In the world about us, the past is distinctly differen[t] [from the future. ...]
we say that the processes going on in the world [around us are asymmetric in time,]
or display an *arrow of time*. Yet this manifest fact [of our] experience is particularly
difficult to explain in terms of the fundamental laws of physics. Newton's laws,
quantum mechanics, electromagnetism, Einstein's theory of gravity, etc., make no
distinction between the past and future – they are *time-symmetric*. Reconciliation of
these profoundly conflicting facts is the topic of this volume. It is an interdisciplinary
survey of the variety of interconnected phenomena defining arrows of time, and their
possible explanations in terms of underlying time-symmetric laws of physics.

Physical Origins of Time Asymmetry

# Physical Origins of
# Time Asymmetry

*Edited by*
J. J. Halliwell
*Imperial College of Science and Technology, London*

J. Pérez-Mercader
*Laboratorio de Astrofísica Espacial y Física Fundamental, Madrid*

W. H. Zurek
*Los Alamos National Laboratory, New Mexico*

 CAMBRIDGE
UNIVERSITY PRESS

Published by the Press Syndicate of the University of Cambridge
The Pitt Building, Trumpington Street, Cambridge CB2 1RP
40 West 20th Street, New York, NY 10011–4211, USA
10 Stamford Road, Oakleigh, Melbourne 3166, Australia

© Cambridge University Press 1994

First published 1994
First paperback edition 1996

Printed in Great Britain at the University Press, Cambridge

*A catalogue record of this book is available from the British Library*

*Library of Congress cataloguing in publication data available*

ISBN 0 521 43328 2 hardback
ISBN 0 521 56837 4 paperback

TAG

# Contents

# Contributors

Andreas J. Albrecht
Theory Group, Blackett Laboratory, Imperial College, London SW7 2BZ, UK.

Julian Barbour
College Farm, South Newington, Banbury, OX15 4JG, UK.

Charles H. Bennett
IBM Corp, Thomas J. Watson Research Centre, P.O. Box 218, Yorktown Heights, NY 10598, USA.

Iwo Bialynicki-Birula
Centre for Theoretical Physics, Lotnikov 32/46, 02-668 Warsaw, Poland.

Carlton M. Caves
Department of Physics and Astronomy, University of New Mexico, Albuquerque, NM 87131, USA.

Thomas M. Cover
Department of Statistics and Department of Electrical Engineering, Durand 121, Stanford University, Stanford, CA 94305, USA.

Paul C.W. Davies
Dept. of Physics and Mathematical Physics, Box 498 GPO, University of Adelaide, South Australia 5001.

Fay Dowker
Department of Physics, University of California, Santa Barbara, CA 93106, USA.

Bryce DeWitt
Department of Physics, University of Texas, Austin, TX 78712, USA.

Murray Gell-Mann
Lauritsen Laboratory, California Institute of Technology, 452-48, Pasadena, CA 91125, USA.

Pedro Gonzalez-Diaz
Instituto de Óptica "Daza de Valdés", CSIC, Serrano 121, 28006 Madrid, Spain.

Robert B. Griffiths
Department of Physics, Carnegie-Mellon University, Pittsburgh, PA 15213, USA.

Jonathan J. Halliwell
Theory Group, Blackett Laboratory, Imperial College, London SW7 2BZ, UK.

ix

James B. Hartle
Department of Physics, University of California, Santa Barbara, CA 93106, USA.

Stephen W. Hawking
Department of Applied Mathematics and Theoretical Physics, Silver Street, Cambridge CB3 9EW, UK.

Bei-lok Hu
Department of Physics and Astronomy, University of Maryland, College Park, MD 20742, USA.

Raymond Laflamme
Theoretical Astrophysics, T-6, Los Alamos National Laboratory, MS B288, Los Alamos, NM 87545, USA.

Joel Lebowitz
Department of Mathematics, Rutgers University, Hill Center - Bush Campus, New Brunswick, NJ 08903, USA.

Seth Lloyd
Complex Systems, T-13, Los Alamos National Laboratory, MS B213, Los Alamos, NM 87545, USA.

Jorma Louko
Department of Physics, University of Wisconsin-Milwaukee, P.O. Box 413, Milwaukee, WI 53201, USA.

Warner A. Miller
Theoretical Astrophysics, T-6, Los Alamos National Laboratory, MS B288, Los Alamos, NM 87545, USA.

Emil Mottola
High Energy Phyiscs, T-8, Los Alamos National Laboratory, MS B285, Los Alamos, NM 87545, USA.

Viatcheslav Mukhanov
Institut für Theoretische Physik, Eidgenössische Hochschule, Hönggerberg, Zürich CH-8093, Switzerland.

Roland Omnès
Laboratoire de Physique Theorique et Hautes Energies, B 210, Université de Paris-Sud, 91405 Orsay, France.

T. Padmanabhan
Inter-University Centre for Astronomy and Astrophysics, Pune University Campus, Ganeshkhind, Pune 411007, India.

Don Page
Department of Physics, 412 Avadh Bhatia Physics Laboratory, University of Alberta, Edmonton T6G 2J1, Canada.

Juan Pablo Paz
Theoretical Astrophysics T-6, Los Alamos National Laboratory, MS B288, Los Alamos, NM 87545, USA.

Juan Pérez-Mercader
Laboratorio de Astrofísica Espacial y Física Fundamental, Apartado 50727, 28080 Madrid, Spain.

Larry Schulman
Department of Physics, Clarkson University, Potsdam, NY 13699-5820, USA.

Benjamin Schumacher
Department of Physics, Kenyon College, Gambier, OH 43022, USA.

Claudio Teitelboim
Centro Estudios Cientificos de Santiago, Casilla 16443, Santiago 9, Chile.

John A. Wheeler
Physics Department, Jadwin Hall, Box 708, Princeton University, Princeton, NJ 08544-0708, USA.

William K. Wootters
Department of Physics, Williams College, Williamstown, MA 01267, USA.

James W. York
Institute of Field Physics, Department of Physics and Astronomy, University of North Carolina, Chapel Hill, NC 27599-3255, USA.

H. Dieter Zeh
Institut für Theoretische Physik der Universität Heidelberg, 69 Heidelberg, Germany.

Wojciech H. Zurek
Theoretical Astrophysics, T-6, Los Alamos National Laboratory, MS B288, Los Alamos, NM 87545, USA.

# Other Participants

Orfeo Bertolami
Instituto Superior Técnico, Departamento de Física, Avenida Rovisco Pais, 1096 Lisboa Cedex, Portugal.

Karel V. Kuchař
Department of Physics, University of Utah, Salt Lake City, Utah 84112, USA.

Glenn Lyons
Department of Applied Mathematics and Theoretical Physics, Silver Street, Cambridge CB3 9EW, UK.

Masahiro Morikawa
Department of Physics, Ochanomizu University, 1-1, Otsuka 2, Bunkyo-ku, Tokyo 112, Japan

José Mourão
Instituto Superior Técnico, Departamento de Física, Avenida Rovisco Pais, 1096 Lisboa Cedex, Portugal.

Alexander Starobinsky
Landau Institute for Theoretical Physics, Academy of Sciences of the USSR, Kosigina 2, Moscow 1173334, USSR.

William G. Unruh
Physics/Cosmology Program CIAR, University of British Columbia, 6224 Agriculture Road, Vancouver V6T 2A6, Canada.

Manuel G. Velarde
UNED-Facultad de Ciencias, Apartado 60.141, Madrid 28080, Spain.

Grigory Vilkovisky
P.N. Lebedev Physics Institute, Leninsky Prospect 53, RU-117924 Moscow, Russia.

# Foreword

The concept of time is perhaps one of the most integrating in human knowledge. It appears in many fields, including philosophy, psychology, biology and, most prominently, in physics, where it plays a central role. It has interested Man of all Ages, and the finest minds from Saint Augustine to Kant to Einstein have paid attention to its meaning, and the mystique shrouding its most notorious property: that of flowing only forward, its irreversibility.

Today, largely thanks to the efforts in popularization of science by some leading physicists, even laymen show an interest in these problems.

More than 25 years ago a meeting on "The Nature of Time" took place at Cornell University. Since then, there has been no forum for discussing these ideas in the light of important developments being made in several areas of physics. Cosmology, general relativity, computer science and quantum mechanics, to mention a few, have undergone a profound evolution in these 25 years.

Identifying these important problem areas, and then fostering and helping their development by bringing together experts in the field, is one area in which our Fundación sees its duty fulfilled.

Fundación Banco Bilbao Vizcaya, in its interest to create the appropriate framework for creative thinking in Frontier Topics in Science, which at the same time affect society at large, and are capable of integrating knowledge, felt that a meeting devoted to a rigorous, deep and exhaustive study of the Physical Origins of Time Asymmetry, was particularly suitable.

We were happy to contribute with our sponsorship to this meeting, where so many of the world's experts came together to discuss such an important and fundamental question. We also hope that through the publication of these minutes the discussions can be brought to the attention of a much wider audience which can profit from what was presented and discussed at the meeting.

José Angel Sánchez Asiain
Presidente
Fundación Banco Bilbao Vizcaya

# Acknowledgements

We would like to express our deep gratitude to NATO, for their initial grant which made the meeting possible, and to the Fundacion Banco de Bilbao Vizcaya (FBBV), for their generous local support.

J. J. H.
J. P.-M.
W. H. Z.

# Introduction

The world about us is manifestly asymmetric in time. The fundamental laws of physics that describe it are not. Can these two undeniable facts be reconciled? This question has been the focus of much attention ever since it was first addressed by Boltzmann in the last century. The fact that it does not belong exclusively to a specific area of academic research makes it a particularly suitable focus for an interdisciplinary meeting. This volume is the outcome of such a meeting.

The meeting, a NATO Advanced Research Workshop, took place in Mazagon, in the Province of Huelva, Spain, from 29 September to 4 October 1991. Its aim was to consider the question of time asymmetry in the light of recent developments in quantum cosmology, quantum measurement theory, decoherence, statistical mechanics, complexity theory and computation.

The case for such a meeting is not hard to make. Although many physicists and scientists from other disciplines have been very interested in the arrow of time, there has not been a meeting on this topic for over twenty years. The last one was at Cornell in the 1960s. In the spring of 1989, and again in the spring of 1990, there were workshops at the Santa Fe Institute on Entropy, Complexity and the Physics of Information. These workshops were of an interdisciplinary nature and cut across many of the fields involved in the time asymmetry workshop reported here. They were very successful, and indicated that workers in the different disciplines involved have many mutual interests and are keen to exchange ideas.

It was of course in the context of statistical mechanics that the question of time asymmetry was first investigated, and this area still remains central to any discussion of the topic. The new areas of investigation, at which this workshop was aimed, are as follows.

1. **Quantum Cosmology.** Cosmologists have appreciated for many years that much of the observed time asymmetry in the universe may be traced back to the particular set of initial conditions with which the universe began. Unfortunately, the great difficulty of modeling the very beginning of the universe has for a long time presented a serious obstacle to further investigation. However, recent

developments in the field of quantum cosmology have led to the construction of a framework in which the beginning of the universe may be discussed in a precise and meaningful way. This has led to a revitalization of interest in the question of the cosmological arrow of time.

2. **Quantum Measurement Theory and Decoherence.** The special needs of quantum cosmology, together with a number of recent experimental advances (the generation of squeezed states, macroscopic quantum effects, mesoscopic systems, and the possibility of interference experiments with large systems) has led in recent years to a resurgence of interest in quantum measurement theory and the interpretation of quantum mechanics. More generally, studies in both quantum cosmology and inflationary universe models have underscored the necessity to acquire a deeper understanding of the emergence of classical properties from quantum systems. Central to discussions of this issue is the distinctly time-asymmetric process of *decoherence* – loss of quantum coherence as a result of interaction with an environment. In particular, the direction of decoherence coincides with, and maybe even defines, an arrow of time. Moreover, a number of recent attempts to quantify the degree of classicality of a decohering system use the notion of information and entropy, providing a link with the fields of complexity and computation.

3. **Complexity.** Many attempts have been made to capture, in precise mathematical terms, the notion of the *complexity* of a physical system. A suitable definition of complexity would provide a rigorous distinction between states of randomness and states of self-organization. Many candidates for a formal measure of complexity have been suggested. They include, for example, the *algorithmic information content* (the length of the shortest computer program required to generate the object in question), the *logical depth* (the execution time required to generate the object by a near-incompressible universal computer progam), and the *thermodynamic depth* (the amount of entropy produced during a state's actual evolution). These efforts are of particular interest in relation to discussions of the arrow of time, in that a measure of complexity might permit one to formalize the intuitive notion that the universe generally evolves from order to disorder, but without having to appeal to the probabilistic considerations involved in the statistical (Gibbs–Shannon) definition of entropy. Indeed, it has been argued by Zurek that in the course of measurements the second law of thermodynamics – and its associated arrow of time – will hold only if the usual definition of entropy is modified to take account of the algorithmic information content (*i.e.*, Kolmogorov complexity) of the acquired data.

4. **Computation.** A comparatively new class of systems which may define an arrow of time has emerged in the field of the physics of computation. This field is concerned with the energetic cost, structural requirements and thermodynamic reversibility of physical systems capable of universal computation. Through the work of Landauer, Bennett and others, it is now established that only

logically *irreversible* operations inevitably dissipate energy, whereas logically reversible ones need not, implying that it is in principle possible to compute at no thermodynamic cost. However, the logically reversible operations may be accomplished only at the expense of cluttering up the memory with an otherwise useless step-by-step record of the computation. This leads to the recognition that — as was anticipated by Szilard more than half century ago — it is actually the erasure of information which results in the increase of entropy. Real systems are in fact always dissipative and the irreversibility is important for stability and error correcting. These considerations may shed considerable light on various long-standing questions in the foundations of statistical mechanics, such as Maxwell's demon. They are also closely related to studies of complexity and tie in with information transfer and processing in biological systems.

Approximately half the papers presented here concern time asymmetry specifically, in the variety of contexts discussed above. The rest concern related topics, reflecting the discussions that took place at the meeting and the participants current interests. The papers are loosely grouped according to the predominant subject they address. The six parts of the book do not fall into a particular order. We will not attempt to summarize or review the papers found here, because the range of topics covered is too broad to describe in a concise and coherent way.

An important part of the workshop was the discussions that took place after each talk. In order to give a flavour of these discussions, transcripts are included after most of the papers. They are neither verbatim nor complete. They have been edited for clarity, and generally just a few of the more interesting and relevant questions and answers are included.

J. J. H.

J. P.-M.

W. H. Z.

## Literature on Time Asymmetry

The meeting at Cornell in the 1960s is described in,

Gold, T. (ed.) (1967) *The Nature of Time*, Cornell University Press, Ithaca.

The first of the two Santa Fe meetings is reported in,

Zurek, W.H. (ed.) (1990) *Complexity, Entropy and the Physics of Information, Santa Fe Institute Studies in the Sciences of Complexity, vol VIII*, Addison-Wesley, Reading, MA.

The following are some general references on time and time asymmetry.

Davies, P. C. W. (1976) *The Physics of Time Asymmetry*, University of California Press, Berkeley, CA.

Denbigh, K.G. (1981) *Three Concepts of Time*, Springer-Verlag, Berlin.

Horwich, P. (1987) *Asymmetries in Time*, MIT Press, Cambridge, MA.
Landsberg, P.T. (ed.) (1982) *The Enigma of Time*, Adam Hilger, Bristol.
Reichenbach, H. (1956) *The Direction of Time*, University of California Press.
Zeh, H. D. (1989) *The Physical Basis of the Direction of Time*, Springer-Verlag, Berlin.

# 1

## Time Today

John Archibald Wheeler

*Physics Department, Princeton University,*
*Princeton, New Jersey 08544, USA, and*
*University of Texas,*
*Austin, Texas 78712, USA*

How come time? It is not enough to joke that "Time is nature's way to keep everything from happening all at once. [1]" A question so deep deserves a deeper look. Let's come back to it, therefore, towards the end of this account, and turn for now to the less intimidating, "How come the asymmetry between past and future?"

In 1939–1945, Richard Feynman and I explored the idea of sweeping out [2], [3] the electromagnetic field from between all the world's charged particles. We knew that Tetrode [4] and Fokker [5] had postulated instead that every charge acts directly on every other charge with a force governed by half the sum of the retarded and advanced potentials. Their considerations and ours led us to the thesis that the force of radiative reaction arises from the direct interaction between the accelerated source charge and the charged particles of the absorber. In this work we encountered so many interesting issues of principle that we went to 112 Mercer Street to consult Einstein about them. He told us of his dialogue with W. Ritz. Ritz [6] had taken the phenomenon of radiative damping to argue that, at bottom, the electrodynamic force between particle and particle must itself be time-asymmetric. Einstein, in contrast, had maintained that electrodynamics is fundamentally time-symmetric, and that any observed asymmetry must, like the flow of heat from hot to cold, follow from the statistics of large numbers of interactions. Einstein's words made us appreciate anew one of our findings, that the half-advanced, plus half-retarded action of the particles of the absorber on the radiating source only then added up to the well-known and observationally tested [7] force of radiative reaction when the absorber contained enough particles completely to absorb the radiation from the source. By this route, Feynman and I learned that the physics of radiation can indeed be regarded as, at bottom, time-symmetric with only the statistics of large numbers giving it the appearance of asymmetry.

The idea that time asymmetry is not the attribute of any individual interaction, but the statistical resultant of large numbers of intrinsically time-symmetric individual couplings is not a new one, of course, but implicit in the great tradition of physics associated with the names of Boltzmann, Kelvin, Maxwell, Smoluchowski, the Ehrenfests, Szilard, Landauer, Bennett and their followers. The essential idea stands

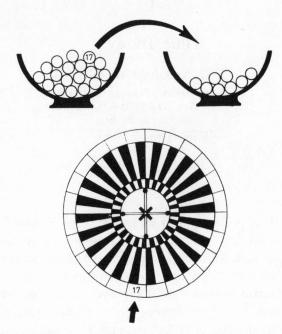

Fig. 1.1. The Ehrenfest double urn in a 1979 rendering. When number seventeen comes up on the roulette wheel, the ball carrying that number is transferred from whichever urn it happens to be in to the other urn. Thus 100 balls, 75 of them initially in the left-hand urn and 25 in the right hand one, gradually approach a 50–50 distribution as "time increases" (more spins of the roulette wheel).

out nowhere more clearly than in the double-urn model of Paul Ehrenfest. (Fig. 1.1). Each spin of the roulette wheel brings up a random number, say seventeen. We look for the ball with that number on it and transfer it from whichever urn it occupies to the other. At the start, there are 75 balls in the left hand urn and only 25 in the right hand urn. Therefore, ball seventeen is three times as likely to lie in the left hand urn as in the right hand one. Thus, the first spin of the roulette wheel is three times as likely to narrow the count discrepancy to 74 and 26 as to raise it to 76 and 24. The numbers of balls in the two urns approach equality with three times the probability that they go the other way. So, too, with further spins of the roulette wheel. Thus the Ehrenfest double-urn models the probabalistic approach of the energy content of two masses to thermal equilibrium: That's how come, we understand, that heat flows from hot to cold. Water and alcohol, once mixed, don't unmix. Past and future don't exchange garments. The one-sided flow of energy from accelerated source charge to multiple far away receptor charges plays out a like scenario.

   As Ehrenfest's double-urn model displays a difference in number of balls in the two urns that undergoes smooth exponential decay plus zigzag up-down fluctuations, so the model of direct half-advanced, half-retarded couplings between particle and particle yields the familiar law of radiative damping with superposed background fluctuations.

Ehrenfest's model invites the question: After a hundred spins of the wheel, will the populations of the two urns by a freak fluctuation ever return to 75 on the left, 25 on the right? The answer, we know, is yes. Moreover, when we make the number of runs sufficiently enormous, there will be a wealth of examples of such a 75–25 outcome following on a 75–25 start. When we pick out a thousand examples of such a "symmetric-ended" hundred-spin run, we find that the average difference in the count of balls in the two urns for these examples follows a hyperbolic cosine curve, with superposed statistical fluctuations. Past and future are symmetrically related.

To select symmetric-ended runs of the Ehrenfest double-urn model, and to average them, is thus to see the analog of temperature following a hyperbolic cosine curve. Do not big bang and big crunch impose a similar selection on the fluctuating development of the microscopic processes going on all the time and everywhere throughout the entire universe? Is it not therefore reasonable, W. J. Cocke asks [8], for such a quantity as the temperature of the background microwave radiation to reach a minimum some billions of years from now and start rising after this "turning of the tide." Cocke's proposal has led to some further discussion [9]: "Even if there is any correlation between statistics and cosmology, it is not a necessary consequence of the reasoning that the statistical tide should turn at the same time as the cosmological tide, nor is it necessary that either time occur exactly half-way between start and stop. There are few model universes easier to analyze in all detail than the Taub model universe [10], its geometry an egg-like deformation of the familiar Friedmann 3-sphere. For the extreme time-asymmetric (large $m'$) case of this model the volume varies with proper time in accordance with a relation which, written parametrically, is [11]

$$\begin{cases} V = 32\pi^2 \ l^3 (m')^2 \sin f \ (1 - \cos f), \\ t = lm'(f - \sin f). \end{cases} \tag{1.1}$$

Only for the special choice of the parameter, $m' = 0$, is the dynamics time symmetric. Moreover, there is no obvious reason why the final value of $n = n''$ in the double-urn experiment should be identified with the initial value; or to spell out the analogy, no obvious reason why the conditions at the big crunch should be in every way identical to those at the big bang.... Despite all ... provisos and caveats, the simplest model makes the greatest appeal in any first survey of ideas. In [that model] the turning of the tide for the statistics is identical in its timing with the transition from expansion to contraction. Also both are mirror-symmetric [in the large, not in detail] with respect to that common time....

"Incentive [9] though the double-urn model is for asking new questions about the universe, it is inadequate for answering them. One shortcoming is evident from the start. The double-urn model is characterized by a single transition rate [the probability per unit time that a ball jump from one urn to the other]. In contrast, the universe is characterized by almost as many transition rates as there are physical processes, from elementary-particle decay rates to the rates of thermonuclear processes in stars, and from the rate of dynamical evolution in star clusters to the rate of decay of turbulence. Nowhere does this limitation of the double-urn model show more conspicuously than in the difficulties it makes for predictions about $\beta$-decay of $^{187}$Re. Which is relevant, the $40 \times 10^9$ a [a = year] half-life for expulsion of the $\beta$-particle or the $10^{-12}$ s time for reducing the expelled $\beta$-particle to thermal equilibrium with its surrounding? Or a complex resultant of these two and many other characteristic times? The predictions of the double-urn model, if one can call them predictions, are utterly different according as one correlates the characteristic decay constant, $\lambda$, of that model with the shorter or the longer of these two times, let alone some unknown third "resultant time constant." In the first case the transition from exponential decay to exponential increase of $^{187}$Re takes place within an extremely short interval of the turning of the tide. To hope to see any evidence of that transition today, at a time when the universe is still expanding, would seem preposterous. However, if the long time of the $\beta$-decay itself is the relevant quantity, then the transition from fall to rise takes place gradually over the whole range from start to end. In this case a significant difference in the effective half-life of $^{187}$Re might be expected as between today and $4.5 \times 10^9$ years ago, when certain stony meteorites were formed."

"Consider first the customary hypothesis that the decay has been exponential ever since the time, $t_{form}$, of the formation of the meteorite, and has continued to have a decay rate, the $\lambda_{apparent}$ [of tradition] equal to that found today,

$$N_{form} \, (^{187}\text{Re}) = N_{now}(^{187}\text{Re}) \exp \, [\lambda_{app} \, (t_{now} - t_{form})]. \tag{1.2}$$

Then the ratio of daughter osmium $^{187}$Os to surviving rhenium in the relevant granules of the meteorite will be

$$R = \frac{N(\text{daughter Os})_{now}}{N(\text{surviving Re})_{now}} = \exp \, [\lambda_{app}\Delta t] - 1."\tag{1.3}$$

In the fifteen years subsequent to these remarks, no anomaly has been confirmed in the apparent half-life, $T_{1/2} = 0.693/\lambda_{app}$, of any long-lived radio nuclide extracted from any ancient source. In brief, no evidence has been discovered from the world of the small for any approach to any turning of the tide of cosmology.

Among all radioactive processes for which a turning of the tide might be conceived, one puts itself forward as especially worthy of investigation, K-meson decay, because it signals (*cf. Fitch, Cronin and colleagues*) [12] a violation of time-reversal invariance.

"Does the degree of violation of T-reversal symmetry change with time? Was the key anomaly measured by Fitch, Cronin and colleagues, that is, was the branching ratio of $K_2^0$-to-two-pion decays relative to all charged modes of decay, greater four billion years ago, in the early days of this planet, than it is today, $2 \times 10^3$? And *will* it be greater as the time of the crunch approaches? The question, put in these latter terms, does not demand an impossibly long time for its answering because a wait of a cosmological order of magnitude should not be necessary for the crunch to manifest itself. Conditions in the vicinity of a black hole are sufficient. After all, black-hole physics first attracted attention [13] because it provides a here-and-now model for the crunch that the simplest cosmology predicts at the end of time. Of course, the Fitch-Cronin equipment [14] is not something that one packs into a briefcase, so it is not easy to imagine some version of it, equipped with a radio transmitter, transported to the vicinity of a black hole and dropped in. But it is difficult to conceive of a more direct test of cosmological flavor to distinguish time symmetry from time asymmetry.

"It is hard enough to measure the CP anomaly in K-meson decay today even with a whole battery of instrumentation at hand. It seems at first sight hopeless to find out what value this critical constant had long ago from any record that can be read in the ancient rocks. However, it also once seemed hopeless to determine whether other physical 'constants' differ appreciably – if only by very small amounts – from their values today. Yet the unexpected discovery [in Gabon, Africa] of the ancient Oklo uranium reactors gave a way [15] in to the analysis of this question of unmatched power. Can some equally powerful and equally unexpected means be found to study the amount of CP = T-asymmetry in K-meson decay long ago? Could that old motto have its place, 'The first step in the answering of a question is the raising of the question'?"

Ever to discover a radioactive transformation rate dependent on time will immediately raise the question: Dependence on proper time? Or dependence on cosmological time? Few illustrations of proper time are better known than cosmic-ray $\mu$-mesons and their decay in flight from point of production in the upper atmosphere to arrival at the ground [16]. No process of transformation is yet known which might be, or which can be predicted to be, responsive to the most natural cosmological time parameter, that is, York's "crunch time," K, the fractional rate of contraction of space per unit advance of proper time, a quantity which starts at $-\infty$ at the big bang, reaches zero at the peak of expansion of any bang-to-crunch model universe, and rises to $+\infty$ at the big crunch.

We thrust a thermometer into a fluid at a chosen point when we want to know its temperature there. Likewise, a pressure gauge measures pressure there. But insert a finger or a sensor at that point and have it feel or read time? No go. Or no go unless and until someone invents a device based on the anomaly in K-meson decay or some other process yet to be discovered that will respond to cosmological or crunch time.

Reflections on time asymmetry have led us from bang-to-crunch cosmology to the question whether a turning in the tide of expansion is accompanied by any such turning in the tide of statistics as would show up in rates of radioactive decay and magnitude of CP violation in K-meson decay. Challenging though this complex of issues about time asymmetry is to analysis and to experiment, we come closer to the mainstream of physics when we ask questions about time itself, unencumbered by the issue of time asymmetry.

## Time as Metronome of Classical Physics

The word Time came, not from heaven, but from the mouth of man, an early thinker, his name long lost. If problems attend the term, they are of our own making. Centuries of analysis have sharpened the meaning of the term and extended its scope. The domain of 'time' has grown from pendulum to sundial and from particle to cosmos.

Whence does time secure its power as metronome whose ticking all things, by magic, heed and obey? However marvelous it was for Columbus and his men and his Arab navigator to see the new sights of the New World, their findings were as nothing compared to the miracles the well-equipped observer discovers around him on first arrival in our own strange world: heat, light, sound, particles, radioactivity, electromagnetic and gravity waves. Not one of these miracles can we explore more readily than time. We hang a stone from a string to make a pendulum. We count its swings to acquire a measure of time of unrivalled simplicity. A second pendulum, identical to the first, gives identical counts for the time from one event to another. Or they would if something did not go wrong with one pendulum. But then, how do we know which one has gone wrong? It is no wonder that Captain James Cook on his voyages of exploration in the South Pacific [17] took three clocks and, when they disagreed, employed the principle of majority rule. What a miracle that in a world of strange sights, sounds and happenings a quantity should present itself so easy to measure and so well-adapted to produce agreement between different observers. Other quantities soon presented themselves to bright-eyed observers, among them, pressure, electric and magnetic fields, even spacetime curvature.

If physics is the science of change, it is impossible to do physics without some understanding of time. Therefore, it is natural that Aristotle defines time as "that aspect of change that enables it to be counted" [18] and it is equally impossible to understand all that time means and is today except in the context of Einstein's spacetime physics.

"The expansion of the empire of time [19] has elevated the concept, human born though it is, to platform upon platform upon platform of authority (Figure 1.2). Regularities of sun and season raised the first foundation. On top of it Newtonian dynamics erected a second and tighter platform; special relativity a third, terraced further in and up; and general relativity stands at the summit, the final level of authority. Not except out of the mouth of Einstein's 1915 and still standard theory of spacetime can one hear the generally agreed account of all that 'time' now means and measures.

"In a single night an empire's magnificent terraced structure of authority can

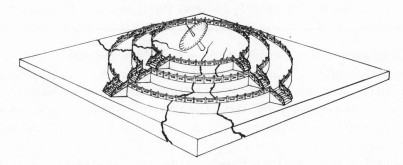

Fig. 1.2. Time rules physics. General relativity counts as its highest and most compactly organized platform of authority. Three cracks suggest that the foundations are falling to pieces and will require reassembly on a new plan. Appreciation is expressed to Ros Lin Chin for making this drawing.

crumble into pieces because of a flaw in the original design. Yet the pieces can be reassembled thereafter into a superior structure according to a quite different guiding principle. The solid state physics of the 1860s, built on density, specific heat, expansion coefficients, elasticity and other 'constants of nature,' had been reconstituted by the 1960s on the plan of wave mechanics, electronic states and modes of excitation.

"Is the structure of all of physics as we see it in this century destined to fall to pieces and be put back together on quite a different plan in the coming century? Is time – ruler of it all today – to be toppled? Fall from primordial concept in the description of nature to secondary, approximate and derived? No longer be fed into physics at the start – as density once was – but pulled out at the end? In other words, should we be prepared to see some day a new structure for the foundations of physics that does away with time?

"Yes; that is the thesis of this account. Yes, because 'time' is in trouble. Yes, because beyond all day-to-day problems of physics, in the profound issues of principle that confront us today, no difficulties are more central than those associated with the concept of 'time,' symbolizing as they do three cracks in the structure of physics:

(i) Time ends in big bang and gravitational collapse.
(ii) 'Past' and 'future' are interlocked in a way contrary to the causal ordering presupposed by 'time.'
(iii) Quantum theory denies all meaning to the concepts of 'before' and 'after' in the world of the very small [19]."

These yet-to-be-explored new quantum-driven features of time show up most sharply when seen against the background of today's standard classical-general-relativity account of time.

*J. A. Wheeler*

## Time Today as Understood in Einstein's 1915 Classical, Still Standard Geometrodynamics

Time as we know it from general relativity is not an entity in itself but is an aspect of the sheet-of-sandpaper picture that physics gives us for spacetime: (1) Events, and (2) the interval, spacelike, time-like, or null, as the case may be, between event and event, and (3) the continuous 4-geometry that we abstract out of these measurements as conceptually economical device to summarize them and to quantify the location of new events.

Wristwatch time for the space traveller or the analogous tick-tick-tick quantity for a particle provide a succession of events between which the intervals are now well defined. Ageing of an electron or a nucleus from A to B lets itself be defined as not time nor space but interval – the currency in which Einstein's classical geometrodynamics operates – added up along the history or world line that leads to B, the here and now of that particle, from A, its place and time of origin in the big bang. What about alternative histories that lead from A to B? The more jogs a history displays in its course, the more it will depart from the smoothness of a geodesic and the smaller, normally, the interval it will rack up. In contrast, the unique world line of maximum integrated interval defines the geodesic AB, and by the length of that geodesic provides one way to ascribe a "time" to the point B – and a maximum possible age for the particle we find there.

Date rocks [20] we can, and with great good consequences for the science of geology. However, to date an electron or any other elementary particle is still beyond our power. General relativity, nevertheless, makes it conceivable [21] that we will someday gain that power provided that the famous asymmetry in K-meson decay [22] or – so suggests V. L. Fitch in a kind personal communication – the related rate of decay of a two-boson system is sensitive to the rate of crunch of 3-geometry. Sensitive? That question of elementary particle physics is fascinating but lies outside the scope of our discussions. 3-geometry and its rate of crunch? To elucidate this phrase is to recall what general relativity teaches about the nature of time:

## Three-Dimensional Geometry Carries Information About Time [23]

To "quantize general relativity" seemed an attractive enterprise in the 1940s and the 1950s. Why did the enterprise seem more difficult [24] the further it was pursued? For no reason so much as this, that it suffered from a false reading of the familiar demand for "covariance": It conceived of the dynamic object as 4-D spacetime. However, spacetime does not wiggle. It is 3-D space geometry that undergoes the agitation. The history of its wiggling registers itself in frozen form as 4-D spacetime. What then is Einstein's classical theory of gravity all about? It is

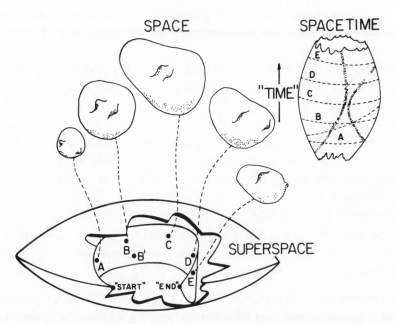

Fig. 1.3. Space, spacetime and superspace. Upper left: Five sample configurations, A, B, C, D, E attained by space in the course of its expansion and recontraction. Below: Superspace and these five sample configurations, each represented by a point in superspace located on one and the same leaf of history that curves through superspace. Upper right: Spacetime, the classical-physics history of space geometry conceived as undergoing a deterministic evolution in time, A, B, C, D, E. Let the representative point move from one location in superspace to another. Then the 3-geometry alters as if alive – a cinema of the deterministic classical dynamics of space. Of these classical concepts, which still continue to make sense in quantum theory? Space, yes. Superspace, yes. Spacetime, no. (Diagram from MTW). [25]

about the dynamics of 3-geometry, or geometrodynamics – the Einsteinian analog of Maxwellian electrodynamics.

Since those early days we have learned that the dynamics of 3-geometry, $^{(3)}G$, both classical and quantum, unrolls in superspace, S (Figure 1.3). Superspace is that infinite-dimensional manifold, each point of which represents one $^{(3)}G$. Two nearby points in superspace represent two 3-geometries that differ only little in shape.

"Time": time as spelled with a "t"? Search about as we may in superspace, nowhere can we catch any sight of it. Of 3-geometries, yes; of time, no. Out of these 3-geometries, however, can we reconstruct time? In classical theory, yes; in quantum theory, no.

Classical theory, plus initial conditions, confronted with the overpowering totality of $^{(3)}G$'s that constitute superspace, picks out that single bent-leaf of superspace which constitutes the relevant classical history of 3-geometry evolving with time. Otherwise put,

(i) Classical geometrodynamics [26] in principle constitutes a device, an algorithm, a rule for calculating and constructing a leaf of history that slices through superspace.

(ii) The $^{(3)}G$'s that lie on this leaf of history are **YES** 3-geometries [YES with respect to the prescribed initial conditions!]; the vastly more numerous $^{(3)}G$'s that do not are **NO** 3-geometries.

(iii) The **YES** $^{(3)}G$'s are the building blocks of the $^{(4)}G$ that is [the relevant] classical spacetime [for this problem, with its specified initial conditions].

(iv) The interweavings and interconnections of these building blocks give the [relevant spacetime; that is, the appropriate] $^{(4)}G$ its existence, its dimensionality and its structure.

(v) In this structure every $^{(3)}G$ has a rigidly fixed location of its own.

(vi) In this sense one can say that the 'many-fingered time' [carried by] each 3-geometry is specified by the very interlocking construction itself. . .[In brief, '3-geometry carries information about time.']" [27]

"How different from the textbook concept of spacetime! There the geometry of spacetime is conceived as constructed out of elementary objects, or points, known as 'events.' Here, by contrast, the primary concept is 3-geometry, in abstracto, and out of it is derived the idea of event. Thus, (1) the event lies at the intersection of such and such $^{(3)}G$'s and (2) it has a timelike relation to (earlier or later than, or synchronous with) some other [nearby] event, which in turn (3) derives from the intercrossings of other $^{(3)}G$'s . . . .

## Fluctuations in Spacetime Geometry

"Quantum theory upsets the sharp distinction between **YES** 3-geometries and **NO** 3-geometries. It assigns to each 3-geometry not a YES or a NO, but a probability amplitude,

$$\Psi(^{(3)}G) \tag{1.4}$$

This probability amplitude [rises to greatest amplitude] near the classically forecast leaf of history and falls off steeply outside a zone of finite thickness extending a little way on either side of the leaf."

Otherwise stated, "quantum fluctuations take place in the geometry. They coexist [28] with the geometrodynamic development predicted by classical general relativity The fluctuations widen the narrow swathe cut through superspace by the classical history of the geometry. In other words, the geometry is not deterministic, even though it looks so at the everyday scale of observation. Instead, at microscopic [or Planck length $L^* = (\hbar G/c^3)^{\frac{1}{2}} \sim 10^{-33}$ cm] scale, it 'resonates' between one configuration and another and another. This terminology means no more and no less than the following:

(i) Each configuration $^{(3)}G$ [of the 3-geometry] has its own probability amplitude $\Psi = \Psi(^{(3)}G)$.

(ii) These probability amplitudes have comparable magnitudes for a whole range of 3-geometries included within the limits $\Delta g \sim L^*/L$, on either side of the classical swathe through superspace.

(iii) This range of 3-geometries is far too variegated on the submicroscopic scale to fit into any one 4-geometry, or any one classical geometrodynamic history.

(iv) Only when one overlooks these small-scale fluctuations and examines the larger-scale features of the 3-geometries do they appear to fit into a single space-time manifold, such as comports with the classical field equations.

"These small-scale fluctuations tell one that something like gravitational collapse is taking place everywhere in space and all the time; that gravitational collapse is in effect perpetually being done and undone; that in addition to the gravitational collapse of the universe, and of a star, one has also to deal with a third and, because it is constantly being undone, most significant level of gravitational collapse at the Planck scale of distances." [29]

"The $^{(3)}G$'s with appreciable probability amplitude are too numerous to be accommodated into any one spacetime. Thus the uncertainty principle declares that spacetime is only an approximate and classical concept. In reality there is no such thing as spacetime. 'Time' itself loses its meaning, and **[at the Planck scale] the words 'before' and 'after' are without application.** These long-known considerations are of importance only at the Planck scale of distances. They all flow out of the straightforward analysis of the dynamics of geometry in the arena of superspace, inescapable conceptual adjunct of general relativity.

Quantum theory demands and physics supplies the correct wave equation to describe how the dynamics of geometry unrolls,

$$-\frac{\nabla^2 \psi}{(\delta^{(3)}G)^2} + ^{(3)}R\psi = 0 \tag{1.5}$$

in abbreviated form [30]; or, properly spelled out, [31]

$$\left( G_{ijkl} \frac{\delta}{\delta \gamma_{ij}} \frac{\delta}{\delta \gamma_{kl}} + \gamma^{\frac{1}{2}} \,^{(3)}R \right) \Psi[^{(3)}G] = 0 \tag{1.6}$$

where

$$G_{ijkl} \equiv \frac{1}{2}\gamma^{-\frac{1}{2}}(\gamma_{ik}\gamma_{jl} + \gamma_{il}\gamma_{jk} - \gamma_{ij}\gamma_{kl}) \tag{1.7}$$

The so-called WDW equation (1.7) transcribes into quantum language the very heart of Einstein's classical geometrodynamics: the classical physics demand that for every probe hypersurface, in whatever way curved, through every event, it is required that [32] the sum of the scalars of extrinsic curvature and intrinsic curvature

$$(\text{Tr}\mathbf{K})^2 - \text{Tr}\mathbf{K}^2 + ^{(3)}R = 0 \tag{1.8}$$

shall be zero. Here, for simplicity, we have annulled the density, $\rho$, of mass-energy that would otherwise appear in a term $16\pi\rho$ on the right hand side of (1.8) where the 0 now stands.

This is not the occasion to recall Élie Cartan's beautiful interpretation of the left hand side of (1.8) in terms of the moment of rotation [33] associated with the spacetime curvature nor the wonderful way this geometric interpretation, together with the mathematical identity that the "boundary of a boundary is zero," lets one understand Eq. (1.8) as geometrodynamics tool to uphold the law of conservation of momentum and energy. In this respect (1.8) is the analog of the central equation of electrodynamics,

$$\nabla E = 4\pi\rho_e \tag{1.9}$$

We annul (1.5), (1.6), and (1.8) because for the sake of simplicity we have excluded from the stage – space – all actors except space geometry itself. Nowhere better than in the concepts just reviewed does today offer us the tools that define time.

The spacetime physics of the past and much of it today off-loads onto particles and light rays the double task to define, by their encounters, those points in spacetime which we call *events* and to measure the **interval** between event and event. Today admits and often prefers the opposite order of ideas: First comes spacetime with all its curvature as characteristic of it as mountains are for landscape or shape is for an automobile fender. One can stretch a rubber sheet over an automobile fender and rule upon it a pattern of intersecting lines labeled $x = \ldots 11 \ldots 12 \ldots 13 \ldots$, $y = \ldots 8 \ldots 9 \ldots 10 \ldots 11 \ldots$. Then one can use the already known spacelike interval, $ds$, or timelike interval, $d\tau$, between event and nearby event to define metric coefficients $g_{ij}(x, y)$ :

$$ds^2 = -(d\tau)^2 = g_{ij}(x, y)dx^i dx^j. \tag{1.10}$$

These metric coefficients, in their dependence on the coordinates $x$ and $y$, are influenced as much by the arbitrary choice of coordinate rulings on the rubber sheet as by the intrinsic shape of the fender itself. Therefore, a dealer in automobile parts would be crazy to use these metric coefficients as tool when he has to distinguish a Lincoln fender from a Mercedes-Benz fender. His own sense of shape is quicker. His motto, if he needs one, reads, "Coordinates, no. Shape, yes." Likewise for the aficionado of landscapes and the Einsteinian with his spacetime geometries.

It is not hard to conceive of a ranch which, as it grows, marks its boundaries of past years by a succession of fences that do not intersect. They "foliate" the landscape. Likewise, the Einsteinian in imagination foliates or slices up spacetime by a sequence of non-intersecting spacelike hypersurfaces. Each of these hyperspaces, or slices, or "3-geometries," is most usefully constructed to be "spacelike" in this sense: The interval between each point and its immediate neighbors in that hypersurface is spacelike. Nothing forbids the construction in imagination of a rebel hypersurface that slices at an angle through the foliated array of hypersurfaces. Typically, it intersects a specified hypersurface of the foliation in a two-dimensional surface. That surface and yet another rebel hypersurface ordinarily intersect in a line. It, cut by a third rebel surface, ordinarily picks out a definite point in spacetime, the lone event that belongs to all four hypersurfaces. Thus within a given spacetime manifold an individual point, and more particularly an instant of time itself, lets itself be fixed by the specification of four intersecting spacelike manifolds. The conclusion is clear. Yes, four-dimensional spacetime can be considered to be put together as a collection of point-like events; but, contrariwise, every point-like event, every instance of a somewhere at an instant, can be considered as selected out of a larger preassigned 4-D spacetime landscape by the intersection of three-dimensional fences running through that landscape. To travel the conceptual road from point to four-dimensional continuum was natural in a day when particles were viewed as the building blocks of everything. Einstein's 1915 and still standard geometrodynamics, however, often makes it more natural to start with spacetime and spacelike slices through spacetime as means to define a point, an event endowed with a place and a time. No label, of course, attaches to the arbitrary point so selected, no number to define time – no number unless, for example, we ourselves number sequentially the slices in the original foliation. Then the time coordinate of the point of intersection lets itself be defined as the numerical label on the selected sheet of the foliation.

Let the first rebel hypersurface be pushed forward through spacetime by a small amount, so that it now intersects a sheet of the foliation that bears a higher index number. We say that attention now attaches to this point of intersection. In other words, by the movement of the rebel hypersurface we have "moved attention forward in time." More generally, we recognize and declare [34] that, "a spacelike hypersurface carries information about time." Make a bulge in it and it will move forward in time. The summit of the bulge will touch tangentially a sheet of the foliation. We can arbitrarily assign a time coordinate to this point of tangency: The number that appears is the number on the sheet of the foliation. Of course there is nothing unique about this assignment of time coordinate. Any monotonic continuous increasing real function of it will serve equally well to orchestrate time.

Geometrodynamics is not accustomed to regarding all of spacetime as laid down in advance and then running different spacelike slices through it. Instead, the space geometry at one instant and some appropriate measure of its time rate of change at that instant are conceived and specified. Then the question is raised how to calculate the subsequent evolution with time of the 3-geometry so as to trace out the whole of the remaining spacetime beyond the specified "initial spacelike hypersurface."

The required initial-value data are conveniently taken to be (1) the 3-geometry $^{(3)}G$, specified in terms of some convenient coordinates $x, y, z$ by metric coefficients [Eq (1.1)] $g_{ij}(x, y, z)$ – out of which the curvature $R(x, y, z)$ intrinsic to the hypersurface follows by differentiation [35] or by appropriate geometric manipulations [36] – and (2) the extrinsic curvature $K(x, y, z)$, telling how the spacelike hypersurface is curved with respect to the embedding 4-geometry. The trace, $K$, of the tensor, $K$, of extrinsic curvature measures the fractional shrinkage in the volume of a cube lying in the initial spacelike hypersurface when each corner of the cube is propagated forward a unit amount of time normal to the hypersurface. The quantity, $K$, sometimes known as the York measure of time [36], goes to positive infinity for a spacelike hypersurface surrounding a region that is collapsing to a black hole, and starts with negative infinity for a 3-sphere hypersurface surrounding the locus of an ideal big bang. In such a history, time runs a natural course from $-\infty$ to $+\infty$.

The law of conservation of momentum and energy, applied to whatever particles and fields – if any – occupy space, demands that at each point of spacetime the Élie Cartan "moment of rotation" [37] defined by the spacetime curvature at that point must equal $16\pi$ times the local density of momenergy [38]; in other words, must vanish in any region unencumbered by fields or particles. The nature of this requirement lets itself be seen in the analogous demand of electrodynamics, that no electric line of force must end in any region of space where there is zero electric charge,

$$\nabla \cdot \mathbf{E} = O \tag{1.11}$$

This one equation seems woefully inadequate to determine the time evolution of all three components of the electric field plus the three components of the magnetic field. However, it is a stronger requirement than appears at first sight. It applies at the point under investigation not only on the chosen spacelike hypersurface through the chosen point, but also for every spacelike hypersurface through the chosen point, whatever its slope, a condition sufficiently powerful to imply the additional three equations of electrodynamics.

Geometrodynamics unfolds its content from an equally compact package, the very heart of Einstein's classical geometrodynamics (GMD). For every probe hypersurface, in whatever way curved, through every event, in space empty of energy and momentum, GMD demands that the sum of the scalars of extrinsic curvature and intrinsic curvature

$$(\mathrm{Tr}\mathbf{K})^2 - \mathrm{Tr}\mathbf{K}^2 + {}^{(3)}\mathrm{R} = 0 \tag{1.12}$$

shall be zero [39].

The term 3-geometry makes sense as well in quantum geometrodynamics as in classical theory. So does superspace. But spacetime does not. Give a 3-geometry, and give its time rate of change. That is enough [under generic circumstances] to fix the whole time-evolution of the geometry; enough, in other words, to determine the entire four-dimensional spacetime geometry, provided one is considering the problem in the context of classical physics. In the real world of quantum physics, however, one cannot give both a dynamic variable and its time rate of change. The principle of complementarity forbids. Given the precise 3-geometry at one instant, one cannot also know at that instant the time-rate of change of the 3-geometry. In other words, given the geometrodynamic field coordinate, one cannot know the geometrodynamic field momentum. If one assigns the intrinsic 3-geometry, one cannot also specify the extrinsic curvature [of that 3-geometry in any purported 4-geometry.]

The uncertainty principle thus deprives one of any way whatsoever to predict or even to give meaning to "the deterministic classical history of space evolving in time." *No prediction of spacetime, therefore no meaning for spacetime*, is the verdict of the quantum principle. That object which is central to all of classical general relativity, the **four-dimensional spacetime** geometry, simply **does not exist**, except in a classical approximation.

These considerations reveal that the concepts of spacetime and time are not primary but secondary ideas in the structure of physical theory. These concepts are valid in the classical approximation. However, they have neither meaning nor application under circumstances where quantum geometrodynamic effects become important. Then one has to forego that view of nature in which every event, past, present, or future, occupies its preordained position in a grand catalog called "spacetime," with the Einstein interval from each event to its neighbor eternally established.. There is no spacetime, **there is no time, there is no before, there is no after**. The question of what happens "next" is without meaning.

That spacetime is not the right way does not mean that there if *no* right way to describe the dynamics of geometry consistent with the quantum principle. Superspace is the key to *one* right way to describe the dynamics.

The hiker looks about him, registers the profile of peaks and valleys silhouetted against the sky, and by comparing it with the contour map he carries in his pocket, knows where he is. Likewise, given the 4-geometry, one can make what spacelike slice one will through that 4-geometry and thereby fix and define a spacelike 3-geometry. Conversely, given the 4-geometry and a 3-geometry compatible with it, one can think in imagination of a spacelike rubber sheet which one pushes forward more in one region of spacetime than in another and whose location one otherwise readjusts with all the care required to reproduce the specified 3-geometry with arbitrary precision. This is the sense in which 3-geometry carries information about time.

In the generic case, the 3-geometry is far from flat. Being the dynamic object in Einstein's geometrodynamics, the 3-geometry itself, $^{(3)}G$, has the familiar dynamic attributes of a field coordinate.

Surely not by limiting our questions to time can we ever expect to understand time. As well close our eyes to all the rest of a huge and complex circuit, as we try to grasp the role of a particular transistor. We are lost until we form some idea of the way of working of the system as a whole. To ask another question, however, which is also hopeless, produces out of the union of no hope and no hope, some hope, in regions of discourse where the two questions may overlap and collaborate and give us some guidance.

Let me therefore counterpose to "How come time?" the question, "How come existence?" and with it, two companion questions which are associated, I think, in the minds of all of us with the idea of existence. (1) "How come the quantum?" absolutely central feature of the physical world as we know it today. And (2) "How come one world out of the registrations of many observer-participants?" This question is especially vivid to those of us who subscribe to something related to Mach's phenomenological view of science, or more precisely, to the idea that what we call "Physics," in the words of Bohr, "is not physics. Physics is what we can say about physics." I don't know any better approach to making headway on this union of two questions, [40] "time" on the one hand, and on the other hand, "How come existence," than four negatives, four nos. " First, no tower of turtles. We should not have one idea, one concept, one principle founded on another, and that on another, and that on still other levels of analysis numberless to man, down into a sunless sea. Instead, we more reasonably expect some closure of thought, some logic loop, some circuit of ideas [41]. And no tower of turtles? No laws! Were we to encounter a law, then we would have to ask where does it come from and with it open up an endless chain of questions. When we accept, however, the demand for no laws, then we also accept no continuum, and therefore, no space and no time.

I have not found any approach to dealing with these issues which offers more comfort than the working hypothesis of "It From Bit": the idea that every "it," every particle, every field of force, even the spacetime continuum itself, derives its function, its meaning, its very existence entirely – even if in some contexts indirectly – from the apparatus-elicited answers to yes or no questions, binary choices, **bits**. This is not to forego admiration for the approach that Murray Gell-Mann and James Hartle have been following [42], to assume quantum mechanics without asking out of what deeper principle it comes and see how far one can make progress on that line with understanding how definiteness comes about. Here in contrast is the idea that quantum theory comes at the bottom from some deeper principle and that we will not really make headway into these deepest questions if we simply take quantum mechanics for granted. Instead, we have to ask "how come quantum mechanics?"

So let me try to spell out some of these question, ideas, and approaches a little more fully by a few pictures and words.

The tablet of granite (Figure 1.4) on which the laws of physics are inscribed was once conceived to endure from everlasting to everlasting. In contrast, there is a moment (the Big Bang) before which there is no "before," and there is a moment (gravitational crunch) after which there is no "after." The laws cannot exist forever and so, not existing forever, they can't be absolutely without exception.

What about this business of "no continuum?" I'm still spelling out these considerations which I put down at the beginning. No continuum. And here I take the words of Willard Van Orman Quine, the mathematician and logician, as my text: "Just as the introduction of the irrational numbers is a convenient myth which simplifies the laws of arithmetic, so. . . physical objects are postulated entities which round out and simplify our account of the flux of existence. . . The conceptual scheme of physical objects is a convenient myth, simpler than literal truth and yet containing

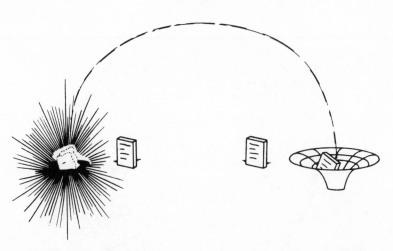

Fig. 1.4. The tablet with the laws of physics inscribed on it – at least in the mind's eye – does not stand from everlasting to everlasting. Therefore here we envisage the "laws" to come into being with the Big Bang and to fade out of existence at gravitational collapse.

that literal truth as a scattered part." At a break in a meeting, I turned around to speak to the colleague sitting just behind me, Quine himself, for whom I have great admiration, and I said, "Van, how come you never speak about the great problems of physics? " "Oh, when I was young, I used to worry about them all the time. But then I solved them to my satisfaction." He later sent me his book, *From a Logical Point of View*, explaining this thesis that the conceptual scheme of physical objects is a convenient myth, simpler than the literal truth and yet containing that literal truth as a scattered part. The quantum: Can we not consider it as our most vivid evidence of this notion of "scattered part?" Could it be that Quine's words lead us to a new and deeper view of physics, "It From Bit," the theme that every particle, every field of force, even the spacetime continuum itself, derives its function, its meaning and its very existence directly or indirectly from apparatus-elicited answers to yes or no questions, bits. That's the thesis which I would propose as way to spell out the concept of "it from bit." As one of many illustrations that one can give of that working hypothesis, let me recall the history of findings about the horizon of a black hole.

In 1970, Demetrios Christodoulou discovered [43] that in every transformation of the Penrose type, that is, in every exchange of energy and angular momentum between an outside agency and a black hole, there is one quantity which either stays constant or always increases, never decreases: the horizon area of the black hole. He discovered that this horizon area has all the attributes of entropy without ever once mentioning the word entropy. Excitement rose when Jacob Bekenstein went on to argue [44] that that area is indeed entropy and that the constant of proportionality between horizon area and any of the usual dimensionless measures of entropy is

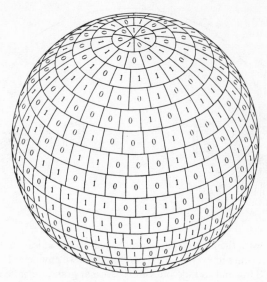

Fig. 1.5. Example of it from bit. The it, the area of the horizon of a black hole, expressed in units of the basic Bekenstein-Hawking area.

the square of the Planck length multiplied by some numerical constant of the order of magnitude of unity. Then Stephen Hawking showed [45] that this constant is $4\pi$ times the logarithm of 2 to the base e. Then Kip Thorne and Wojtek Zurek went on to consider the union of radiation, particles and what have you, to make a black hole [46]. To give a complete description of what goes in takes an enormous amount of information, but it is information that can be expressed in a quantum form, and therefore one can from it sort out and list on page after page the different entitities and quantum states of what goes in to make a given black hole, a black hole of a given horizon area. To list not the information itself but to tell on which page in this great dictionary to look for the information, we need to tell the page number. The Bekenstein-Hawking number gives us, not that page number itself, but the number of binary digits in that page number (Figure 1.5). That finding provides one example of the concept of it from bit in action.

The book edited by Zurek, published a year ago, entitled Complexity, Entropy and the Physics of Information contains further discussion of this point and these ideas. The encounter of Heisenberg and Bohr that led to the principle of indeterminism and then to the principle of complementarity excited Bohr's old and much admired professsor at the University of Copenhagen, Harald Høffding. Professor Høffding invited Bohr and Heisenberg around in the evening to explain to him what this bruhaha on indeterminism and complementarity was all about. Regarding the double slit experiment (Fig. 1.6) he asked, "Where can the electron be said to **be** in

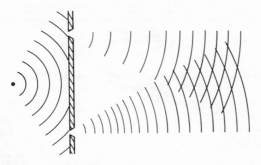

Fig. 1.6. The double slit experiment. Not shown is the detector at the right that registers electron arrival.

its travel from the point of entry to the point of detection?" Bohr's reply deserves to be quoted: "**To be**? **To be**? What does it **mean** 'to *be*'?"

When we turn from electron to photon we recognize that the photon has no existence in the atom before it is emitted. It has no existence in the detector before the detector goes off. But any talk about what the photon is doing between the point of production and the point of reception is, as we know, simply mere talk. To put it in a more dramatic form, the photon is a great smoky dragon. (Figure 1.7). The point of entry of the photon is indicated by the tail. And the point of reception is indicated by the mouth of the great smoky dragon biting the one counter or the other, but in between all is cloud.

Put it yet another way. We used to think that the world exists "out there" independent of us, we the observer safely hidden behind a one-foot thick slab of plate glass (Figure 1.8), not getting involved, only observing.

However, we've concluded in the meantime that that isn't the way the world works. In fact we have to smash the glass, reach in, install a measuring device. But to install that equipment in that place prevents the insertion in the same place at the same time of equipment that would measure the momentum. We are inescapably involved in coming to a conclusion about what we think is already there. Surely we will never understand how come Time until we understand how come the quantum. Whence the necessity for quantum mechanical observership? Does not the necessity for observership reveal itself in this central feature of nature, that we have no right to say that the electron is at such-and-such a place until we have installed equipment, or done the equivalent, to locate it. What we thought was there is not there until we ask a question. No question? No answer!

I'm afraid everyone has heard of my game of Twenty Questions in its surprise version too often to want to hear it again, but let me only just recall–I come into the room, I say "Is it edible?" and my friend that I asked thinks, thinks, thinks, and says finally, "No." Then I ask the next person, "Is it mineral?" Thinks and thinks. "Yes." And so it goes. But before my permitted twenty questions run out, I must

Fig. 1.7. Smoky dragon as symbol of Bohr's elelmentary quantum phenomenon. This drawing was made by Field Gilbert of Austin, Texas, 1983.

arrive at some word. "Is it 'cloud'?" And the respondent that I ask this time thinks and thinks and thinks. I couldn't understand why the delay because supposedly the word already existed in the room, and all he'd have to do is give an answer "yes" or "no." But instead he thinks. Finally he says "yes" and everybody bursts out laughing. They explain to me that when I came in there was no word in the room, that everybody could answer my questions as he wished with one small proviso – that if I challenged, and he couldn't supply a word compatible with all the previous answers, he lost and I won. So it was just as difficult for them as it was for me. Thus the word was brought into being by the choice of questions asked and answers given. The game reminds us of how we find out about the electron, its position or momentum. What we get depends partly on our choice of question asked. However, we don't have the whole voice in it. "Nature" has the rest of the answer.

This example seems to say, that what we call "reality" is a strange thing. Did not the universe exist for some billions of years before there was any life around to observe it? From all we've discovered about cosmology, we are led to believe that starting with the big bang the universe expanded and reached its present dimension

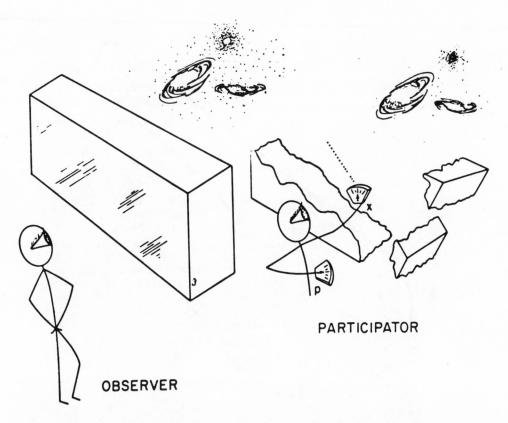

**PARTICIPATOR**

**OBSERVER**

Fig. 1.8. Quantum mechanics evidences that there is no such thing as a mere "observer (or register) of reality." The observing equipment, the registering device, "participates in the defining of reality." In this sense the universe does not sit "out there."

giving rise on the way to the growth of life and mind. We who use observing devices in the here and now give a tangible meaning to the polarization of a photon given out from a quasar at a time long before there was any Earth or any life. So in this sense, we are participators in giving a tangible meaning to those early days. What, then, about the portions of this immense universe that nobody has yet observed? There are billions of years yet to come, there are billions upon billions of living places yet to be inhabited and we come to the fantastic dream that out of observer-participancy at that enormous scale, we can hope to get this whole vast show.

Preposterous? Yes. Crazy? Yes. But surely some day, testable. Put the same thought in yet other language: existence as a meaning circuit. Thus, physics gives rise to light, pressure and sound – means of communication (Figure 1.9); gives birth to chemistry and biology and thus to communicators; and communicators communicating give rise at last to meaning, meaning [47] as the joint product of all

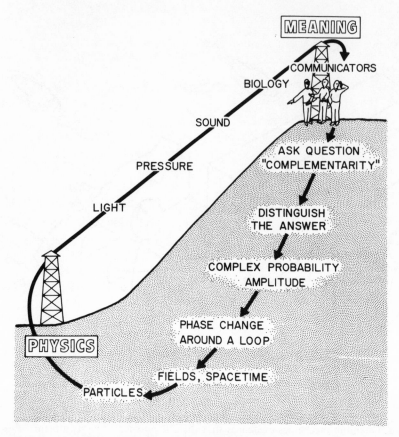

Fig. 1.9. World viewed as a self-synthesizing system of existences. Physics gives light and sound and pressure – tools to query and to communicate. Physics also gives chemistry and biology and, through them, observer- participators. They, by way of the devices they employ, the questions they ask, and the registrations that they communicate, put into action quantum-mechanical probability amplitudes and thus develop all they do know or ever can know about the world.

the information exchanged between those who communicate, meaning as compatible with that old idea that meaning is agreement.

In a double-slit electron-interference experiment of the type proposed by Aharonov and Bohm [48], the interference fringes experience a phase shift proportional – so it is customary to say – to the flux of magnetic field through the domain bounded by the two electron paths. We reverse the language when we turn to the "It From Bit" interpretation of nature. We speak of the magnetic field, and, by extension, spacetime and all other fields, and the whole world of particles built upon these fields, as having no function, no significance, no existence, except insofar as they affect wave phase, affect a 2-slit interference pattern, or, more concretely, affect the counting rate of elementary quantum phenomena. Fields and particles give physics and close the loop.

Out of meaning, Wootters has pointed out, we come to an understanding of the machinery of complementarity, so central to the machinery of quantum theory [49].

That is to say on the one hand we have the asking of a question but with it we have the distinguishing of an answer. The statistician and geneticist, R. A. Fisher, already in the 1920s before the advent of modern quantum theory, taught us [50] that the concept of probability amplitude is more fundamental than probability itself in distinguishing between one population and another. Thus the relevant quantity in a kind of Hilbert space of different populations (Figure 1.10) is not the probability of blue eyes, grey eyes and brown eyes, but the square root of that probability.

The angle between the probability amplitude for those two populations in that real Hilbert space measures the distinguishability of those two populations. One finds himself forced to probability amplitude by the very concept of distinguishability. What about the other side of the story of the quantum, the idea of complementarity? Stückelberg long ago [51], Saxon [52] more recently, and Wooters in more detail [53] have explained how complementarity demands more than real probability amplitudes: complex probability amplitudes.

Out of complex probability amplitudes we found the concept of phase change around a loop, and on that foundation the concept of fields and spacetime, from there to what we call particles, and from there back to physics. So we don't have a tower of turtles to deal with but a logic loop, a meaning loop. (Figure 1.9).

Phase change around a loop in electricity and magnetism as calculated by differentiation from the vector potential leads to the familiar local Maxwell field: But going the other way, integration of the vector potential around a loop, we can come to flux and hence the equally familiar Faraday line of force. [54]

"The Faraday line of force supplies a happy analog for the Ashtekar loop. Both lines and loops put at the center of attention, not the local field (obtained by differentiation of the appropriate potential) but the integral of the relevant potential around a loop. In electromagnetism this idea has become familiar:

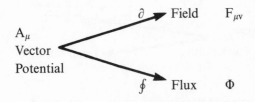

Moreover, the magnetic flux $\Phi$ expresses itself in direct physical terms as well by one or other familiar measuring techniques as by it-from-bit definition *á la* Aharonov and Bohm [55] (see Figure 1.6, supplemented by embraced magnetic flux).

"Ashtekar invented the analogous loop-integral method [56] to deal with geometry:

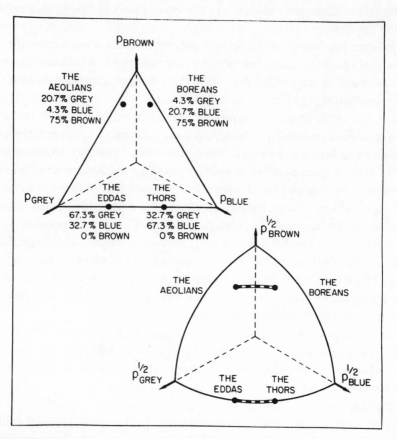

Fig. 1.10. From probabilities to probability amplitudes as tool for determining distinguishability (Aeolians vs. Boreans or Eddas vs. Thors). Triangle above: probabilities of gray, blue and brown eyes for tribes plotted in three-dimensional probability space. Quarter-sphere below, Hilbert space: same information with axes now measuring "probability amplitudes." The angle (dashed arcs) between two points in this Hilbert space measures the distinguishability of the two populations. W. K. Wootters is thanked for assistance in preparing this diagram.

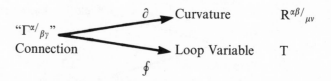

Here the connection, differentiated, gives curvature, whereas integrated around a loop it gives a two-index loop variable T. This connection, however, as signified by the quotes, is not the one familiar in texts of relativity and it is not normally a real-valued quantity. To give a little impression of its character it may be enough to note that electromagnetism admits a similar complex "connection" built by combining magnetic potential A with the imaginary unit times the electric field E.

If the writings of Oliver Heaviside and his followers could make complex-valued quantities familiar to every engineer dealing with electrical machinery, the work nowadays being published on the loop representation, and since carried out by at least four groups, may well provide similar enlightenment to all concerned with the dynamics of geometry. We can measure the magnetic flux through the region between the two branches of the electron beam by measuring the shift it causes in the pattern of interference fringes and via a like fringe shift determine a certain surface integral of spacetime curvature. In the case of geometry, through the loop representation introduced by Ashetekar and other colleagues, we can proceed likewise in analyzing geometry. Instead of going from the connection by differentiation to the local curvature we deal instead with an integral, a loop integral which, however, has some connection with the idea of a knot and a knot-class. All this is background for dealing with the issue of time. In all the history of physics there is no more solidly founded approach to the origin of the concept of time than what general relativity (GMD) gives us.

GMD tells us not to impose our familiar concept of spacetime, for example, as a God-given necessity, but to start with the concepts of a space that starts small, gets bigger and shrinks, and these 3-geometries as 3-dimensional space slices through that spacetime. Then we can use the location of a 3-geometry itself in the enveloping 4-geometry as an indicator of what time is, or where it is. This is the BSW concept that 3-geometry in itself carries information not only about space geometry but also about time. If we think of the different 3-geometries we get by slicing this particular spacetime in different ways, we think of those different geometries including lumps, bumps and wiggles as indicators in a great map with one point for each, then we can give that great map the fancy name of superspace. Each point in superspace gives us one conceivable stage in the dynamic development of 3-geometry. A leaf of superspace gives us a collection of 3-geometries that may or may not represent a classical history of space geometry undergoing its deterministic dynamic evolution. But nowhere in this picture is there any time directly to be seen. There is only 3-geometry. Yet this is what classical physics in its highest incarnation tells us is at bottom the nature of time. In place of the deterministic leaf of history cutting through superspace, quantum theory gives us a thickened leaf, a collection of 3-geometries more vast than we can fit into any one spacetime. In consequence, at the smallest distances, no matter how symbolically we want to portray conditions, we have fluctuations in the geometry which are so great at the Planck scale that the very concepts of before and after lose their significance, their meaning, their application. Time, in this sense, is not the be-all and end-all of the scheme of physics. To put it more quantitatively, we have fluctuations (1) in the position of a harmonic oscillator given by a simple Gaussian wave function, (2) in a collection of harmonic oscillators given by a product of Gaussians (3) for the electromagnetic field or, better, for the magnetic field in the ground state a functional, built like the product of Gaussians, which tells us that the smaller the scale of distances we consider, the more markedly fluctuations show up and (4) in the case of geometry, the fluctuations at the Planck level of distances, as estimated by similar reasoning, tell us that the very concepts of before and after simply have no meaning or application at the Planck scale.

To deal with the quantum dynamics of geometry, one can write down the relevant equation symbolically in the form shown in Eq. (1.2) or in proper detail, as Bryce DeWitt showed, in the form spelled out in Eq. (1.3). Of all ways to describe the quantum mechanics of geometry, however, I don't know any that holds out more promise of new insights than the Ashtekar loop formulation and I don't know any account of it all which is clearer, or provides a more helpful injection of physical ideas on top of the mathematics than the recent account by Carlo Rovelli [56]. Miller [57] and Kheyfets [58] have shown how this account can be introduced and derived from the wonderful "boundary of a boundary" principle for understanding what general relativity is all about, the principle given us by Cartan's formulation of Einstein's equations. Out of it come these loops and a translation of a wave equation into the language of loops and one comes then to the concept that what counts is not the loop which has a location in space, but something which rises above space and does not require space for its definition for picturing but a loop which distinguishes itself from another loop by its so-called knot class–an object which lets itself be translated into the language of bits, digits.

So we have the very great question: Does this mathematical machinery allow of the elementary apparatus-elicited yes or no response to a yes or no question? And if so, does this bit response give us the opportunity to begin to see not only the quantum without introducing the quantum but space and time without space and time? So in reply to the original question which I stated at the beginning, "How Come Time?" I come back again to another question "How Come the Quantum?"

## Summary

In summary, Time today has as many different meanings as there are theoretical frameworks for doing physics. For most everyday purposes we employ the Newtonian concept of time, except that instead of comparing three simple pendulums we compare three vibrating-quartz-crystal watches and in case of disagreement we adopt the principle of majority rule. When we deal with a high speed mu meson, we accept special relativity and figure the spacetime interval between start and decay. When we push resolution to the limit, we ditch the spacetime continuum, we accept the Ashtekar loop representation of quantum gravity, we envisage spacetime as fabric rather than continuum, we recognize the element of spacetime to be, not a spacetime event but the interstice between two threads of warp and two of woof, a domain which in everyday language has dimensions of the order of the square of the Planck length, $1.616 \times 10^{-33}$ cm. In short, time is clothed in a different garment for each role it plays in our thinking.

For any deep advance in our understanding of time we would seem to depend most of all on a new insight, yet to be won, on "How come the quantum?" To win that insight we have less than a decade to the centenary of Planck's 7 October 1900 discovery.

As we meet on Time at Mazagon there sits about 26 kilometers to the west of us, also on the south coast of Spain, the little town of Palos. Christopher Columbus tells us "On the third of August, 1492, a Friday, I left Palos and stood out to sea half an hour before sunrise." From weighing anchor on that day until his landfall on the twelfth of October he demanded of his crew nine weeks of sail, "West: nothing to the northward, nothing to the southward."

We've started later in the morning, we are to have less than one week here to try out our ship before we venture on an even longer voyage. We have no compass and we have a more difficult voyage to accomplish because we lack any plan to find "How Come Time?" – except to try to discover "How Come the Quantum?" Bon Voyage!

## References

[1] Graffito from the men's room of the Pecan Street Cafe, Austin, Texas (1976).

[2] Wheeler, J. A. and Feynman, R. P., Interaction with the Absorber as the Mechanism of Radiation. *Rev. Mod. Phys.* **17**, 157–181 (1945).

[3] Wheeler, J. A. and Feynman, R. P, Classical Electrodynamics in Terms of Direct Interparticle Action, *Rev. Mod. Phys.* **21**, 425–433 (1949).

[4] Tetrode, H, *Zeits. Physik*, **10**, 317 (1922); See also G. N. Lewis, *U.S. Nat. Acad. Sci. Proc.* **12**, 22 (1926).

[5] Fokker, A. D., *Zeits. Physik*, **58**, 386 (1929); *Physica*, **9**, 33 (1929) and **12**, 145 (1932).

[6] Ritz, W. and A. Einstein, *Zeits. Physik*, **13**, 145 (1909).

[7] Partridge, B., The Absorber Theory of Radiation and the Future of the Universe, *Nature*, **244**, 263 (1973).

[8] Cocke, W.J., "Statistical time symmetry and two-time boundary conditions in physics and cosmology," *Phys. Rev.*, **160**, 1165–1170 (1967).

[9] Wheeler, J. A., Frontiers of Time, North Holland, Amsterdam, 451-467; in G. Toraldo di Francia, ed., Problems in the Foundation of Physics, *Proceedings of the International School of Physics, Enrico Fermi, Course 72*, Italian Physical Society, Bologna, 395–497 (1979).

[10] Taub, A. H., "Empty spacetimes admitting a 3-parameter group of motions," *Ann. of Math.*, **53**, 472-490 (1969); and C. W. Misner and A. H. Taub, *Sov. Phys. JETP*, **28**, 122–133 (1951).

[11] JAW, Ref. 9, p. 460.

[12] Christenson, J., J. Cronin, V. Fitch and R. Turlay, hereafter referred to as CCFT, "Evidence for the $2\pi$ Decay of the $K_2^0$ Meson," *Phy. Rev. Lett.* **13**, 138–140 (1964).

[13] Wheeler, J., "Geometrodynamics and the Issue of the Final State," in *Relativity, Groups and Topology: Lectures Delivered at Les Houches during the 1963 Session of the Summer School of Theoretical Physics, University of Grenoble*, eds. C. DeWitt and B. DeWitt, Gordon and Breach, New York and London, 317–522 (1964).

[14] CCFT, Ref. 12

[15] Petrov, Y. "The Oklo Natural Nuclear Reactor," *Sov. Phys. Usp.*, **20**, (11), 937-943 (1977); Shlyakhter, A. "Direct Test of the Time-Independence of Fundamental Nuclear

Constants Using the Oklo Natural Reactor," *Atomki Report A1*, Leningrad, 40–56 (1983); Shlyakhter, A., "Direct test of the constancy of fundamental nuclear constants," *Nature*, **264**, (11), 340 (1976); Dyson, F., "Variations in Constants," in J. Lannutti and P. Williams, eds. *Current Trends in the Theory of Fields*, Amer. Inst. Phys., New York, 163–168 (1978).

[16] Taylor, E. and J. Wheeler. See example, discussion of time-stretching with mu-mesons in *Spacetime Physics: Introduction to Special Relativity, 2nd ed.*, W. H. Freeman, N.Y., 23 (1992).

[17] Cook, Capt. James, Article by J. Beaglehole in C. C. Gillispie, ed., *Dictionary of Scientific Biography*, 3, C. Scribner's Sons, NY, 396–397, (1981).

[18] Owen, G., Article on Aristotle in C. C. Gillispie, ed., *Dictionary of Scientific Biography*, **1**, C. Scribner's Sons, NY, 253 (1981).

[19] Wheeler, J., text quoted from (1982) *Frontiers of Time*, in "Physics and Austerity: Law Without Law," transl. into Chinese by Fang Li Zhi, Anhui Science and Technology Publications, Anhui, China, 40 (1982).

[20] C. T. Harper, ed., (1973) *Geochronology: Radiometric dating of rocks and minerals*, Dowden, Hutchinson & Ross, Inc., Stroudsberg, Pa. (1973); and D. York and R. Farquhar, *The Earth-age and Geochronology*, Pergamon, Oxford (1972).

[21] Qadir, A. and J. Wheeler, Late Stages of Crunch, in Y. S. Kim and W. W. Zachary, eds., *Proceedings of the Internatioal Symposium on Spacetime Symmetries, Nuclear Physics 3, Proceedings Supplements*, N. Holland, Amsterdam, 345–348 (1989).

[22] CCFT, Ref. 12.

[23] Baierlein, R., H. Sharp and J. Wheeler, hereafter referred to as BSW, "Three-dimensional geometry as carrier of information about time," *Phys. Rev.* **126**, 1864–1865 (1962).

[24] For one perspective on how the problem of quantization looked when time was treated as no more than another coordinate, see for example James L. Anderson, "Q-number Coordinate Transformations and the Ordering Problem in General Relativity," pp. 389-392 in M. E. Rose, ed., *Eastern Theoretical Physics Conference, University of Virginia, October 26-27, 1962*. The reference to I owe to the kindness of Prof. Anderson who also cites for a recent discussion J. L. Friedman, *Phys. Rev.*, **D37**, 3495 (1988); and N. C. Tsamis and R. P. Woodard, *Phys. Rev.*, **B36**, 3691 (1987).

[25] Misner, C., K. Thorne and J. Wheeler, hereafter referred to as MTW, Gravitation, W. H. Freeman, San Francisco, now in New York, pp. 1183 (1973).

[26] MTW, pp.1184–85.

[27] BSW, Ref. 23.

[28] MTW, pp. 1193–94.

[29] MTW, p. 1183.

[30] Wheeler, J., Einstein-Schroedinger Equation, p. 276 ff., esp. p. 278 in "Superspace and the Nature of Quantum Geometrodynamics," pp. 242-307 in *Battelle Rencontres* (1968); *1967 Lectures in Mathematics and Physics*.

[31] DeWitt, B., Quantum theory of gravity, I, *Phys. Rev.* **160**, 1113–1148 (1967); Quantum theory of gravity, II, The manifestly covariant theory, *Phys. Rev.* **162**, 1195-1239 (1967); Quantum theory of gravity, III, Application of the covariant theory, *Phys. Rev.* **162**, 1239–1256 (1967).

[32] MTW., Eqs 11 and 12, pp. 421–423.

[33] Cartan, É., hereafter referred to as ÉC, Leçons sur la Géométrie des Espaces de Riemann, Gauthier-Villars, Paris, France, Box 8.2, Ex. 21.4 endpapers (1928 and 1946) hereafter referred to as ÉC; Misner, C. W. and J. A. Wheeler, "Conservation laws and the boundary of a boundary," 338–351 in Shelest, V., ed., *Gravitatsiya: Problem i Perspektivi: pamyati Alekseya Zinovievicha Petrova posvashaetsya*, Naukova Dumka,

Kiev (1972); and J. Wheeler, "The Boundary of a Boundary: Where the Action is!", pp. 109–121, in *A Journey Into Gravity and Spacetime*, Scientific American Library, New York, (1990).

[34] BSW, Ref. 23.

[35] MTW, p. 218–240.

[36] MTW, p. 244–302.

[37] JAW, "Geometrodynamic Steering Principle Reveals the Determiners of Inertia," in *Int. J. Mod. Phys.* **A**, 3, No. 10, 2207-2247, World Scientific, Singapore (1988).

[38] ÉC, Ref. 33.

[39] JAW, "The Boundary of a Boundary: Where the Action is!," *Journey Into Gravity and Spacetime*, Scientific American Library, New York, Chap. 7 (1990).

[40] MTW, Chap. 21.

[41] JAW, "World as System Self-synthesized by Quantum Networking," in *I.B.M. Jour. of Res. and Dev.*, **32**, 4–15 (1988).

[42] M. Gell-Mann and J. Hartle, "Quantum Mechanics in the Light of Quantum Cosmology," in *Proc. 3rd Int. Symp. Found. of Quan. Mech.*, Physical Society of Japan, Tokyo, 321–343 (1989).

[43] Christodoulou, D., "Reversible and Irreversible Transformations in Black Hole Physics," *Phys. Rev. Lett.*, **25**, 1596–1597 (1970).

[44] Bekenstein, J. D., "Generalized Second Law of Thermodyamics in Black-hole Physics," *Phys. Rev.*, **D8**, 3292-3300 (1973).

[45] Hawking, S. W., "Black Holes and Thermodynamics" in *Phys. Rev.*, **13**, 101–197 (1976).

[46] Zurek, W. H., and K. S. Thorne. "Statistical Mechanical Origin of the Entropy of a Rotating, Charged Black Hole," *Phys. Rev. Lett.*, **20**, 2171–2175 (1985).

[47] Føllesdal, D., "Meaning and Experience," 25–44, in S. Guttenplan, ed., *Mind and Language*, Oxford, Clarendon (1975).

[48] Aharonov, Y. and D. Bohm, *Phys. Rev.*, **115**, 485–491 (1959).

[49] Wootters, W. K., hereafter referred to as WKW, "The Acquisition of Information from Quantum Measurements." Ph.D. dissertation, University of Texas at Austin (1980). Available from University Microfilms, Inc., Ann Arbor, MI 48106.

[50] Fisher, R. A. "On the Dominance Ratio," *Proc. Roy. Soc. Edin.*, **42**, 321–341 (1922); Fisher, R. A., *Statistical Methods and Statistical Inference*, New York, Hafner, 8–17 (1956)

[51] Stueckelberg, E.C.G. "Théorme unitarité de S," *Helv. Phys. Acta*, **25**, 577-580 (1952) and "Quantum Theory in Real Hilbert Space," *Helv. Phys. Acta*, **33**, 727-752 (1960).

[52] Saxon, D. S., *Elementary Quantum Mechanics*, Holden, San Francisco, (1964).

[53] WKW, Ref. 49.

[54] Miller, W., A. Ashtekar and C. Rovelli, "A loop representation for the quantum Maxwell field," in *Class and Quan. Grav.*, **9**, 1121–1150 (1992).

[55] Aharonov and Bohm, Ref. 48.

[56] Rovelli, C., "Ashtekar's formulation of general relativity and loopspace non-perturbative quantum gravity: a report," in *Class and Quan. Grav.*, **8**, 1613–1675 (1991).

[57] Miller, W. and A. Kheyfets, "The Boundary of a Boundary Principle in Field Theories and the Issue of Austerity of the Laws of Physics," *J. Math. Phys.*, **32**, 3168–3175 (1991) and, "É Cartan Moment of Rotation in Ashtekar's Formulation of Gravity." *J. Math. Phys.* **33**, 2242–2248 (1992).

[58] Kheyfets, Ref. 57.

# Part One

Information, Computation and Complexity

# 2

# Complexity in the Universe

Charles H. Bennett

*IBM Research Division, T. J. Watson Research Center*
*Yorktown Heights, NY 10598, USA.*

**Abstract**

Nontrivial, "complex" or "organized" states of a physical system may be characterized as those implausible save as the result of a long causal history or evolution. This notion, formalized by the tools of the theory of universal digital computers, is compared to other notions of complexity, and an attempt is made to sketch open problems in the computation theory and statistical physics whose resolution would lead to a better fundamental understanding of "self-organization" in the universe.

The manifest complexity of many parts of the universe, especially living organisms and their byproducts, was formerly thought to be an expression of divine creativity, but is now widely believed to result from a general capacity of matter, implicit in known physical laws, to "self-organize" under certain conditions.

As a rough illustration of the essential ideas of self-organization, consider a sealed aquarium illuminated by a light source and containing a dead mixture of prebiotic chemicals. After a long time this system will fluctuate or otherwise find its way into a "live" macrostate containing, for example, fish and green plants. This live state will be somewhat stabilized relative to dead states by the light, which enables the organisms to grow, reproduce and metabolically defend themselves against thermal and other degradation. In an aquarium of ordinary size, dead states would still be overwhelmingly more probable even in the presence of the light, because spontaneous biogenesis would probably be far less likely than a "gambler's ruin" ecological fluctuation in which, say, the fish died of starvation after eating all the plants. But the larger the aquarium, the less likely will be a simultaneous extinction everywhere within it, and the more likely a spontaneous biogenesis somewhere within it. Finally, if the aquarium were the size of the earth, it might spend most of its time alive, as suggested by paleontological evidence of one dead-to-live and no live-to-dead transitions so far.

To elevate this kind of thinking from a truism that everyone agrees with but no one really understands, to a level of provable or refutable conjectures in statistical physics, we need a more rigorous and mathematical definition of "complexity," the quantity that supposedly increases when a self-organizing system organizes itself. As might be expected, the problem of defining complexity is itself complex, and there are many satisfactory definitions of different kinds of complexity. Below we compare a number of candidates for a definition of complexity, dismissing most of them as unsuitable to our purpose, without meaning to disparage their appropriateness in other contexts. For further details see [2][4][5].

An object might be considered complex if it has *complex behavior or function*, for example if it is able to grow, reproduce, adapt, or evolve in an appropriate environment. Even if it were possible to find mathematical definitions of these properties, we believe a more structural and less functional definition is needed to understand self-organization, because even functionally inert objects, such as a dead body, a book, or a fossil, subjectively can be said to be complex, and would not plausibly be found in a universe lacking some sort of self-organization, or, God forbid, divine intervention.

A more mathematical property related to complex function is *computational universality,* the ability of a system to be programmed through its initial condition to simulate any digital computation. Originally demonstrated for computer-like models such as Turing machines and deterministic cellular automata, computational universality has subsequently been demonstrated for models more closely resembling those studied in mechanics and statistical mechanics, e.g. the hard sphere gas in an appropriate periodic potential [10], noisy cellular automata in 1 and 3 dimensions [12] [13], systems of partial differential equations [17] and even a single classical particle in a finitely complicated box [16]. The ability of universal systems to simulate one another entails that the dynamics of any one of them encodes, in a straightforward manner, the dynamics of any other, and indeed of any process whose outcome can be determined by logical deduction or numerical simulation. For example, one can readily find an initial condition for Moore's particle [16] which will enter a designated region of space if and only if white has a winning strategy in chess, and another initial condition that will do so if and only if the millionth decimal digit of $\pi$ is a 7. Computational universality therefore now appears to be a property that realistic physical systems can have; moreover if a physical system does have that property, it is by definition capable of behavior as complex as any that can be digitally simulated.

However, computational universality is an unsuitable criterion of complexity for our purposes because it is a functional property of systems rather than a structural property of states. In other words it does not distinguish between a system merely capable of complex behavior and one in which the complex behavior has actually occurred. The complexity measure we will ultimately advocate, logical depth, is closely related to the notion of universal computation, but it allows complexity to increase as it intuitively should in the course of a "self-organizing" system's time development.

*Thermodynamic potentials,* such as entropy or free energy, measure capacity for irreversible change, but do not agree with subjective complexity. A human body is more complex than a vat of nitroglycerine, but has lower free energy. Similarly a bottle of sterile nutrient solution has higher free energy, but lower subjective complexity, than the bacterial culture it would turn into if inocculated with a single seed bacterium. The growth of bacteria following inocculation is a thermodynamically irreversible process analogous to crystallization of a supersaturated solution inocculated with a seed crystal. Each is accompanied by a decrease in free energy, and, even in the absence of a seed, is vastly more probable than its reverse: the spontaneous melting of a crystal into a supersaturated solution, or the spontaneous transformation of bacteria into high-free-energy nutrients. The unliklihood of a bottle of sterile nutrient transforming itself into bacteria is therefore not a manifestation of the second law, but rather of a putative new "slow growth law" which forbids complexity, however it is defined, to increase quickly, but allows it to increase slowly, e.g. over geological time in biogenesis. This example also illustrates the non-additivity of subjective complexity. One bacterium seems much more complex than none, but only slightly less complex than the bottle full of descendants it can quickly give rise to.

*Algorithmic Information Content,* also called Algorithmic Entropy, Algorithmic Complexity, or Solomonoff-Kolmogorov-Chaitin Complexity [20][7][8], formalizes the notion of amount of information necessary to uniquely describe a digital object $x$. A digital object means one that can be represented as a finite binary string, for example, a genome, an Ising microstate, or an appropriately coarse-grained representation of a point in some continuum state space. The algorithmic entropy $H(x)$ of such an object is defined as the negative base-2 logarithm of the object's *algorithmic probability*, $P(x)$. This in turn is defined as the probability that a standard universal computer $U$, randomly programmed (for example by the proverbial monkey typing at a binary keyboard with two keys), would embark on a computation yielding $x$ as its sole output, afterward halting. The algorithmic probability $P(x)$ may be thought of a weighted sum of contributions from all programs that produce $x$, each weighted according to the negative exponential of its binary length, which is the probability that the monkey will type that particular program and so cause it to be executed. An *algorithmically random* string is defined as one of maximal information content, nearly equal to the length of the string (even if a string has no regularities permitting it to be produced with higher probability, any $N$–bit string can be generated with probability at least $2^{-(N+O(\log N))}$ by a "print program" in which the monkey essentially types the string out verbatim, along with instructions, of length $O(\log N)$, directing the computer to pass these $N$ bits on directly to the output and then halt).

Turning now to the sum of $P(x)$ over *outputs,* this sum $\sum_x P(x)$ is not equal to unity as one might first suppose, because, as is well known, an undecidable subset of all universal computations fail to halt, and so produce no output. Therefore $\sum_x P(x)$ is an uncomputable irrational number less than 1. This number, called

Chaitin's Omega [7], has many remarkable properties [14], such as the fact that its uncomputable digit sequence is a maximally compressed form of the information required to *solve* the halting problem.

Despite being defined in terms of a particular universal computer, algorithmic probability is machine-independent up to an multiplicative constant (and algorithmic entropy up to an additive constant), because of the ability of universal computers to simulate one another (programs for one machine can be adapted to run on another by prefixing each program with a constant string, directing the second machine to simulate the first).

Though very differently defined, algorithmic entropy is typically very close to ordinary statistical entropy $-\sum p \log p$ in value. To take a simple example, it is easy to show that almost all $N$–bit strings drawn from a uniform distribution (of statistical entropy $N$ bits) have algorithmic entropy nearly $N$ bits. More generally, in any concisely describable ensemble of digital objects, e.g. a canonical ensemble of Ising microstates at a given temperature, the ensemble average of the objects' algorithmic entropy closely approximates the whole ensemble's statistical entropy [20] [1]. In the case of continuous ensembles, the relation between algorithmic and statistical entropy is less direct because it depends on the choice of coarse-graining. Zurek [21] discusses some of the conceptual issues involved.

For this reason algorithmic information is best thought of as a measure of randomness, not subjective complexity, being maximal for coin-toss sequences, which are among the least organized subjectively. Typical organized objects, on the other hand, precisely because they are partially constrained and determined by the need to encode coherent function or meaning, contains less information than random sequences of the same length; and this information reflects not their organization, but their residual randomness.

For example, the algorithmic information content of a genome represents the extent to which it is underdetermined by the constraint of viability. The existence of noncoding DNA, and the several percent differences between proteins performing apparently identical functions in different species, make it clear that a sizable fraction of the genetic coding capacity is given over to transmitting such "frozen accidents", evolutionary choices that might just as well have been made otherwise.

A better way of applying statistical or algorithmic information to the definition of organization is to use it to characterize the correlations typical of organized or complex objects: two parts of such an object taken together typically require fewer bits to describe than the same two parts taken separately. This difference, the *mutual algorithmic information* between the parts, is the algorithmic counterpart of the non-additivity of statistical or thermodynamic entropy between the two parts, the amount by which the entropy of the whole falls short of the sum of the entropies of the two parts. In many contexts, e.g., communications through a noisy channel, mutual information can be viewed as the "meaningful" part of a message's information, the rest being meaningless information or "noise".

A body is said to have long range order if even arbitrarily remote parts of it are correlated. However, crystals have long range order but are not subjectively very complex. Organization has more to do with the *amount* of long-range correlation, i.e., the number of bits of mutual information between remote parts of the body. Although we will ultimately recommend a different organization measure (logical depth), remote mutual information merits some discussion, because it is characteristically formed by nonequilibrium processes, and can apparently be present only in small amounts at thermal equilibrium.

If two cells are taken from opposite ends of a multicellular organism, they will have a large amount of mutual information, if for no other reason than the presence in each cell of the same genome with the same load of frozen accidents. As indicated earlier, it is reasonably certain that at least several per cent of the coding capacity of natural genomes is used to transmit frozen accidents, and hence that the mutual information between parts of a higher organism is at least in the hundred megabit range. More generally, mutual information exists between remote parts of an organism (or a genome, or a book) because the parts contain evidence of a common, somewhat accidental history, and because they must function together in a way that imposes correlations between the parts without strictly determining the structure of any one part. An attractive feature of remote mutual information for physical systems is that it tends to a finite limit as the fineness of coarse-graining is increased, unlike simple information or entropy in a classical system.

Since mutual information arises when an accident occurring in one place is replicated or propagated to another remote place, its creation is an almost unavoidable side effect of reproduction in a probabilistic environment. Another obvious connection between mutual information and biology is the growth of mutual information between an organism and its environment when the organism adapts or learns.

Further support for remote mutual information as an organization measure comes from the fact that systems stable at thermal equilibrium, even those with long range order, exhibit much less of it than nonequilibrium systems. Correlations in systems at equilibrium are generally of two kinds: short range correlations involving a large number of bits of information (e.g. the frozen-in correlations between adjacent lattice planes of an ice crystal, or the instantaneous correlations between atomic positions in adjacent regions of any solid or liquid), and long range correlations involving only a few bits of information. Typical of these latter correlations are infinite-range correlations associated with order parameters such as magnetization and crystal lattice orientation and phase. Even when these order parameters are continuous, they convey only a few bits of information, owing to the thermal and zero-point disorder which causes the lattice orientation, say, of an $N$–atom crystal to be well-defined only to about $\log N$ bits precision. Besides involving much less information, remote correlations at equilibrium differ qualitatively from the non-equilibrium ones discussed earlier: equilibrium correlations, in a system with short-range forces, must be propagated through an intervening medium, while nonequililbrium ones (e.g.

between the contents of two newspaper dispensers in the same city) need not pass through the intervening medium but are instead typically propagated through a V-shaped path in spacetime connecting the random origin of the information at an earlier time with two separated copies of it at a later time.

Despite these advantages, we believe remote mutual information is an unsatisfactory complexity measure because large quantities of it can be produced rapidly, by subjectively trivial nonequilibrium processes, in violation of the slow growth law. For example, by pulverizing a piece of glass with a hammer, one can produce a kind of 3-dimensional jigsaw puzzle of atomically complementary random fracture surfaces, with a non-additivity of entropy, between two specimens of the powder, proportional to the area of complementary surface between them. A greater non-additivity could be produced by enzymatically replicating, and then stirring, a solution of random, biologically meaningless DNA molecules to produce a kind of jigsaw puzzle soup, two spoonfuls of which would have macroscopically less than twice the entropy of one spoonful. In both these examples, the mutual information is formed by nonequilibrium processes and would decay if the system were allowed to approach a state of true thermal equilibrium, e.g. by annealing of the separated fracture surfaces.

A conspicuous feature of many nontrivial objects in nature and mathematics is the possession of a *fractal or self-similar or hierarchical structure,* in which a part of the object is identical to, or is described by the same statistics as, an appropriately scaled image of the whole. This often beautiful property is too specialized to be an intuitively satistfactory criterion of complexity because it is absent from some subjectively complex objects, such as the decimal expansion of pi, and because, on the other hand self-similar structures can be produced quickly, e.g. by deterministic cellular automata, in violation of the slow growth law. Even so, the frequent association of self-similarity with other forms of organization deserves comment. In some cases, self-similarity is a side-effect of computational universality, because a universal computer's ability to simulate other computers gives it in particular the ability to simulate itself. This makes the behavior of the computer on a subset of its input space (e.g., all inputs beginning with some prefix *s* that tells the computer to simulate itself) replicate its behavior on the whole input space.

*Logical Depth,* the plausible number of computational steps in an object's causal history, is the complexity measure we chiefly recommend. A logically deep object, in other words, is one containing internal evidence of having resulted from a long computation, or from a dynamical process requiring a long time for a computer to simulate. Thus a fossil is deep because it is plausible only as a byproduct of a long evolution, unlike the complementary fracture surfaces in the broken glass example above, which are plausible as the result of a short evolution.

To formalize this notion, we consider the distribution of *running times* of computations by which the standard universal computer might produce the digital output $x$. Let $P_t(x)$ be the probability that the standard universal computer, randomly programmed by monkeys as before, would produce the output $x$ by a computation that halts in time $\leq t$. Thus $P_t(x)$, for each $x$, is a monotonically increasing function of $t$, approaching in the long time limit $P_\infty(x) = P(x)$, i.e. the ordinary time-unbounded algorithmic probability discussed before. A digital object $x$ is said to be "$t$ deep with $b$ bits confidence" iff $P_t(x)/P(x) < 2^{-b}$, in other words, if all but a fraction $< 1/2^b$ of the monkey computations that produce $x$ take more time than $t$ to do so. Inasmuch as the set of universal computations producing $x$ may be regarded as a fairly-weighted microcosm of all causal or logical processes by which $x$ could have arisen, for an object to be $t$ deep with $b$ bits confidence means that the complementary null hypothesis, that $x$ originated by a process of fewer than $t$ steps, can be rejected at the $2^{-b}$ confidence level, i.e. as less likely tossing $b$ consecutive tails with a fair coin. The confidence parameter $b$ may seem a nuisance, but it is a necessary part of the idea. Since there are many ways of computing any output $x$, we can make no absolutely certain assertions about how $x$ originated based on intrinsic evidence, only assertions at some level of statistical confidence. As in ordinary statistical discussions, we will sometimes omit mention of the confidence parameter, assuming it to have been set at a value that is safe and conservative in the given context.

Thus defined, depth can be shown to be machine-independent and to obey the slow growth law to within a polynomial in the computation time and an additive constant plus a term of order $\log b$ in the confidence parameter [5]. This imprecision is unfortunately characteristic of the theory of computation times, which typically differ by a small polynomial between one universal machine and another (e.g. one machine may require time $t^2 + 4t + 23$ to simulate what another can do in time $t$).

Algorithmically random strings, of maximal information content (nearly equal to their length) are shallow because the fast-running print program mentioned above contributes a significant fraction of their rather low algorithmic probability. At the other extreme, trivial nonrandom strings such as '0000000...' are also shallow, because though their algorithmic probability is high, a great deal of it can be accounted for by small fast programs of the form "FOR I=1 TO N; PRINT '0'; NEXT I;". On the other hand a string such as the second million digits of pi, which looks random and is not the output of any known small fast program, but is the output of a small slow program (Compute pi, throw away the first million digits, and print the next million), has the possibility of being deep. (This remains unproven, though. See below for a discussion of provably deep strings.)

Returning to the realm of physical phenomena, we note that use of a universal computer frees the notion of depth from excessive dependence on particular physical processes (e.g., prebiotic chemistry) and allows an object to be called deep only if there is no shortcut path, physical or non-physical, to reconstruct it from a concise description. An object's logical depth may therefore be less than its chronological age. For example, old rocks typically contain physical evidence (e.g., isotope ratios) of the time elapsed since their solidification, but would not be called deep if the aging process could be recapitulated quickly in a computer simulation. Intuitively, this means that the rocks' plausible history, though long in time, was rather uneventful, and therefore does not deserve to be called long in a logical sense.

Although a deep object cannot quickly be made from a shallow one (slow growth rule) a deep object can be quickly made by juxtaposing *two* shallow objects, if these are correlated in a deep way. To see this, let $x$ be a deep string and $r$ be a random string of the same length, generated by coin tossing. Both $r$ and the string $y$ obtained by XORing $r$ and $x$ bit by bit are uniformly distributed over the space of $N$–bit strings, and so both are with high probability algorithmically random and therefore shallow. However the concatenation string $ry$, from which $x$ can quickly be made, is deep because of the deep correlation between $r$ and $y$.

In nature, something like the reverse of this process is more common: a deep object, interacting with its surroundings, typically contaminates them and makes them deep too. For example, outside our hotel, I found this beer-can pull-tab on the ground. I would say that a beer-can pull-tab, although a trivial and worthless byproduct of biological evolution, is so a priori implausible except as a byproduct some such evolution that it probably made the ground it was on nearly as deep as the civilization that produced the beer.

Although time (machine cycles) is the resource closest to the intuitive notion of computational work, space (i.e. memory) is also important because it corresponds to a statistical mechanical system's number of particles or degrees of freedom. The maximum relevant time for a system with $N$ degrees of freedom is of order $2^N$, the Poincaré recurrence time; and the deepest state such a system could relax to would be one requiring time $2^N$, but only memory $N$, to compute from a concise description.

Unfortunately, it is not known that any space-bounded physical system or computer can indeed produce objects of such great depth (exponential in $N$). This uncertainty stems from the famous open P=?PSPACE question in computational complexity theory [11], i.e., from the fact that it is not known whether there exist computable functions requiring exponentially more time to compute than space. In other words, though most complexity theorists suspect otherwise, it is possible that the outcome of every exponentially long computation or physical time evolution in a space-bounded system can be predicted or anticipated by a more efficient algorithm using only polynomial time.

A widely held contrary view among complexity theorists today, considerably stronger than the mere belief that P is not equal to PSPACE, is that there are "cryptographically strong" pseudorandom number generators [6][15], whose successive outputs, on an $N$-bit seed, satisfy all polynomial time (in $N$) tests of randomness. The existence of such generators implies that space-bounded universal computers, and therefore any physical systems that mimic such computers, can after all produce exponentially deep outputs.

Deep mathematical objects can be shown to exist without invoking any unproven assumptions by diagonal arguments similar to that used to prove the existence of uncomputable functions. For example, for appropriate values of $N$ (greater than a few thousand, say, to be safely larger than overhead in program size required to combine simple subroutines or program one simple machine to simulate another), the algorithm

```
By exhaustive simulation of all possible computations running
less than $2^N$ steps, find and print out the lexicographically
first $N$-bit string $x$ whose algorithmic probability, from
computations running less than $2^N$ steps, is less than $2^{-N/2}$
```

defines specific $N$–bit string that by construction is $2^N$ deep with about $N/2-\log N-c$ bits confidence, where $c$ is the number of bits required to program the above algorithm in machine language. The string must exist because there are too many $N$–bit strings for them all to have time-bounded algorithmic probability as great as $2^{N/2}$, and of the ones that do not, there must be a first.

Though such constructions establish the existence of deep objects, actual execution of the algorithm would use so much space and time (exponential and double-exponential in $N$, respectively) as to be utterly nonphysical.

It is worth noting that neither algorithmic information nor depth is an effectively computable property. This limitation follows from the most basic result of computability theory, the unsolvability of the halting problem, and reflects the fact that although we can prove a string nonrandom (by exhibiting a small program to compute it) we cannot in general prove it random. A string that seems shallow and random might in fact be the output of some very slow running small program, which ultimately halts but whose halting we have no means of predicting. This open-endedness is a necessary feature of the scientific method: at any time some phenomena will always be incompletely understood, so they appear more random and less deep than than they really are.

The uncomputablilty of depth is no hindrance in the present theoretical setting where we assume a known cause (e.g., a physical system's initial conditions and equations of motion) and try to prove theorems about the depth of its typical effects. Here it is usually possible to set an upper bound on the depth of the effect by first showing that the system can be simulated by a universal computer within a time $t$ and then invoking the slow growth rule to argue that such a computation, deterministic or probabilistic, is unlikely to have produced a result much deeper than $t$. On the other hand, proving lower bounds for depth, e.g., proving that a given deterministic or probabilistic cause certainly or probably leads to a deep effect, though always possible in principle, is more difficult, because it requires showing that no equally simple cause could have produced the same effect more quickly.

Aside from its nonspecific usefulness in clarifying intuition, the notions of complexity discussed here raise potentially decidable questions in statistical physics and the theory of computation concerning necessary and sufficient conditions for the production of complexity, especially logical depth.

In the theory of computation the relation of depth to classic unproved conjectures in time and space complexity has been mentioned.

In statistical physics, the role of dissipation in generating and stabilizing complexity is a major problem area. The need for dissipation to produce and stabilize remote non-additive entropy in locally interacting systems has already been mentioned and is fairly well understood. Concerning depth, one may ask in general how dissipation can help error-correcting computation to proceed despite the locally destructive effects of noise.

One obvious way dissipation assists in error-correction is by allowing compression (many-to-one mapping) of a system's information-bearing degrees of freedom, which, in making the error, have undergone a one-to-many mapping. Another way dissipation may help is by exempting systems from the Gibbs phase rule which applies to equilibrium systems with short-ranged interactions [3]. In typical $d$–dimensional equilibrium systems of this sort, barring symmetries or accidental degeneracy of parameters such as occurs on a coexistence line, there is a unique thermodynamic phase of lowest free energy. The nucleation and growth of this most stable state renders equilibrium systems ergodic and unable to store information reliably in the presence of "hostile" (i.e. symmetry-breaking) noise. Since they forget their initial conditions, such systems cannot be programmed by them, and so cannot be computationally universal. Analogous dissipative systems, because they have no defined free energy in d dimensions, are exempt from this rule. A $d + 1$ dimensional free energy can be defined, but varying the parameters of the $d$ dimensional model does not in general destabilize one phase relative to another [9].

One may ask what other properties besides irreversibility a system needs to take advantage of the exemption from Gibbs phase rule. Known examples, such as Toom's cellular automaton rules [19], lack rotation symmetry, but it is not known whether this is necessary.

Conversely one can ask to what extent equilibrium systems (e.g. quasicrystals) can be caused to have computationally complex ground states, even though they remain subject to the Gibbs phase rule [18].

Finally one can ask whether dissipative processes such as turbulence, that are not explicitly computational or genetic or error-correcting, can still generate large amounts of remote non-additive entropy. Do they generate logical depth? Does a persistent hydrodynamic phenomenon such as Jupiter's Great Red Spot contain internal evidence of a nontrivial dynamical history leading to its present state, or is there no systematic objective difference between a the red spot of today and that of a century ago?

## Acknowledgements

These ideas are the outcome of about 25 years of thoughtful discussions and suggestions from from Ray Solomonoff, Gregory Chaitin, Rolf Landauer, Dexter Kozen, Gilles Brassard, Leonid Levin, Peter Gacs, Stephen Wolfram, Geoff Grinstein, Tom Toffoli, Norman Margolus, and many others. Some of this work was done while the author was visiting California Institute of Technology as a Sherman Fairchild Scholar.

## Discussion

**Schulman**  Is this "nonadditive entropy" information, entropy, or something you'd measure with a calorimeter?

**Bennett**  Both. It can be expressed as a nonadditivity of algorithmic information, or as a nonadditivity of thermodynamic entropy that could be measured, in the case of the DNA soup, by integrating along a reversible calorimetric path in which the duplicated DNA was reversibly restored to to its non-duplicated state by a carefull reversal of the action of the copying enzymes.

**Lebowitz**  What is the relation between the complexity of the beer can top and that of the Alhambra, or between a Rembrandt painting and a child's crayon drawing?

**Bennett**  The Alhambra is deeper, but maybe not much. Both contain evidence of the general scope of biological and cultural evolution, but the Alhambra may contain evidence of additional causal processes not necessary to produce beer cans, and not likely side-effects of any beer-can-producing civilization.

**Miller**  With regard to your notion of logical depth, what do you mean by "very long"? For what message length does your definition become well-defined?

**Bennett**  The messages need to be longer than the number of bits required to program one simple universal computer to simulate another, or to program the fairly simple algorithms implicit in the proof of the slow growth law, typically a few thousand bits, for depth to be reasonably robust.

**Cover**  You mentioned that algorithmic complexity is computer-independent. Is that true also of logical depth?

**Bennett**  Less so. [As noted above in the printed version of the talk,] logical depth, being based on time complexity, suffers from the polynomial slop typical of time complexity results that attempt to be machine-independent over a reasonably broad range of machines.

**Unruh**  Isn't the function $H(x)$ undefined since you can't know if the random program won't stop.

**Bennett**  This makes $H(x)$ uncomputable, but it is still well defined.

**Gell-Mann**  1. I believe it would be helpful to include, in the list of systems characterized by what they can do rather than what they are, COMPLEX ADAPTIVE SYSTEMS that can adapt or evolve. 2. Charlie and his friends are typically interested in long messages, for which additive constants and polynomial functions may not matter much. If one cares about systems described by shorter messages, then it is desirable to know from the beginning the describing system, the nature of the language it employs, the coarse-graining of what it is describing, and so forth. Only in that way can absolute quantities be defined, if at all. 3. Although it is not relevant to Charlie's argument, it should be noted that between a schema like DNA and a "phenotypic" object like a human being, a large amount of partly random information is introduced in the course of development, so that the individuality of a human being is much greater than that of the DNA.

**Lloyd**   Is the beer can pull tab as complex as the civilization that produced it?

**Bennett**   It depends on how much of the world's history was plausibly necessary to produce the beer can. It also depends on whether one defines depth using a purely serial machine such as a Turing machine, or a moderately parallel one, capable of simulating, without having to slow down, all the parallel dynamical processes going on in our civilization. In the latter case, the depth of civilization is only greater to the extent that it contains objects not plausible as byproducts of a beer-can-producing civilization, since plausible byproducts could be simulated at no extra cost. In the former case, the difference may be greater, reflecting the extent to which civilization contains evidence of causal processes not plausibly *necessary* to produce beer cans.

**Wootters**   In your definition of algorithmic entropy, is there a reason that you used the monkey formulation rather than the length of the shortest program that produces the desired output sequence?

**Bennett**   Including all the other programs besides the shortest makes only an additive constant difference in algorithmic entropy, but is necessary in the definition of depth, where the other programs, besides the shortest, help to determine the significance parameter. Also it is possible, though not proven, that there may be objects that are "deterministically deep but probabilistically shallow", in other words, objects that have a high fast probability, but no single small fast program.

**Zurek**   First a comment, then a question. Andy Albrecht was wondering about the relevance of such algorithmic considerations to the issue of "coarse grainings," and you have implicated me. I do not want to go into details here, so let me only mention that one way in which algorithmic randomness is helpful in this context is that it can be used to help clarify the well-known problem of the "simplicity" of coarse-grainings. It is often argued that a choice of coarse-graining is a privilege of the observer and, therefore, the entropy is defined with respect to it has an observer-dependent value. This is certainly true. Nevertheless algorithmic randomness could be used to prove that the observers which can communicate with ease will also agree on which coarse-grainings are simple. Therefore their estimates of entropy will agree to very high accuracy. Now for the question. Could you comment on the "thermodynamic depth" which has also been proposed as a measure of complexity?

**Bennett**   I meant to. Thermodynamic depth differs from the complexity measures I have been emphasizing here in that it depends on the history rather than just the state. The thermodynamic depth of the history of the igneous rock would, as I understand, be large, reflecting the large amount of dissipation that occurred in that history, whereas the logical depth of the rock is small, because of the ability to short-circuit this long history by a short computation.

**Albrecht**   What about operating system–dependence? There can be an operating system that prints out the human genome every time you press "H"?

**Bennett**  If the operating system is treated as part of the program (external data fed into the computer) there is no problem. If it is treated as part of the computer, then that computer (with the whole human genome built in) could not fairly be called a simple computer. Even if one perversely decided to use it as the standard universal computer, algorithmic entropies defined on it would not differ from those defined on a simple Turing machine by more than a rather large additive constant, the information content of the human genome.

**Albrecht**  What is the complexity of the system after the bacteria have died?

**Bennett**  Lower. More specifically it depends on how soon after they have died. Immediately afterward, it is probably pretty deep. When the bacteria have all decayed to an equilibrium mixture of carbon dioxide, and water, etc., they are shallow again.

**Albrecht**  So complexity need not increase monotonically like entropy?.
**Bennett**  That is correct.

**Teitelboim**  Is our universe deeper than any other conceivable universe?

**Bennett**  I don't know. I guess that there might be other universes with less wasted motion than ours, more efficient computations and less forgetting of deep things that have been computed before, but on the other hand my remark about deep objects contaminating their environment suggests that not much depth is ever destroyed.

# References

[1]  C.H. Bennett, "The Thermodynamics of Computation— a Review", *Internat. J. Theoretical Phys.* **21**, 905-940 (1982).
[2]  C.H. Bennett, "Information, Dissipation, and the Definition of Organization", in *Emerging Syntheses in Science*, David Pines ed., Addison-Wesley (1987).
[3]  C.H. Bennett and G. Grinstein *Phys. Rev. Lett.* **55**, 657-660 (1985).
[4]  C.H. Bennett, "On the Nature and Origin of Complexity in Discrete, Homogeneous, Locally-Interacting Systems", *Foundations of Physics* **16**, 585-592 (1986).
[5]  Charles H. Bennett "Logical Depth and Physical Complexity" in *The Universal Turing Machine – a Half-Century Survey*, edited by Rolf Herken Oxford University Press 227-257 (1988).
[6]  M. Blum and S. Micali "How to Generate Cryptographically Strong Sequences of Pseudo Random Bits," *SIAM J. on Computing* **13**, 850-864 (1984).
[7]  G. Chaitin, "A Theory of Program Size Formally Identical to Information Theory", *J. Assoc. Comput. Mach.* **22**, 329-340 (1975).
[8]  G. Chaitin, "Algorithmic Information Theory", *IBM J. Res. Develop.* **21**, 350-359, 496, (1977). Cf also *Information, Randomness, and Incompleteness—Papers on Algorithmic Information Theory*, (World Scientific Press, Singapore 1987).
[9]  E. Domany and W. Kinzel, *Phys. Rev. Lett.* **53**, 311 (1984).
[10]  E. Fredkin and T. Toffoli, *Internat. J. of Theo. Phys.* **21**, 219, 1982.
[11]  M. Garey and D. Johnson, *Computers and Intractability, a Guide to NP Completeness* (Freeman, 1979).
[12]  P. Gacs, *Journal of Computer and System Sciences* **32**, 15-78 (1986)
[13]  P. Gacs and J. Reif, *Proc. 17th ACM Symposium on the Theory of Computing*, 388-395 (1985).
[14]  Gardner, Martin, "Mathematical Games" *Scientific American* 20-34 (November 1979)
[15]  L. Levin "One-Way Functions and Pseudorandom Generators" *ACM Symposium on Theory of Computing*, (1985).

[16]  C. Moore, *Phys. Rev. Lett.* **64**, 2354 (1990).

[17]  S. Omohundro, *Physica* (Amsterdam) **10D**, 128-134 (1984).

[18]  Charles Radin, *J. Math. Phys.* **26**, 1342 (1985) and *J. Stat. Phys.* **43**, 707 (1986).

[19]  A. L. Toom in *Adv. in Probability* **6** (Multicomponent Systems, ed. by R. L. Dobrushin), Dekker, New York (1980), pp. 549-575.

[20]  A.K. Zvonkin and L.A. Levin, "The Complexity of Finite Objects and the Development of the Concepts of Information and Randomness by Means of the Theory of Algorithms", *Russ. Math. Surv.* **256** 83-124 (1970).

[21]  W.H. Zurek, *Phys. Rev.* **A 40**, 4731-4751 (1989)

# 3

# Information, Entropy, and Chaos

Carlton M. Caves[†]

*Center for Laser Studies*
*University of Southern California*
*Los Angeles, CA 90089-1112, U.S.A*

## 3.1 Introduction

How are information and entropy connected? In two ways. First, there is a trade-off between information about a physical system and the system's entropy: acquire $n$ bits of information, and thereby reduce the entropy by $nk_B \ln 2$ or, equivalently, increase the free energy by $nk_B T \ln 2$, an increase that can be extracted as work. This trade-off holds only in a statistical sense, but in that sense it has a precise mathematical formulation. Second, there is a free-energy cost associated with information. The cost is not incurred in acquiring information or in processing it (reversibly). The bill must be paid only when information is erased (Landauer 1961). The minimum erasure cost comes to $nk_B T \ln 2$ and thus cancels the work made available by acquiring the $n$ bits of information (Bennett 1982).

The first of these connections is a purely mathematical relation between information and entropy; physics is introduced by the second connection, which relates information to free energy and work. The two connections together suggest regarding the erasure cost as a *negative* contribution to the *total* free energy. Equivalently, as Zurek (1989a, 1989b) has proposed, one should enlarge the concept of entropy to include both the ordinary entropy and a contribution from the information "we" have about a physical system. Information having been granted a physical status, one can talk about a "physics of information."

To avoid excessive use of "we" and "one," I use the following construct: a physical system is observed by a "memory," which gathers, stores, and manipulates information about the system. The primary difficulty in understanding the physics of information arises from a tension between two ways of viewing the system and the memory (Zurek 1989b). From the "outside" the system and the memory are both viewed as physical systems, described in the standard language of statistical physics. On this outside view connections between information and entropy are nearly trivial, because ordinary entropy is formally identical to Shannon's *statistical*

† Present address: Department of Physics and Astronomy, University of New Mexico, Albuquerque, NM87131-1156, USA

measure of information (Shannon and Weaver 1949; Gallager 1968). From the inside, however, the memory sees itself as storing a certain number of bits, which record something about the system. The connection to entropy is not so clear, because the memory wants an *absolute* measure of information, simply a count of the one-bit registers it uses. The close connection between information and entropy has been appreciated for some time, notably by Szilard (1929) in a seminal paper, by Brillouin (1962) in his identification of information with negentropy, and by Jaynes (1983) in a lifetime of work elucidating the Bayesian view of probabilities, but the connection has been obscured by tension between the outside view, where a statistical measure of information suffices, and the inside view, where an absolute measure of information is required.

The necessary absolute measure of information comes from the theory of computation and is called *algorithmic information* (Solomonoff 1964; Kolmogorov 1965; Chaitin 1966). The algorithmic information of an object is, crudely speaking, the length (in bits) of the shortest complete description of the object. The shortest description is the one of interest because it sets the minimum erasure cost. The notion of algorithmic information is made precise by letting the object be a binary digit string and by letting a complete description be a computer program that computes the string. Formally, the algorithmic information $I(q)$ of a binary string $q$ is the length (in bits) of the *shortest* program which, when presented to a *universal* computer, causes the computer to compute $q$ and then to *halt*. This article does not review any details of algorithmic-information theory; the reader is referred to papers (Zvonkin and Levin 1970; Chaitin 1987) in the computer-science literature and to discussions (Zurek 1989a, 1989b; Caves 1990a) with a more physical bent. It should be noted that algorithmic information is computer-dependent, so results in algorithmic-information theory are typically proved to within a computer-dependent additive constant.

Consider now a situation where there are $\mathcal{J}$ alternatives—in physical language a system that has $\mathcal{J}$ states. A typical scenario in the physics of information involves the following: the memory, starting with some background information, observes the system, processes or uses the new information, and ultimately erases information to get back to its background state of knowledge. Two kinds of algorithmic information enter into this scenario. The first is the algorithmic background, or prior information $I_0$. Typically the background information is the knowledge that the alternatives occur with probabilities $p_j$. The *algorithmic background information* $I_0$ is then the amount of information required to generate a description of the system together with a list of the alternatives and their probabilities. Given the background information, the memory needs additional information to pick out a particular alternative. The second kind of information, the *conditional algorithmic information* $\Delta I_j$, is the additional information, beyond the background information, required to generate a complete description of—i.e., to specify—alternative $j$. One of the principal results of algorithmic-information theory is that the sum $I_0 + \Delta I_j$ is (to within a computer-dependent constant) the joint algorithmic information $I_j$ to generate both the background information and a description of alternative $j$.

The physics of information is mainly concerned with conditional algorithmic information, that being the information that is gathered by observation, processed and used by the memory, and ultimately erased. In this article a $\Delta$ serves as a reminder that conditional algorithmic information is the difference between two other algorithmic informations, which apply to different states of knowledge about the system. Conditional algorithmic information is always defined relative to some background information, which must be specified, because what information is regarded as background and what as foreground can change.

The first connection between information and entropy is a purely mathematical statement (Zvonkin and Levin 1970; Bennett 1982; Zurek 1989b; Caves 1990a),

$$-\sum_{j=1}^{\mathcal{J}} p_j \log p_j \leq \sum_{j=1}^{\mathcal{J}} p_j \Delta I_j \leq -\sum_{j=1}^{\mathcal{J}} p_j \log p_j + O(1) \,, \tag{3.1}$$

where $O(1)$ denotes a computer-dependent additive constant.† The quantity on both the left and the right of the double inequality (3.1) is Shannon's statistical measure of the information in $\mathcal{J}$ alternatives which have probabilities $p_j$. The quantity in the middle is the average conditional algorithmic information of the alternatives. Both the left and right inequalities are consequences of coding theory, the left being a strict inequality, but the right including a computer-dependent constant.

For the rest of this paper, I use, following Zurek (1989a), "physicist's notation"— $\lesssim$, $\simeq$, $\gtrsim$—to denote approximate equality and inequality within a computer-dependent additive constant. With this convention the double inequality (3.1) can be converted to an approximate equality

$$\sum_{j=1}^{\mathcal{J}} p_j \Delta I_j \simeq -\sum_{j=1}^{\mathcal{J}} p_j \log p_j \,, \tag{3.2}$$

which says that the *average* conditional algorithmic information is very nearly Shannon's statistical information. This approximate equality is the crucial link between the absolute information of the inside view and the statistical information of the outside view. Notice that it holds only on the average: there can be and generally are algorithmically simple alternatives for which $\Delta I_j \ll -\sum p_j \log p_j$, but such simple alternatives are atypical.

It is perhaps surprising that the approximate equality (3.2) holds for *any* probabilities $p_j$, but this can be so because the conditional algorithmic informations $\Delta I_j$ are defined relative to background information that includes the probabilities. In this article, I use Eq. (3.2) repeatedly in the case of minimal background information, just the information necessary to describe the system and to list alternatives—and

---

† Throughout this article, log denotes the base-2 logarithm, which is appropriate for measuring information in bits, whereas ln is reserved for the natural logarithm.

thus to assign *uniform* probabilities $p_j = 1/\mathscr{J}$. Specialized to this case, Eq. (3.2) involves an unweighted average of conditional algorithmic informations, each defined relative to minimal background information:

$$\frac{1}{\mathscr{J}} \sum_{j=1}^{\mathscr{J}} \Delta I_j \simeq \log \mathscr{J} . \qquad (3.3)$$

The unweighted average is the conditional algorithmic information of a *typical* alternative. To estimate the information needed to specify a typical alternative, relative to minimal background information, one need only count the alternatives and take a logarithm.

The second connection between information and entropy is *Landauer's* (1961, 1985, 1988) *principle: to erase a bit of information in an environment at temperature T requires dissipation of energy* $\geq k_B T \ln 2$. On the outside view Landauer's principle is a consequence of the Second Law. A memory that stores one bit has equal probabilities to be in two distinct physical configurations, which correspond to its one-bit record, and thus it has an entropy of one bit or $k_B \ln 2$ in thermodynamic units. To erase the memory means to take it reliably from either configuration to some standard configuration, its entropy thus going to zero. The Second Law requires that the entropy of the environment increase by at least $k_B \ln 2$, which means that energy $\geq k_B T \ln 2$ must be dissipated into the environment. Conservative dynamics cannot take two distinct memory configurations to a single final configuration, but dynamical considerations alone do not assign an erasure cost—the Second Law does that.

Justified on the outside view, Landauer's principle acquires its significance on the inside view, with its absolute measure of information. Here a curious subtlety crops up. If a memory "knows" the internal physical configuration that corresponds to a one-bit record, it can design a conservative erasure mechanism. Knowing its internal configuration amounts to having a copy of its own record. Erasing the original record leaves the copy, which still has one bit of entropy on the outside view, so the Second Law does not require dissipation. Thus, when one talks about a memory gathering and processing information, one means that the memory "knows" what is recorded in its memory, but does not know the corresponding physical configuration.

The best way to gain familiarity with the information-entropy connection is to put it to work exorcising Maxwell demons. Bennett (1982, 1987, 1988b) realized that Landauer's principle is the magic formula for dealing with demons. Zurek (1989a, 1989b) used Bennett's idea as the cornerstone in the physics of information. I summarize here a general free-energy analysis taken from Caves (1990b).

Consider a physical system in thermal equilibrium at temperature $T$. The system has states labeled by $j$, with energy $E_j$ and multiplicity $\mu_j$ (corresponding to finer grainings of classical phase space or to quantum degeneracies). In thermal equilibrium the probability for the system to be in state $j$ is

$$p_j = \frac{\mu_j}{Z_0} e^{-E_j/k_B T} , \tag{3.4}$$

where $Z_0 \equiv \sum \mu_j e^{-E_j/k_B T}$ is the partition function. In equilibrium the system's mean energy, entropy (in bits), and free energy are given by

$$E_0 = \sum_j p_j E_j , \tag{3.5}$$

$$H_0 = -\sum_j p_j \log(p_j/\mu_j) , \tag{3.6}$$

$$F_0 = E_0 - k_B T \ln 2\, H_0 = -k_B T \ln 2 \log Z_0 . \tag{3.7}$$

The algorithmic background information $I_0$ appropriate to equilibrium is the information necessary to generate a description of the system together with a list of its states and their thermal probabilities.

Suppose now that the system occupies state $j$. The energy $E_j$, entropy $H_j = \log \mu_j$ (in bits), and free energy

$$F_j = E_j - k_B T \ln 2\, H_j = -k_B T \ln 2\, \log\left(\mu_j e^{-E_j/k_B T}\right) \tag{3.8}$$

of this state correspond to changes from equilibrium $\Delta E_j = E_j - E_0$, $\Delta H_j = H_j - H_0$, and

$$\Delta F_j = F_j - F_0 = \Delta E_j - k_B T \ln 2\, \Delta H_j = -k_B T \ln 2 \log p_j \equiv W_j^{(+)} . \tag{3.9}$$

The change in free energy is the amount of work $W_j^{(+)}$ available as the system returns isothermally to equilibrium from state $j$. If the demon/memory observes that the system occupies state $j$, the amount of information it gathers is the conditional algorithmic information $\Delta I_j$ to specify state $j$, relative to the background information. The minimum cost for erasing this information is

$$W_j^{(-)} = k_B T \ln 2\, \Delta I_j . \tag{3.10}$$

The demon/memory can operate an engine cycle in which it observes the state of the system, extracts work $W_j^{(+)}$ as the system returns to equilibrium, and pays an erasure cost $W_j^{(-)}$ to return to its background state of knowledge. The *net* work extracted on the average,

$$\overline{W} = \sum_j p_j \left( W_j^{(+)} - W_j^{(-)} \right) = k_B T \ln 2 \left( -\sum_j p_j \log p_j - \sum_j p_j \Delta I_j \right) \leq 0 , \tag{3.11}$$

is guaranteed not to be positive by the strict left inequality of the double inequality (3.1). Though the demon/memory cannot win, the soft right inequality implies that it can come close to breaking even, at least in principle. This free-energy analysis is framed carefully in terms of averages, for which there is a rigorous statement (3.1) of information-entropy balance. Such care is too cumbersome for the rest of this article, where I generally work in terms of "typical" quantities instead of averages.

It is instructive to write the average work extracted in other ways:

$$\overline{W} = \sum_j p_j(\Delta F_j - k_B T \ln 2 \, \Delta I_j) = -k_B T \ln 2 \sum_j p_j(\Delta H_j + \Delta I_j) \simeq 0 \,. \qquad (3.12)$$

*These expressions combine Landauer's principle with an average information-entropy balance that comes from the approximate equality (3.2); they summarize neatly the connections between information and entropy.* The first form of $\overline{W}$ suggests defining a *total* free energy

$$\mathscr{F} \equiv F - k_B T \ln 2 \, I = E - k_B T \ln 2 \, (H + I) \,, \qquad (3.13)$$

which recognizes the Landauer erasure cost by including it as a negative contribution to total free energy. This total free energy is equivalent to Zurek's (1989a, 1989b) proposal that the *total* entropy (in bits),

$$\mathscr{S} \equiv H + I \,, \qquad (3.14)$$

should include both ordinary entropy $H$ and an algorithmic contribution $I$. To avoid confusion, I use the following nomenclature throughout the rest of this article: $\mathscr{S}$ is called the total entropy; the ordinary entropy $H$ of statistical physics (in bits) is called *statistical entropy*, since it is formally identical to Shannon's statistical information.

The philosophy underlying this article is that *to say that a system occupies a certain state implies that one has the information to generate a description of that state.* What else could "say" mean? Compelling though this philosophy may be, it has no physical content without Landauer's principle, which connects information to free energy and work. A consistent theme throughout this article is that questions in the physics of information must be referred to free energy and work.

The information-entropy balance that exorcises Maxwell demons is by now cut and dried. Where might one learn something new? One place to look is the algorithmic background information, because there information-entropy balance does not apply. The deepest part of the background information, which contains the primitive notions and higher-level language that allow us to get started describing physical systems, is probably a "can of worms," because we don't know how to quantify this information. I set aside this deep part of the background information and concentrate on a part of the background information—the ability to generate a list of states and their probabilities—that can be quantified easily.

One can estimate the size of this part of the algorithmic background information for a *typical* probability distribution just by counting the number of distributions that can be assigned to $\mathscr{J}$ alternatives. This count is finite because the individual probabilities $p_j$ are given to a finite accuracy $\delta p$. The probability simplex for $\mathscr{J}$ alternatives is a $(\mathscr{J} - 1)$-dimensional "tetrahedron" with edges of length $\sqrt{2}$. Hence, the volume of the probability simplex is

$$\mathscr{V} = \frac{\sqrt{\mathscr{I}}}{(\mathscr{I}-1)!} . \qquad (3.15)$$

At resolution $\delta p$ each probability distribution occupies a volume

$$\delta v = \sqrt{\mathscr{I}}(\delta p)^{\mathscr{I}-1} \qquad (3.16)$$

on the probability simplex. Thus the number of probability distributions at this level of resolution is

$$\mathscr{N} = \frac{\mathscr{V}}{\delta v} = \frac{1}{(\mathscr{I}-1)!(\delta p)^{\mathscr{I}-1}} , \qquad (3.17)$$

corresponding to information to specify a typical distribution,

$$I_0 \simeq \log \mathscr{N} = \mathscr{I} \log\left(\frac{e/\mathscr{I}}{\delta p}\right) + \frac{1}{2}\log\left(\frac{(\mathscr{I}\delta p)^2}{2\pi\mathscr{I}}\right) + O\left(\frac{1}{\mathscr{I}}\right) \underset{\mathscr{I}\gg 1}{\sim} \mathscr{I}\log\left(\frac{e/\mathscr{I}}{\delta p}\right) . \qquad (3.18)$$

The asymptotic $\mathscr{I} \gg 1$ form can be interpreted to mean that the number of bits needed to specify each probability is $\log((e/\mathscr{I})/\delta p)$, which indicates that only the probability accuracy beyond $e/\mathscr{I}$ is important.

Although Eq. (3.18) is the information to specify a *typical* probability distribution, there are distributions that can be specified with far less information. If the number of alternatives is a typical integer $\mathscr{I}$, the *uniform* distribution can be specified by the $I_0 \simeq \log \mathscr{I}$ bits needed to give $\mathscr{I}$. Moreover, if the number of alternatives is an algorithmically simple integer like $\mathscr{I} = 2^N$, the uniform distribution can be specified by the $I_0 \simeq \log N = \log\log \mathscr{I}$ bits needed to give the integer $N$. The probability distributions of statistical physics are all of this latter type, with $I_0 \ll \log \mathscr{I}$. For macroscopic systems, typical distributions are literally impossible to specify, and even those distributions with $I_0 \simeq \log \mathscr{I}$ cannot be specified in practice.

A small fraction of the algorithmic information (3.18) of a typical probability distribution overwhelms the statistical entropy, which is bounded above by $\log \mathscr{I}$. Of course, it matters not how big $I_0$ is so long as it remains background information, never erased. The rest of this article explores when and how it might move to the foreground, where it weighs in the balance of free energy and work. In this regard, what immediately comes to mind is dynamics—classical dynamics, where the system state is an evolving probability distribution on phase space, and quantum dynamics, where the system state is an evolving pure state, whose specification requires not just probabilities, but phases as well. The idea (Zurek 1989b) is that algorithmic information can increase—and, hence, total free energy can decrease—as a system evolves under purely conservative evolution. This idea motivates the theme of this article, taken up in Section 3 and encapsulated in the following question: *How does one lose the ability to extract work as a system evolves?*

Before turning to dynamics, however, I consider in Section 2 the general question of how much information is needed to specify a phase-space pattern or a quantum-mechanical pure state. There are so many phase-space patterns and so many pure states that it takes an enormous amount of information to specify a typical one. A typical phase-space pattern or a typical pure state has *total* entropy much bigger than the entropy of thermal equilibrium. Operationally, this means that work can be extracted as a system is transformed from thermal equilibrium to a typical phase-space pattern or a typical pure state.

Motivated by these considerations, I ask in Section 3 whether dynamical evolution with an algorithmically simple Hamiltonian, starting from an algorithmically simple initial state, can access algorithmically complex phase-space patterns or algorithmically complex pure states. The answer seems to be no, at least for reasonable times. For classical non-mixing evolution (Zurek 1989b) and for quantum-mechanical evolution, the information needed to specify the evolved state increases logarithmically in time, much too slow to be of interest. For chaotic classical evolution, sensitivity to initial conditions is manifested as a linear increase in algorithmic information. It turns out for macroscopic systems that this linear increase, though much faster than logarithmic, is still too slow to be of interest on the scale set by the equilibrium entropy.

The analysis in Section 2 indicates, however, that the phase-space pattern that evolves under classical chaotic evolution or the pure state that evolves under quantum evolution is immersed in a sea of superficially similar, but exceedingly complex states, whose total entropy is enormous, much bigger than the equilibrium entropy. The system can be nudged into one of these complex states by even the tiniest (energy-conserving) perturbation. This hyper-sensitivity to perturbation means that for slightly perturbed chaotic or quantum evolution, the best strategy for extracting work from the system is not to keep track of the perturbed state, but rather to average over the perturbation and to put up with the consequent standard increase in statistical entropy. There is no such hyper-sensitivity to perturbation for non-mixing classical evolution, which occurs within the space of possible initial states. Thus hyper-sensitivity to perturbation provides a link between classical chaos and quantum dynamics. Indeed, it suggests a reversal of conventional wisdom: quantum mechanics is not somehow deficient relative to statistical physics because it lacks chaotic sensitivity to initial conditions; rather chaos elevates classical mechanics to a level of sensitivity to perturbation which is essential for statistical physics and which is inherent in quantum dynamics.

### 3.2 Entropy of Phase-Space Patterns and Quantum Mechanical Pure States

Throughout this article the discussion is couched in terms of the statistical physics problem of an isolated (fixed-energy), macroscopic (many-degrees-of-freedom) system. The energy is assumed to lie within a narrow interval of width $\delta E$ centered on a fiducial energy $E_0$. Despite the focus on macroscopic systems, the mathematical results can be specialized to systems with only a few degrees of freedom, although some conclusions might be different. Within this general framework I consider both classical and quantum-mechanical systems. Some of the material of this section is also discussed in Caves (1993).

### 3.2.1 Phase-Space Patterns

Consider first a classical Hamiltonian system that has $f = 2D \gg 1$ degrees of freedom. The system's dynamics transpire within a $2D$-dimensional phase space. The energy constraint picks out an allowed $2D$-dimensional phase-space volume $\mathscr{V}_0$. Using the grid provided by some convenient set of canonical coordinates, divide this volume into $\mathscr{J} = q_i^{2D}$ coarse-grained cells of identical volume $\mathscr{V}_i = \mathscr{V}_0/\mathscr{J}$. Here $q_i$ is crudely the number of cells along each phase-space dimension. I assume that $\log q_i \geq 1$; moreover, although $q_i$ can be quite large, $\log q_i$ is utterly negligible compared to $2D$. Reasonable values might lie in the range $3 \lesssim \log q_i \lesssim 100$.

If the system is in thermal equilibrium, it is described by the microcanonical ensemble—identical probabilities $p_j = 1/\mathscr{J}$ for each of the coarse-grained cells. The equilibrium statistical entropy relative to the coarse graining is

$$H_i = \log(\mathscr{V}_0/\mathscr{V}_i) = \log \mathscr{J} = 2D \log q_i \,. \tag{3.19}$$

In contrast, if the system is constrained to occupy a single coarse-grained cell, its statistical entropy relative to the coarse graining is zero, and the change in statistical entropy relative to equilibrium is

$$\Delta H = -H_i = -\log \mathscr{J} \,. \tag{3.20}$$

What about algorithmic information? The background information appropriate to equilibrium consists of the information necessary to generate a physical description of the system and to generate a list of coarse-grained cells. The physical description of the system is summarized in the Hamiltonian and associated boundary conditions. Let $I_H$ be the number of bits required to specify the Hamiltonian and the boundary conditions, typically less than a few thousand bits even for a macroscopic system.†️ To generate a list of coarse-grained cells requires giving the number of cells $\mathscr{J}$ and the energy $E_0$ and energy range $\delta E \ll E_0$ in some energy unit. There is a natural unit $\tilde{E} = \delta E/\mathscr{J}$, in which the energy range is the same as the number of cells. Specifying the number of cells $\mathscr{J} = 2^{H_i}$ is equivalent to specifying $H_i$, which requires about $\log H_i$ bits. Specifying $E_0$ in the natural unit means the following: first, give the significant digits (the digits out to accuracy $\delta E$), this requiring about $\log(E_0/\delta E)$ bits; second, give the number of trailing zeroes after the significant digits, this being

---

† It is easy to imagine—but essentially impossible to describe—systems for which $I_H \gtrsim 2D$: the Hamiltonian might be algorithmically complex, as in a spin glass, or the boundary conditions might be algorithmically complex, as in a gas confined to a container that has an algorithmically complex boundary. Such systems have a level of background complexity not contemplated in the present discussion.

the same as specifying the energy range in natural units and thus the same as specifying $\mathcal{J}$.

Putting all this together yields an amount of background information

$$I_B \simeq I_H + \log(E_0/\delta E) + \log H_i = I_H + \log(E_0/\delta E) + \log 2D + \log \log q_i \qquad (3.21)$$

(Caves 1990a), which is utterly negligible on the scale set by $2D$, as it must be if we are to be able to generate a statistical description of the macroscopic system. Notice that if $\delta E/E_0 \sim 1/\sqrt{f}$, as in an approximation to the canonical ensemble, the second and third terms in Eq. (3.21) are comparable.

Given this minimal background information, a memory needs additional information

$$\Delta I \simeq \log \mathcal{J} = H_i \qquad (3.22)$$

to specify a typical cell. There are cells for which the additional information is essentially zero—those with simple "addresses" in the coordinate grid—but they are atypical. Thus the typical change in total entropy, relative to physical equilibrium and to the "equilibrium" background information, is

$$\Delta \mathcal{S} = \Delta H + \Delta I \simeq 0 . \qquad (3.23)$$

As in the analysis of a Maxwell demon in the Introduction [cf. Eq. (3.12)], the information-entropy balance (3.23) means that for a typical coarse-grained cell (or on the average), there is no work available when the erasure cost of the memory's record is included.

Imagine using the same coordinate grid to divide each coarse-grained cell into $n = q^{2D}$ fine-grained cells, for a total of $\mathcal{K} = n\mathcal{J}$. Here $q$ is the number of fine-grained cells along each dimension of a coarse-grained cell. I make the same assumptions for $\log q$ as for $\log q_i$—i.e., $1 \leq \log q \ll 2D$, with a typical range of 3 to 100. The equilibrium statistical entropy relative to the fine graining is

$$H_0 = \log\left(\frac{\mathcal{V}_0}{\mathcal{V}_i/n}\right) = \log \mathcal{K} = H_i + \log n = 2D(\log q_i + \log q) , \qquad (3.24)$$

and the statistical entropy of a coarse-grained cell is

$$H = \log n = 2D \log q . \qquad (3.25)$$

Consider now filling $n$ fine-grained cells—i.e., identical probabilities $p_k = 1/n$ for $n$ cells and probability zero for all others—to create a pattern on phase space at the scale of the fine graining. The pattern fills the same volume $\mathcal{V}_i$ as a coarse-grained cell—it has statistical entropy $H = \log n$ relative to the fine graining—and thus has the same change in statistical entropy, $\Delta H = H - H_0 = -H_i$, relative to equilibrium as does one of the coarse-grained cells. The number $\mathcal{N}$ of such patterns is the number of ways of choosing $n$ cells out of $\mathcal{K}$ cells, without regard to order:

$$\mathcal{N} = \frac{\mathcal{K}!}{n!(\mathcal{K} - n)!} . \qquad (3.26)$$

Why bother with these patterns? Because they have the same reduction in statistical entropy as a coarse-grained cell—i.e., the same free-energy increase. Thus they yield the same amount of work $k_B T \ln 2 (-\Delta H)$ as the system "expands" isothermally to fill the energy-allowed volume of phase space. The total-entropy change, including the information to specify a pattern, is a very different story, however, because there are so many patterns.

The appropriate background information now is the information needed to generate a description of the system, a list of the coarse-grained cells, and a list of the fine-grained cells. (Also needed is information to generate a list of the patterns made up of fine-grained cells, but this requires essentially no information beyond that necessary to generate a list of the cells themselves.) The further information beyond the background information (3.21) at the coarse-grained level is the number $\log q$ necessary to specify $n = 2^{2D \log q}$. This requiring about $\log \log q$ bits, the amount of equilibrium background information at the fine-grained level is

$$
\begin{aligned}
I_B &\simeq I_H + \log(E_0/\delta E) + \log H_i + \log \log q \\
&= I_H + \log(E_0/\delta E) + \log 2D + \log \log q_i + \log \log q .
\end{aligned} \tag{3.27}
$$

Given the background information, the further information necessary to specify a typical pattern is

$$
\Delta I \simeq \log \mathcal{N} \underset{\substack{n \gg 1 \\ \mathcal{K} - n \gg 1}}{\sim} n \log\left(\frac{\mathcal{K}}{n}\right) + (\mathcal{K} - n) \log\left(\frac{\mathcal{K}}{\mathcal{K} - n}\right) \underset{n/\mathcal{K}=1/\mathcal{J} \ll 1}{\sim} n \log \mathcal{J} = n H_i . \tag{3.28}
$$

The second asymptotic form can be understood in the following way. Given a list of the fine-grained cells, a pattern can be specified by giving the positions of the $n$ filled cells. Since the filled cells are sparse $(n/\mathcal{K} = 1/\mathcal{J} \ll 1)$, typically separated by about $\mathcal{J}$ unfilled cells, an efficient approach is to give the number of unfilled cells between successive filled cells. This takes about $\log \mathcal{J}$ bits per filled cell, or $n \log \mathcal{J}$ bits for the whole pattern.

If $n$ is large, the additional information (3.28) is enormous on the scale set by the statistical entropy $H_i$. There are patterns with much smaller $\Delta I$: for example, the original coarse-grained cells typically have a comparatively small $\Delta I \simeq \log \mathcal{J}$, and those with simple phase-space addresses have still smaller $\Delta I$. All of the simple patterns are, however, highly atypical. For the typical pattern the change in total entropy, relative to physical equilibrium and to equilibrium background information, is dominated by the information needed to specify the pattern:

$$
\Delta \mathcal{S} = \Delta H + \Delta I \simeq \log(\mathcal{N}/\mathcal{J}) \underset{\substack{n \gg 1 \\ n/\mathcal{K}=1/\mathcal{J} \ll 1}}{\sim} n \log \mathcal{J} . \tag{3.29}
$$

No information-entropy balance here! This is a huge entropy increase (a free-energy decrease). *A typical phase-space pattern has much higher total entropy than a state of thermal equilibrium.* There is no work available from a typical pattern when one takes into account the cost of the record needed to specify it. Just the opposite, in fact: it should be possible to extract work $k_B T \ln 2 \, \Delta\mathscr{S}$ as the system is transformed from equilibrium to a typical pattern.

Before addressing this provocative assertion, it should be noted how the phase-space patterns are related to the Introduction's discussion of the number of probability distributions. Let $A$ be the set of probability distributions on the fine-grained cells which have entropy change $\Delta H = -H_i$ within resolution $\delta H$ and which have probabilities given to accuracy $\delta p$. The phase-space patterns are a subset of $A$. Perhaps the discussion in this subsection should be given in terms of the number $\mathscr{N}_A$ of distributions in $A$, but this number is difficult to evaluate. Since $\mathscr{N}_A > \mathscr{N}$, the conclusions reached here might be more dramatic if given in terms of $\mathscr{N}_A$. How much more dramatic is unclear, and the number of patterns is already more than big enough to make the point. Thus it is considerably easier, both here and in Section 3, to deal with phase-space patterns instead of more general probability distributions on phase space.

To see how to extract work as the system goes from thermal equilibrium to a typical pattern, first group the patterns into phase-space "partitions." Each partition contains $\mathscr{J}$ non-intersecting patterns, which partition phase space into $\mathscr{J}$ equal volumes $\mathscr{V}_i$ and which together fill the energy-allowed volume $\mathscr{V}_0$. It is possible to group the patterns in such a way that each pattern belongs to one and only one partition, so that there are

$$\frac{\mathscr{N}}{\mathscr{J}} = \frac{(\mathscr{K}-1)!}{(n-1)!\,(\mathscr{K}-n)!} \tag{3.30}$$

partitions in all. The proof that such a grouping is possible was given by Baranyai (see Brouwer and Schrijver 1979). For each partition, imagine fashioning a "template" that, when placed on the system, prevents it from moving one pattern to another within that partition. Finally, introduce an auxiliary system that has $\mathscr{N}/\mathscr{J}$ states, each of which corresponds to one of the partitions.

Initially, both the system and the auxiliary system are in equilibrium, with uniform probabilities over their respective states. A memory observes the state of the auxiliary system, thereby reducing the auxiliary system's statistical entropy by $\log(\mathscr{N}/\mathscr{J})$ bits and collecting (on average) $\simeq \log(\mathscr{N}/\mathscr{J})$ bits of information. Using its record of the auxiliary system's state, the memory extracts work $k_B T \ln 2 \, \log(\mathscr{N}/\mathscr{J})$ as the auxiliary system returns to equilibrium. Using the same record again, the memory selects the corresponding system template. After applying that template to the system, it observes which of the $\mathscr{J}$ patterns the system occupies, thereby collecting (on average) an additional $\simeq \log \mathscr{J}$ bits of information. At this stage the auxiliary system is back in equilibrium and is irrelevant to the discussion. The system is constrained to occupy a particular pattern, the memory stores (on average) $\simeq \log(\mathscr{N}/\mathscr{J}) + \log \mathscr{J} = \log \mathscr{N}$ bits, which are *required* to specify the pattern, and the

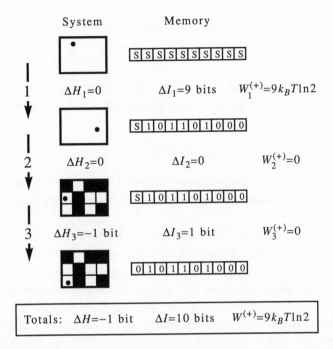

Fig. 3.1. A single molecule (the system), initially free to roam its entire container, is transformed in three steps to occupy a particular pattern consisting of 6 out of $\mathcal{K} = 12$ boxes. A memory with 10 binary registers, each initially in a standard state (denoted by "S"), stores the 10 bits $\left[\log \mathcal{N} = \log\left(12!/(6!)^2\right) = \log 924 = 9.85\right]$ needed to specify the pattern of boxes. Given at each step are the change in the system's statistical entropy, the change in the number of bits of memory used, and the work extracted. After the three-step transformation, the system's statistical entropy has changed by $\Delta H = -1$ bit, the memory stores $\Delta I = 10$ bits, and work $W^{(+)} = k_B T \ln 2 (\Delta H + \Delta I) = 9 k_B T \ln 2$ has been extracted.

memory has extracted work $\simeq k_B T \ln 2 \Delta \mathcal{S}$, as promised. Notice that the memory's record splits naturally into two parts: a huge number of bits, $\log(\mathcal{N}/\mathcal{J})$, for which partition and a conventional $\log \mathcal{J}$ bits for which pattern within the partition.

It is instructive to present a more explicit example of this procedure, modeled on the famous Szilard (1929) engine. In this example the patterns are formed in configuration space instead of in phase space. This makes it easier to see how to extract work from a pattern and to imagine applying the template constraints. In addition, the example dispenses with the auxiliary system, work being extracted directly from the memory. The example is depicted in Fig. 3.1.

Consider a rectangular container of volume $V$ in which there is a single molecule. The molecule is in equilibrium with the walls of the container, which are at temperature $T$. Let a Cartesian grid divide $V$ into $\mathcal{K}$ boxes, all of volume $V/\mathcal{K}$. Imagine coloring half of the boxes black, thereby creating a pattern of $n = \mathcal{K}/2$ black boxes and a complementary pattern of $n = \mathcal{K}/2$ white boxes. The total number of possible patterns—black and white—is

$$\mathcal{N} = \frac{\mathcal{K}!}{\left( (\mathcal{K}/2)! \right)^2} \, , \tag{3.31}$$

so the information required to specify a typical pattern, relative to minimal background information that can generate a description of the container and the division into boxes, is

$$\Delta I \simeq \log \mathcal{N} \underset{\mathcal{K} \gg 1}{\sim} \mathcal{K} \, . \tag{3.32}$$

The asymptotic $\mathcal{K}$ bits mean that an efficient way to specify a pattern for $\mathcal{K} \gg 1$ is to give a binary string of length $\mathcal{K}$, each entry specifying black or white for a particular box.

There are $\mathcal{N}/2$ pairs of complementary black and white patterns, each of which divides $V$ into two equal volumes. For each pair of complementary patterns, imagine constructing a *physical* partition that prevents the molecule from moving between black and white boxes. Suppose that such a partition constrains the molecule and that one observes that the molecule occupies the black boxes—not to be confused with observing *which* black box! The change in statistical entropy relative to having no constraints is $\Delta H = -1$ bit. To extract the available work $k_B T \ln 2$, one could move all the black boxes to one side of the container and all the white boxes to the other side, separating the two by a partition down the middle; then, knowing the molecule is on the black side, one could extract work $k_B T \ln 2$ as the molecule pushed a piston into the white side.

On the other hand, for a typical pattern the change in total entropy, relative to equilibrium and the minimal background information, is

$$\Delta \mathcal{S} = \Delta H + \Delta I \simeq \log \mathcal{N} - 1 = \log(\mathcal{N}/2) \, . \tag{3.33}$$

One should be able to extract work $k_B T \ln 2 \, \Delta \mathcal{S}$ as the molecule is taken from equilibrium to occupation of a particular pattern. To see how this is done, introduce a memory that has $\log \mathcal{N}$ binary registers to store a pattern record. Starting with the molecule in equilibrium, free to roam the entire container, and with the memory in a standard state, storing no information, the transformation is carried out in three steps (see Fig. 3.1).

1. All memory registers except the first—$\log \mathcal{N} - 1$ registers in all—are allowed to randomize, with extraction of work (inverse of erasing)

$$W_1^{(+)} = k_B T \ln 2 \left( \log \mathcal{N} - 1 \right) , \tag{3.34}$$

   after which the memory stores $\Delta I_1 = \log \mathcal{N} - 1$ bits.
2. The memory uses its $\Delta I_1$-bit record to select one of the $\mathcal{N}/2$ partitions and applies that partition to the container.

3. The memory observes whether the molecule is in the black part or the white part of the container and records $\Delta I_3 = 1$ bit for black or white in its first memory register, thereby changing the molecule's statistical entropy by $\Delta H_3 = -1$ bit.

After step 3 the molecule is constrained to occupy a particular pattern, the memory stores

$$\Delta I = \Delta I_1 + \Delta I_3 = \log \mathcal{N} \text{ bits,} \tag{3.35}$$

which are *necessary* to record which pattern the molecule occupies, and work $W_1^{(+)} = k_B T \ln 2 \, \Delta \mathcal{S}$ has been extracted. The amount of extracted work shows that the final configuration of the system (after step 3) has lower free energy than the standard state (before step 1)—lower by $k_B T \ln 2 \log(\mathcal{N}/2)$—and thus higher total entropy by $\log(\mathcal{N}/2)$. An alternative transformation lets all $\log \mathcal{N}$ memory registers randomize, with extraction of work $k_B T \ln 2 \log \mathcal{N}$, of which $k_B T \ln 2$ is used to "compress" the system into the pattern stored in the memory's record.

Suppose that after step 3 the memory observes which box the molecule occupies, thereby collecting on average $\log(\mathcal{K}/2)$ bits that balance the molecule's further reduction in statistical entropy. Should one conclude that this configuration has a total-entropy change $\Delta \mathcal{S} = \log(\mathcal{N}/2)$? No, because the memory's $\log(\mathcal{N}/2)$ bits for which partition are excess information, irrelevant to specifying which box the molecule occupies. To have a *minimal* record, the memory should erase the excess bits at a cost that just cancels the extracted work.

It is useful to consider two engine cycles based on this example. The first cycle starts with the same standard state (before step 1), proceeds through steps 1–3, and adds two further steps to get back to the standard state.

4. The system returns to its standard (equilibrium) state with extraction of work $W_4^{(+)} = k_B T \ln 2$ (as described above).
5. The memory returns to its standard state (all registers empty) at erasure cost $W_5^{(-)} = k_B T \ln 2 \Delta I = k_B T \ln 2 \log \mathcal{N}$.

The net work extracted, $W_1^{(+)} + W_4^{(+)} - W_5^{(-)}$, is zero. In this cycle the background information is the information needed to generate a description of the system at the level of division into boxes. The information gathered by the memory during the cycle includes both the $\log \mathcal{N} - 1$ bits for choosing a partition and the 1 black-or-white bit from observing the molecule.

The second kind of cycle uses the result of step 1 as the standard state. Step 4 is the same, but in step 5 only the first memory register needs to be erased—at cost $k_B T \ln 2$—to return the memory to its standard state. In this cycle the $\log \mathcal{N} - 1$ bits for choosing a partition become background information, which tells the memory how to partition the container into two equal volumes. The cycle is just a fancy Szilard (1929) engine, with the container partitioned in a bizarre way instead of by inserting a partition down the middle.

These two cycles emphasize that the change in total entropy must be defined relative to some initial system state *and* to some initial background information. The difference between the two cycles is precisely whether the $\log(\mathcal{N}/2)$ which-partition bits are background information, to be carried forward from cycle to cycle, or foreground information, collected afresh during each cycle and erased to return to the initial background information. The total-entropy increase when the molecule is confined to a typical pattern is real, but it must be understood as an increase relative to physical equilibrium *and* to minimal background information.

Return now to the general discussion of phase-space patterns. Why is one pattern complex and another simple? Why are the original coarse-grained cells relatively simple, whereas other patterns occupying the same phase-space volume are exceedingly complex? Only because the coarse-grained cells arise from the gridding of phase space, which is generated by some set of canonical coordinates that are presumed to be simple. Choose different canonical coordinates for the gridding, and arrive at a different set of simple patterns. My own view is that this relativity of complexity is inevitable: the phase-space description of dynamics doesn't just pop into existence; it must be built up, presumably from the set of canonical coordinates used for the Hamiltonian. How these preferred coordinates arise and in what sense they are simple are questions that must be addressed to the deep part of the background information—the "can of worms" mentioned in the Introduction. Fortunately, this interesting and diverting issue is unimportant for the discussion here. What matters is that the overwhelming number of patterns are complex, not which are simple and which are complex.

The complex patterns arise because of the fine-grained structure that underlies any gridding of phase space. Fine graining permits an explosion in the number of patterns at the statistical-entropy level set by a coarser graining. Recall the philosophy that underlies this article: to say that a system is transformed to a particular pattern implies that one has the information necessary to generate a description of that pattern; that information having a Landauer erasure cost, it must be included in the total entropy. Nonetheless, the discussion in this subsection makes clear that most of the information to specify a typical pattern—specifically, the $\log(\mathcal{N}/\mathcal{J})$ bits of which-partition information—is never stored in the system in any sense and cannot be gotten by observing the system. The most information that can be gotten by observing the system is $\log \mathcal{J}$ bits at the coarse-grained level and $\log \mathcal{K}$ bits at the fine-grained level. The which-partition information is *information about how one chooses to observe the system*—i.e., about how to partition phase space into patterns.

Even though the templates partition phase space into patterns with the coarse-grained volume, they can be used to get the $\log \mathcal{K}$ bits that are available at the fine-grained level. Successive observations of the system with different templates reveal intersections of patterns within the different templates and thus eventually isolate the system in one of the fine-grained cells.

Is there any way that complex patterns might be produced naturally, instead of being forced on the system by observation? The dynamical evolution of chaotic Hamiltonian systems comes immediately to mind, because the evolution probes finer and finer scales on phase space and thus creates patterns on finer and finer scales. This possibility is taken up in Section 3.

### 3.2.2 Quantum-Mechanical Pure States

The quantum-mechanical discussion is surprisingly similar to the preceding classical discussion, yet it is simpler and more compelling, because the complex "patterns" are the pure states that arise from the Hilbert-space structure of quantum mechanics. The fundamental difference is that these quantum "patterns" do not arise from any underlying fine graining, there being no structure "underneath" the pure states.

Consider now an isolated quantum system. The energy eigenstates within the allowed energy range $\delta E$ span a $\mathcal{J}$-dimensional Hilbert space. The quantum statistical entropy when the system is in equilibrium—i.e., when it is spread uniformly over the energy eigenstates and thus is described by the microcanonical density operator $\hat{\rho} = \hat{1}/\mathcal{J}$—is

$$H_0 = -\text{tr}(\hat{\rho} \log \hat{\rho}) = \log \mathcal{J} . \tag{3.36}$$

In contrast, if the system occupies any pure state, the statistical entropy is $H = 0$, corresponding to a statistical-entropy change relative to equilibrium,

$$\Delta H = H - H_0 = -\log \mathcal{J} . \tag{3.37}$$

What about algorithmic information? The appropriate background information consists of the information necessary to generate a physical description of the system and to generate a list of the energy eigenstates within $\delta E$. The discussion leading to Eq. (3.21) applies with "coarse-grained cell" replaced by "energy eigenstate" and with $H_i$ replaced by $H_0$. In particular, the natural energy unit $\tilde{E} = \delta E / \mathcal{J}$ takes on added significance as the mean spacing between energy eigenstates. The upshot is an amount of "equilibrium" background information (Caves 1990a)

$$I_B \simeq I_H + \log(E_0/\delta E) + \log H_0 . \tag{3.38}$$

The additional information required to specify a typical pure state can be estimated in the following way. Introduce an orthonormal basis $|j\rangle$, $j = 1,\ldots,\mathcal{J}$. Any pure state state $|\psi\rangle$ can be expanded in this basis as

$$|\psi\rangle = \sum_j c_j |j\rangle = \sum_j \sqrt{p_j} e^{i\phi_j} |j\rangle . \tag{3.39}$$

To specify a typical pure state requires giving $\mathcal{J}$ complex amplitudes $c_j = \sqrt{p_j} e^{i\phi_j}$. Equivalently, one can give $\mathcal{J}$ probabilities $p_j$, or $\mathcal{J}$ real amplitudes $a_j = \sqrt{p_j}$, and $\mathcal{J}$ phases $\phi_j$. Suppose one gives the probabilities to accuracy $\delta p$ and the phases to accuracy $\delta\phi$. One expects that only the probability resolution beyond $e/\mathcal{I}$ is important, as in Eq. (3.18), so the number of significant digits in each probability is $\sim \log\big((e/\mathcal{I})/\delta p\big)$. The phases are defined modulo $2\pi$, so the number of digits in each phase is $\sim \log(2\pi/\delta\phi)$. Thus one estimates the additional information required to specify a typical pure state as

$$\Delta I \sim \mathcal{I}\left[\log\left(\frac{e/\mathcal{I}}{\delta p}\right) + \log\left(\frac{2\pi}{\delta\phi}\right)\right] . \tag{3.40}$$

For macroscopic systems, the additional information (3.40) is enormous on the scale set by the statistical entropy $H_0$. There are pure states with much smaller $\Delta I$: for example, a typical energy eigenstate has a comparatively small $\Delta I \simeq \log \mathcal{I}$, and there are certainly energy eigenstates that have $\Delta I$ near zero. The reason the energy eigenstates are relatively simple is that the equilibrium background information includes the Hamiltonian, which can be used to compute the energy eigenstates and eigenvalues. Some of the relativity of complexity that arises in the classical analysis thus seems to be absent from quantum mechanics. The energy eigenstates are relatively simple because of the Hamiltonian description of a quantum system. They provide a simple orthonormal basis for constructing all the other pure states—and, hence, the whole Hilbert space—by superposition. The special role of the energy eigenstates is highlighted by the dynamical analysis in Section 3.

Nothing in this discussion implies that *all* superpositions of energy eigenstates have the typical $\Delta I$ of Eq. (3.40); there are superpositions that are algorithmically simple. For example, the coherent states of a harmonic oscillator are defined in terms of a single complex number $\alpha$, which determines all the energy amplitudes; the information to specify a typical coherent state is determined largely by the accuracy to which one gives the real and imaginary parts of $\alpha$.

One can put the estimate (3.40) for $\Delta I$ on a firmer footing by introducing a metric on Hilbert space and using it to calculate the volume of Hilbert space. The natural distance between two pure states, $|\psi\rangle$ and $|\psi'\rangle$, is the Hilbert space-angle (Wootters 1981)

$$s = \cos^{-1}\big(|\langle\psi'|\psi\rangle|\big) . \tag{3.41}$$

In the coordinates that arise from the expansion (3.39), the line element for this metric is

$$ds^2 = \sum_{j=1}^{\mathcal{I}}(da_j^2 + a_j^2\,d\phi_j^2) - \left(\sum_{j=1}^{\mathcal{I}}a_j^2\,d\phi_j\right)^2 = \sum_{j=1}^{\mathcal{I}}\left(\frac{dp_j^2}{4p_j} + p_j\,d\phi_j^2\right) - \left(\sum_{j=1}^{\mathcal{I}}p_j\,d\phi_j\right)^2 . \tag{3.42}$$

The normalization condition,

$$\sum_{j=1}^{\mathcal{I}}p_j = \sum_{j=1}^{\mathcal{I}}a_j^2 = 1 , \tag{3.43}$$

reduces the dimension by 1, making a $(\mathscr{J} - 1)$-sphere in the real amplitudes; the irrelevance of overall phase changes introduces a degeneracy in the metric, which is manifested in the squared one-form at the end of the line element and which further reduces the dimension by 1. The result is a metric on the $2(\mathscr{J} - 1)$-dimensional space of Hilbert-space rays, called projective Hilbert space (Gibbons 1992).

The volume of projective Hilbert space can be computed to be

$$\mathscr{V} = \frac{\pi^{\mathscr{J}-1}}{(\mathscr{J}-1)!} \tag{3.44}$$

(Gibbons 1992). The desired resolution in Hilbert space is described by introducing a volume per pure state, $\delta v$. At this level of resolution, the number of pure states becomes

$$\mathscr{N} = \mathscr{V}/\delta v\,, \tag{3.45}$$

which leads immediately to the additional information, relative to the equilibrium background information, to specify a typical pure state:

$$\Delta I \simeq \log \mathscr{N} = \log(\mathscr{V}/\delta v)\,. \tag{3.46}$$

Percival (1992) has introduced $\log \mathscr{N}$ as an entropy measure, which he calls "Hilbert entropy," and he has pointed out that the enormous size of $\log \mathscr{N}$ means that a typical pure state cannot be prepared—too much information is required! Here I go further: Landauer's principle assigns an energy cost to $\Delta I$, and thus I include it in the total entropy of a typical pure state, in accordance with Zurek's proposal.

The proper quantum state to associate with a resolution volume $\delta v$ is not really a pure state, but rather the density operator obtained by taking an unweighted average of the pure states within $\delta v$. If the resolution volume is a sphere of radius $R \ll 1$, one finds that

$$\mathscr{N}^{-1} = \delta v/\mathscr{V} = R^{2(\mathscr{J}-1)}\,; \tag{3.47}$$

moreover, the statistical entropy of the density operator is

$$H \simeq \frac{\mathscr{J}-1}{\mathscr{J}} R^2 \log\left(\frac{e\mathscr{J}}{R^2}\right) \underset{\mathscr{J}\gg1}{\sim} R^2 \log \mathscr{J}\,. \tag{3.48}$$

The statistical-entropy change $\Delta H = H - H_0$ is thus not quite $-\log \mathscr{J}$, but the difference is unimportant for the discussion in this subsection.

The resolution volume $\delta v$ clearly has something to do with the contemplated resolutions for probabilities and phases. One would like to let $\delta v$ be the volume of a "box" that has sides $\delta p$ in each probability dimension and sides $\delta \phi$ in each phase dimension, but according to the metric (3.42), this volume depends on location in Hilbert space. The natural place to calculate this volume is at a state that has uniform probabilities $p_j = 1/\mathscr{J}$, because there all the probability dimensions are equivalent and all the phase dimensions are equivalent. This leads to a resolution volume

$$\delta v = \mathscr{J} \left( \frac{\delta p \, \delta \phi}{2} \right)^{\mathscr{J}-1} , \qquad (3.49)$$

where the leading factor of $\mathscr{J}$ arises from the degeneracy due to overall phase changes. [Notice that in the limit $\mathscr{J} \gg 1$ a spherical resolution volume has the same volume as a probability-phase box if $R^2 = (\delta p/(e/\mathscr{J}))(\delta \phi/2\pi)$.] The number of pure states is

$$\mathscr{N} = \frac{1}{\mathscr{J}!} \left( \frac{2\pi}{\delta p \, \delta \phi} \right)^{\mathscr{J}-1} , \qquad (3.50)$$

and the additional information to specify a typical pure state becomes

$$\Delta I \simeq \log \mathscr{N} = \mathscr{J} \log \left( \frac{e/\mathscr{J}}{\delta p} \frac{2\pi}{\delta \phi} \right) + \frac{1}{2} \log \left( \frac{(\mathscr{J}\delta p)^2}{2\pi \mathscr{J}} \frac{(\delta \phi/2\pi)^2}{\mathscr{J}^2} \right) + O \left( \frac{1}{\mathscr{J}} \right)$$

$$\underset{\mathscr{J} \gg 1}{\sim} \mathscr{J} \left[ \log \left( \frac{e/\mathscr{J}}{\delta p} \right) + \log \left( \frac{2\pi}{\delta \phi} \right) \right] . \qquad (3.51)$$

[cf. Eq. (3.18)], in accordance with the estimate (3.40).

The entropy change of a typical pure state, relative to equilibrium and to equilibrium background information, is dominated by the enormous algorithmic information contribution:

$$\Delta \mathscr{S} = \Delta H + \Delta I \simeq \log(\mathscr{N}/\mathscr{J}) \underset{\mathscr{J} \gg 1}{\sim} \mathscr{J} \left[ \log \left( \frac{e/\mathscr{J}}{\delta p} \right) + \log \left( \frac{2\pi}{\delta \phi} \right) \right] . \qquad (3.52)$$

*A typical pure state has much higher total entropy than a state of thermal equilibrium.* One should be able to extract work $k_B T \ln 2 \, \Delta \mathscr{S}$ as the quantum system is transformed from equilibrium to a typical pure state. Before discussing the required transformation, it should be emphasized that the large amount of information needed to specify a typical pure state is just another way of saying that projective Hilbert space can accommodate an enormous number of pure states: if one asks that the states be 0.2 radian apart—i.e., $R \simeq 0.1$—which corresponds to giving just over three bits of information for each probability and each phase, then there are $\mathscr{N} \simeq 10^{2(\mathscr{J}-1)}$ states in projective Hilbert space.

The general discussion of how to do this proceeds exactly as in the classical case, except that the patterns and the phase-space partitions of the classical analysis become pure states and orthonormal bases in the quantum case. I assume that it is possible to group the pure states into orthonormal bases in such a way that each pure state belongs to one and only one basis, of which there are thus $\mathscr{N}/\mathscr{J}$ in all.[†] The templates of the classical analysis are replaced by measuring devices, each of which is designed to distinguish the pure states in one of the orthonormal bases.

---

† I know of no proof that this grouping can be done, nor do I know how to formulate a precise conjecture, since the division of projective Hilbert space into patches of volume $\delta v$ might mean that one can't insist on exact orthogonality. The grouping seems reasonable, nonetheless, and it is possible for the example in Fig. 3.2.

With these replacements, the procedure for extracting work $k_B T \ln 2 \, \Delta \mathscr{S}$ is the same as in the classical analysis. At the end of the procedure the quantum system occupies a particular pure state, and the memory stores (on average) $\simeq \log \mathscr{N}$ bits, which are *required* to specify the pure state.

Again the memory's record splits naturally into $\log(\mathscr{N}/\mathscr{J})$ bits for specifying which measurement basis and $\log \mathscr{J}$ bits for which pure state is found within the chosen basis. That most of the information goes into specifying how to observe the system is easier to swallow in quantum mechanics, since we are used to the idea that there are many incompatible ways of observing a quantum system. Nonetheless, it should be emphasized that, just as for the classical system, the which-basis information is never stored in the quantum system and cannot be gotten by observing the system. The most information that can be gotten by observing the system is the $\log \mathscr{J}$ bits for which state within a basis.

It is instructive to present a more explicit example, which the reader will recognize as a re-write of the example of Fig. 3.1. Re-write though it is, however, the quantum example is less artificial and thus more compelling than the classical example. Consider an unpolarized photon incident on a polarizing beamsplitter, as depicted in Fig. 3.2. The preferred axis of the beamsplitter can be set at any angle $\theta_p$ relative to the vertical, in which case the beamsplitter separates orthogonal linear polarizations at angles $\theta_p$ and $\theta_p + \pi/2$. By observing the photon's output direction, one infers its linear polarization. Thus the beamsplitter and a following one-bit observation transform an initially unpolarized photon to a particular state of linear polarization, specified by angle $\theta$ relative to the vertical, with a change in statistical entropy $\Delta H = -1$ bit. The range of polarization angles $\theta$ runs from 0 to $\pi$, so the relevant range of beamsplitter angles $\theta_p$ runs from 0 to $\pi/2$.

Suppose one wishes to prepare a linearly polarized photon with polarization angle $\theta$ specified to accuracy $\delta\theta$. The number of polarization angles is thus

$$\mathscr{N} = \pi/\delta\theta \, . \tag{3.53}$$

The additional information to specify a typical angle, relative to minimal background information that can generate a list of angles, is

$$\Delta I \simeq \log \mathscr{N} = \log(\pi/\delta\theta) \, , \tag{3.54}$$

which is the number of digits in the binary representation of $\theta/\pi$. For a typical polarization angle the total-entropy change is

$$\Delta \mathscr{S} = \Delta H + \Delta I \simeq \log \mathscr{N} - 1 = \log(\mathscr{N}/2) \, . \tag{3.55}$$

Start with an unpolarized photon and with a $\log \mathscr{N}$-bit memory in its standard state. Carry out the transformation to a linearly polarized photon in three steps (see Fig. 3.2).

1. All memory registers except the first—$\log \mathscr{N} - 1$ registers in all—are allowed to

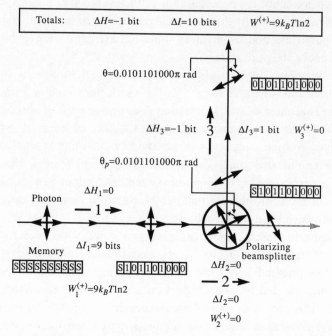

Fig. 3.2. An unpolarized photon is transformed in three steps to a particular linear polarization (the state of linear polarization is depicted in a plane that is rotated by 90° so that it lies in the plane of the paper). The final polarization is specified by an angle $\theta$ (relative to the vertical), which is given to 10 binary digits (accuracy $\delta\theta = 2^{-10}\pi$). A memory with 10 binary registers, each initially in a standard state (denoted by "S"), stores the 10 angle bits. Given at each step are the change in the photon's statistical entropy, the change in the number of bits of memory used, and the work extracted. After the three-step transformation, the photon's statistical entropy has changed by $\Delta H = -1$ bit, the memory stores $\Delta I = 10$ bits, and work $W^{(+)} = k_B T \ln 2 (\Delta H + \Delta I) = 9 k_B T \ln 2$ has been extracted.

randomize, with extraction of work $W_1^{(+)} = k_B T \ln 2 (\log \mathcal{N} - 1)$. The memory's first register remains in the standard state, and the remaining registers store a binary string $r$ of length $\Delta I_1 = \log \mathcal{N} - 1$ bits.

2. The memory sets the polarizing beamsplitter at angle $\theta_p = 0.0r\,\pi$, after which the photon passes through the beamsplitter.

3. The memory observes the photon's output direction, recording in its first register the $\Delta I_3 = 1$ bit for which orthogonal polarization—0 for angle $\theta_p$ and 1 for angle $\theta_p + \pi/2$. The observation changes the photon's statistical entropy by $\Delta H_3 = -1$ bit.

After step 3 the photon has linear polarization with angle $\theta = \theta_p = 0.0r\,\pi$ or with angle $\theta = \theta_p + \pi/2 = 0.1r\,\pi$, the memory stores the polarization angle modulo $\pi$ as the binary string $0r$ or $1r$ of length $\Delta I = \Delta I_1 + \Delta I_3 = \log \mathcal{N}$ bits, and work

$W_1^{(+)} = k_B T \ln 2 \Delta \mathscr{S}$ has been extracted, confirming that the final state has lower free energy than the initial state.

The complexity of typical pure states is a consequence of the Hilbert-space setting of quantum mechanics, which means that the number of pure states far exceeds the number of distinguishable states that make up an orthonormal basis. This is reminiscent of the explosion in the number of phase-space patterns, where the number of patterns far exceeds the number of non-intersecting patterns that make up a partition. There is, however, a crucial difference: successive observations with different phase-space templates eventually obtain all the information necessary to isolate the system in a single fine-grained cell, after which the results of further such observations are wholly predictable and thus yield no information; successive quantum-mechanical observations with different measurement bases—e.g., polarizing beamsplitters at 45°—never become predictable and thus continue to yield fresh information. The new information supersedes the old, never revealing any structure beneath the pure states (no hidden variables!). So what? When one grants to information a physical status, as in this article, the bottomless well of quantum information becomes both intriguing and disturbing. Indeed, the ability of quantum systems to manufacture information without end—quantum rolls of the dice—is the central mystery of quantum mechanics, stated in information-theoretic terms. Perhaps the approach sketched here might allow a glimpse behind the veil of quantum reality.

Pursuing elusive quantum reality is not, however, my purpose. How might one produce a complex pure state naturally, instead of by observation? Just as for classical systems, temporal evolution comes to mind. Can dynamical evolution transform a simple initial state to a complex final one? That question, for classical and quantum systems, occupies Section 3. Answering it provides insight into the connection between chaos and quantum mechanics and into the relation of each to irreversibility and statistical physics.

### 3.3 Complexity in Hamiltonian Dynamical Evolution

Consider a classical or a quantum system that is prepared in a simple initial state, "simple" meaning that the reduction in statistical entropy far exceeds the additional information needed to specify the state, so that nearly all the reduction in statistical entropy is available as work. Among all possible states, the simple states are atypical. We need not concern ourselves here with how the system gets into an atypical state (presumably by drawing on an external supply of free energy). The fact is that for macroscopic systems, simple initial states are our only concern, because complex initial states require in practice too much information to prepare.

A more precise formulation uses the language developed in Section 2. For a classical system the initial state is a coarse-grained cell, and for a quantum system it is a pure state. In either case the change in statistical entropy, relative to physical equilibrium, is $\Delta H = -\log \mathscr{J}$. Simplicity means that the additional information needed to specify the initial state, relative to equilibrium background information, satisfies

$$\Delta I \equiv I_{IS} \ll \log \mathscr{J}. \tag{3.56}$$

The change in total entropy,

$$\Delta \mathscr{S} = \Delta H + \Delta I \simeq -\log \mathscr{J}, \tag{3.57}$$

is dominated by the reduction in statistical entropy, almost all of which can be translated into work.

In this section I ask what happens when the constraints that keep the system in the initial state are removed. Experience answers that there is no longer any work available, an answer formalized by saying that the statistical entropy increases to its maximum—i.e., by $\log \mathscr{J}$. Hamiltonian dynamics answers that the statistical entropy is constant. A way out of this classic conundrum has been proposed by Zurek (1989b): the information necessary to specify the evolved state is not constrained by Hamiltonian dynamics and can increase with time, thereby allowing the *total* entropy, including the algorithmic contribution, to increase.

To investigate this proposal, I refer entropy changes now not to physical equilibrium and equilibrium background information, but rather to the initial state and the information necessary to specify it. Thus the background information $I_0$ includes both the equilibrium background information $I_B$—see Eq. (3.21) for the classical coarse-grained background information and Eq. (3.38) for the quantum background information—and the additional information $I_{IS}$ to specify the initial state:

$$I_0 = I_B + I_{IS}. \tag{3.58}$$

Zurek's proposal looks dicey as soon one realizes that the background information already contains the information necessary to evolve the system forward in time— the Hamiltonian in $I_B$ and the initial conditions in $I_{IS}$. Zurek realized, however, that one is interested in specifying the system state—and thus in being able to extract work—*at a particular time*; that time must be specified in addition to the Hamiltonian and the initial conditions.

Zurek (1989b) applied this idea to an ergodic, but non-mixing classical Hamiltonian system, which visits all the coarse-grained cells in a recurrence time $\mathscr{J}\tau$, where $\tau$ is the characteristic time to hop from one cell to another. The time $t$ must be given to accuracy $\tau$, requiring about $\log(t/\tau)$ binary digits, so the additional information to specify the system state at time $t$ is typically

$$\Delta I \simeq \log(t/\tau). \tag{3.59}$$

This example illustrates that the algorithmic contribution to total entropy can increase, but the logarithmic increase is too painfully slow to be of interest. The algorithmic contribution approaches the equilibrium entropy only after a recurrence time.

*How does one lose the ability to extract work as an isolated system evolves from a simple initial state?* That is the theme explored in this section. The jumping-off point is Zurek's suggestion that the algorithmic information to specify the system's evolved state can increase in time. This suggestion is applied to classical chaotic systems and to quantum systems, where, in contrast to a non-mixing classical system, the evolved state "lives" in a space that is much bigger than the space of initial states. The tentative conclusion is that simple states do *not* evolve to complex states on reasonable time scales, but rather that the evolved state is immersed in a sea of complex configurations into which it can be perturbed by tiny perturbations.

### 3.3.1 Globally Chaotic Classical Hamiltonian Evolution

Consider a chaotic classical Hamiltonian system (Lichtenberg and Lieberman 1983; MacKay and Meiss 1987; Schuster 1988; Rasband 1990). The interest here being macroscopic systems, the best I can do is to model crudely the chaotic dynamics, in the hope that the model captures the key features. I assume that the system is *globally* chaotic within the energy-allowed volume of phase space. I characterize the sensitivity to initial conditions by a single quantity, the Kolmogorov entropy $K$ (Kolmogorov 1959; Lichtenberg and Lieberman 1983; Schuster 1988), which I assume to have the same value everywhere within the energy-allowed volume.

I assume further that there are $D = f/2$ positive Lyapunov exponents, all of roughly the same size $\lambda$. The Kolmogorov entropy being the sum of the positive Lyapunov exponents (Pesin 1977), one has $\lambda = K/D$. For each positive Lyapunov exponent, there must be a companion negative exponent, to ensure conservation of phase-space volume.† For a gas, $\lambda = K/D$ is roughly the collision rate for a single molecule; appropriating this language generally, I refer to $D/K$ as the collision time.

I idealize the pattern that evolves from an initial coarse-grained cell as a $D$-dimensional "sheet" which, thrown into the air, bends and folds back on itself. Relative to the original phase-space coordinates (or, equivalently, to the initial coarse-grained cells), the $D$ "longitudinal" dimensions along the sheet expand as $e^{\lambda t} = e^{Kt/D}$, whereas the "thickness" of the sheet decreases as $e^{-\lambda t} = e^{-Kt/D}$ in each of the $D$ "transverse" dimensions. To resolve this phase-space pattern, one must grid phase space on a scale at least as fine as the thickness of the sheet. Thus, at time $t$, I introduce a fine graining in which each coarse-grained cell has $q = e^{\lambda t} = e^{Kt/D}$ fine-grained cells along each of its dimensions. Each coarse-grained cell contains $n = q^{2D} = e^{2Kt}$ fine-grained cells. At this level of fine graining the pattern at time $t$ is a set of $n$ filled fine-grained cells, which connect to make a $D$-dimensional sheet, one cell thick in each of the transverse dimensions. To match the use of bits

---

† Energy conservation implies that there are at least two vanishing Lyapunov exponents. Ignoring this, as I do, makes essentially no difference to the discussion of macroscopic systems.

for information, I find it convenient to use 2-foldings instead of $e$-foldings for the Kolmogorov entropy. Thus I define $K' = K/\ln 2$, so that

$$q = 2^{K't/D} \qquad \text{and} \qquad n = q^{2D} = 2^{2K't} . \tag{3.60}$$

No matter how fine the graining, one cannot keep track of the evolved pattern exactly, and thus one loses the ability to extract all of the work inherent in the initial coarse-grained cell. The fine graining introduced here is just fine enough to resolve the evolved pattern, there being about a factor of 2 of slop in each of the $D$ transverse dimensions. Effectively, the phase-space volume increases by a factor $\simeq 2^D$, corresponding to a statistical-entropy increase

$$\delta H \simeq D . \tag{3.61}$$

Entropy changes smaller than $\delta H$ are too small to be of interest at this level of resolution.

What is the additional algorithmic information, beyond the initial background information (3.58), required to specify the pattern at time $t$? One must describe the contemplated level of fine graining, which requires $\simeq \log(K't/D)$ bits to give the number $q = 2^{K't/D}$; this unimportant contribution is the additional $\log\log q$ bits of fine-grained background information in Eq. (3.27). The important contribution comes from specifying the time. If $\tau$ is a characteristic transition time between coarse-grained cells, then $2^{-K't/D}\tau$ is the characteristic transition time between fine-grained cells—i.e., the time for the evolved pattern to move across its thickness. The time must be given to this fine-grained accuracy, which requires $\simeq \log(t/2^{-K't/D}\tau)$ bits. Putting this together, one finds that the additional information to specify the pattern at time $t$, relative to background information (3.58), is

$$\Delta I \simeq \log\left(\frac{K't}{D}\right) + \log\left(\frac{t}{\tau}\right) + \frac{K't}{D} . \tag{3.62}$$

The second term is the logarithmic increase found for a non-mixing ergodic system. The dominant term is the last one. Its linear increase in $t$ makes it much larger than the logarithmic terms, yet the $D$ in the denominator means that is still too slow to be of interest for macroscopic systems. It becomes important on the scale $\delta H \simeq D$ when $K't/D \simeq D$, which in gas language means about $D$ collision times, far too long to be of interest.

These considerations indicate that *chaotic evolution via an algorithmically simple Hamiltonian does not take a simple initial state to a complex final state, at least for reasonable times*. What about the complexity of chaotic evolution? A first step in addressing this question is to review the information-theoretic interpretation (Farmer 1982a, 1982b) of the Kolmogorov entropy. Divide the interval from zero to $t$ into a sequence of closely spaced times. At the first time, partition the evolved pattern into pieces defined by intersections with the coarse-grained cells. At the second and subsequent times, sub-partition it further in the same way. At time $t$ the pattern is partitioned into many pieces, each of which corresponds to a unique sequence of coarse-grained cells—i.e., a coarse-grained trajectory that begins in the initial coarse-grained cell. The Kolmogorov entropy $K'$ is the asymptotic (in $t$) *rate* in bits/sec at which one must supply new information to pick out a particular coarse-grained trajectory. Crudely speaking, $K't$ is the number of bits of initial data needed to predict a coarse-grained trajectory out to time $t$, and $2^{K't}$ is the total number of coarse-grained trajectories at time $t$.

In the model used here, the number of coarse-grained trajectories can be estimated by saying that each trajectory corresponds to an intersection of the $D$-dimensional evolved sheet with a coarse-grained cell. The typical such intersection contains $q^D = 2^{K't}$ fine-grained cells, leading to $\simeq n/2^{K't} = 2^{K't}$ coarse-grained trajectories in all. This successful estimate builds confidence that the crude model captures the key features of chaotic evolution.

The information $K't$, already huge at one collision time, quantifies the difficulty in predicting a chaotic trajectory or, equivalently, the sensitivity to initial conditions of chaotic dynamics. Yet *predicting a trajectory has nothing directly to do with extracting the work inherent in the initial coarse-grained cell.* To predict the coarse-grained trajectory out to time $t$ typically requires an additional $K't$ bits of initial background information (information balanced by a $K't$-bit reduction in statistical entropy) beyond the background information for the problem considered here. To extract the work inherent in the initial coarse-grained cell, however, one needs not any particular trajectory, but rather the entire *pattern* at time $t$. The linear term $K't/D$ in $\Delta I$ is a manifestation of the difficulty in predicting chaotic dynamics, yet even though the pattern is made up of the trajectories, specifying the entire pattern requires much less information—by a factor $1/D$—than specifying a trajectory within the pattern.

Although the sheet that evolves from a simple initial coarse-grained cell is itself relatively simple, the discussion in Section 2 encourages us to consider a *typical* $D$-dimensional phase-space sheet consisting of $n = q^{2D}$ fine-grained cells. The typical sheet is not nearly as complex as a typical arbitrary phase-space pattern, because the fine-grained cells must connect to make a $D$-dimensional sheet, but it is exceedingly complex nonetheless. I view this typical $D$-dimensional sheet as coming from a perturbation of the evolved sheet. For this reason I give the information to specify a typical sheet relative to the fine-grained background information *and* to information that locates and orients the sheet in phase space, because this information is already available in the evolved sheet. The remaining complexity of a typical sheet resides in its transverse undulations along the longitudinal dimensions. Thus, to estimate the number of sheets, one can start with a flat $D$-dimensional grid of $n$ cells and move cells in the $D$ transverse dimensions to produce transverse undulations.

Consider first fine-scale undulations, which can be gotten, while maintaining connectivity, by restricting the candidates for moving to those cells—$n/2^D$ in number— that lie at the "all-even" positions in the grid. This restriction means that each candidate cell is surrounded by a buffer of stationary cells. The possibilities for each candidate cell are to stay put or to move to a neighboring cell in the $D$ transverse dimensions. There being three possibilities for each transverse dimension—move "up," stay put, or move "down"—each candidate has a total of $3^D$ possibilities. The number of sheets that can be produced by this procedure is

$$\mathcal{N} = \left(3^D\right)^{n/2^D} , \tag{3.63}$$

which leads to the information to specify a typical sheet, relative to fine-grained background information and to location-orientation information,

$$\Delta I \simeq \log \mathcal{N} = \frac{n}{2^D} D \log 3 . \tag{3.64}$$

Suppose now that the undulations in each longitudinal dimension are on the scale of $q_s \geq 2$ fine-grained cells. To estimate the number of sheets, divide the flat grid into "blocks" that have $q_s$ cells on a side, there being $n/q_s^D$ such blocks. The center cell of each block can move up to $q_s/2$ units in any transverse dimension, maintaining connectivity by "dragging along" surrounding cells in the block so as to leave the boundary of the block unmoved. Each center cell having $(q_s + 1)^D$ possibilities, the number of sheets is

$$\mathcal{N} = \left((q_s + 1)^D\right)^{n/q_s^D} , \tag{3.65}$$

corresponding to information to specify a typical sheet, relative to fine-grained background information and to location-orientation information,

$$\Delta I \simeq \log \mathcal{N} = \frac{n}{q_s^D} D \log(q_s + 1) . \tag{3.66}$$

The finest longitudinal structure corresponds to $q_s = 2$.

It is useful to define a parameter

$$\alpha \equiv q_s/q , \tag{3.67}$$

which measures the scale of the longitudinal structure in units of the coarse-grained cells. In these units the largest possible scale for the longitudinal structure is roughly $q_i$, the number of coarse-grained cells along each phase-space dimension, so one has the condition

$$\alpha \lesssim q_i . \tag{3.68}$$

In terms of $\alpha$ the information to specify a typical sheet becomes

$$\Delta I \sim \left(\frac{q}{\alpha}\right)^D D \log(\alpha q) = \left(\frac{2^{K't/D}}{\alpha}\right)^D D \left(\frac{K't}{D} + \log \alpha\right) . \tag{3.69}$$

In making the estimate (3.69), two assumptions are required: (i) the longitudinal extent of the sheet—$q^2$ fine-grained cells or $q$ coarse-grained cells—must be at least as big as the longitudinal scale, i.e.,

$$q \geq \alpha \quad \Longleftrightarrow \quad K't/D \geq \log \alpha ; \tag{3.70}$$

(ii) $q_s \geq 2$, i.e.,

$$q \geq 2/\alpha \qquad \Longleftrightarrow \qquad K't/D \geq 1 - \log \alpha . \qquad (3.71)$$

Within a few collision times of satisfying the appropriate condition (3.70) or (3.71), the information (3.69) is enormous on the scale set by $\delta H \simeq D$ or even by $H_i = 2D \log q_i$. Although the sheet that evolves from a simple initial condition is algorithmically simple, it is immersed in a sea of patterns which, though apparently similar, are extremely complex. The complexity of the chaotic dynamics manifests itself in that the evolved pattern "lives" in a space that is far larger than the original space of coarse-grained cells. This is to be contrasted with non-mixing evolution, which occurs within the space defined by the initial coarse graining.

The evolved sheet lies very "close" to patterns of much higher complexity. Just how close is made dramatically clear by considering a pattern that is identical to the evolved sheet, except that *one* fine-grained cell is moved to a neighboring cell in the transverse dimensions. The number of such patterns is

$$\mathcal{N} = n(3^D - 1) = 2^{2K't}(3^D - 1) , \qquad (3.72)$$

so the information needed to specify a typical pattern of this sort, relative to the information to describe the evolved sheet, is

$$\Delta I \simeq \log \mathcal{N} \underset{D \gg 1}{\sim} 2K't + D \log 3 , \qquad (3.73)$$

where $2K't$ is the number of bits needed to specify which fine-grained cell is moved and $D \log 3$ is the number of bits needed to specify which neighboring cell it moves to. The $2K't$ bits are significant on the scale of $\delta H \simeq D$ after just one collision time, and they overwhelm the statistical-entropy reduction $H_i = 2D \log q_i$ after $\ln q_i$ collision times—this from the tiniest perturbation into the sea of complex patterns!

A statistical description of a perturbation that changes just one cell says that each of the $\mathcal{N}$ patterns has equal probability $1/\mathcal{N}$. The number of fine-grained cells that the system can occupy includes the $n$ cells in the evolved sheet and the $\mathcal{N}$ neighboring cells (which are in one-to-one-correspondence with the patterns), for a total of $\mathcal{N} + n = 3^D n$ cells. Within each pattern the system actually occupies one fine-grained cell, the probability that a particular fine-grained cell is occupied being $1/n$. Averaged over all patterns, the probability that a given neighboring cell is occupied is $1/\mathcal{N}n$, and the probability that a cell in the evolved sheet is occupied is $(1/n)(1 - 1/n)$. Calculating the statistical entropy $H$ of this set of probabilities and converting to a statistical-entropy change relative to the $\log n$ bits for a coarse-grained cell, one finds

$$\Delta H = H - \log n = \frac{1}{n} \log \mathcal{N} - \left(1 - \frac{1}{n}\right) \log\left(1 - \frac{1}{n}\right) \underset{n \gg 1}{\sim} \frac{1}{n} \log \mathcal{N} \simeq \frac{1}{n} \Delta I . \quad (3.74)$$

This is an unimportant increase in statistical entropy, negligible on the scale $\delta H \simeq D$. For a perturbation to neighboring cells to produce a significant change in statistical entropy, it must be able to move *all* cells to a neighboring cell. This gives a statistical-entropy change $\Delta H = D \log 3$, but if the cells' moves are statistically independent, the information to specify a typical perturbed pattern (generally not connected) becomes $\Delta I \simeq \log(3^D)^n = n\Delta H$.

That $\Delta I \gg \Delta H$ is a formal way of saying that the perturbation "accesses" a space of perturbed sheets that is much bigger than the space of the original coarse-grained cells. Indeed, for any particular perturbation, the ratio $\Delta I / \Delta H$ ($\sim n$ in the two examples) is a useful measure of this discrepancy in size. The information $\Delta I$ is the logarithm of the number of patterns accessed by the perturbation, whereas the statistical-entropy increase $\Delta H$ is the logarithm of the number of coarse-grained-cell volumes accessed by the perturbation. Non-mixing evolution provides a useful contrast, illustrating what happens when evolution does not open up a bigger space of possibilities: if a perturbation moves the evolved coarse-grained cell to a neighboring one, both the additional information needed to specify a typical neighboring cell and the statistical-entropy increase on averaging over the perturbation are given by $\log 3^D = D \log 3$.

In the presence of a perturbation, one can choose between two strategies, which have grossly different costs. One can keep track of the perturbed phase-space sheet at a resolution that allows one to extract almost all the work $k_B T \ln 2 \, H_i$ available from the initial coarse-grained cell. The typical information needed to do this is $\Delta I$; the corresponding erasure cost, $k_B T \ln 2 \, \Delta I$, exceeds the work, even for tiny perturbations. Put differently, the *total* entropy of a typical perturbed sheet exceeds the *total* entropy of the equilibrium state. Alternatively, one can give up keeping track of the perturbed sheet and average over the perturbation. The resulting increase $\Delta H$ in statistical entropy has a free-energy penalty $k_B T \ln 2 \, \Delta H$ that is negligible compared to the erasure cost for $\Delta I$. The most efficient strategy in terms of work—i.e., most efficient in keeping the total entropy down—is to average over the perturbation rather than to keep track of the perturbed sheet. The ratio $\Delta I / \Delta H$ quantifies just how desirable this strategy is.

We thus come to a tentative bottom line: *for slightly perturbed chaotic evolution, the work-efficient strategy is to average over the perturbation and to put up with the resulting conventional increase in statistical entropy.* A first step in investigating this bottom line is to study chaotic Hamiltonian systems that are subjected to small energy-conserving perturbations. The appropriate analytical tool is the stochastic Liouville equation (and associated "Fokker-Planck equation" on the space of phase-space distributions), whose average gives a master equation for the evolution of an *averaged* phase-space distribution. The algorithmic information $\Delta I$ for a typical perturbed phase-space distribution is given by a statistical information on the space of phase-space distributions, whereas the statistical-entropy increase $\Delta H$ comes from the standard statistical entropy of the averaged phase-space distribution.

Since this program has not been carried out, I guess the results. The guess has been confirmed in a detailed analysis (Schack and Caves 1992) of a perturbed version of the baker's map, a prototype for classical chaos. Suppose the perturbation is modeled as phase-space diffusion with diffusion constant $\mathcal{D}$ and correlation length $\alpha$, both in units of coarse-grained cells. Introduce a dimensionless diffusion strength

$$\beta \equiv \sqrt{\frac{2K/D}{\mathcal{D}}} , \qquad (3.75)$$

whose inverse measures the amount of diffusion, in coarse-grained cells, during half a collision time. I assume a weak perturbation so that $\beta \gg 1$. There is a critical time $t_c$ at which the thickness of the unperturbed sheet is $\beta^{-1}$:

$$2^{-K't_c/D} = \beta^{-1} \qquad \Longleftrightarrow \qquad K't_c/D = \log \beta . \qquad (3.76)$$

Before $t_c$, the sheet evolves more or less as it does in the absence of the perturbation. After $t_c$, however, the diffusion becomes effective on the scale of the sheet thickness. Averaged over the diffusion, the sheet continues to expand in the longitudinal dimensions, but its thickness remains fixed at $2^{-K't_c/D} = \beta^{-1}$ coarse-grained cells or $q/\beta = 2^{K'(t-t_c)/D}$ fine-grained cells. The volume of the averaged sheet increases as $2^{K'(t-t_c)}$, leading to a statistical-entropy increase

$$\Delta H \simeq K'(t - t_c) . \qquad (3.77)$$

Estimating the number of perturbed sheets that lie within the averaged sheet is essentially the same problem as estimating the number of $D$-dimensional sheets that have transverse undulations on a longitudinal scale $q_s = \alpha q = 2^{K'(t-t_c)/D}\alpha\beta$, the scale set in this case by the correlation length of the diffusion. One starts with the unperturbed sheet, aggregates the fine-grained cells into $n/q_s^D$ blocks that have $q_s$ cells on a side, and estimates the number of perturbed sheets by considering the number of ways to move the center cell of each block in the transverse dimensions. If the scale of the longitudinal structure exceeds the thickness of the averaged sheet, i.e.,

$$q_s \gtrsim q/\beta \qquad \Longleftrightarrow \qquad \alpha\beta \gtrsim 1 , \qquad (3.78)$$

then the distance a center cell can move in any transverse dimension is limited by the thickness of the sheet to $2^{K'(t-t_c)/D}/2 \lesssim q_s/2$ fine-grained cells. There being thus about $2^{K'(t-t_c)}$ possibilities per center cell, the number of perturbed sheets becomes

$$\mathcal{N} \simeq \left(2^{K'(t-t_c)}\right)^{n/q_s^D} , \qquad (3.79)$$

and the information needed to specify a typical perturbed sheet, beyond that needed to describe the unperturbed sheet, is

$$\Delta I \simeq \log \mathcal{N} \underset{n \gg 1}{\sim} \frac{n}{q_s^D} K'(t - t_c) = 2^{K'(t-t_c)} \left(\frac{\beta}{\alpha}\right)^D K'(t - t_c) . \qquad (3.80)$$

For this estimate to hold, two conditions must be satisfied: (i) longitudinal extent of the unperturbed sheet bigger than the diffusion correlation length, i.e.,

$$K't/D \gtrsim \log \alpha \tag{3.81}$$

[Eq. (3.70)], and (ii) $t \gtrsim t_c$, i.e.,

$$K't/D \gtrsim \log \beta . \tag{3.82}$$

As soon as these conditions are met—i.e., as soon as the diffusion becomes effective—the information to specify a typical perturbed sheet explodes exponentially, and the information-to-statistical-entropy ratio goes as

$$\frac{\Delta I}{\Delta H} \underset{n \gg 1}{\sim} 2^{K'(t-t_c)} \left( \frac{\beta}{\alpha} \right)^D , \tag{3.83}$$

consistent with the preceding discussion. It should be emphasized that the exponential explosion of $\Delta I$ is a consequence of having longitudinal structure on a scale smaller than the longitudinal extent of the unperturbed sheet [Eq. (3.81)]; in contrast, $\Delta H$, which involves an average over the perturbed sheets, is indifferent to the scale of the longitudinal structure. The exponential explosion stops once there is more than one fold in the evolved sheet within each phase-space volume defined by the perturbation correlation length.

Suppose, for example, that $\beta \simeq \alpha$. Then at one collision time after the critical time, $\Delta H \simeq D/\ln 2$ becomes significant, but $\Delta I \sim e^D D/\ln 2$ is enormous—so enormous, at least for a macroscopic system, as to merit further discussion, which is postponed till Section 4. One other situation deserves mention. If $\beta \lesssim \alpha$, the same reasoning yields an estimate $\Delta I \sim K'(t - t_c)$ during the interval $t_c \lesssim t \lesssim (D/K') \log \alpha$, so that $\Delta I/\Delta H \sim 1$ during that interval. The statistical-entropy increase during this interval is limited to $\Delta H \simeq K'(t - t_c) \lesssim D \log(\alpha/\beta) \lesssim D \log(q_i/\beta) \lesssim H_i/2$, which means, crudely speaking, that even in the extreme case of large correlation length ($\alpha \simeq q_i$) and strong diffusion ($\beta \simeq 1$), the exponential growth of $\Delta I$ sets in before the increase in statistical entropy can account for the equilibrium entropy.

### 3.3.2 *Quantum-mechanical evolution*

One can already surmise that a simple pure quantum state remains relatively simple as it evolves under Hamiltonian evolution, at least for reasonable times. The discussion in Section 2 indicates, however, that the evolved state "lives" in an enormous space of complex pure states. Tiny perturbations access the surrounding complex pure states and thus produce a huge information-to-statistical-entropy ratio. This sensitivity to small perturbations is just the well known sensitivity to destruction of quantum coherence.

To see how this works out, recall that we are interested in energies within an interval of width $\delta E = \hbar \delta \omega$ centered on a fiducial energy $E_0 = \hbar \omega_0$, where $\delta \omega$ and $\omega_0$ are the corresponding frequency-interval width and fiducial frequency. Denote the energy eigenvalues within the allowed interval by $E_j = \hbar(\omega_0 + \omega_j)$, $j = 1, \ldots, \mathcal{J}$ ($|\omega_j| \leq \delta\omega/2$), with corresponding energy eigenstates $|j\rangle$. Let the initial system state have the energy-eigenstate expansion

$$|\psi(0)\rangle = \sum_{j=1}^{\mathcal{J}} a_j|j\rangle \,, \tag{3.84}$$

where, without loss of generality, we can assume that the amplitudes $a_j = \sqrt{p_j}$ are real. Recall that I am assuming that the initial state is simple in the sense of Eq. (3.56). The underlying idea is that the amplitudes associated with the initial state are localized in some simple cell of the corresponding classical "phase space."

The system state at time $t$ is

$$|\psi(t)\rangle = e^{-i\omega_0 t} \sum_{j=1}^{\mathcal{J}} a_j e^{-i\phi_j}|j\rangle \,, \qquad \phi_j \equiv \omega_j t \,. \tag{3.85}$$

Suppose the phase $\phi_j$ is known to some finite accuracy. Let $\tilde{\phi}_j$ be the phase truncated to this accuracy, and let $\delta\phi_j$ be the resulting phase error, i.e.,

$$\phi_j = \tilde{\phi}_j + \delta\phi_j \,. \tag{3.86}$$

Written explicitly in terms of the errors, the system state becomes

$$|\psi(t)\rangle = e^{-i\omega_0 t} \sum_{j=1}^{\mathcal{J}} a_j e^{-i\tilde{\phi}_j} e^{-i\delta\phi_j}|j\rangle \,. \tag{3.87}$$

The truncated phases correspond to a "fiducial" approximation to the system state,

$$|\tilde{\psi}\rangle \equiv e^{-i\omega_0 t} \sum_{j=1}^{\mathcal{J}} a_j e^{-i\tilde{\phi}_j}|j\rangle \,. \tag{3.88}$$

The phase errors lead to loss of coherence in the energy-eigenstate basis, described formally by averaging $|\psi(t)\rangle\langle\psi(t)|$ over the errors to get a system density operator.

Before doing the average, notice that the phase error

$$\delta\phi_j = \delta\omega_j t + \omega_j \delta t \tag{3.89}$$

arises from two sources, errors in the eigenfrequency and errors in the time. Using only the background information, one can make the eigenfrequency error as small as desired, because the background information contains the Hamiltonian and the boundary conditions necessary to compute the eigenfrequencies to any accuracy. In contrast, the time error depends on information beyond the background information—specifically, on how accurately the time is specified. If the time is given to accuracy $\tau$ ($|\delta t| \leq \tau/2$), then the appropriate density operator is

$$\hat{\rho} = \int_{-\tau/2}^{\tau/2} \frac{d\delta t}{\tau} |\psi(t)\rangle\langle\psi(t)| = \sum_{j,k} a_j a_k e^{-i(\tilde{\phi}_j - \tilde{\phi}_k)}|j\rangle\langle k| \left( \frac{\sin[(\omega_j - \omega_k)\tau/2]}{(\omega_j - \omega_k)\tau/2} \right) \,, \tag{3.90}$$

where the factor in large parentheses describes loss of coherence.

The density operator (3.90) is hard to work with, so I take a slightly different tack here. The phase error at the extreme eigenfrequencies lies in the interval defined by

$$|\delta\phi_j| \le (\delta\omega/2)(\tau/2) \equiv \delta/2 \,. \tag{3.91}$$

At eigenfrequencies within the allowed range the phase error is smaller. This is overkill. One can maintain roughly the same level of coherence by relaxing the requirements on the eigenfrequency error so that all phases $\phi_j$ are given to accuracy $\delta$, with $\delta\phi_j$ lying in the interval (3.91). To keep the density operator (3.90) nearly pure, thus enabling one to extract almost all the work available from a pure state, one must have

$$\delta = \delta\omega\,\tau/2 \lesssim 1 \,. \tag{3.92}$$

With this new tack the appropriate density operator becomes

$$
\begin{aligned}
\hat{\rho} &= \int \frac{d\delta\phi_1 \dots d\delta\phi_{\mathscr{I}}}{\delta^{\mathscr{I}}} |\psi(t)\rangle\langle\psi(t)| \\
&= \sum_j p_j |j\rangle\langle j| + \left(\frac{\sin^2(\delta/2)}{(\delta/2)^2}\right) \sum_{j\neq k} a_j a_k e^{-i(\tilde{\phi}_j - \tilde{\phi}_k)} |j\rangle\langle k| \\
&= \mu^2 \sum_j p_j |j\rangle\langle j| + (1-\mu^2)|\tilde{\psi}\rangle\langle\tilde{\psi}| \,,
\end{aligned}
\tag{3.93}
$$

where each phase error is integrated from $-\delta/2$ to $+\delta/2$ and where, for the last form, the decoherence factor in large parentheses is replaced in favor of

$$\mu^2 \equiv 1 - \frac{\sin^2(\delta/2)}{(\delta/2)^2} \underset{\delta\lesssim 1}{\simeq} \frac{\delta^2}{12} = \frac{(\delta\omega\,\tau)^2}{48} \,. \tag{3.94}$$

Notice that even for $\delta = 1/2$ ($\tau = 1/\delta\omega$), $\mu^2 \simeq 0.02$ is still small.

I make one further simplification. For quantum systems with a chaotic classical limit, there is evidence (Haake, Kús, and Scharf 1987; Sanders and Milburn 1989; Życzkowski 1990) that the energy eigenstates typically have amplitudes to be all over phase space and that the expansion (3.84) of an initial state that is largely confined within a chaotic region of phase space contains roughly equal contributions from nearly all energy eigenstates. Throughout the remainder of this subsection, I make the extreme simplifying assumption that all the energy amplitudes are identical, with value

$$a_j = \sqrt{p_j} = 1/\sqrt{\mathscr{I}} \,. \tag{3.95}$$

I return to the significance of this assumption below.

With this assumption the density operator (3.93) becomes

$$\hat{\rho} = (\mu^2/\mathscr{I})\hat{1} + (1-\mu^2)|\tilde{\psi}\rangle\langle\tilde{\psi}| = \eta\left(\hat{1} - |\tilde{\psi}\rangle\langle\tilde{\psi}|\right) + \xi|\tilde{\psi}\rangle\langle\tilde{\psi}| \,, \tag{3.96}$$

where

$$\xi \equiv 1 - \frac{\mathscr{J}-1}{\mathscr{J}}\mu^2 \qquad \text{and} \qquad \eta \equiv \frac{\mu^2}{\mathscr{J}}. \qquad (3.97)$$

It is easy to calculate the quantum statistical entropy of this density operator (its diagonal form has probability $\xi$ for state $|\tilde{\psi}\rangle$ and probability $\eta$ for each of $\mathscr{J}-1$ orthogonal states):

$$\delta H = -\text{tr}(\hat{\rho}\log\hat{\rho}) = -(\mathscr{J}-1)\eta\log\eta - \xi\log\xi \underset{\mathscr{J}\gg 1}{\sim} \mu^2\log\mathscr{J} = \mu^2 H_0. \qquad (3.98)$$

This statistical-entropy increase quantifies how much work is lost as a consequence of tracking the time (or the phases) with finite accuracy; it sets the scale for weighing the importance of other entropy increases. Notice that for large $\mathscr{J}$, $\mu^2$ is the lost fraction of the original work.

The information needed to specify the fiducial pure state (3.88), beyond the background information (3.58), is the time $t$ to accuracy $\tau$:

$$\Delta I \simeq \log(t/\tau). \qquad (3.99)$$

The logarithmic increase in $\Delta I$, identical to that for a non-mixing ergodic system, indicates that quantum evolution is not sensitive to initial conditions and thus not chaotic in the usual sense. The algorithmic information (3.99) approaches the equilibrium entropy only after a time $t = \mathscr{J}\tau = 2\delta\tau_R$, where $\tau_R \equiv \mathscr{J}/\delta\omega$ is a quantum recurrence time. *Quantum-mechanical evolution under an algorithmically simple Hamiltonian takes simple initial states to simple final states, at least for reasonable times.*

Quite a different picture emerges when one considers perturbed quantum evolution. The discussion in Section 2 indicates that the evolved pure state is immersed in a sea of complex pure states, into which it can be easily perturbed. Just how "close" these complex states are is illustrated by considering a perturbed pure state that is identical to the fiducial state, except that *one* phase is shifted by one unit of $\delta$. The additional information needed to specify the perturbed state, relative to the information to describe the fiducial state, is

$$\Delta I \simeq \log\mathscr{J} + 1 = H_0 + 1, \qquad (3.100)$$

where the $\log\mathscr{J}$ bits specify which of the $\mathscr{J}$ phases is changed and the 1 bit specifies whether that phase increases or decreases. This tiniest perturbation requires information on the scale of the equilibrium entropy $H_0$, but described statistically, it leads to a statistical entropy that is negligible compared to $\delta H$.

For a perturbation to produce a significant change in statistical entropy, it must have the potential to change *all* phases. Consider then a perturbation that can change each phase independently by up to $(m-1)/2$ units in either direction, each phase thus having $m$ possibilities. Statistically such a perturbation is equivalent to replacing $\delta$ by $\Delta \equiv m\delta$. For $\Delta \lesssim 1$ and $\mathcal{J} \gg 1$, the statistical-entropy change, relative to a pure state, is

$$\Delta H \sim \frac{\Delta^2}{12} \log \mathcal{J} \,. \tag{3.101}$$

In contrast, the number of perturbed phase patterns is

$$\mathcal{N} = m^{\mathcal{J}} = (\Delta/\delta)^{\mathcal{J}} \,, \tag{3.102}$$

so the information needed to specify a typical perturbed phase pattern, relative to the information to specify the fiducial state, is

$$\Delta I \simeq \log \mathcal{N} = \mathcal{J} \log(\Delta/\delta) \,. \tag{3.103}$$

The information-to-statistical-entropy ratio,

$$\frac{\Delta I}{\Delta H} \sim \frac{\mathcal{J}}{\log \mathcal{J}} \frac{\log(\Delta/\delta)}{\Delta^2/12} \,, \tag{3.104}$$

plays a dual role, just for chaotic dynamics. The information $\Delta I$ is the logarithm of the number of phase patterns accessed by the perturbation (relative to phase accuracy $\delta$), whereas the statistical-entropy increase $\Delta H$ is, roughly speaking, the logarithm of the number of *orthogonal* states accessed by the perturbation. The large value of $\Delta I / \Delta H$ indicates that the perturbation accesses a space of pure states that is much bigger than the dimension of Hilbert space, and it also expresses the quantum bottom line: *for slightly perturbed quantum evolution, the work-efficient strategy is to average over the perturbation, rather than to keep track of the perturbed pure state.*

To investigate this bottom line, a first step is to study quantum systems that are subjected to small energy-conserving perturbations. The appropriate analytical tool is the stochastic Schrödinger equation (and associated "Fokker-Planck equation" on projective Hilbert space), whose average gives a master equation for the system density operator. In its simplest form, this program can be carried through in a few lines. Under an energy-conserving stochastic Hamiltonian, each energy eigenvalue $E_j = \hbar(\omega_0 + \omega_j + \delta\omega(t))$ acquires a stochastic component $\delta\omega(t)$, and each phase in the evolved state (3.85),

$$\phi_j = \omega_j t + \delta\phi(t) \,, \tag{3.105}$$

acquires a stochastic component

$$\delta\phi(t) = \int_0^t dt' \, \delta\omega(t') \,. \tag{3.106}$$

The fiducial state $|\tilde{\psi}\rangle$ is given by Eq. (3.88) with $\tilde{\phi}_j \equiv \omega_j t$.

Suppose that the frequency perturbations $\delta\omega_j(t)$ are independent stationary white-noise processes with common diffusion constant $\mathscr{D}$, so that

$$\overline{\delta\omega_j(t)\delta\omega_k(t')} = \mathscr{D}\delta_{jk}\delta(t - t') \,, \tag{3.107}$$

where an overbar denotes an average with respect to the stochastic processes. The phase perturbations $\delta\phi_j(t)$ are independent Wiener processes, each of which has a zero-mean Gaussian probability distribution,

$$p(\delta\phi_j) = \frac{1}{\sqrt{2\pi\mathscr{D}t}} \exp\left(-\frac{(\delta\phi_j)^2}{2\mathscr{D}t}\right), \tag{3.108}$$

with variance

$$\overline{\delta\phi_j^2(t)} = \mathscr{D}t. \tag{3.109}$$

An immediate consequence is that

$$\overline{e^{-i\delta\phi_j(t)}} = e^{-\mathscr{D}t/2}. \tag{3.110}$$

Averaged over the perturbation, the density operator at time $t$ is

$$\hat{\rho}(t) = \overline{|\psi(t)\rangle\langle\psi(t)|} = \sum_j p_j|j\rangle\langle j| + e^{-\mathscr{D}t}\sum_{j\neq k} a_j a_k e^{-i(\tilde{\phi}_j - \tilde{\phi}_k)}|j\rangle\langle k|. \tag{3.111}$$

The decoherence factor is an exponential decay $e^{-\mathscr{D}t}$. Calculation of the statistical entropy proceeds just as in the calculation of $\delta H$, with $\mu^2$ now given by

$$\mu^2 = 1 - e^{-\mathscr{D}t} \underset{\mathscr{D}t \ll 1}{\simeq} \mathscr{D}t. \tag{3.112}$$

The result is a statistical entropy

$$\Delta H = -\text{tr}(\hat{\rho}\log\hat{\rho}) = -(\mathscr{J} - 1)\eta\log\eta - \xi\log\xi \underset{\mathscr{J} \gg 1}{\sim} (1 - e^{-\mathscr{D}t})\log\mathscr{J}. \tag{3.113}$$

How much information is required to specify a typical perturbed pure state with phase accuracy $\delta$, thereby enabling one to extract nearly all the work available from a pure state? According to Eq. (3.2), the *average* algorithmic information is the statistical information derived from the phase probability distributions (3.108), discretized at the phase accuracy $\delta$. Thus, provided $\delta \ll \sqrt{\mathscr{D}t}$, the average information to specify a perturbed pure state, relative to the information to specify the fiducial state, is

$$\Delta I \simeq -\int d\delta\phi_1 \dots d\delta\phi_{\mathscr{J}}\, p(\delta\phi_1)\dots p(\delta\phi_{\mathscr{J}})\log\left(\delta^{\mathscr{J}}p(\delta\phi_1)\dots p(\delta\phi_{\mathscr{J}})\right)$$

$$= \mathscr{J}\log\left(\frac{\sqrt{2\pi e\mathscr{D}t}}{\delta}\right), \tag{3.114}$$

which has the obvious interpretation that $\log\left(\sqrt{2\pi e\mathscr{D}t}/\delta\right)$ is effectively the number of bits needed to specify each phase. The information-to-statistical-entropy ratio,

$$\frac{\Delta I}{\Delta H} \sim \frac{\mathscr{J}}{\log\mathscr{J}}\frac{\log\left(\sqrt{2\pi e\mathscr{D}t}/\delta\right)}{1 - e^{-\mathscr{D}t}}, \tag{3.115}$$

is enormous as soon as the discretization of the Gaussians can be trusted. A detailed analysis (Schack and Caves 1993) of a perturbed version of a quantized baker's map confirms the main conclusion of this simple analysis—that the algorithmic information to keep track of the perturbed pure state grows much faster than the statistical entropy that follows from averaging over the perturbation.

This simple analysis is valuable primarily for illustrating a general procedure: quantum statistical entropy is calculated from the density operator in the standard way, whereas the average algorithmic information for a perturbed pure state is a statistical information calculated from a discretized probability distribution on projective Hilbert space. Otherwise, little more is learned from this diffusion analysis than from a perturbation into precisely $m = \Delta/\delta$ phases.

The discussion in this section suggests a link between chaotic classical dynamics and quantum dynamics. In both cases the state that evolves from a simple initial state remains simple, at least for reasonable times, but it is immersed in a sea of algorithmically complex states, into which it can be easily perturbed. As a consequence, even a tiny perturbation leads to such enormous algorithmic cost that the work-efficient strategy is to average over the perturbation and to put up with the resulting conventional increase in statistical entropy. This hyper-sensitivity to perturbations is the relevant property for statistical physics. For quantum dynamics the hyper-sensitivity is a consequence of the Hilbert-space setting of quantum mechanics, which opens up the enormous space of pure states as soon as quantum evolution begins. For chaotic classical dynamics the hyper-sensitivity arises as the chaotic evolution probes finer and finer scales in phase space and thus opens up an enormous space of phase-space patterns. From this point of view, chaos promotes classical dynamics to a level of sensitivity to perturbations which is inherent in quantum evolution. For quantum mechanics one gets a new twist on the sensitivity to destruction of quantum coherence; for classical chaos, a different way to think about sensitivity to initial conditions.

There is, however, a possible deeper connection. If, in contrast to the assumption in this subsection, the initial quantum state has substantial amplitudes for $\mathscr{J}' < \mathscr{J}$ energy eigenstates, then the results of this subsection for energy-conserving perturbations hold roughly with $\mathscr{J}$ replaced by $\mathscr{J}'$. The information to specify a typical perturbed state is then roughly $\mathscr{J}'$ times a logarithmic factor. There is a critical value $\mathscr{J}' \simeq \log \mathscr{J}$ below which the perturbed information cannot become significant on the scale set by the equilibrium statistical entropy $\log \mathscr{J}$. The potential connection to classical chaos arises from studies of quantum systems that have chaotic classical limits (Haake, Kús, and Scharf 1987; Sanders and Milburn 1989; Życzkowski 1990). In the classical "phase space" there are typically regions of regular motion embedded in regions of chaotic motion. If the initial quantum state is largely isolated within a chaotic region, it typically has amplitudes for many energy eigenstates, but if it is isolated within a regular region of phase space, it typically has amplitudes for only a few energy eigenstates. The potential connection to the explosion of algorithmic information and the consequent hyper-sensitivity to perturbations deserves further investigation.

### 3.4 Conclusion and Discussion

This article sketches a tentative view of irreversibility. The question: *How does one lose the ability to extract work as a physical system evolves away from a simple initial state?* The tentative answer: *Keeping track of the evolved state under slightly perturbed chaotic or quantum dynamics has such an enormous algorithmic cost that the work-efficient strategy is to average over the perturbation and to put up with the attendant increase in statistical entropy.* The key idea: *Classical chaotic dynamics and quantum dynamics both open up an enormous space of states—classical phase-space patterns and quantum-mechanical pure states—which even a tiny perturbation can access.*

Sketched is the word—if anything remains to be done, almost everything remains to be done. As was emphasized in Section 2, most of the information $\Delta I$ needed to specify a typical phase-space pattern or a quantum-mechanical pure state cannot be stored in the system. In the examples in Section 2, most of this information resides in how a memory chooses to observe the system. For the perturbations considered in Section 3, most of this information must be information about the particular way that a "reservoir," whatever it is, perturbs the system. There are several consequences.

If one has the information $\Delta I$ to specify the perturbed state, one can extract not only the full amount of work from the system, but also additional work from the reservoir, the total work just balancing the erasure cost $k_B T \ln 2 \Delta I$. There is a balance between information and statistical entropy in the end, with the reservoir included. *To get the desired work from the system, one must collect enough information to extract an enormously greater amount of work from the reservoir*—not much consolation if one intended to get work from the system. This statement reformulates the present point of view in system-reservoir language.

The typical size of $\Delta I \sim \left( 2^{H_i} \text{ (classical chaotic) or } 2^{H_0} \text{ (quantum)} \right) \sim 2^{10^{23}}$ for a macroscopic system is far too large, even for a perturbation involving the entire Universe. For macroscopic systems it is simply impossible to have perturbations with the properties described in Section 3. For example, a realistic energy-conserving quantum perturbation cannot be white up to frequency $\tau^{-1} = \delta\omega/2\delta$—there will be correlations on time scales considerably larger than $\tau$—and, more importantly, a realistic perturbation cannot perturb all phases independently—there will be considerable correlation between the phases of different energy eigenstates. Using realistic perturbations does not affect the main conclusion that $\Delta I / \Delta H$ becomes large, provided that the perturbation senses the size discrepancy between the space the evolved state "lives" in and the original state space—i.e., provided that there is structure, averaged away in calculating the statistical entropy, in the perturbed classical phase-space patterns or in the pattern of perturbed quantum phases. What this means for the properties of a perturbing reservoir and its interaction with the system is a topic for further work.

Interaction—there's a word that has not occurred in this article. The initial step in investigating the present point of view, as mentioned above, is to add a stochastic piece to the system Hamiltonian and to frame the analysis in terms of stochastic Liouville or stochastic Schrödinger equations. A more realistic further step is to model the perturbation as an interaction with a reservoir. There are techniques for deriving system master equations and thus calculating the growth of statistical entropy, but evaluating the information needed to specify the evolved system state presents a knottier problem. Each realization of a stochastic Hamiltonian leads to a system state, but for a fully interacting reservoir model, correlations between the system and reservoir make it difficult to assign system states. The right question to ask is the following: how much information must be gathered from the reservoir— more precisely, at what *rate* must information be gathered from the reservoir—to retain the ability to extract the full amount of work from the system? It might be difficult to formulate this question precisely, particularly for quantum mechanics, where one must deal with the many different, incompatible ways to observe from the reservoir.

Perhaps the most obvious criticism of the present point of view can be phrased as follows: "Maybe it's true that the evolved state is algorithmically simple because it can computed from a short program—essentially the Hamiltonian, the initial condition, and the time—but it's a truth with no content. To extract work from the evolved state, one must actually know it—not just the ingredients that go into computing it—and to know it, one must *perform* the computation, a whale of a job that is essentially impossible." There are two responses. The first (easy) response is that the computational complexity of the evolved state—i.e., the number of steps needed to compute it or some equivalent measure of computational difficulty—has no in-principle free-energy cost, so it cannot be included within the work-oriented analysis of this article.

The second (more productive) response is that this criticism might be an invitation to reformulate the present point of view. To set this reformulation within a larger context, it is useful to introduce Bennett's (1986, 1988a, 1990) logical depth. The logical depth of an "object" is measured by the computational complexity of the shortest program that can generate the object. An object with large logical depth is necessarily algorithmically simple, but it can be generated only by a very long computation. Section 3 is all about objects with large logical depth: the phase-space sheets that evolve from simple coarse-grained cells via the chaotic dynamics of a simple Hamiltonian and the pure states that evolve from simple initial pure states via quantum evolution under a simple Hamiltonian.

Logical depth captures a notion of interesting complexity. A long computation time can be seen as a long history that builds into an object such exquisitely subtle structure that its underlying simplicity is difficult to uncover. It seems reasonable that any object with such exquisite structure lies "close" to objects which, though superficially similar, are algorithmically complex and into which it can be "easily" perturbed. Perhaps large logical depth is equivalent to the physical property of hyper-sensitivity to perturbation developed in this article. Specifically, that the algorithmically simple evolved state in chaotic or quantum dynamics is difficult to compute might be equivalent to the property that it can be nudged into algorithmically complex states by tiny perturbations.

I perceived a consensus at this Workshop that the cosmological arrow of time arises because simple initial conditions evolve to complex final conditions. I'm not convinced. If the initial state of the Universe is simple and if the Universe evolves under a simple Hamiltonian, how did the present complexity arise? Not from evolution, which only makes simplicity difficult to discern, and by definition not from external perturbations. The information content of the Universe is almost entirely a consequence of quantum rolls of the dice. But then where did *that* information come from?

### Acknowledgements

I thank W. G. Unruh for challenging me to come up with the examples in Figs. 3.1 and 3.2, A. S. Lane for helping me think about patterns in high-dimensional spaces, and R. Schack for reading the manuscript carefully and suggesting improvements.

### Discussion

**Omnès:** In a derivation of classical physics from quantum physics, I investigated chaotic systems [Omnes, R. 1990 *Ann. Phys. (N.Y.)*, **201**, 354–447]. After chaos is established down to cells of quantum-mechanical size, classical physics is no longer consistent with quantum mechanics. However, if successive measurements, made at intervals $\Delta t \sim D/K$, each locate the system within a coarse-grained cell, the statistical predictions of classical physics and quantum physics agree up to negligible errors depending on the coarse graining. Thus one should expect the same information at large times in the classical and the quantum cases.

**Caves:** This question deals with successive observations of the system, a situation that I have argued is not relevant for statistical physics. It is, nonetheless, interesting to speculate on how the growth of algorithmic information in classical chaotic and quantum dynamics might be given a unified description. An initial coarse-grained cell in phase space is described quantum mechanically as an incoherent superposition of $n_Q = \mathcal{V}_i/h^D = q_Q^{2D}$ pure states. Classical chaotic evolution lasts until the thickness of the evolved sheet is of quantum size, i.e., until a time $t_Q$ given by $2^{K't_Q/D} = q_Q$. During the interval $0 \leq t \leq t_Q$ of chaotic evolution, the time must be specified to accuracy $\tau(t) = 2^{-K't/D}\tau$. For $t \geq t_Q$ quantum evolution takes over, with $\tau(t)$ becoming a constant time accuracy

$$\tau_Q \equiv \tau(t_Q) = 2^{-K't_Q/D}\tau = \frac{\tau}{q_Q} \quad \Longleftrightarrow \quad \delta\omega\,\tau_Q \simeq \frac{\delta E\,\tau/h}{q_Q} \sim 1,$$

where $\delta E\,\tau/h$ is the ratio of a classical action to the quantum of action. With these definitions the dominant contribution to the information needed to specify the evolved state becomes

$$\Delta I \simeq \log\left(\frac{t}{\tau(t)}\right) = \begin{cases} \log(t/\tau) + K't/D\,, & 0 \le t \le t_Q\,, \\ \log(t/\tau_Q)\,, & t \ge t_Q\,. \end{cases}$$

Although this argument is little more than units, it indicates how a detailed analysis might proceed. Especially interesting would be a study of the sensitivity to perturbations at the classical-quantum transition.

# References

Bennett, C. H. (1982) The thermodynamics of computation—a review. *Int. J. Theor. Phys.*, **21**, 905–940.

Bennett, C. H. (1986) On the nature and origin of complexity in discrete, homogeneous, locally-interacting systems. *Found. Phys.*, **16**, 585–592.

Bennett, C. H. (1987) Demons, engines and the Second Law. *Scientific American*, **257**(5), 108–116.

Bennett, C. H. (1988a) Dissipation, information, computational complexity and the definition of organization. In *Emerging Syntheses in Science*, Ed. D. Pines, Addison-Wesley, Redwood City, California, pp. 215–231.

Bennett, C. H. (1988b) Notes on the history of reversible computation. *IBM J. Res. Develop.*, **32**, 16–23.

Bennett, C. H. (1990) How to define complexity in physics, and why. In *Complexity, Entropy and the Physics of Information*, Ed. W. H. Zurek, Addison-Wesley, Redwood City, California, pp. 137–148.

Brillouin, L. (1962) *Science and Information Theory*, 2nd Ed., Academic, New York.

Brouwer, A. E. and Schrijver, A. (1979) Uniform hypergraphs. In *Packing and Covering in Combinatorics*, Ed. A. Schrijver, Mathematisch Centrum, Amsterdam, pp. 39–73.

Caves, C. M. (1990a) Entropy and information: How much information is needed to assign a probability? In *Complexity, Entropy and the Physics of Information*, Ed. W. H. Zurek, Addison-Wesley, Redwood City, California, pp. 91–115.

Caves, C. M. (1990b) Quantitative limits on the ability of a Maxwell demon to extract work from heat. *Phys. Rev. Lett.*, **64**, 2111–2114.

Caves, C. M. (1993) Information and entropy. *Phys. Rev.* **E47**, 4010–4017.

Chaitin, G. J. (1966) On the length of programs for computing finite binary sequences. *J. Assoc. Computing Machinery*, **13**, 547–569.

Chaitin, G. J. (1987) *Information, Randomness, and Incompleteness: Papers on Algorithmic Information Theory*, World Scientific, Singapore.

Farmer, J. D. (1982a) Information dimension and the probabilistic structure of chaos. *Z. Naturforsch.*, **37a**, 1304–1325.

Farmer, J. D. (1982b) Dimension, fractal measures and chaotic dynamics. In *Evolution of Order and Chaos*, Ed. H. Haken, Springer, New York.

Gallager, R. G. (1968) *Information Theory and Reliable Communication*, Wiley, New York.

Gibbons, G. W. (1992) States and density matrices. In *Complex Geometry and Mathematical Physics*, special issue of *J. Geom. Phys.* **8**, 147.

Haake, F., Kús, M., and Scharf, R. (1987) Classical and quantum chaos for a kicked top. *Z. Phys.*, **B65**, 381–395.

Jaynes, E. T. (1983) *Papers on Probability, Statistics, and Statistical Physics*, Ed. R. D. Rosenkrantz, Kluwer, Dordrecht, Holland.

Kolmogorov, A. N. (1959) *Dokl. Akad. Nauk. SSSR*, **124**, 754.

Kolmogorov, A. N. (1965) Three approaches to the quantitative definition of information. *Problemy Peredachi Informatsii*, **1**(1), 1–7. English translation: *Prob. Inform. Transmission*, **1**, 1–7 (1965).

Landauer, R. (1961) Irreversibility and heat generation in the computing process. *IBM J. Res. Develop.*, **5**, 183–191.

Landauer, R. (1985) Fundamental physical limitations of the computational process. In *Computer Culture: The Scientific, Intellectual, and Social Impact of the Computer*, Ed. H. Pagels, *Ann. N.Y. Acad. Sci.*, **426**, 161–170.

Landauer, R. (1988) Dissipation and noise immunity in computation and communication. *Nature*, **355**, 779–784.

Lichtenberg, A. J. and Lieberman, M. A. (1983) *Regular and Stochastic Motion*, Springer, New York.

MacKay, R. S. and Meiss, J. D., Eds. (1987) *Hamiltonian Dynamical Systems*, Hilger, Bristol.

Percival, I. C. (1992) Quantum Records. In, *Quantum Chaos – Quantum Measurements*, Eds. P. Cvitanović, I. Percival and A. Wirzba, Kluwer, Dordrecht, Holland, pp. 199–204.

Pesin, Ya. B. (1977) Characteristic Lyapunov exponents and smooth ergodic theory. *Usp. Mat. Nauk*, **32**(4), 55–112. English translation: *Russ. Math. Surveys*, **32**(4), 55–114 (1977).

Rasband, N. (1990) *Chaotic Dynamics of Nonlinear Systems*, Wiley, New York.

Sanders, B. C. and Milburn, G. J. (1989) The effect of measurement on the quantum features of a chaotic system. *Z. Phys.*, **B77**, 497–510.

Schack, R. and Caves, C. M. (1992) Information and entropy in the baker's map. *Phys. Rev. Lett.* **69**, 3413–3416.

Schack, R. and Caves, C. M. (1993) Hypersensitivity to perturbations in the quantum baker's map. *Phys. Rev. Lett.* **71**, 525–528.

Schuster, H. G. (1988) *Deterministic Chaos*, 2nd Ed., VCH, Weinheim, Germany.

Shannon, C. E. and Weaver, W. (1949) *The Mathematical Theory of Communication*, University of Illinois, Urbana, Illinois.

Solomonoff, R. J. (1964) A formal theory of inductive inference. Parts I and II. *Information and Control*, **7**, 1–22 and 224–254.

Szilard, L. (1929) On the decrease of entropy in a thermodynamic system by the intervention of intelligent beings. *Zeit. Phys.*, **53**, 840–856. English translation: In *Quantum Theory and Measurement*, Eds. J. A. Wheeler and W. H. Zurek, Princeton University, Princeton, New Jersey, 1983, pp. 539–548.

Wootters, W. K. (1981) Statistical distance and Hilbert space. *Phys. Rev.*, **D23**, 357–362.

Zurek, W. H. (1989a) Thermodynamic cost of computation, algorithmic complexity and the information metric. *Nature*, **341**, 119–124.

Zurek, W. H. (1989b) Algorithmic randomness and physical entropy. *Phys. Rev.*, **A40**, 4731–4751.

Zvonkin, A. K. and Levin, L. A. (1970) The complexity of finite objects and the development of the concepts of information and randomness by means of the theory of algorithms. *Usp. Mat. Nauk*, **25**(6), 85–127. English translation: *Russ. Math. Surveys* **25**(6), 83–124 (1970).

Życzkowski, K. (1990) Indicators of quantum chaos based on eigenvector statistics. *J. Phys.*, **A23**, 4427–4438.

# 4

# Demonic Heat Engines

## Benjamin W. Schumacher

*Department of Physics*
*Kenyon College*
*Gambier, OH 43022, USA*

## 4.1 Demons and the Second Law

One of the many equivalent formulations of the Second Law of Thermodynamics is a limit on the efficiency of heat engines (Fermi 1936). A heat engine does work by exploiting a temperature difference between two parts of its environment; according to the Second Law, its efficiency $\eta$ is bounded by

$$\eta \leq 1 - \frac{T_2}{T_1} \tag{4.1}$$

where $T_1$ and $T_2$ are the high and low working temperatures, respectively. The maximum efficiency is recognizable as the efficiency of a Carnot engine, in which an ideal gas is employed as a working fluid in a reversible cycle of adiabatic and isothermal processes.

Recently, new insight into the Second Law has been obtained by re-examining a famous thought-experiment, Maxwell's demon (Maxwell 1871). The demon is a microscopic being that observes the motions of individual molecules in an equilibrium system and then uses this information to do work, thus violating the Second Law. An important simple example of this idea was proposed by Szilard (Szilard 1929). Szilard's demon operates on a one-molecule gas confined in a chamber. The demon first locates the molecule, then introduces a moveable partition into the chamber. Since the demon knows which side of the partition the molecule is on, it may expand the one-molecule gas by moving the partition, obtaining work.

Szilard claimed the quantum-mechanical process of measurement by which the demon observes the gas molecule will require the demon to expend an energy $k_B T$, just offsetting the work obtained. (The shorthand $k_B$ represents $k_B \ln 2$ throughout my discussion.) In other words, Szilard supposed that the acquisition of information in a quantum measurement necessarily involves dissipation, and that this dissipation is just sufficient to rescue the Second Law.

More recently, Bennett (Bennett 1982) and Zurek (Zurek 1989) have both pointed out that the *acquisition* of information is not necessarily irreversible, even in quantum mechanics. They view Maxwell's demon as a kind of computer, which acquires and processes information during its operation. Bennett has shown that computers may function in a completely reversible way, and that the only computer operation that requires energy dissipation is the *erasure* of information, which costs at least $k_B T$ per erased bit. According to Bennett, therefore, the demon fails because it must *erase* information to operate in a closed cycle, and the erasure of $N$ bits of information will dissipate an energy $N k_B T$. Zurek points out that, even if the demon retains its stored information without erasing anything, the Second Law may be salvaged by including an additional term in the entropy reflecting the *algorithmic randomness* of the demon's memory state. On average, each bit in the computer's memory adds $k_B$ to the total entropy.

## 4.2 A Demonic Engine

Maxwell's demon fails in its task because it operates in an environment that is in equilibrium. The energy *value* of acquiring information is equal to the energy *cost* of free memory in which to store the information.

On the other hand, suppose the demon is placed in an environment that is not in equilibrium? It should then be possible for a demon to accomplish work by extracting energy from its surroundings. For example, we might consider a demon placed between two thermal reservoirs of different temperature. The energy value of information (about the warm reservoir) would then exceed the energy cost of free memory (relative to the cool reservoir). How efficiently can the demon function as a heat engine?

Let us begin by considering a version of Maxwell's demon which operates in a completely reversible way, and which we might call a "standard" demon. It consists of three parts: a one-molecule Szilard gas, a reversible computer, and a reversible energy storage device (or "battery"). These subsystems are coupled to each other and to the environment in the following ways:

- The Szilard gas can be thermally coupled to an external thermal reservoir.
- The Szilard gas is mechanically coupled to the battery, so that work performed by the gas may be reversibly stored in the battery and energy from the battery may be used to do work upon the gas.
- Memory elements in the computer may be reversibly coupled to the Szilard gas. The volume of the gas cell may be divided into two equal halves, which we will denote A and B. The molecule may be in either half. A particular memory element has exactly two states, 0 and 1. The coupling connects the location of the molecule with the state of the memory element in the following way:

$$
\begin{array}{ccc}
A\ 0 & \longrightarrow & A\ 0 \\
A\ 1 & \longrightarrow & B\ 0 \\
B\ 0 & \longrightarrow & B\ 1 \\
B\ 1 & \longrightarrow & A\ 1
\end{array}
$$

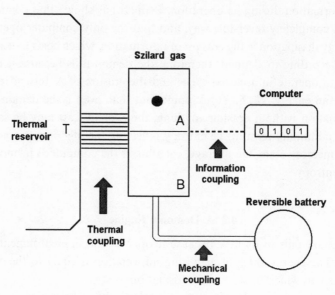

Fig. 4.1. The standard demon.

Note that if the memory element is initially 0 (blank), this coupling causes the memory element to record the location of the molecule in the code 0 = A, 1 = B.

We may also imagine that the demon may exchange energy from the battery and information from the computer with other systems. A schematic of the standard demon is shown in Figure 4.1.

A standard demon connected to a thermal reservoir at a temperature $T$ has two operating cycles, called READ and ERASE. In the READ cycle, the demon first locates the molecule in the Szilard gas and then allows the gas to expand isothermally. This will increase the amount of information in the computer's memory by one bit and add a mean energy of $k_B T$ to the battery. In the ERASE cycle, the demon first compresses the Szilard gas isothermally into region A, then transfers the contents of one bit of its memory to the gas using the information coupling. Note that the coupling leaves the demon's memory in the state 0, so that the bit has been erased; however, the initial compression of the gas requires an average energy $k_B T$ from the battery.

Clearly, a standard demon operating in a READ/ERASE double cycle connected to a thermal reservoir can gain no net energy on average, since the ERASE cycle costs as much energy as the READ cycle provides. On the other hand, suppose our system included *two* demons connected to a different thermal reservoirs and able to exchange information and energy. The information acquired in the READ cycle of one demon may now be disposed of in the ERASE cycle of the other demon, perhaps at a reduced cost. Leftover energy could be used to lift weights, etc.

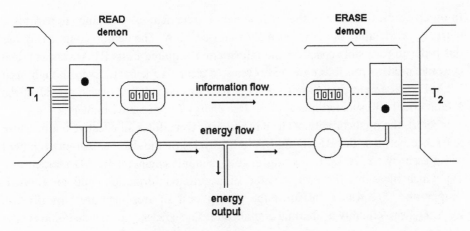

Fig. 4.2. A demonic heat engine.

A diagram of this arrangement is shown in Figure 4.2. This pair of demons functions as a *demonic heat engine*, exploiting the temperature difference between the reservoirs to do work by acquiring and erasing information. In the demonic heat engine, the READ cycle is performed on the thermal reservoir at the higher temperature $T_1$; then information and mechanical energy is reversibly transfered to the second demon, which performs the ERASE cycle on the cooler reservoir at temperature $T_2$.

The demonic heat engine has aquired an energy of $k_B T_1$ in each cycle, but it has been forced to use some of this energy $(k_B T_2)$ to dispose of unwanted information. It is thus easy to see that the demonic heat engine has a thermodynamic efficiency

$$\eta = 1 - \frac{T_2}{T_1}$$

which is, of course, simply the efficiency of an ideal ordinary heat engine. The READ/ERASE cycle is thermodynamically equivalent to the Carnot cycle.

### 4.3 Remarks

In an ordinary heat engine, it is easy to identify the heat flow between the two thermal reservoirs. In the demonic heat engine, however, there are separate transfers of information and mechanical energy. This indeed constitutes a flow of heat provided we adopt Zurek's suggestion and count the algorithmic entropy of the transfered bits as "entropy flow".

The demonic heat engine makes use of a temperature difference between two reservoirs; but other demonic engines could make use of other non-equilibrium degrees of freedom in their environments. Suppose, for example, that some order in the thermal reservoir caused the measured positions for a standard demon's Szilard gas to follow a highly correlated sequence: AAAAAABBBBBBBAAAABBBBB ....

In other words, the effect of the environment on the demon's "sensing" apparatus is partially predictable, since A is usually followed by A. The demon could easily use this order to reversibly *compress* the information acquired in its READ cycles; then it would need to run fewer ERASE cycles to clear its memory. Even though each ERASE cycle might cost as much energy ($k_B T$) as each READ cycle provided, the demon could still succeed in acquiring free energy from its environment.

In short, the predictability of the environment constitutes an effective temperature difference between the READ and ERASE cycles of the demon. An environment will present two effective temperatures to a demonic engine: a READ temperature $T_R$, which measures the energy value of acquired information, and an ERASE temperature $T_E$, which measures the energy cost of free memory. In general, $T_R \geq T_E$, with equality at thermal equilibrium. The efficiency of the demonic engine will be

$$\eta \leq 1 - \frac{T_E}{T_R}. \tag{4.2}$$

In a sense, we ourselves are demonic engines. We gain access to energy through the information we possess about our environment. The effective READ temperature of our environment is astronomically high: even the sum of the genetic, cultural, and sensory information available to human beings is small compared to thermodynamic entropies, yet the free energy that this makes available to human beings is enormous. The ERASE temperature is a few hundred Kelvins, but free energy is so easily available that we typically store and process information in thermodynamically "wasteful" ways.

I have one final observation that is more directly germane to the subject of this workshop. In the demonic heat engine, the demon at the warm reservoir transfered the contents of its memory to the demon at the cool reservoir, clearing its own memory cells. Suppose instead that the first demon sent over a *copy* of its information, retaining its own record afterward. The ERASE cycles of the cool reservoir demon would not now restore the demonic heat engine to its initial state, and more erasure would be required. In other words, for the demonic heat engine to function at its ideal efficiency, the efficiency at which the total entropy of the entire system remains constant over time, *no copy can be made of the transfered information*.

So the *increase* in entropy in a demonic heat engine is related to the *copying* of information. If information is copied within the demon—or indeed, if an external observer copies information from the demon's memory by examining it—entropy increases during the operation of the demonic heat engine.

Since this talk comes near to the beginning of the workshop, let me venture a modest prediction. I predict that most of the suggestions made here about the physical origin of time asymmetry will boil down to Nature's enormous propensity for *making copies* of information. Why Nature should be so constituted is a deeper mystery, and a much harder problem.

*Note added after the workshop*: I am not certain to what degree my prediction has been fulfilled in the discussions of decoherence and quantum cosmology that have dominated much of our agenda. It is arguable that the "leaking away" of quantum phase relations into the external environment that is so crucial in models of decoherence is an instance of the uncontrolled copying of information in Nature.

## Discussion

**Davies**  You are using the concept of temperature to describe a non-equilibrium system. What exactly do you mean?

**Schumacher**  I mean nothing more than the ratio between the energy transfer and entropy change of a system: $\Delta E = T\Delta S$. To be careful, we should call this an "effective temperature".

**Bennett**  (1) Szilard almost understood it. At the end of his paper, he did a calculation in which entropy only increased during erasure. (2) The representation of the demon you use [on your transparencies] is that used by Larry Gronik in his cartoon about Maxwell's Demon in the July 91 issue of Discover magazine.

**Schumacher**  You are quite right about Szilard. As for my cartoon demon, I got it from the figure on the Underwood deviled ham can.

**Gell-Mann**  So-called "creation scientists" sometimes argue that biological evolution is impossible because it allegedly violates the second law of thermodynamics. Some real scientists occasionally claim that biological evolution and related phenomena are instances of local exceptions to the second law.

If we take into account the environment, with its regularity, and the algorithmic complexity term in the entropy (where it is appropriate), we see that evolution that adapts, more or less, to those environmental regularities represents an example of the second law and not an exception.

**Schumacher**  Yes, I agree.

**Lloyd**  Two points: First, the amounts of information to which we are accustomed are negligible compared to normal thermodynamical quantities. All the libraries in the world contain much less information than that in a gallon of gas.

Second, our normal heat engines are already "demonic": their ability to do work depends crucially on their ability to get and process information.

**Schumacher**  This is a very good point. All heat engines must process information. This is true even of the Carnot engine: it must know which of the heat reservoirs is hot and which is cold. Otherwise it might act as a refrigerator!

**Davies**  Following from Seth Lloyd's remarks: Is the fact that a copy of Encyclopaedia Britannica costs much more than a gallon of gasoline a reflection of the fact that the information content of the former has greater "value" (or depth) than the latter?

**Schumacher**  Perhaps the encyclopaedia is valuable because the information is strongly correlated to the rest of the universe. It contains information about the composition of air, how to obtain gasoline, how to build an internal combustion engine, etc. One could imagine a very clever demonic engine which read the encyclopaedia, then went off and used the information to obtain lots and lots of energy. This would correspond in my scheme to an extremely high READ temperature—aquisition of a relatively small amount of information carries with it access to a huge amount of energy.

**Omnès**  Would you please elaborate upon what you said last, i.e., nature is constantly making copies of itself.

**Schumacher**  I have in mind, among other things, the phenomenon of decoherence, about which so much is said at this conference. Information about quantum correlations gets copied into many environmental degrees of freedom, beyond any hope of recovery, and so phase relationships between branches of the wave functions are effectively destroyed. But analogous things happen in classical systems as well.

**Hartle**  Does the measure of irreversibility supplied by the "multiplication of copies" agree quantitatively with entropy augmented by complexity, or disagree with it?

**Schumacher**  I am not sure. I have not really proposed a quantitative measure of irreversibility based on copying; however, that is a very nice idea. I would be surprised if it did not work out.

**Zeh**  Is the fact that "Nature likes to make copies" not essentially equivalent to the retardation of radiation (which carries away information)? Philosophers like to call it the "fork of causality".

**Schumacher**  Yes, certainly they are related. I would rather say that the retardation of radiation is a crucial example of Nature's propensity for copying information.

**Albrecht**  Is not the ability of nature to make all these copies in turn related to the out-of-equilibrium state of the universe? This is reflected in the presence of heat baths at two different temperatures in the apparatus you describe, or, for example, in the availability of blank tape which a demon could put to good use.

**Schumacher**  Yes, I think that you are right. If you try to copy something onto a tape that already contains random bits (using some reversible writing scheme, such as the XOR function), you just change the random bits to another set of random bits. You have not made the tape any harder to erase. If copying of information does occur, there must be lots of "blank tape" in the universe—that is, the universe must be far from equilibrium.

**Starobinsky**  If you really want to develop further your proposal that the source of irreversibility in nature lies in unconscious multiple copying of information, you have to take into account unavoidable errors that appear in this process. As a result, only approximate copies will be produced which can equally well be considered as containing new information. To check if they are identical to an original and to correct errors (if desired) will cost more energy and will produce more entropy. We all encounter this problem when typing our papers or sending them by e-mail.

**Schumacher**  Errors in copying will, of course, only make things worse for erasing, and so contribute to the irreversibility of the process. In this sense, copying errors would be a source of "friction" in a demonic heat engine, reducing the actual efficiency still further from the ideal efficiency.

## References

Bennett, C. H. (1982) The thermodynamics of computation—a review. *International Journal of Theoretical Physics*, **21**, 905–940.

Fermi, E. (1936) *Thermodynamics*, Dover, New York.

Maxwell, J. C. (1871) *Theory of Heat*, Longmans, London.

Szilard, L. (1929) On the decrease of entropy in a thermodynamic system by the intervention of intelligent beings. (In German) *Zeitschrift für Physik*, **53**, 840–856. Translation reprinted in Wheeler, J. A. and Zurek, W. H., eds., (1983) *Quantum Theory and Measurement*, Princeton University Press, Princeton.

Zurek, W. H. (1989) Algorithmic randomness and physical entropy I. *Physical Review* A **40**, 4731–4751.

# 5

# Which Processes Satisfy the Second Law?

Thomas M. Cover[†]

*Durand Bldg Rm 121*
*Stanford University*
*Stanford, CA 94301, USA*

## 5.1 Introduction

The second law of thermodynamics states that entropy is a nondecreasing function of time. One wonders whether this law is built into the physics of the universe or whether it is simply a common property of most stochastic processes. If the latter is the case, we should be able to prove the second law under mild conditions.

Thus motivated, we will reverse the usual physical development and put the emphasis on stochastic processes, physically generated or otherwise, and attempt to determine the family of processes for which the second law holds. In the course of this treatment we will suggest that relative entropy and conditional entropy are natural notions of what is meant by entropy in the second law. Certainly, the second law is true under milder conditions as we shift to these definitions.

We shall concern ourselves here, primarily, with discrete time finite state Markov processes. To the extent that the physical universe is Markovian, our comments will apply to physics. Here we should be aware that coarse graining (lumping of states) of a Markov chain may destroy Markovity. Also, while the Schrödinger wave function seems to evolve in a Markovian manner, the associated probabilities do not. Thus Markovity is a strong assumption.

We shall use Shannon entropy throughout. We ask whether the second law of thermodynamics is true of all finite state Markov processes. We shall find, somewhat surprisingly, that it is only true of doubly stochastic Markov processes. Equivalently, the second law is only true of Markov processes for which the equilibrium distribution is uniform over the finite state space. We will find that a slight change in the statement of the second law suffices to cover all Markov chains. Instead of the statement, "entropy always increases," we may substitute the more general statement that "relative entropy (of the current distribution with respect to the stationary distribution) decreases."

An interesting discussion of time symmetry and the second law can be found in Mackey (1992). The development of the second law from the physical standpoint is

† Email: cover@isl.stanford.edu.

argued in Van Kampen (1990), Wehrl (1978) and Tisza (1966), where good histories of the subject can be found. The consequences of the second law for Maxwell's Demon can be found in the collection of papers by Leff and Rex (1990). A probabilistic investigation of the behavior of entropy for stochastic processes can be found in Kullback (1959), Renyi (1961), Csiszar (1967), Fritz (1971), and Cover and Thomas (1991).

## 5.2 Entropy and its Interpretations

Let $X$ be a random variable drawn according to a probability mass function $p(x)$ over the finite set of outcomes $\mathcal{X}$. Shannon entropy is defined as

$$H(X) = -\sum p(x) \log p(x).$$

We shall sometimes denote this as $H(p)$. Here the entropy $H$ has the interpretation that it is the minimal expected number of yes-no questions required to determine the outcome $X$. It can also be shown that it is the minimum expected number of fair coin flips required by a random number generator to generate an outcome $X$ with the desired distribution. Also of importance is the conditional entropy $H(X|Y)$ which can be written as

$$H(X|Y) = -\sum_{x,y} p(y)p(x|y) \log p(x|y) = \sum_y p(y)H(X|Y = y).$$

It can be shown, by writing $\log p(x, y) = \log p(x) + \log p(y|x)$, that $H(X, Y) = H(Y) + H(X|Y)$. The strict concavity of the logarithm and Jensen's inequality immediately yield the result

$$H(X|Y) \le H(X),$$

with equality if and only if $X$ and $Y$ are independent. Thus, conditioning always reduces entropy. In fact, the reduction in entropy is strict unless the conditioning random variable $Y$ is independent of $X$.

Some other interpretations of the entropy $H$ are as follows:

*Descriptive complexity:*

$$H(X) \le El(X) < H(X) + 1$$

where $El(X)$ is the minimum expected number of bits in the description of $X$.

*Asymptotic equipartition theorem:*
If $X_1, X_2, \ldots$ be a discrete valued ergodic random process. Then the actual probability of the sequence of outcomes $X_1, \ldots, X_n$ is close to $2^{-nH}$ where $H$ is

the entropy rate of the process defined by $H = \lim_{n \to \infty} H(X_1, \ldots, X_n)/n$. More precisely,

$$p(X_1, \ldots, X_n) = 2^{-nH + o(n)},$$

where $o(n)/n$ converges to 0 as $n \to \infty$, with probability 1. The number of such "typical" sequences is approximately $2^{nH}$. This result allows one to interpret $2^H$ as the volume of the effective support set of $X$.

*Kolmogorov complexity:*

Let $K(x) = \min_{p:U(p)=x} l(p)$ be the minimum program length for a computer $\mathcal{U}$, which causes the computer to print $x$ and halt. Let $X_1, X_2, \ldots$ be an ergodic process with entropy rate $H$. Then

$$E \frac{K(X_1, X_2, \ldots, X_n)}{n} \to H, \qquad \text{as } n \to \infty.$$

Thus Kolmogorov complexity and Shannon entropy are asymptotically equal for ergodic processes. See Cover and Thomas (1991) for a proof for independent identically distributed processes.

*Number of microstates:*

For rough statements of the second law, it often suffices to take the logarithm of the number of microstates corresponding to a given macrostate in order to characterize the entropy of that macrostate. Implicit in this is that the probability is uniformly distributed over the microstates. (See our remarks about the asymptotic equipartition theorem.) In any case, the critical calculation of the number of microstates usually involves a multinomial coefficient which can be shown to be equal to

$$\binom{n}{np_1, np_2, \ldots, np_m} = 2^{nH(p_1, p_2, \ldots, p_m) + o(n)}.$$

Another important quantity for this discussion will be the relative entropy $D(p\|r)$, sometimes known as the Kullback-Leibler information, or the information for discrimination. The relative entropy $D(p \| r)$ between two probability mass functions $p(x)$ and $r(x)$, $p(x) \geq 0$, $\sum p(x) = 1$, $r(x) \geq 0$, $\sum r(x) = 1$, is defined by

$$D(p \| r) = \sum_x p(x) \log \frac{p(x)}{r(x)}.$$

The relative entropy is always nonnegative, as shown in the following theorem:

**Theorem 1.** $D(p \| r) \geq 0$ *with equality if and only if* $p(x) = r(x)$ *for all* $x$.

**Proof.** Let $A$ be the support set of $p(x)$. We use Jensen's inquality and the strict concavity of the logarithm to show

$$-D(p \parallel r) = \sum_A p(x) \log \frac{r(x)}{p(x)} \leq \log \sum_A p(x) \frac{r(x)}{p(x)} = \log \sum_A r(x) \leq \log 1 = 0.$$

The interpretations of relative entropy are as follows:

*Likelihood ratio:*
The relative entropy is the expected log likelihood ratio between distributions $p$ and $r$.

*Hypothesis testing exponent:*
The probability of error in a hypothesis test between distribution $p$ and distribution $r$ for independent identically distributed observations drawn according to one of these, has a probability of error given to first order in the exponent by $P_e = e^{-nD}$. Thus $D$ is the degree of difficulty in distinguishing two distributions.

*Redundancy:*
If one designs an optimal description for distribution $r$ when in fact distribution $p$ is true, then instead of requiring $H(p)$ bits for the description, the random variable requires $H(p) + D(p \parallel r)$ bits, as given in the following expression:

$$H + D \leq E_p l(X) < H + D + 1.$$

*Large deviation theory:*
Also, relative entropy arises in large deviation theory. The probability of physical data appearing to have macrostate $r$ when in fact observations are drawn according to $p$ is $e^{-nD(r \parallel p) + o(n)}$.

## 5.3 General Results about Increase in Entropy

Although we shall eventually argue that the entropy increase is not true for general Markov chains, there are a number of preliminary general results about the increase of entropy which agree with intuition.

First, it makes sense that for any stochastic process whatsoever, Markov or not, in equilibrium or not, the entropy of the present state given the past increases as the amount of information about the past decreases. This is due to the fact that conditioning always reduces entropy. This is embodied in the following theorem.

This theorem proves that the conditional entropy of the present given the far past increases as the past recedes, but simple examples exist for which the conditional entropy of the process at time $n$ given the fixed past up to time 0 may actually decrease.

Let $X_m^n$ denote $(X_m, X_{m+1}, X_{m+2}, \ldots, X_n)$ throughout this discussion.

**Theorem 2.** *For all stochastic processes, $H(X_0 | X_{-\infty}^{-n})$ is monotonically nondecreasing.*

*The apparently similar quantity $H(X_n|X^0_{-\infty})$ does not generally increase with n, but it does increase if the process is stationary.*

**Proof.** Conditioning reduces entropy. Thus

$$H(X_0|X^{-n}_{-\infty}) = H(X_0|X^{-(n+1)}_{-\infty}, X_{-n}) \le H(X_0|X^{-(n+1)}_{-\infty}),$$

proving the first statement. In the second statement, periodic processes provide counterexamples to the monotonicity of $H(X_n|X^0_{-\infty})$, but the additional assumption of stationarity yields

$$H(X_{n+1}|X^0_{-\infty}) = H(X_n|X^{-1}_{-\infty}) \ge H(X_n|X^0_{-\infty}),$$

establishing the increase of $H(X_n|X^0_{-\infty})$ for stationary processes.

We can now demonstrate a nice symmetry property of the conditional entropy of the present, given the past and given the future, for all stationary processes, Markov or otherwise. (A stationary process is in equilibrium.)

**Theorem 3.** $H(X_0|X_{-1}, X_{-2}, \ldots, X_{-n}) = H(X_0|X_1, X_2, \ldots, X_n)$ *for all stationary processes, Markov or otherwise. Also, for all stationary processes,*

$$H(X^{-1}_{-n}|X_0) = H(X^n_1|X_0).$$

**Proof.** By stationarity, $H(X_{-n}, \ldots, X_{-1}, X_0) = H(X_0, X_1, \ldots, X_n)$. Then the chain rule yields

$$H(X^{-1}_{-n}|X_0) + H(X_0) = H(X^n_1|X_0) + H(X_0),$$

thus proving the second assertion. The first assertion is proved similarly.

**Remark.** The fact that the entropy of the present given the n-past is equal to the entropy of the present given the n-future is somewhat surprising in light of the fact that the statement is true even for time-irreversible processes. Consider, for example, a Markov chain with transition matrix

$$P = \begin{bmatrix} .1 & .9 & 0 \\ 0 & .1 & .9 \\ .8 & 0 & .2 \end{bmatrix}.$$

Here it is clear that one can determine the direction of time by looking at the sample path. Nonetheless, the entropy of the present given a chunk of the future is equal to the entropy of the present given the corresponding chunk of the past.

## 5.4 Relative Entropy Always Decreases

We now state a theorem about relative entropy which shows the monotonic increase of relative entropy for all Markov chains, stationary or not. From this we will derive, by application, the second law of thermodynamics, which holds for doubly stochastic Markov chains. Versions of the following theorem appear in Kullback

(1959), Cover and Thomas (1991), Van Kampen (1990), Fritz (1973), Csiszár (1967), Renyi (1961), and the survey by Wehrl (1978).

**Theorem 4.** *Let $\mu_n$ and $\mu'_n$ be two probability mass functions on the state space of a finite state Markov chain at time n. Then $D(\mu_n \parallel \mu'_n)$ is monotonically decreasing. In particular, if $\mu$ is the unique stationary distribution,*

$$D(\mu_n \parallel \mu) \searrow 0.$$

Before proving this we need a definition and a lemma. We first define a conditional version of the relative entropy.

**Definition:** Given two joint probability mass functions $p(x, y)$ and $q(x, y)$, the *conditional relative entropy* $D(p(y|x) \parallel q(y|x))$ is the expected value of the relative entropies between the conditional probability mass functions $p(y|x)$ and $q(y|x)$ averaged over the probability mass function $p(x)$. More precisely,

$$D(p(y|x) \parallel q(y|x)) = \sum_x p(x) \sum_y p(y|x) \log \frac{p(y|x)}{q(y|x)}.$$

**Lemma:** *(Chain rule for relative entropy.)*

$$D(p(x, y) \parallel q(x, y)) = D(p(x) \parallel q(x)) + D(p(y|x) \parallel q(y|x)).$$

**Proof:** Write $p(x, y)/q(x, y) = p(x)p(y|x)/q(x)q(y|x)$ and expand $D(p(x, y) \parallel q(x, y))$.

**Proof of Theorem 4:** Let $\mu_n$ and $\mu'_n$ be two probability mass functions on the state space of a Markov chain at time $n$, and let $\mu_{n+1}$ and $\mu'_{n+1}$ be the corresponding distributions at time $n+1$. Let the corresponding joint mass functions be denoted by $p$ and $q$. Thus $p(x_n, x_{n+1}) = p(x_n)r(x_{n+1}|x_n)$ and $q(x_n, x_{n+1}) = q(x_n)r(x_{n+1}|x_n)$, where $r(\cdot|\cdot)$ is the probability transition function for the Markov chain. Then by the chain rule for relative entropy, we have two expansions:

$$
\begin{aligned}
D(p(x_n, x_{n+1}) \parallel q(x_n, x_{n+1})) &= D(p(x_n) \parallel q(x_n)) + D(p(x_{n+1}|x_n) \parallel q(x_{n+1}|x_n)) \\
&= D(p(x_{n+1}) \parallel q(x_{n+1})) + D(p(x_n|x_{n+1}) \parallel q(x_n|x_{n+1})).
\end{aligned}
$$

Since both $p$ and $q$ are derived from the Markov chain, the conditional probability mass functions $p(x_{n+1}|x_n)$ and $q(x_{n+1}|x_n)$ are both equal to $r(x_{n+1}|x_n)$ and hence $D(p(x_{n+1}|x_n) \parallel q(x_{n+1}|x_n)) = 0$. Now using the non-negativity of $D(p(x_n|x_{n+1}) \parallel q(x_n|x_{n+1}))$, we have

$$D(p(x_n) \parallel q(x_n)) \geq D(p(x_{n+1}) \parallel q(x_{n+1}))$$

or

$$D(\mu_n \parallel \mu'_n) \geq D(\mu_{n+1} \parallel \mu'_{n+1}).$$

Consequently, the distance between the probability mass functions is decreasing with time $n$ for any Markov chain.

Finally, if we let $\mu'_n$ be any stationary distribution $\mu$, then $\mu'_{n+1} = \mu'_n = \mu$. Hence

$$D(\mu_n \parallel \mu) \geq D(\mu_{n+1} \parallel \mu),$$

which implies that any state distribution approaches the stationary distribution as time passes. The sequence $D(\mu_n \parallel \mu)$ is a monotonically non-increasing non-negative sequence and must therefore have a limit. It can be shown that the limit is actually 0 if the stationary distribution is unique.

We now specialize this result to obtain the result for entropy increase for Markov chains.

**Theorem 5.** *Consider a finite state Markov chain. Then $H(X_n) \nearrow$ for any initial distribution on $X_0$ if and only if the Markov transition matrix is doubly stochastic, i.e., if and only if the stationary distribution for the Markov chain is uniform.*

**Proof.** Let $m$ denote the number of states. We note that

$$D(\mu_n \parallel \mu) = \sum_x \mu_n(x) \log \frac{\mu_n(x)}{(1/m)} = -H(\mu_n) + \log m.$$

Thus monotonic decrease in $D$ induces a monotonic increase in $H$. Moreover, if the stationary distribution is unique, then $D \searrow 0$ by Theorem 4, and $H(X_n) \nearrow \log m$.

To see that there are initial distributions for which the entropy decreases when the doubly stochastic conditions are not satisfied, let the initial state $X_0$ have the uniform distribution. Then, $H(X_0) = \log m$, which is the maximum possible entropy. As time goes on, $H(X_n)$ will converge to $H(\mu)$, the entropy of the stationary distribution. Since the stationary distribution is not uniform for this example, the entropy must decrease at some time.

### 5.5 Stationary Markov Chains

A number of entropy-increase or second law theorems are true if the process is already in equilibrium. This seems strange since a process in equilibrium is stationary and the entropy will remain constant. However, the appropriate entropy is the conditional entropy of the future given the present. That is, if one cuts into a process in equilibrium and observes its state, the conditional uncertainty of the future will grow with time.

**Theorem 6.** *If $X_n$ is stationary Markov chain, then the entropy $H(X_n)$ is constant, and*

$$H(X_n|X_1) \nearrow$$

*with $n$.*

**Proof.** Stationarity implies the marginal distributions are the same; thus $H(X_n)$ is constant. To prove monotonicity, we use conditioning and Markovity to show

$$H(X_{n+1}|X_1) \geq H(X_{n+1}|X_2, X_1) = H(X_{n+1}|X_2) = H(X_n|X_1),$$

where the first inequality follows from conditioning, the second from Markovity, and the last from stationarity.

## 5.6 Time Asymmetry

It is intriguing that a time asymmetric law like the second law of thermodynamics arises from a time symmetric physical process. This is not so puzzling if one believes that the initial conditions are extraordinary – for example, if one starts in a low entropy state. Thus, the time asymmetry comes from the asymmetry between the initial and final conditions.

However, if the process is in equilibrium, then the entropy is constant. Nothing could be more time symmetric than that. However, the conditional entropy $H(X_n|X_0)$ of a state at time $n$ given the present, is monotonically increasing as observed in Theorem 6. There is no asymmetry in this because the conditional uncertainty $H(X_{-n}|X_0)$ of the past given the present is also monotonically increasing. In short, the observation of the state of a process in equilibrium at time 0 yields an amount of information about the past and about the future which monotonically dissipates with time. Thus there is symmetry: conditional entropy increases in both directions of time.

A true time asymmetry arises when we consider relative entropy. We have observed that the relative entropy distance $D(\mu_n \parallel \mu'_n)$ between two probability mass functions on the state space decreases with time for Markov chains. This is true even for transition matrices $r(x_{n+1}|x_n)$ that generate time reversible Markov chains. Here, then, is an apparent paradox. Why can't we reverse time, argue that the reversal of a Markov process is also Markov, and conclude that $D(\mu_n \parallel \mu'_n)$ decreases?

The answer is that the time reversed processes, although Markov, do not have the same transition matrices, *i.e.* $p(x_n|x_{n+1}) \neq q(x_n|x_{n+1})$, so the argument in the proof of Theorem 4 does not apply. We conclude that there is indeed a time-asymmetric behavior $(D(\mu_n \parallel \mu'_n) \searrow)$ even for Markov processes generated from time symmetric physical laws.

## 5.7 The Relation of Time-Discrete and Time-Continuous Markov Chains

It should be pointed out that the study of time-continuous and time-discrete Markov chains may lead to different statements about the second law of thermodynamics. In a time-continuous Markov chain, one has intensities $\lambda_{ij}$ for the Poisson rate at which transitions take place from state $i$ to state $j$. A typical condition (Yourgrau et al. 1982) for the $\dot{H}$ theorem to hold, for example, would be the microscopic reversibility condition $\lambda_{ij} = \lambda_{ji}$.

A time-discrete Markov chain can be thought of as being generated by a time-continuous Markov process where the states are labeled $X_1, X_2, X_3, \ldots$ as the transitions occur. Thus in our formalism

$$H(X_n) = H(X(t)|N(t) = n),$$

where $N(t)$ denotes the number of transitions that have taken place in the continuous-time Markov chain. It may well be that $H(X(t))$ increases while

$$H(X_n) = H(X(t)|N(t) = n)$$

does not increase with $n$. Thus, conditioning on the number of events may change the qualitative statement of the second law for such processes.

The discrete-time analysis in this paper deals with the event-driven rather than absolute time driven idea of time.

## 5.8 Summary of Relevant Results

We now gather these results in increasing order of generality.

The entropy $H(X_n)$ increases for a finite state Markov chain with an arbitrary initial distribution if the stationary distribution is uniform.

The conditional entropies $H(X_n|X_1)$ and $H(X_{-n}|X_0)$ increase with time for a stationary Markov chain.

The conditional entropy $H(X_0|X_n)$ of the initial condition $X_0$ increases for any Markov chain.

The relative entropy $D(\mu_n\|\mu)$ between a distribution $\mu_n$ and the stationary distribution $\mu$ decreases with time for any Markov chain.

The conditional entropy $H(X_n|X_0, X_{-1}, \ldots)$ of a process at time $n$ given the past up to time zero increases for any stationary process.

The conditional entropy $H(X_0|X_{-n}, X_{-(n+1)}, \ldots)$ of the present given the past increases as the past recedes for *all* processes.

## 5.9 Conclusions

While the second law of thermodynamics is only true for special finite-state Markov chains, it is universally true for Markov chains that the relative entropy distance of a given distribution from the stationary distribution decreases with time. This leads to a natural general statement of the second law for Markov chains: relative entropy decreases. To specialize this to the statement that entropy increases, one must restrict oneself to Markov chains with a uniform stationary distribution or to certain Markov chains with a suitable low entropy initial condition.

Finally, the second law of thermodynamics says that uncertainty increases in closed physical systems and that the availability of useful energy decreases. If one can make the concept of "physical information" meaningful, it should be possible to augment the statement of the second law of thermodynamics with the statement, "useful information becomes less available." Thus the ability of a physical system to act as a computer should slowly degenerate as the system becomes more amorphous and closer to equilibrium. A perpetual computer should be impossible.

## Discussion

**Albrecht**   Relative entropy sounds like it has something to do with coarse graining. Can one think of it as being relative to a particular description or parameterization of phase space?

**Cover**   Yes. Although the coarse graining is arbitrary, the natural distribution to place on the coarse graining is the stationary or equilibrium distribution. Then, the relative entropy distance of the current distribution from this stationary distribution is monotonically decreasing in time for any Markovian process. Consequently, the difficulty in measuring the difference of the current distribution from the equilibrium distribution increases with time. One should be aware, however, that coarse graining can destroy Markovity unless the partition is adroitly chosen.

**Lloyd**   It is a commonly claimed feature of the psychological arrow of time that we know more about the past than we do about the future. How do you square this with the result that the amounts of information about the past and future that a system has in the present are equal in a stationary process?

**Cover**   The key ingredient is that the statement holds in general only for processes in equilibrium. The theorem you are referring to states that the future is as uncertain as the past, conditioned on the present, for stationary Markov processes. That is to say, where we are going is conditionally as uncertain as how we got to where we are, at least for Markov processes in equilibrium. This intriguing symmetry is true even for time-asymmetric Markov chains. Apparently, this symmetry in information about past and future given the present follows entirely from the stationary Markovian assumption of the process and not from the time symmetry. That's the main point.

## References

Cover, T., and Thomas, J. (1991) *Elements of Information Theory*, Wiley, New York.

Csiszar, I. (1967) Information type measures of difference of probability distributions. *Studia Sci. Math. Hung.*, **2**, 299-318.

Fritz, J. (1973) An information-theoretical proof of limit theorems for reversible Markov processes. *Transactions of the Sixth Prague Conference on Information Theory, Statistical Decision Functions, Random Processes*, Verlag Dokumentation, Munich, 183-197.

Kullback, S. (1959) *Information Theory and Statistics*, Wiley, New York.

Leff, H., and Rex, A., eds., (1990) *Maxwell's Demon: Entropy, Information, Computing*, Princeton University Press, Princeton, New Jersey.

Mackey, M. (1992) *Time's Arrow: The Origins of Thermodynamic Behavior*, Springer-Verlag, New York.

Renyi, A. (1961) On measures of entropy and information. *Proc. 4th Berkeley Symposium on Math. Stat. and Probability*, **1**, 547-561, Berkeley.

Tisza, L. (1966) *Generalized Thermodynamics*, MIT Press, Cambridge, Mass.

Van Kampen, N.G., (1990) *Stochastic Processes and Chemistry*, North Holland, 1990.

Wehrl, A. (1978) General properties of entropy. *Rev. Mod. Phys.*, **50**, 221.

Yourgrau, W., Van der Merwe, A., Raw, G. (1982) *Treatise on Irreversible and Statistical Thermophysics*, Dover Publications, New York.

# 6

# Causal Asymmetry from Statistics

Seth Lloyd[†]

*California Institute of Technology*
*Pasadena, CA 91125, USA*

## 6.1 Introduction

Aristotle's analysis of the world in terms of cause and effect formed a cornerstone of Renaissance thought. Efficient cause explained the sinfulness of the world as an effect of Adam's fall, while final cause justified religious and political institutions as 'caused' by God's intentions for the future of mankind. The Enlightenment's questioning of religious and social institutions robbed arguments by final cause of much of their force. Hume denied final cause, and regarded efficient cause as arising simply from the human habit of calling 'cause' the first in time of two events that occur in constant conjunction (Hume 1739). By the beginning of the current century, the intellectual status of causal reasoning had receded to the point that Russell could write (Russell 1929), "The law of causality, I believe, like much that passes muster among philosophers, is a relic of a bygone age, surviving, like the monarchy, only because it is erroneously supposed to do no harm."

Russell's conviction of the anachronistic nature of causal law came from well–established 19th century ideas in physics: the fundamental description of the world was given by Hamiltonian evolution; the behaviour of Hamiltonian systems over a given time interval was determined equally by conditions given at the beginning or at the end of the interval; the particular Hamiltonians that seemed to describe physical systems were quadratic in momentum, and so were invariant under the transformation $t \longrightarrow -t$. In this view, the underlying dynamics of the world were completely time symmetric, and temporal and causal asymmetry were merely artifacts of our inability to perceive the true microscopic workings of the world around us, an inability that forced us to rely on statistical descriptions.

The greater the explanatory power of a prevalent physical model, the greater is the temptation to regard that model's picture of the world as fundamental. Russell's dismissal represents a low water mark in causal and statistical reasoning. In the twentieth century, the successes of statistical and quantum mechanics have

† This work supported in part by the U. S. Department of Energy under Contract No. DE-AC0381-ER40050

lent legitimacy to probabilistic reasoning, to the extent that most physicists now regard the deterministic evolution of classical mechanics as an approximation, and the stochastic natures of quantum mechanics and of chaos as fundamental. With the growing importance of statistical techniques has come a resurgent interest in causality and its relationship to statistics, exemplified by the work of Reichenbach (Reichenbach 1956). Causal reasoning, like the monarchy, survived both Russell and his remark.

In the present work, the only requirement made of a cause is that under some circumstances, variation in the outcome of the cause produces a correlated variation in the outcome of the effect. In essence, the cause–effect relationship can be thought of as a one–way communications channel through which the cause sends information that the effect receives. The basic idea here is that causal connection implies the possibility of statistical correlation, while the absence of causal connection implies the absence of correlation. If two events are correlated, then either one has an effect on the other, or the two have a common cause in their past. By extending these ideas to many events, one arrives at a method for deriving patterns of statistical dependence and independence from the causal relations between the events.

## 6.2 Correlation and Information

This programme may be made formal as follows. Suppose that the statistical information possessed about a set of events $A, B, C, \ldots$ takes the form of a joint probability distribution $p(abc\ldots)$ over the possible outcomes $a, b, c, \ldots$ of $A, B, C, \ldots$. From this joint distribution, one can derive various marginal distributions, such as $p(ab) \equiv \sum_{c\ldots} p(abc\ldots)$, the probability for $A$ to have outcome $a$ and $B$ to have outcome $b$, and conditional probability distributions, such as $p(a|b) \equiv p(ab)/p(b)$, the probability that the outcome of $A$ is $a$ given that the outcome of $B$ is $b$.

Two events $A$ and $B$ are correlated if fixing the outcome of $B$ can change the probabilities for the outcome of $A$, that is, if $p(a|b) \neq p(a)$ for some $a, b$. Note that $p(a|b) \neq p(a)$ if and only if $p(b|a) \neq p(b)$: correlation is a symmetric relationship. On the one hand, if you throw a rock at a window, there is a better than normal chance that the window will break. On the other hand, if a window breaks, there is a better than normal chance that someone has thrown a rock at it. The degree of correlation between events can be measured using information theory (Shannon and Weaver 1949). Define $I(A) \equiv -\sum_a p(a) \log_2 p(a)$; $I(A)$ is the average number of bits of information required to specify the outcome of $A$. If the outcome of $B$ is fixed, the average number of bits required to specify the outcome of $A$ is $I(A|B) \equiv \sum_b p(b)\left(-\sum_a p(a|b) \log_2 p(a|b)\right) = I(AB) - I(B)$, where $I(AB) = -\sum_{ab} p(ab) \log_2 p(ab)$ is the average number of bits required to specify the outcome of both $A$ and $B$. The degree of correlation between $A$ and $B$ can be measured by how much knowledge of the outcome of $B$ reduces the amount of information required to specify the outcome of $A$: the corresponding quantity,

$I(A;B) \equiv I(A) - I(A|B) = I(A) + I(B) - I(AB)$, is called the mutual information between $A$ and $B$. Note that $I(A;B) = I(B;A)$: the amount that one finds out about $A$ by knowing $B$ is equal to the amount that one finds out about $B$ by knowing $A$. Note also that $I(A;B) \geq 0$, with equality if and only if $p(a|b) = p(a)$ for all $a, b$. If $I(A;B) = 0$, then $A$ and $B$ are said to be independent: knowing the outcome of $B$ imparts no knowledge about the outcome of $A$.

Mutual information was originally defined in communications theory, to measure the capacity of communication channels. Regarding the cause–effect relationship as such a channel, one may take the mutual information between cause and effect as a measure of the amount of information that the effect is receiving from the cause. In fact, this is strictly true only when an event has only one cause. When an effect has more than' one cause, just as when a communications channel has more than one input, one must be more careful in measuring the amount of information transmitted from cause to effect.

## 6.3 Causal Models

Directed graphs will be used to model the causal relationships between events. So, for example, $A \longrightarrow B$ will indicate that $A$ has an effect on $B$ in the sense given above, that variation in the outcome of $A$ induces a correlated variation in the outcome of $B$ under some circumstances. Similarly, $A \longrightarrow B \longrightarrow C$ indicates that $A$ has an effect on $B$, $B$ has an effect on $C$, and that $A$ may have an effect on $C$, but only through its effect on $B$. That is, $A \longrightarrow B$ indicates that $A$ has a *direct* effect on $B$, unmediated by any of the other variables about which we possess statistical information.

### 6.3.1 Models with Two Variables

Causal models imply the existence of independence relationships between the events in the model. Consider a model, $A \quad B$, for the events $A$ and $B$: in this model, there is no causal connection whatsoever between $A$ and $B$. Since there is no casual connection between $A$ and $B$, there should be no correlation between $A$ and $B$: that is, the model, $A \quad B$ implies that $I(A;B) = 0$. Similarly, the models, $A \longrightarrow B$ and $B \longrightarrow A$, both imply that $I(A;B)$ need not be equal to zero.

At first sight, it might be thought that $A \longrightarrow B$ should imply that $I(A;B)$ is strictly greater than zero. This is not the case. Consider, for instance, an exclusive $OR$ gate, whose output is equal to 1 when exactly one of its inputs is 1, and equal to 0 otherwise. Label the inputs $A, C$ and the output $B$. Suppose that each of the four possible combinations for the inputs $A$ and $C$, $00, 01, 10$, and $11$, have equal probability. It is easy to verify that even though variation in the input $A$ can cause a correlated variation in the output $B$, in the absence of knowledge of the value of $C$, $A$ and $B$ are uncorrelated. The reason for this lack of correlation is that the value of $C$ determines the way in which variation in $A$ induces variation in $B$: if $C = 0$, then variation in $A$ produces a perfectly correlated variation in $B$ — when $A$ is 0, $B$ is 0, when $A$ is 1, $B$ is 1; if $C = 1$, then variation in $A$ produces a perfectly anticorrelated variation in $B$ — when $A$ is 0, $B$ is 1, when $A$ is 1, $B$ is 0. If both values for $C$ are equally probable, then the correlation between $A$ and $B$ when $C = 0$ is counterbalanced by the anticorrelation when $C = 1$, and $A$ is not correlated with $B$, even though it is a cause of $B$. If $C = 0$ and $C = 1$ are not equally probable, however, $A$ and $B$ will be correlated. $A \longrightarrow B$ generally implies that $I(A;B) > 0$; special conditions are required on the other causes of $B$ to destroy this correlation.

For causal models containing only two variables, the situation can be summarized as follows: $A \quad B$ implies that $I(A;B) = 0$; $A \longrightarrow B$ and $B \longrightarrow A$ generally imply that $I(A;B) > 0$, but are also consistent with $I(A;B) = 0$. If the statistics for the outcomes of $A$ and $B$ give $I(A;B) > 0$, then they rule out the model $A \quad B$. $I(A;B) > 0$ does not imply that either $A \longrightarrow B$ or $B \longrightarrow A$ is the case, however. Correlation between $A$ and $B$ can also be explained by the existence of a common cause $\mathscr{X}$ that lies outside the set of events about which we possess statistics: $A \longleftarrow \mathscr{X} \longrightarrow B$ can also give $I(A;B) > 0$. At the level of two events, no causal asymmetry can be derived from statistical relations. Two events are either correlated or not, and correlation gives no clue as to which is cause and which is effect, or whether both are effects of some common cause.

### 6.3.2 Three and More Variables

At this point, it may seem that an elaborate notation has been introduced merely to state the obvious: events whose outcomes are correlated are causally connected, either by standing in a cause–effect relationship to each other or by possessing some common cause. The usefulness of introducing directed graphs to model causal situations arises when one desires to keep track of the causal relationships between more than two variables. In addition, with more than two variables, the asymmetric nature of the cause–effect relationship gives rise to recognizable statistical patterns.

The simplest causal asymmetry to have a statistical signature is the 'causal fork' (Reichenbach 1956). Consider two causes that have a common effect: $A \longrightarrow B \longleftarrow C$, for example, the exclusive $OR$ gate of the previous section. In this causal model, the only connection between $A$ and $C$ is the fact that they have a common effect. But there is no reason why two events should be correlated simply because they have a common effect. Therefore, $A \longrightarrow B \longleftarrow C$ implies $I(A;C) = 0$. Now suppose that the outcome of $B$ is fixed to 0. By the rule for an exclusive $OR$ gate, if $A$ is 0 then $C$ must be 0, if $A$ is 1 then $C$ must be 1: fixing the outcome of $B$ implies perfect correlation between the outcomes of $A$ and of $C$. So $A \longrightarrow B \longleftarrow C$ implies that $I(A;C) = 0$, but $I(A;C|B)$ need not equal 0. Now consider the same model, but with the direction of the cause–effect relationships reversed: $A \longleftarrow B \longrightarrow C$. Here $B$ is a common cause of both $A$ and $C$. Since $A$ and $C$ have a common cause, there is no reason to expect that $I(A;C) = 0$. In addition, since the only causal connection between $A$ and $C$ is their common cause $B$, fixing the value of $B$ should destroy the correlation between $A$ and $C$: $I(A;C|B) = 0$. That is, $A$ and $C$ are correlated because variation in the outcome of $B$ induces a correlated variation in the outcomes of $A$ and $C$. In the absence of any variation in $B$, there is no correlation between $A$ and $C$. An example is a logic gate that has one input, $B$, and two outputs, $A = B$, and $C = 1 - B$; here $A$ and $C$ are perfectly anticorrelated, but fixing the value of $B$ destroys that correlation.

The two causal models with the cause–effect relationships reversed imply entirely different statistical independence relations. Having a common effect does not induce correlation between events, while having a common cause does. Controlling for the outcome of a common effect can make the outcomes of its causes correlated, while controlling for the outcome of a common cause has the opposite result. Reichenbach identified the different independence relationships implied by common causes and by common effects as forming the basis for our identification of causal asymmetries in the world around us. This asymmetry in causation is responsible for the primary psychological arrow of time, our belief that we can change the future, but not the past. The difference between common cause and common effect implies that correlations between events in the present are to be ascribed to common causes in the past. In particular, the correlation between our memories of past actions, and those events that our actions have affected, lead us to identify our past actions as a common cause of those present events and present memories. However, there is no reason why a future choice of action should generate correlation between present events and present state of mind. Therefore, insofar as our choices are designed to effect a positive correlation between our desires and the state of affairs in the world around us, this correlation lies in the future; the past is beyond our control.

The general rule for deriving independence relations between events in causal models with many variables can now be presented. The following result is due to Pearl (Pearl 1988), and results from applying the two rules given above: 1) two events whose only causal connection is a common effect should not be correlated unless the outcome of their common effect is fixed; 2) a common cause tends to induce correlation between its effects unless its outcome is fixed. Suppose that a set of events $\mathbf{Y}$ have their outcomes fixed. Consider a path within the causal model. This path is capable of inducing correlation between the events that make up its endpoints if all the common effects along the path either belong to the set $\mathbf{Y}$ or have some descendant in $\mathbf{Y}$, and if no other events in the path belong to $\mathbf{Y}$. Such a path is called *open*. A path that is not open is called *closed*. If there is no open path between $A$ and $B$ given $\mathbf{Y}$, then $I(A;B|\mathbf{Y}) = 0$.

That is, a path is capable of generating correlation between its endpoints given $\mathbf{Y}$ if none of the common or intermediate causes — places where correlation is generated or propagated — along the path are fixed by fixing $\mathbf{Y}$, and if all the common effects — places where the propagation of correlation breaks down —

have been fixed to some degree, thus allowing correlation between their causes. So, for example, in the path $A \longrightarrow B \longleftarrow C \longrightarrow D \longleftarrow E$, fixing the outcomes of $B$ and $D$ allows correlation between $A$ and $E$, while fixing the outcomes of $C$ and $D$ allows no correlation between $A$ and $E$.

The general rule for deriving conditional independence relations from causal models can now be given. Consider a causal model, given by a directed graph whose vertices represent both the events about which statistics are available, and also 'hidden' events about which no statistical information is available. Consider three non–overlapping sets of events, $\mathbf{X}$, $\mathbf{Y}$ and $\mathbf{Z}$, taken from the events within the model about which statistics are available. $\mathbf{X}$ and $\mathbf{Z}$ are conditionally independent given $\mathbf{Y}$, $I(\mathbf{X}; \mathbf{Z}|\mathbf{Y}) = 0$, if there are no open paths between events $X \in \mathbf{X}$ and $Z \in \mathbf{Z}$ given $\mathbf{Y}$. If the causal model correctly represents the cause–effect relationships between events, then the conditional independence relationships derived from the model must hold. Statistically significant deviations from the implied independence relations falsify the model.

In contrast, the existence of independence relationships above and beyond those implied by the causal model does not falsify the model. Two events may possess an open path between them, and still have statistically independent outcomes, as in the exclusive $OR$ gate above. However, in the absence of detailed knowledge of the actual form of causal influence, one generally expects the conditional independence relationships implied by the causal model to be the only ones actually present in the data. For an open path not to generate correlation between its endpoints requires special sorts of causal influence along the path; in the exclusive $OR$ gate example, variation in one of the inputs fails to generate a correlated variation in the output if the values 0 and 1 for the other input are equal. Any deviation from equiprobability on the part of the second input will allow correlation between the other input and the output. A 'generic' open path generates correlation between its endpoints.

## 6.4 Bayesian Networks

Given these methods for deriving conditional independence relations from causal models, one can ask, When do two causal models imply the same set of independence relationships? The answer to this question is particularly simple when one restricts one's attention to causal models whose graphs contain neither directed loops nor unobserved variables. Such models, represented by directed, acyclic graphs, are called Bayesian Networks (Pearl 1988).

The first point to note is that two directed, acyclic graphs that have different links (ignoring the directionality of those links) imply different conditional independence relations. Any two events that are not directly linked can be made independent by fixing the values of some set of events, while no conditional independence relation can be derived between two events that are directly linked. So two Bayesian

Networks that imply the same set of independence relations must have links in the same places.

The second point to note is that two Bayesian Networks with links in the same places, but with different unlinked common effects, imply different independence relations. An unlinked common effect is one such as $A \longrightarrow B \longleftarrow C$, where the two causes have no direct link between them. An acyclic model that contains $A \longrightarrow B \longleftarrow C$ always implies some conditional independence relation between $A$ and $C$ in which the value of $B$ is *not* fixed. Any other set of directions for the links, $A \longrightarrow B \longrightarrow C$, $A \longleftarrow B \longleftarrow C$, or $A \longleftarrow B \longrightarrow C$, gives conditional independence between $A$ and $C$ *only* if $B$ is fixed. So two Bayesian Networks that imply the same set of independence relations must have links in the same places, and the same set of unlinked common effects.

It is straightforward to verify that once the positions of the links and the unlinked common effects are given, the direction of the remainder of the links may be varied in any way that does not produce a directed loop or a new unlinked common effect — all Bayesian Networks obtained by such a process imply the same set of conditional independence relationships. So two Bayesian Networks imply the same set of conditional independence relations if and only if they share the same link locations and the same unlinked common effects. This result was derived by the author and used in analyzing financial data (1986–87); the same result was derived independently by Pearl and Verma (Pearl, Verma 1990). Similar results can be derived if the no directed loop and no unobserved event restrictions are relaxed.

## 6.5 Causal Asymmetry in Physical Systems

The notion of causality in physics is the same as that assumed as a basis for causal modelling: variation in the outcome of the cause produces a correlated variation in the outcome of the effect. The methods of modelling causal systems given here, unsurprisingly, give correct results when applied to physical systems. As an example, the requirement in electrodynamics that the source–free part of the incoming electromagnetic field vanish is a way of realizing the requirement that correlated variation between the motions of charged particles be caused by the motions of charged particles in the past. Some subtleties arise, however.

Since the causal models considered here correspond to directed graphs, to apply them to physical systems, one must either discretize the physical system, or look at causal models that contain hidden variables. In addition, causal models are Markovian in nature, defining causal relationships in terms of conditional probabilities, while nondissipative physical systems are generally characterized by Hamiltonian mechanics, a very particular type of Markov process. The deterministic nature of Hamiltonian mechanics implies the existence of statistical independence relations above and beyond those simply implied by the causal structure of a system. For example, controlling for the initial or final state of a Hamiltonian system completely determines the system's trajectory, and destroys the statistical correlation between all variables, whether common causes or common effects. As a result, for Hamiltonian systems in the absence of noise, causal asymmetry cannot generally be derived from statistics. Whenever noise is introduced, however, as in Brownian motion, or in other systems described by master equations, application of the above methods generally results in a unique set of statistical independence relations. In such cases, statistical relations can be used completely to determine the causal structure of the system.

The methods described in this article are classical in nature, and do not take into account quantum mechanics. Quantum mechanical correlations can violate Bell's inequality: that is, the correlation between two spins in the Bohm version of the Einstein–Podolsky–Rosen *gedanken* experiment is of a form that cannot be reduced to zero by controlling for the value of a common cause (Bell 1964, Bohm 1951, Einstein, Podolsky, Rosen 1935). This result does not invalidate the present work, however. The correlation between the spins is still due to a common cause in the past: this common cause — the $S$-wave state out of which the spins arise — is fundamentally non–classical, and is not an 'event' whose outcomes can be assigned probabilities. Such quantum–mechanical causes can be included in causal models as hidden common causes. The resulting causal models, though expressed in terms of classical probabilities, are perfectly consistent both with quantum mechanics and with experiment. Bell's inequalities tell us only that such common causes cannot be resolved by experiment: hidden quantum–mechanical causes will always remain hidden.

## 6.6 Conclusion

The methods presented here are useful in ruling out causal models that predict independence relations that are not realized by the data. If hidden common causes can be ruled out *a priori*, then in many occasions, one and only one causal model is consistent with the data. If common causes cannot be ruled out, then although one can still rule out causal models that are inconsistent with the data, no unique model can be derived from the data; one can always postulate a model with many different hidden causes (e.g., a conspiracy theory), that explains the correlations in the data as accurately as a model with few or no hidden causes. In such cases a further principle, such as Occam's razor, must be introduced to identify the most plausible causal model.

In closing, it should be noted that when applied to the universe as a whole, the causal models presented here require a particular type of initial condition. If correlation in the present is to be ascribed to common causes in the past, then at the unique moment at which there was no past, there should be no correlation. Although one must be careful about extrapolating classical methods back to a quantum initial condition, the present work implies that in addition to being in a state of low entropy, the universe began in a state with no correlation between spacelike separated points beyond that required by the Heisenberg uncertainty principle. In fact, Euclidean quantum gravity calculations point to just such an initial condition (Halliwell 1994, Laflamme 1994).

## References

Bell, J.S. (1964) On the Einstein Podolsky Rosen paradox. *Physics*, **1**, 195–200.

Bohm, D. (1951) *Quantum Theory*, Prentice Hall, Englewood Cliffs.

Einstein, A., Podolsky, B., and Rosen, N. (1935) Can quantum–mechanical description of physical reality be considered complete? *Physical Review*, **47**, 777–780.

Halliwell, J.J. (1994) This volume.

Hume, D. (1739) *A Treatise of Human Nature*, John Noon, London. Reprinted 1906, Clarendon Press, Oxford.

Laflamme, R. (1994) This volume.

Pearl, J. (1988) *Probabilistic Reasoning in Intelligent Systems*, Morgan Kauffman, San Mateo.

Pearl, J., and Verma, T.S. (1990) Equivalence and synthesis of causal models, Technical Report R-150, Cognitive Systems Laboratory, University of California, Los Angeles.

Reichenbach, H. (1956) *The Direction of Time*, University of California Press, Berkeley.

Russell, B. (1929) *Mysticism and Logic*, Norton, New York.

Shannon, C.E., and Weaver, W. (1949) *The Mathematical Theory of Communication*, University of Illinois Press, Urbana.

# Part Two

Statistical Origins of Irreversibility

# 7

# Stirring up Trouble

P. C. W. Davies

*Department of Physics and Mathematical Physics*
*The University of Adelaide, South Australia*

**Abstract**

The nature and origin of time asymmetry is reviewed. Some persistent misconceptions are dispelled by drawing careful distinctions between the flow of time and time asymmetry, and between coarse-graining and the microscopic asymmetry associated with environmental noise. The significance of branch systems is stressed. The manner in which the expansion of the universe produces a time asymmetry is briefly discussed.

## 7.1 Does Time Flow?

You have to accept the idea that subjective time with its emphasis on the now has no objective meaning... the distinction between past, present and future is only an illusion, however persistent.

A. Einstein

Discussions of 'the arrow of time' frequently conflate three distinct concepts. The first is of these is the assertion that time 'passes'. This so-called flow or flux of time was made explicit in Newton's definition of time (time "flows equably, without relation to anything external" – a notion from which Newton derived the idea of 'fluxions', or calculus, for the rate of change of dynamical quantities). The flux of time is the source of many metaphors: time's winged chariot, the river of time, the fleeting arrow... It is based on the subjective impression that we are being carried along on a tide of events, bourne irresistibly into the future. Alternatively, we observe helplessly as future events are bourne down upon us. Either way, the notion that time *passes*, or moves in any way, is utterly mysterious to the physicist. Time, like space, simply *is*. To assert that time flows at one second per second is a vacuous statement.

The second, related, concept is that time is divided into 'the past', 'the present' and 'the future'. In popular parlance, the present, or 'now' is continually shifting towards the future, so that some events which are in 'the future' eventually come about, or 'happen', and are thereby consigned to 'the past'.

There is a temptation to attach a metaphorical arrow to this flux of time, an arrow which points in the direction of the motion of the now, i.e., from past to future. For many people, this is what is meant by the arrow of time. But as the above quotation of Einstein illustrates, notions of a divided, kinetic time are usually regarded by physicists as illusory. On the other hand, and by contrast, physicists do recognize that the events in the physical world often display a distinct asymmetry – an asymmetry *in* time between past and future. It is important to make a sharp distinction between an asymmetry *of* the world *in* time from an intrinsic asymmetry *of* time itself. The meaning of the former concept is that successive physical states may be ordered with a distinct directionality.

It is helpful to give this idea an operational meaning. Imagine a movie film taken of a sequence of events. Usually this sequence would 'look wrong' if played in reverse. Consider, for example, a film sequence of an egg that rolls off a table to the floor and smashes. You do not need to be a physicist to know, on seeing a film of a smashed egg reconstituting itself, that you have been fooled. It is important to realize, however, that the asymmetry of the set of physical states here is a structural property of the set, and does not depend on the movie actually being played (either forwards or backwards – to make time 'run') in order to manifest that asymmetry. The film could be chopped up into frames, and the frames stacked up: the asymmetry would still be evident. If the frames were shuffled, we could make a passable attempt at re-ordering them correctly.

One may still use the arrow metaphor to denote this asymmetry of the world in time. Using an arrow to denote an asymmetry is quite usual. For example, a compass needle has an arrow pointing north to represent the north-south asymmetry of the Earth's magnetic field. It is conventional to have the arrow of time point from past to future. However, this need not imply that the arrow, or time itself, is actually moving from past to future, any more than the north-directed compass arrow implies that the compass, or the Earth's magnetic field, is moving north.

It is in this latter sense that the arrow of time can be meaningfully studied by physicists, who treat time as a co-ordinate, not as a flowing entity. Nor do physicists divide time up into past, present and future. The terms 'past' and 'future' are used instead to denote directionality *along* the time co-ordinate, not as names for *regions* of time, just as north and south are directions across the Earth's surface, and not geographical places or regions. Actually this is not quite true. People do speak of 'the Deep South' (or, in the case of Queensland, the 'Deep North'). But this is merely a *façon de parler*. They really mean 'a region towards the south of most of the rest of the United States' (or Australia), not to a place which *is* 'the' South.

The foregoing distinctions between various arrows of time have not always been clearly made, with the result that the subject, already confusing enough, has been further muddled. Some scientists, not content with dismissing the flow of time as an illusion, yet realizing that physics-as-we-know-it makes no provision for time to flow, conclude that the physicists must have left something out. Some key quality about time is missing in the standard theory. What is needed, they argue, is an 'extra ingredient' to give time its elusive flux. Some have sought this ingredient in quantum mechanics (e.g., Landau & Lifschitz 1968), others in cosmology (e.g., Gold 1962) and others in departures from the known laws of dynamics (e.g., Mehra & Sudarshan 1972). These attempts to 'graft on' a mysterious extra quality to bestow a flowing motion upon a time which is simply 'there' is reminiscent of attempts to explain the vital quality of living organisms by adding a 'life force' or 'elan vital' to otherwise lifeless matter.

Historically, reluctance to accept the flow of time as an illusion has led to the following reasoning:

1. A flow of time requires or implies a time asymmetry.
2. Therefore time cannot flow in reversible systems (i.e., those symmetric in time).
3. Time does flow, so those systems that seem reversible are not really so.
4. Therefore we must search for subtle effects that introduce an irreversibility to all systems 'through the back door'.

Some scientists in the Prigogine school have gone so far as to suggest that where there is no asymmetry there is no time. (For a recent review of Prigogine's philosophy, see Coveney & Highfield 1990.)

## 7.2 The Reversibility Paradoxes

To clarify these matters, let us go back to the original discussion of irreversibility in thermodynamics, i.e., Boltzmann's theorem applied to a box of gas. Imagine a rigid impermeable box divided in two by a rigid impermeable movable membrane, the left half of the box containing inert gas A and the right gas B. At some moment the membrane is removed and the gases start to diffuse into one another. After a while equilibrium is reached with the gases inextricably mixed and each at a uniform density throughout the box. The process seems irreversible: we do not expect to encounter mixtures of gases spontaneously unmixing themselves.

The initial state is clearly more ordered than the final state, so the transition represents a rise in entropy in accordance with the second law of thermodynamics. An explanation for this rise is readily forthcoming: it is caused by the various intermolecular collisions, which are random, and so shuffle the molecules into a jumbled state.

This simple explanation has often been regarded as paradoxical. If the underlying laws of dynamics that the molecules obey are time-symmetric, how can the system as a whole display a time asymmetry? The resolution is well known (Davies 1974, Zeh 1989). A box of gas that is permanently isolated from its environment has no inbuilt arrow of time. If an isolated box of gas is in a low-entropy state at time $t$, then it will very probably be in a state of higher entropy at $t + \epsilon$. But (because of

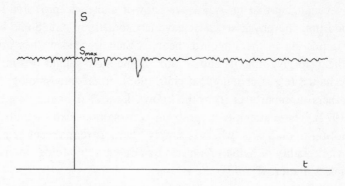

Fig. 7.1.

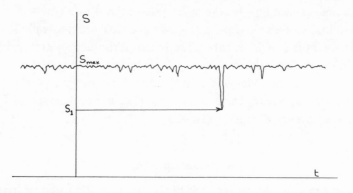

Fig. 7.2.

the underlying time symmetry) it was very probably also in a higher entropy state at $t - \epsilon$ too.

The situation is depicted in Fig. 7.1, which graphs the entropy $S$ of the gas as a function of time. Mostly the entropy remains close to the maximum value corresponding to thermodynamic equilibrium. But random fluctuations about equilibrium occur, which cause many small excursions from $S_{max}$. Larger fluctuations also occur, but much less often. This graph displays no arrow of time.

Suppose we are to select a low-entropy state at random. Physically this could be achieved by waiting until the entropy fluctuates to the desired value $S_1$, whereupon a bell rings to tag the state. Graphically (Fig. 7.2) this corresponds to selecting the value $S_1$ on the entropy axis and proceeding to the right (or the left!) until the entropy curve is encountered. The point $P$ of encounter is very likely to be close to the minimum value of a fluctuation rather than, say, half-way down a somewhat larger fluctuation. The reason: because the larger fluctuation is far less probable than the minimal one necessary to reach the value $S_1$.

## 7.3 Branch Systems

What does the foregoing thought experiment with the bell tell us? It suggests that an arbitrarily selected low-entropy state of a permanently and completely isolated system is very probably close to a minimum in the entropy curve, and hence a point of time symmetry. The fact that the entropy almost certainly subsequently rises does not bestow an arrow of time on the system, because it is also almost certainly the case that the entropy previously fell. It should be noted that the famous 'condition of molecular chaos', wherein there are no correlations between the positions and momenta of molecules, is necessary to prove Boltzmann's theorem that the entropy of the gas will rise. This condition can be proved to be valid only at a minimum point of time symmetry in the entropy curve, and can therefore be applied equally in both time directions.

It is clear that the situation with the bell differs from the experiment in which the membrane is removed in a crucial respect. In the membrane experiment the box of gas does not remain permanently isolated from the outside world. Instead the external experimenter tampers with the system, by removing the membrane at $t = 0$. Thus the gas did not achieve its initial low-entropy state at $t = 0$ as a result of a stupendously rare fluctuation from equilibrium. It achieved it because the system was constructed to have a membrane separating two chambers *prior to the box becoming isolated from the rest of the universe.*

In principle an experimenter could observe an isolated box of gas for an enormous duration. At some stage, a fluctuation will cause the two gases to spontaneously unmix and retreat to opposite ends of the box, at which point a membrane could be inserted (and subsequently removed). But that is not how the gases achieved their unmixed state in the case discussed. In the real world, physical systems are not isolated for all time: this is a fictional idealization. They may be (quasi-) isolated for a period of time, but they arise initially as branch systems that separate off from the main environment. In the case under discussion, the box of separated gases plus membrane would have been made that way by a laboratory technician, who separated the contents of the box off from the wider environment by deliberate manipulation.

The importance of branching for an understanding of the arrow of time was first emphasised by Hans Reichenbach (1956). If the system concerned forms at $t = 0$ as a low-entropy branch system, and its entropy subsequently rises, the question of whether the entropy previously fell (thus maintaining the time symmetry) does not now arise: the system simply did not exist as a branch system prior to $t = 0$.

One can still ask why, nevertheless, the entropy invariably rises after $t = 0$. Reichenbach's analysis of branch systems suggests a plausible answer. If the membrane is removed at random, i.e. at an arbitrary moment, then the situation is analogous to randomly selecting a low-entropy state in a permanently isolated system – the 'bell' method discussed above. And as we have seen, that method almost certainly selects a state near the minimum of the entropy curve where molecular chaos prevails and the entropy subsequently rises. If the membrane is not removed at random, then the experimenter has the possibility of waiting (a long time!) until a rare chance fluctuation which further reduces the state of the gas is about to develop, and to remove the membrane at that precise moment. In this case the entropy would fall after the membrane is removed (e.g. the gases might further retreat to opposite ends of the box, leaving a space between them). There is a finite, but exceedingly remote, chance that such a reduction would occur even if the membrane were removed at random. In the overwhelming majority of cases, however, the entropy will rise.

This line of reasoning seems to resolve the reversibility paradox for the sort of branch systems that occur in daily life. If, for example, I add milk to my tea the two fluids soon diffuse into one another. And if I stir the tea with a spoon I help the process along. I do not find that stirring serves to retard the mixing, or to unmix the fluids. The reason for this is, surely, that the process of adding the milk, or initiating stirring, happens at an arbitrary moment as far as the molecular evolution of the fluids are concerned. The teacup contents serve as a (quasi-isolated) branch system, formed at the moment of adding the milk, and there is no reason to suppose that there exists a molecular conspiracy between the tea molecules and the milk molecules, or the spoon. In the absence of such conspiratorial correlations, there is a very high probability that the entropy will rise, i.e. mixing will occur.

My discussion has focused on 'laboratory' examples, in which the experimenter interferes with the system of interest. The basic argument remains intact, however, in the case that the branch systems form spontaneously in the natural world. The crucial element is the lack of correlation between the macroscopic process of separation, or creation, of a branch system as an identifiable system, and the microscopic degrees of freedom that make up that system. Only in the case of 'conspiratorial' correlations, wherein a branch system separates off at precisely that moment when it is about to undergo a very rare entropy-decreasing fluctuation, will the arrow of time be 'wrong'. For this to occur it would be necessary for the conspiracy to be programmed into the cosmic initial conditions. We can conclude from the fact that reversed-arrow branch systems never seem to occur that the initial conditions of the universe do not contain the requisite conspiratorial correlations.

## 7.4 Environmental Mixing

In spite of the plausibility of the foregoing account, there is a powerful school of thought that it is totally misconceived. All the preceding discussion has been based on the familiar definition of entropy in statistical mechanics, derived through coarse-graining and the application of a probability measure. Some physicists (e.g. Blatt (1959)) have argued that a real physical property such as time asymmetry cannot derive from the voluntary relinquishing of information about the physical system by the observer: How can throwing away information lead to knowledge about something? If it is true that, in principle, an observer could have complete knowledge of the microstates of a closed and isolated physical system, then the entropy of that system would remain constant. And this property could be checked if, by some clever device, we were able to instantaneously reverse all molecular motions and get the gas to return exactly to its initial state. How can the fact that *in practice* we never have such complete information, 'create' an arrow of time? Does this not imply that the arrow is actually subjective?

I believe that this objection is misconceived, and stems from the erroneous assumption, mentioned in Section 1, that there exists a mysterious extra quality of time (itself derived from a conflation of the 'flow' of time with time asymmetry) which has to be 'found' in some physical mechanism. The fact that coarse-graining is a 'semi-subjective' mechanism rules it out, therefore, as a candidate in this misconceived search. Those who subscribe to this line of argument then go on to discuss a possible acceptable mechanism, also having to do with the fact that permanently totally isolated boxes of gas are fictional idealizations, but now referring not to the act of separating off the system from the wider environment as a branch system, but to the fact that complete isolation is impossible. No physical walls are able to totally isolate a gas from the effects of external forces. Even if some way could be found to shield a gas from electromagnetic and nuclear interactions with the outside world, gravitational influences would surely penetrate. As these fluctuating external forces (due, for example, to the movement of matter in the Andromeda galaxy) are random and in principle unknowable (some of them are entering our particle horizon for the first time) they introduce a genuine stochasticity in the gas, and not merely one due to our decision to coarse-grain. If we tried the molecular-reversal device mentioned above, and looked carefully, we would see that this 'environmental noise' factor destroyed the precise microscopic reversibility of the molecular motions, and would prevent the gas finding its way back to exactly the right initial state. And the longer the gas is allowed to evolve, and suffer random external noise, the less accurate is the reversal trick likely to be.

One way of contrasting the two situations discussed above is in terms of information. In a truly isolated system all the information about the microscopic degrees of freedom remains within the box, albeit scrambled up and in a not-very-accessible form. Coarse-graining amounts to our neglecting scrambled information. But it is still there really. How then (so the objection goes) can anything truly irreversible have taken place? But by contrast, environmental noise coming through the walls of the box is genuinely information-destroying; or one can think of the information as leaking out through the walls of the box and disappearing for good into the universe. This is real irreversibility.

While it is undeniably true that environmental noise represents true irreversibility, I believe it is clearly false to claim that our everyday experience of the arrow of time is due to that process. When I stir my tea, the mixing of the tea and milk is due to the stirring, not to random gravitational jiggles from Andromeda or wherever.

(Such jiggles will happen too, but they exercise a negligible effect in comparison with the spoon.)

It may be countered that the irreversible mixing to which I allude is only a pseudo-irreversibility. But so what? If the purpose of a theory is to explain the arrow of time as we experience it in the real world (which includes the coarse-graininess of our senses) then the explanation is forthcoming without the need for environmental noise. The explanation fails only if one insists on the line of reasoning itemized in Section 1: namely, assume that the arrow of time (flow of time?) is some mysterious, objectively real, extra property of time, then attribute this property to environmental noise, and go on to say that, having found it, this property 'explains' the 'everyday' arrow of time such as displayed in teacup phenomena.

As remarked, there is a close analogy between this chimera of an extra property of time and the notion of a 'life force'. Living organisms differ from inanimate matter by being alive. Yet no individual atom of an organism gives evidence of being alive, or behaving in any way differently from a similar atom in a nonliving system, just as no individual atom displays an arrow of time. Yet the assemblage of atoms making up an organism collectively possess the quality of life, just as an assemblage of atoms collectively possess the quality of irreversible change, or an arrow of time. But we no more need to appeal to 'real irreversibility' than to a 'real life force'. It would be quite wrong to argue that I am not alive because none of my atoms are alive, or that the quality of being alive is subjective, or illusory, unless we can discover a life force. The fact that, in principle, one could know the exact microscopic arrangements of all the (nonliving) atoms that make up a bacterium does not mean that the bacterium can be considered 'really' inanimate. We have to face the fact that there do exist legitimate holistic, macroscopic, collective, organizational properties of physical systems, and that these properties are perfectly 'real' and in need of explanation. In the case of the arrow of time (though perhaps not yet of life), we have such an explanation.

In its quantum incarnation, environmental noise becomes environmental decoherence. This process replaces the 'collapse of the wave function'. Whilst decoherence certainly provides another example of an arrow of time, it is clearly not necessary to suppose that the wavefunction of a box of gas (or the contents of a teacup) should have decohered before we can say it exhibits the phenomenon of an arrow of time. In other words, almost all examples of the arrow of time in daily life are not attributable to wave function decoherence, but to the fact that they are branch systems created at random, i.e. without any correlation between the macroscopic process of separation from the main environment and the microscopic processes going on in the gas (or whatever).

## 7.5 Origin of the Arrow

A full explanation of the arrow of time must account for two things: (i) the nature of the arrow, and (ii) its origin. The foregoing sections have been directed towards dispelling some misconceptions about its nature. In this section I shall sketch a theory of its origin.

As we have seen, in daily life the arrow manifests itself in the way that quasi-isolated branch systems separate off from the main environment in uncorrelated less-than-maximum entropy states. Such an explanation supposes, of course, that the wider environment is in a less-than-maximum entropy state itself. So long as we confine attention to subsystems, the question of why their initial states are able to possess less-than-maximum entropy can always be explained by appealing to the wider environment. An examination of realistic branch systems usually shows that they emerge as the result of a chain, or hierarchy, of branchings which, if traced back, expand out into wider and wider regions of the universe. Thus most of the important time-asymmetric phenomena on Earth are driven by the thermodynamic disequilibrium that exists in the vicinity of the sun, while the sun's own disequilibrium can be traced back to its nuclear constitution, which in turn takes us back to the big bang. Eventually, the origin of the arrow of time always refers back to the cosmological initial conditions. There exists an arrow of time only because the universe originated in a less-than-maximum entropy state.

At first sight, this seems to conflict with our knowledge of the early universe. The cosmic microwave background radiation is accurately Planckian, and indicates that the early universe was in a state of thermodynamic equilibrium, rather than disequilibrium. How, then, has it achieved its less-than-maximum entropy state today? The paradox is easily resolved: The expansion of the universe has caused it to depart from equilibrium.

Does this transition from equilibrium to disequilibrium not constitute a violation of the second law of thermodynamics? No. What has happened is depicted in Fig. 3. At some time around one second, the material content of the universe was in a state of equilibrium, having the maximum possible entropy for the constraints at that time. As the universe expanded, however, the maximum possible entropy rose. The actual entropy also rose, but less fast. In particular, the relaxation time for nuclear processes to allow the cosmological material to keep pace with the changing constraints (due to the expansion) was much longer than the expansion time, so the material began to lag further and further behind equilibrium conditions (equilibrium meaning in the nuclear case that this material is in the form of the most stable element - iron). Hence an 'entropy gap' opened up. The continuing expansion of the universe serves to try and widen that gap slightly (though now through other processes than nucleosynthesis), while physical processes such as starlight production serves to try and narrow it.

It is important to realise that the crucial effect of the expansion was in the early universe - hence the sudden widening of the gap early on. Today it seems likely (though I haven't checked) that the gap is narrowing: the universe produces copious quantities of entropy at a rate which I imagine is faster than the (now rather feeble) expansion raises the maximum possible entropy. The actual entropy will presumably

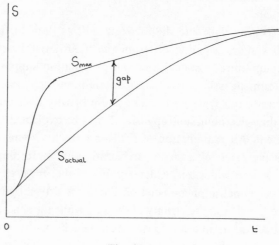

Fig. 7.3.

asymptote towards the maximum possible entropy in the very far future.

What we call the arrow of time is driven by the aforementioned entropy gap, so we can legitimately say that an explanation for the origin of the arrow of time can be traced back to the expansion of the universe. Do not fall for the trap, as some have done, of supposing that the *continuing* expansion provides the notorious 'extra ingredient' needed to explain the 'real' arrow; and especially don't conclude that if the expansion reverses, so must the arrow. The correct conclusion is this: the arrow of time observed in daily life owes its origin to the fact that the expansion of the universe during the first few minutes was much faster than the nucleosynthesis relaxation time. As a result, the majority of the cosmological material remains trapped for aeons in the form of light elements such as hydrogen, which is (from the nuclear point of view) metastable - it would prefer to be iron. It is from this metastable stuff that stars like our sun are busily creating thermodynamic disequilibria in their vicinities, and thereby driving all the 'everyday' time-asymmetric processes.

A fully satisfactory explanation for the origin of the arrow still has to explain why the universe is expanding in the first place. The expansion mode belongs to the gravitational dynamics of the universe, so to trace the argument back further we must examine the combined gravitational-matter system. Clearly this system as a whole did not start in a state of thermodynamic disequilibrium (unlike the matter part alone). Evidently the gravitational field at the time of the big bang was in a much-less-than-maximum entropy state. The entropy of the gravitational field is still ill-understood, and it seems likely that a full explanation of the fact that the universe started out in a highly improbable state gravitationally will have to await a satisfactory theory of quantum gravity.

## Discussion

**Omnès**   I wonder whether one should look for *the* cause of time asymmetry, i.e. a unique reason for it. It seems to me that there are three arrows of time: logical (for instance in the treatment of information), thermodynamical and cosmological. It is important to understand why they coincide but each of them constitutes a separate physical problem.

**Davies**   I believe there is strong evidence to link the thermodynamic and cosmological arrows of time, precisely because the former requires a consideration of initial conditions, and the latter is the only subject that has attempted to present a theory of initial conditions. However, the informational arrow is more subtle. There is a crude link with the thermodynamic arrow because information is negative entropy. But there are other aspects to this problem. For example, alongside rising entropy there is the property of rising organizational complexity in the universe. This is most conspicuous in biology. Here one deals not with information as such, but the quality of information – what has been defined by computer scientists as *depth*. I do not believe we understand the arrow of increasing depth, nor how it relates to the other arrows.

**Halliwell**   A comment. I wonder whether or not there is an aspect of Reichenbach's approach that you may not have covered. My understanding of Reichenbach is that he was trying to explain the thermodynamic arrow of time. The thermodynamic arrow of time is the fact that in the world about us, there exist a vast number of quasi-isolated systems in states of low, increasing entropy. For example, the universe contains a vast number of stars in states of low entropy, with increasing entropy as they radiate. It seems to me that it is *this* feature of the universe that branch systems are designed to explain. The initially low entropy of the universe is of course *part* of the explanation, but it is only through a lengthy series of branching processes that this is connected to the low entropy of the large number of systems about us, and hence to the thermodynamic arrow of time. Any comments?

**Davies**   The low initial entropy of the universe is a necessary, but not sufficient condition to explain the unidirectionality of entropy change in branch systems. The 'bell' argument is very suggestive that random low-entropy branch systems are created in a state of molecular chaos, but it stops far short of a proof. The problem as I see it is that the very concept of branch system is rather nebulous. Any attempt to tighten up this point would, I think, need to consider the fact that the universe has evolved through a series of self-organizing instabilities, leading to what Freeman Dyson has called 'hang-ups', where subsystems get 'hung up' for long durations in quasi-stable, quasi-isolated states. The reason for the existence of these hang-ups clearly involves aspects of the laws of physics (e.g. the values of some constants) as well as initial conditions. In summary, I think you are right to identify a gap in the argument here.

## References

Blatt, J. M. (1959) *Prog Theor Phys* **22**, 745.
Coveney, P. & Highfield. R, (1990) *The Arrow of Time* (W H Allen).
Davies, P. C. W. (1974) *The Physics of Time Asymmetry* (University of California Press).
Gold, T. (1962) *Amer J Phys* **30**, 403.

Landau, L. D. & Lifschitz, E. M. (1968) *Statistical Physics* (2nd edition; Pergamon Press) p. 31.

Mehra, J. & Sudarshan, E. C. G. (1972) *Nuovo Cimento* **11B**, 215.

Reichenbach, H. (1956) *The Direction of Time*, University of California Press.

Zeh, H-D. (1989) *The Direction of Time* (Springer-Verlag).

# 8

# Time's Arrow and Boltzmann's Entropy

Joel L. Lebowitz

*Departments of Mathematics and Physics*
*Rutgers University, New Brunswick, NJ 08903, USA*

*Time past and time future*
*What might have been and what has been*
*Point to one end, which is always present.*

T.S. Eliot in Four Quartets

Dedicated to my teachers, Peter Bergmann and Melba Phillips, who taught me statistical mechanics and much, much more.

## 8.1 Introduction

Let me begin by declaring my premises: for the purpose of this article, my motion of time is essentially the Newtonian one – time is real and the basic laws of physics are time reversible, they connect the states of a physical system, possibly of the whole universe, at different instants of time. This of course does not take account of relativity, special or general, and is therefore certainly not the whole story. Still I believe that the phenomenon we wish to explain, namely the time asymmetric behavior of macroscopic objects, would be for all practical purposes the same in a non-relativistic universe. I will therefore focus here on idealized versions of the problem, in the simplest context, and then see how far the answers we get go towards its solution.

## 8.2 Acknowledgements

The analysis I present here is certainly not novel. It is based in large part on various sections in the books of R. Feynman [1], O. Penrose [2], R. Penrose [3], and D. Ruelle [4]. I have also benefitted greatly from discussions and arguments with many colleagues: those with Yakir Aharonov, Gregory Eyink, Oliver Penrose, Eugene Speer and especially Shelly Goldstein have been particularly useful to me. I have also learned much from the other participants at this conference. Since their contributions also appear in this volume I make no explicit reference to their lectures. Finally, I want to thank the organizers of the conference for a splendid meeting.

### 8.3 The Problem

Given these premises I will start by formulating the problem concerning the origin of the distinction, i.e. asymmetry, between past and future in a non-relativistic universe. Now this distinction is so obvious in all our immediate experiences that it is often quite hard to explain just what exactly is the problem which needs an explanation. On the other hand it is equally hard, once the question has been formulated, to answer it in a way that puts an end to the discussion once and for all. There appears to be no way to convince some people, including sometimes one's self, that the problem has really been resolved, once and for all, by Boltzmann and that there is no need to worry about it (and hold conferences about it) again and again. Let me quote Schrödinger (selectively) in one of his many discourses about this problem [5]: "the spontaneous transition from order to disorder is the quintessence of Boltzmann's theory ... This theory really grants an understanding and does not ... reason away the dissymmetry of things by means of an a priori sense of direction of time variables... No one who has once understood Boltzmann's theory will ever again have recourse to such expedients. It would be a scientific regression beside which a repudiation of Copernicus in favor of Ptolemy would seem trifling." Schrödinger continues however, "... nevertheless objections to the theory have been raised again and again in the course of the past decades and not (only) by fools but (also) by fine thinkers. If we ... eliminate the subtle misunderstandings ... we ... find ... a significant residue ... which needs exploring ... "

I will come back to the "significant residue" after I formulate the basic problem in an idealized setting: Consider an isolated macroscopic system evolving in time, as exemplified by the schematic snapshots of a gas in the four frames in Fig. 8.1. The dots in this figure represent schematically the density profile of the gas at different times during the undisturbed evolution of the system and the question is to identify the time order in which the sequence of snapshots were taken. The "obvious" answer, based on experience is: time increases from left to right – any other order is clearly impossible. Now it would be very simple and nice if this answer could be justified from the laws of nature. But this is not the case, for the laws of physics, as we know them, tell a different story: if the sequence going from left to right is a permissible one, so is the one going from right to left. This is most easily seen in classical mechanics and so I shall use this language for the present. I believe that the situation is similar in quantum mechanics and will discuss that later.

### 8.4 Mathematical Formulation

The complete microscopic (or micro) state of an isolated mechanical system of $N$ particles can be represented by a point $X$ in its phase space $\Gamma$, $X = (\underline{r}_1, \underline{v}_1, \underline{r}_2, \underline{v}_2, \ldots, \underline{r}_N, \underline{v}_N) \in \Gamma$, $\underline{r}_i$ and $\underline{v}_i$ being the position and velocity of the $i$th particle. The Hamiltonian time evolution of this micro state is described by a flow $T_t$, i.e., as $t$

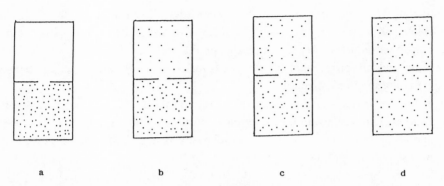

Fig. 8.1. Snapshots of macroscopic density profiles of an isolated container of gas at four different times.

varies between $-\infty$ and $+\infty$, $T_t X$ traces out the trajectory of a point $X \in \Gamma$, with $T_{t_1} T_{t_2} = T_{t_1+t_2}$ and $T_0$ the identity. Thus if $X(t_0)$ is the micro state at time $t_0$ then the state at time $t_1$ is given by

$$X(t_1) = T_{t_1-t_0} X(t_0).$$

Consider now the states $X(t_0)$ and $T_\tau X(t_0) = X(t_0 + \tau)$, $\tau > 0$. If we reverse (physically or mathematically) all velocities at time $t_0 + \tau$, we obtain a new microscopic state, which we denote by $RX(t_0 + \tau)$. We now follow the evolution for another interval $\tau$ to get $T_\tau R T_\tau X(t_0)$. The time reversible nature of the Hamiltonian dynamics then tells us that the new micro state at time $t_0 + 2\tau$ is just the state at $X(t_0)$ with all velocities reversed, i.e. $T_\tau R T_\tau X(t_0) = RX(t_0)$.

Let us return now to our identification of the sequence in Fig. 8.1. The snapshots clearly do not specify uniquely the microscopic state $X$ of the system; rather they represent macroscopic states, which we denote by $M$. To each macro state $M$ there corresponds a set of micro states making up a region $\Gamma_M$ in the phase space $\Gamma$. Thus if we were to divide the box in Fig. 8.1 into say a million little cubes then the macro state $M$ in each frame could simply specify the number $N_j$ of particles in cube $j$, $j = 1, \ldots, 10^6$. In order to make the volume of $\Gamma_M$ finite let us assume that we are also given the total energy of this gas which, like the total particle number $\Sigma N_j = N$, does not change from frame to frame.

Clearly this specification of the macroscopic state contains some arbitrariness, but this need not concern us right now. What is important is that the snapshots shown in the figure contain no information about the velocities of the particles so that if $X \in \Gamma_M$ then also $RX \in \Gamma_M$. (The technical reader might worry that we have left out the velocity field from the macro description in Fig. 8.1. However, since the figure is meant to be solely illustrative I will continue to use only the particle density). Now we see the problem with our definite assignment of a time order to the snapshots in the figure: going from a macro state $M_1$ at time $t_1$, to another macro state $M_2$ at time $t_2 = t_1 + \tau$, $\tau > 0$, means that there is a micro state $X \in \Gamma_{M_1}$ for which $T_\tau X = Y \in \Gamma_{M_2}$, *but* then also $RY \in \Gamma_{M_2}$ and $T_\tau RY = RX \in \Gamma_{M_1}$. Hence the snapshots depicting $M_\alpha$, $\alpha = a, b, c, d$, in Fig. 8.1 could as far as the laws of mechanics (which for the moment we take to be the laws of nature) go, correspond to time going in either direction.

It is thus clear that our judgement of the time order in Fig. 8.1 was not based on the dynamical laws of evolution alone; they permit either order. Rather it was based on experience: one direction is common and easily arranged, the other is never seen. *But why should this be so?*

## 8.5 Boltzmann's Answer

The answer given by Boltzmann's statistical theory starts by associating to each macroscopic state $M$ and thus to each phase point $X$ (through the $M(X)$ which it defines) a "Boltzmann entropy", defined (up to multiplicative and additive constants) as

$$S_B(M) = \log |\Gamma_M|$$

where $|\Gamma_M|$ is the phase space volume associated with the macro state $M$, i.e. $|\Gamma_M|$ is the integral of the Liouville volume element $\prod_{i=1}^{N} dr_i \, dv_i$ over $\Gamma_M$. Boltzmann's stroke of genius was to see the connection between this microscopically defined entropy $S_B(M)$ and the thermodynamic entropy $S_{eq}$, which is a macroscopically defined, operationally measurable (up to additive constants), extensive function of macroscopic systems in *equilibrium*. Thus when the gas in Fig. 8.1 is in equilibrium at a given energy $E$ and volume $V$,

$$S_{eq}(E, V, N) \equiv N s_{eq}(e, v) \simeq S_B(M_{eq}), \quad e = E/N, \ v = V/N, \tag{8.1}$$

where $M_{eq}(E, V, N)$ is the macro state observed when the system is in equilibrium at a given $E$ and $V$. By $\simeq$ we mean that for large $N$, when the system is really macroscopic, the equality holds up to negligible terms when both sides of (8.1) are divided by $N$ and the additive constant, which is independent of $e$ and $v$, is suitably fixed. We require here that the number of cells used to define $M_{eq}$ should grow more slowly than $N$. For a lucid discussion of this point see Chapter V in Oliver Penrose's book [2].

Boltzmann's great insight at once gave a microscopic interpretation of the mysterious thermodynamic entropy of Clausius as well as a natural generalization of entropy to nonequilibrium macro states $M$. Even more important, it gave a plausible explanation of the origin of the second law of thermodynamics – the formal (if restricted) expression of the time asymmetric evolution of macroscopic states occurring in nature. It is certainly reasonable to expect that when a macroscopic constraint is lifted in a system in equilibrium, as when a seal which was confining the gas to half the box in Fig. 8.1 is removed, the dynamical motion of the microscopic phase point will "more likely" wander into newly opened regions of $\Gamma$ for which $\Gamma_M$ is large than into those for which $\Gamma_M$ is small. Thus if we monitor the time evolution of the macro state $M(t)$ (short for $M(X(t))$) we expect it to change in such a way that $S_B(M(t))$ will "typically" increase as time increases.

In particular the new macroscopic equilibrium state $M_{eq}$ which will be reached by the system as it evolves under the new, less constrained, Hamiltonian in the bigger box can then be "expected" to be one for which $\Gamma_M$ has the largest phase space volume, i.e. $S_B(M_{eq}) \geq S_B(M)$ for all $M$ consistent with the remaining constraints. Note that when we take the system in Fig. 8.1 to be macroscopic, say one mole of gas in a one liter container, the ratio of $\Gamma_{M_{eq}}$ of the unconstrained system and the one constrained to the bottom half of the container (roughly $|\Gamma_{M_d}|/|\Gamma_{M_a}|$) is of order $10^{10^{20}}$. We thus have "almost" a mechanical derivation of the second law.

The eventual attainment of a macroscopic equilibrium state of the gas in the larger box is expressed by the zeroth law of thermodynamics: it certainly depends on the microscopic dynamics having some reasonable ergodic behavior. The precise requirements are unknown but are probably very mild since we are interested in systems with very large $N$. The latter is required in any case for the whole business of thermodynamic equilibrium to make sense (see discussion later) and should be generally sufficient for producing adequate ergodic behavior of real systems, including deterministic chaos with its attendant sensitive dependence on initial conditions and small pertubations. We shall assume this to be the case and not discuss it further. Very large $N$ also takes care of the objection against Boltzmann involving Poincaré recurrence times.

## 8.6 Mathematical Elaboration

Boltzmann's ideas are, as Ruelle [4] says, at the same time simple and rather subtle. They introduce notions of probability as indicated by the use of such words as "likely", "expected", etc, into the "laws of nature" which, certainly at that time, were quite alien to the scientific outlook. Physical laws were supposed to hold without any exceptions, not just almost always. Thus it is no wonder that many well known objections were raised against Boltzmann's ideas by his contemporaries (Loschmidt, Zermelo, ..., see S. Brush [6] for a historical account) and that, as Schrödinger wrote in 1954, objections have continued to be raised in the past decades. It appears to me, in 1992, that although ideas of probability, unpredictability and chaos are now part of the general scientific and even popular world outlook, Boltzmann's seminal ideas are still not universally accepted or understood. Let me try to explain how I understand them. To do this I will formulate a bit more precisely the nature of the probability distribution to which the notions of "likely", "expected", etc. used in the previous paragraph refer. (One can actually make these ideas very precise mathematically but that does not necessarily illuminate the physical reality – in fact it can obscure it.)

Let us first consider the gas in Fig. 8.1a to be in equilibrium in the bottom half of the box – excluded from the other half by a wall. Observations and analysis show that the exact values of certain types of phase functions $f(x)$ such as the number of particles contained in some subset of the volume, their kinetic energy in that region, or the force exerted by the gas on some region of the wall, will fluctuate in time about some stationary mean value. The single and multi-time statistics of such observations (obtained by independent repetitions of a specified experiment or situation) will be stationary in time – that is more or less what is meant by the system being in equilibrium. Furthermore the relative magnitude of these fluctuations will decrease as the size of the region increases. A quantitative theory of this behavior – both averages and fluctuations, including time statistics – can be obtained by the use of the Gibbs microcanonical ensemble [2]. This ensemble assigns probabilities to finding the micro state $X$ in a phase space region $\Delta$, *consistent with the specified constraints*, proportional to the volume of $\Delta$.

These probabilities can be interpreted either subjectively or as a statement about empirical statistics. Whatever the interpretation it is important that such ensembles are, by Liouville's theorem, time invariant under the dynamics. For our purposes we shall regard these probabilities as representing the fraction of time (over a sufficiently long time period) which the system spends in $\Delta$ since, for the reasons discussed earlier, we can assume that macroscopic systems are effectively ergodic. For such systems the microcanonical ensemble is the only stationary measure for which the probability density is absolutely continuous with respect to the projection of Liouville measure on the energy surface in $\Gamma$ [7]. (The central role played by Liouville measure is to be noted but, aside from its intuitive "obviousness" and experimental validation, it is only partially justified at present [even assuming ergodicity]. Perhaps the best justification is that in the classical limit of quantum statistical mechanics equal Liouville volume corresponds to equal weight for each quantum state. There are also some interesting stability arguments for singling out these Gibbs ensembles.)

The microcanonical ensemble thus provides a quantitative measure for the fraction of time a typical equilibrium trajectory will spend in untypical regions of the phase space, e.g. in regions where $S_B(M)$ differs significantly from its maximal value. The fractions of time $S_B$ will be decreasing or increasing are the same – in fact the behavior of $M(t)$ is entirely symmetric around local minima of $S_B(t)$, c.f. [8], p. 249.

Suppose now that we compute the time evolution of a microscopic phase point, which is "typical" of such a microcanonical ensemble, when the constraining partition is removed. Then it can be proven in some model systems, and is "believed" to be true for systems with realistic potentials obeying Newtonian laws, that the time evolution of the momentum and energy density will be described to a "high degree of accuracy" by the appropriate time asymmetric macroscopic equations, e.g.

Navier-Stokes type equations of hydrodynamics [8,9]. A particular consequence of these equations is a detailed prediction of how $S_B(M(t))$ increases monotomically in time. This is of course more than is needed for just having $S_B$ increase but I want to discuss it a bit here becauses it fleshes out Boltzmann's statistical ideas for more general systems than the dilute gases which can be treated by Boltzmann's kinetic equation.

The requirement that we start with an equilibrium state is actually too restrictive. We can also start with a nonuniform macroscopic density profile, such as the state $M_b$ given in Fig. 8.1, and consider micro states typical of local equilibrium type ensembles consistent with $M_b$. We then find again evolution towards states like $M_c$ for subsequent times.

In the above statement the words in quotes have the following meanings: "typical" behavior is that which occurs with large probability with respect to the given initial ensemble, i.e. the set of points $X$ in the ensemble for which the statement is true comprise a region of the energy surface whose volume fraction is very close to one, for $N$ very large; "believed" means that the basic ingredients of a mathematical proof are understood but an actual derivation is too difficult for our current mathematical technology; "high degree of accuracy" means that the hydrodynamic equations become exact when the ratio of microscopic to macroscopic scales goes to zero – that is in the so called hydrodynamic scaling limit, c.f. [8,9].

The main ingredient in this analysis is first and foremost the very large number of microscopic events contributing to the macroscopic evolution. Since the direct influence between the particles of the system takes place on a microscopic scale the macroscopic events satisfy, for realistic interactions (and here is where the gap in our mathematics is greatest), a "law of large numbers" which means that there is very little dispersion about specified deterministic behavior, i.e. that we are able to derive macroscopic laws not just for averages over ensembles but for individual systems – with probability approaching one when the micro-macro scale separation becomes large.

## 8.7 Typical versus Averaged Behavior

Having results for typical micro states rather than averages is not just a mathematical nicety but goes to the heart of the problem of deriving observed macroscopic behavior – we do not have ensembles when we carry out observations like those illustrated in Fig. 8.1. What we need and can expect to have is typical behavior. This also relates to the distinction (unfortunately frequently overlooked or misunderstood) between irreversible and chaotic behavior of Hamiltonian systems. The latter, which can be observed in systems consisting of only a few particles, will not have a uni-directional time behavior in any particular realization. Thus if we had only a few hard spheres in the box of Fig. 8.1, we would get plenty of chaotic dynamics and very good ergodic behavior (mixing, K-system, Bernoulli) but, we could not tell the time order of any sequence of snapshots. To summarize: when a constraint is lifted from a macroscopic system in equilibrium at some time $t_0$ then in the overwhelming majority of cases, i.e. with probability approaching one with respect to the micro-canonical ensemble, the micro state $X(t_0)$ will be such that the subsequent evolution of $M(t)$ will be governed by irreversible macroscopic laws.

We may regard the above (I certainly do) as the mathematical elaboration (and at least partial proof) of Boltzmann's original ideas that the observed behavior of macroscopic systems can be understood by combining dynamics with phase-space volume considerations.

This may be a good point to compare Boltzmann's entropy – defined for a micro state $X$ of a macroscopic system – with the more usual entropy $S_G$ of Gibbs, defined for an ensemble density $\rho(X)$ by

$$S_G(\{\rho\}) = - \int \rho(X)[\log \rho(X)]dX.$$

If we now take $\rho(X)$ to be the generalized microanonical ensemble associated with a macro state $M$,

$$\rho_M(X) \equiv \begin{cases} |\Gamma_M|^{-1}, & \text{if } X \in \Gamma \\ 0, & \text{otherwise} \end{cases}$$

then clearly,

$$S_G(\{\rho_M\}) = \log |\Gamma_M| = S_B(M).$$

It is a consequence of this equality that the two entropies agree with each other (and with the macroscopic thermodynamic entropy) for systems in local equilibrium, up to negligible terms in system size.

It is important to note however that the time evolutions of $S_B$ and $S_G$ are *very* different. As is well known, $S_G(\{\rho\})$ does not change in time when $\rho$ evolves according to the $T_t$ evolution, while $S_B(M)$ certainly does. In particular even if we start a system in a state of local thermal equilibrium, such as $M_a$ in Fig. 8.1, $S_G$ would equal $S_B$ only at that initial time. Subsequently $S_B$ would typically increase while $S_G$ would not change with time. $S_G$ would therefore not give any indication that the system is evolving towards equilibrium. This is connected with the fact discussed earlier that the micro state of the system does not remain typical of the local equilibrium state as it evolves under $T_t$. Clearly the relevant entropy for understanding the time evolution of macro systems is $S_B$ and not $S_G$. Unfortunately this point is often missed in many discussions and leads to unnecessary confusion. The use of $S_G$ in nonequilibrium situations is often a convenient technical tool but is not related directly to the behavior of an individual macroscopic system.

## 8.8 Irreversibility and Macroscopic Stability

Coming back now to the time ordering of the macro states in Fig. 8.1 we would say that the sequence going from left to right is typical for a phase point in $\Gamma_{M_a}$.

The sequence going from right to left on the other hand while possible is highly untypical for a phase point in $\Gamma_{M_d}$. The same is true if we compare *any pair* of the *macro states* in that sequence: left to right is typical, right to left atypical for the $\Gamma_M$ representing the *initial state*. Experience tells us that there is no "conspiracy" – if we do something to a macroscopic system and then leave it isolated, its future behavior is that of a "typical" phase point in the appropriate $\Gamma_M$. This then determines our time ordering of the snapshots.

Mechanics itself doesn't of course rule out deliberately creating an initial micro state, by velocity reversal or otherwise, for which $S_B(t)$ would be decreasing as $t$ increases and thus make the sequence in Fig. 8.1 go from right to left – it just seems effectively impossible to do so in practice. This is presumably related to the following observations: the macroscopic behavior of a system with micro state $Y$ in the state $M_b$ *coming* from $M_a$ which is typical with respect to $\Gamma_{M_a}$, i.e. such that $T_\tau Y$ is typical of $\Gamma_{M_a}$, $S_B(M_a) < S_B(M_b)$, is *stable* against perturbations as far as its future is concerned but very *unstable* as far as its *past* (and thus of the future behavior of $RY$) is concerned [10]. That is the *macro state* corresponding to $T_t Y$ is stable for $t > 0$ but unstable for $t < 0$. (I am thinking here primarily of situations where the equations describing the macroscopic evolution, e.g. the Navier-Stokes equations, are stable. In situations, such as the weather, where the forward macroscopic evolution is chaotic, i.e. sensitive to small perturbations, [4], all evolutions will still have increasing Boltzmann entropies in the forward direction. For the backward evolution of the micro states however the unperturbed one has decreasing $S_B$ while the perturbed ones have [at least after a very short time] increasing $S_B$. So even in "chaotic" regimes the forward evolution of $M$ is much more stable than the backward one.)

This behavior can be understood intuitively by noting that a random perturbation of $Y$ will tend to make the micro state more typical and hence will not interfere with the unperturbed behavior of increasing $S_B$ for all $t > 0$ while the forward evolutions of $RY$ is towards smaller phase space volume which requires "perfect aiming". It is somewhat analogous to those pinball machine type puzzles where one is supposed to get a small metal ball into a particular small region. You have to do things just right to get it in but almost anything you do gets it out into larger regions. For the macroscopic systems we are considering the disparity between relative sizes of the comparable regions in the phase space is unimaginably larger.

The difference between the stability of the macroscopic evolution in the forward, entropy-increasing, direction and its instability in the reverse direction is very relevant to understanding the behavior of systems which are not completely isolated – as is the case in practice with all physical systems. In the direction in which the motion is stable this lack of complete isolation interferes very little with our ability to make predictions about macroscopic behavior. It however almost completely hampers our ability to actually observe "back motion" following the application of some type of velocity reversal as in the case of spin echo experiments. After a *very short* time in which $S_B$ decreases the outside perturbations will make it increase again [4]. The same happens also in computer simulations where velocity reversal is easy to accomplish but where roundoff error plays the role of outside perturbations.

## 8.9  Remaining Problems

### 8.9.1  Significant Residue

I now turn to the "significant residue" of Schrödinger. As for the "subtle misunderstandings" I can only hope that they will be taken care of. The point is that when we consider a local equilibrium corresponding to a macroscopic state like $M_b$, and compute, via Newton's equations, the antecedent macro state of a typical micro state $X \in \Gamma_{M_b}$, we also get a macro state like $M_c$ and not anything resembling $M_a$. This is of course obvious and inevitable: since the local equilibrium ensemble corresponding to the macro state $M_b$, at some time $t_o$, gives equal weight to micro states $X$ and $RX$ it must make the same prediction for $t = t_o - \tau$ as for $t = t_o + \tau$. (The situation would not be essentially changed if our macro state also included a macroscopic velocity field.)

We are thus apparently back to something akin to our old problem: Why can we use statistical arguments based on phase space volume (e.g. local equilibrium type ensemble) considerations to make predictions about the future behavior of macroscopic systems but not to make retrodictions? Now in the example of Fig. 8.1 if indeed the macro state $M_b$ came from $M_a$, and we take its micro-state at that *earlier time* to be typical of equilibrium with a constraining wall, i.e. of $\Gamma_{M_a}$, then its micro state corresponding to $M_b$ is *untypical* of points in $\Gamma_{M_b}$: by Liouville's theorem the set of all such phase points has at most volume $|\Gamma_{M_a}|$ which is much smaller than $|\Gamma_{M_b}|$. Nevertheless its *future* but not its *past* behavior, as far as macro states are concerned, will be similar to that of typical points taken from $\Gamma_{M_b}$. It is for this reason that we can use autonomous equations, like the diffusion equation, to predict *future* behavior of real macroscopic systems without worrying about whether their micro states are typical for their macro states. They will almost certainly not be so after the system has been isolated for some time – although in the real world the inevitable small outside perturbations might in fact push the system towards typicality – certainly if we wait long enough, i.e. we are in an equilibrium macro state.

The above analysis thus explains why, if shown only the two snapshots $M_b$ and $M_c$ and told that the system was isolated for some time interval which included the time between the two observations, our ordering would be $M_b$ before $M_c$ and not vice versa. This would in fact be based on there being an initial state like $M_a$, with even lower entropy than $M_b$, for which the micro state was typical. From such an initial state we get a monotone behavior of $S_B(t)$ with the time ordering $M_a$, $M_b$ and $M_c$. If on the other hand we *knew* that the system in Fig. 8.1 had been "completely" isolated for a very long time, compared to the hydrodynamic relaxation time of the system before the snapshots in Fig. 8.1 were taken then (in this *very very very* unlikely case) we would have no basis for assigning an order to the sequence since, as already mentioned, fluctuations from equilibrium are typically symmetric about times in which there is a local minimum of $S_B$. In the absence of any knowledge about the history of the system before and after the sequence we use our experience to deduce that the low entropy state $M_a$ was the initial prepared state [11].

The origin of low entropy initial states poses no problem in "laboratory situations" such as the one depicted in Fig. 8.1. In such cases systems are prepared in states of low Boltzmann entropy by "experimentalists" who are themselves in low entropy states. Like other living beings they are born in such states and maintained there by eating low entropy foods which in turn are produced by plants using low entropy radiation coming from the sun, etc., etc. But what about events in which there is no human participation, e.g. if instead of Fig. 8.1 we are given snapshots of a meteor and the moon before and after their colision? Surely the time direction is just as obvious as in Fig. 8.1.

To answer this question along the Boltzmann chain of reasoning leads more or less inevitably (depending on considerations outside our domain of discourse) to a consistent picture with an initial "state of the universe" having a very small value of its Boltzmann entropy, i.e. an initial macro state $M_o$ for which $|\Gamma_{M_o}|$ is a very small fraction of the "total available" phase space volume. Roger Penrose, in his excellent chapter on the subject of time asymmetry [3], takes that inital state, the macro state of the universe just after the "big bang", to be one in which the energy density is uniform. He then estimates that $|\Gamma_{M_o}|/|\Gamma_{M_f}| \sim 10^{-10^{123}}$, where $M_f$ is in the state of the "final" crunch, with $|\Gamma_{M_f}| \sim$ total available volume. This is a sufficiently small number (in fact much smaller than necessary) to produce all we observe. The initial "micro state of the universe" can then be taken to be typical of $\Gamma_{M_o}$.

In R. Penrose's analysis the low value of $S_B(M_o)$, for a universe with a uniform density, compared to $S_B(M_f)$ is due to the vast amount of the phase space corresponding to macro states with black holes, in which the gravitational energy is very negative. I do not claim to understand the technical aspects of this estimate, which involves the Bekenstein-Hawking formula for the entropy of a black hole; it certainly goes beyond the realm of classical mechanics being considered here. The general idea, however, that the gravitational energy, which scales like $N^2$ for a star or galaxy, can overwhelm any non-gravitational terms, which scale like $N$, seems intuitively clear.

### 8.10 The Cosmological Initial State Problem

I hope that I have convinced you that, as Schrödinger says, "Boltzmann's theory ... really grants an understanding ...". It certainly gives a plausible and consistent picture of the evolution of the unvierse following some initial low entropy state $M_o$.

The question of how $M_o$ came about is of course beyond my task (or ability) to answer. That would be, as Hawking puts it, "knowing the mind of God" [12]. Still, as R. Penrose has pointed out, it would be nice to have a theory which would force, or at least make plausible, an initial $M_o$ so special that its phase space volume $|\Gamma_{M_o}|$ is infinitesimally small compared to the proverbial needle in the haystack, see Fig. 7.19 in [3]. He and others have searched, and continue to do so, for such a theory. While these theories properly belong to the, for me, esoteric domain of quantum cosomolgy there is, from a purely statistical mechanical or Boltzmannian point of view, a naturalness to a spatially homogeneous initial state $M_o$. Such an $M_o$ would indeed be an equilibrium state in the absence of gravity. It is therefore tempting to speculate that "creation" or the big bang was "just" the turning on of gravity, but I am told by the more knowledgeable that this is quite unreasonable. The initial state problem is thus very much open. It is by far the oldest open problem.

Within the context of special (or singular) origin theories of which the big bang is a special example, widely accepted as the truth, there is nothing, not even time, before the intial state. There is an alternate suggestion, dating back to much before the advent of black holes or the big bang theory, in which one doesn't have to assume a special singular creation. Boltzmann speculated that a low entropy "initial state" may have arisen naturally as a fluctuation from an "equilibrium universe." This is in some ways a very appealing minimal hypothesis requiring no beginning or end or special creation. All you have to do is wait "long enough" and you will get any state you want, assuming that a microcanonical ensemble and some mild form of ergodicity exist for the universe as a whole. This requires, at the minimum, some short range regularization of gravity. We shall not worry however about such "technical details" since, as we shall argue next, such a hypothesis is very implausible for other entirely conceptual reasons.

While the obvious objection to this hypothesis, that such a fluctuation is enormously unlikely, can be countered by the argument that if indeed the history of the microstate of the universe is typical of trajectories in $\Gamma$ then, without waiting for some huge flucuation, we humans would not be here to discuss this problem, there remains a more serious objection. As pointed out by Schrödinger and others and particularly by Feynman [1], the actual "size" of the observed ordered universe is too large by orders and orders of magnitude for what is needed. A fluctuation producing a "universe" the size of our galaxy would seem to be sufficient for us to be around. In fact using *purely* phase space volume arguments the "most likely" fluctuation scenario of how I come to be here to discuss this problem is one where only "I" or even only my consciousness really exists, i.e. one in which the smallest region possible is out of equilibrium – and this happened just this instant. While irrefutable as an academic debating position this is, of course, even more in conflict with our observed macro state (e.g. our memories). Merely accepting that what we observe and deduce logically from our marvelous scientific instruments about the world is really there, the idea of a recent fluctuation seems "ridiculous" and therefore makes the whole fluctuation from equilibrium scenaria seem highly implausible. In fact Feynman after discussing the problem in some detail concludes (in the tape of his lecture) that "...it is necessary to add to the physical laws the hypothesis that in the past the universe was more ordered, in the technical sense (smaller $S_B$), than it is today – to make sense, and to make an understanding of the irreversibility" [1].

I should say however that, even after rejecting the "fluctuation from equilibrium" scenario, the evidence or argument present for any particular, minimal $S_B$ initial state of the universe, is not entirely without difficulty. Let me present the problem in the form of a question: given that $|\Gamma_{M_o}|$ is so extraordinarily small compared to the available $|\Gamma| \sim |\Gamma_{M_f}|$ and hence that *every* point in $\Gamma_{M_o}$ is atypical of $\Gamma$, how can we rule out an initial micro state which is itself atypical of $\Gamma_{M_o}$ – *whatever* the original $M_o$ was? This could correspond to a scenario in which $S_B$ first decreased, reaching a minimum value at some long ago time. Since we do not assume this scenario that the trajectory of the universe is typical of the whole $\Gamma$ – in fact we permit a singular initial condition as in big bang theories – some of the objections to a long ago fluctuation from equilibrium scenario are not so telling. Also in this type of scenario we need not assume any symmetry about the minimum entropy state.

The alternatives come down to this: if we accept that the Boltzmann entropy was minimal for the initial state $M_o$, then the initial micro state can be assumed to be typical of $\Gamma_{M_o}$, while in a universe in which $S_B$ first decreased and only then increased, the initial micro state would have to be *atypical* with respect to $M_o$. It seems to me that there is a strong rationale for not accepting such an additional improbable beginning without being forced to it by some observational considerations. This seems to be the point which Schrödinger tried to illustrate with his prisoner story [13]. The concluding moral of that story in which the poor prisoner, who has (very probably) missed his chance for freedom by not being willing to trust probabilities in his favor after realizing that the initial state he was dealing with had to be an unlikely one, is "Never be afraid of dangers that have gone by! It is those ahead which matter."

## 8.11 Quantum Mechanics

The analysis given above in terms of classical mechanics can be rephrased, formally at least, in terms of quantum mechanics. We make the following correspondences:

(i) micro state $X \Leftrightarrow$ wave function $\psi(r_1, \ldots, r_N)$

(ii) time evolution $T_t X \Leftrightarrow$ unitary Schrödinger evolution $U_t \psi$

(iii) velocity reversal $RX \Leftrightarrow$ complex conjugation $\bar{\psi}$

(iv) phase space volume of macro state $|\Gamma_M| \Leftrightarrow$ dimension of projector on macro state $M$.

This correspondence clearly preserves the time symmetry of classical mechanics. It does not however take into account the non-unitary or "wave function collapse" (measurement) part of quantum mechanics, which on the face of it appears time-asymmetric. In fact this theory "is concerned exclusively with the prediction of probabilities of specific outcomes of future measurements on the basis of the results of earlier observations. Indeed the reduction of the wave packet has as its operational contents nothing but this probablistic connection between successive observations." The above quote is taken from an old article by Aharonov, Bergmann and Lebowitz (ABL) [14] which to me still seems reasonable now. In fact I will now quote the whole abstract of that article:

"We examine the assertion that the "reduction of the wave packet," implicit in the quantum theory of measurement introduces into the foundations of quantum physics a time-asymmetric element, which in turn leads to irreversibility. We argue that this time asymmetry is actually related to the manner in which statistical ensembles are constructed. If we construct an ensemble time symmetrically by using both initial and final states of the system to delimit the sample, then the resulting probability distribution turns out to be time-symmetric as well. The conventional expressions for prediction as well as those for "retrodiction" may be recovered from the time-symmetric expressions formally by separating the final (or the initial) selection procedure from the measurements under consideration by sequences of "coherence destroying" manipulations. We can proceed from this situation, which resembles prediction, to true prediction (which does not involve any postselection) by adding to the time-symmetric theory a postulate which asserts that ensembles with unambiguous probability distributions may be constructed on the basis of preselection only. If, as we believe, the validity of this postulate and the falsity of its time reverse result from the macroscopic irreversibility of our universe as a whole, then the basic laws of quantum physics, including those refering to measurements, are as completely time symmetric as the laws of classical physics. As a by-product of our analysis, we also find that during the time interval between two noncommuting observations, we may assign to a system the quantum state corresponding to the observation that follows with as much justification as we assign, ordinarily, the state corresponding to the preceding measurement."

I interpret the ABL analysis as showing that one can conceptually and usefully separate the measurement formalism of conventional quantum theory into two parts, a time symmetric part and a second-law type asymmetric part – which can be traced back, using Boltzmann type reasoning, to the initial low entropy state of the universe. (Of course it is not clear how to discuss meaningfully the concept of measurment in the context of the evolution of the universe as a whole.)

I believe that my colleagues agree with this interpretation of our work. Aharonov in particular has emphasized and developed further the idea described in the last sentence of the Abstract. He assigns two wave functions to a system – one coming from the past and one from the future measurement. It is not clear to me whether this will lead to new insights into the nature of time. Aharonov does think so and there are others too who feel that there are new fundamental discoveries to be made about the nature of time [11]. While this is certainly something interesting to think about, it definitely goes beyond my introductory premises so I will not pursue this further here.

## 8.12 Concluding Remarks

The reader who has gotten to this point will have noticed that my discussion has focused almost exclusively on what is usually referred to as the thermodynamic arrow of time and on its connection with the cosmological arrow. I did not discuss the asymmetry between advanced and retarded electromagnetic potentials or "causality" [11]. It is my general feeling that these and other arrows, like the one in the wave packet reduction discussed in the last section, are all manifestations of Boltzmann's general principle, and of the low entropy initial state of the universe. For this reason I also agree with most physicists that there would be no change in the monotone increase of entropy if and when the universe stops expanding and starts contracting.

Let me close by noting the existence of many well-known and some obscure connections between "entropy" and degree of order or organization in various physical and abstract systems far removed from the simple gas in Fig. 8.1. It is my feeling that, at least when dealing with physical objects containing many microscopic constituents, e.g. macroscopic or mesoscopic systems, the distinction between Boltzmannian and Gibbsian entropies, made earlier for simple systems, is always important and needs to be explored. I am therefore suggesting that there is interesting work to be done on obtaining more refined definitions of such concepts for complex systems like a Rembrandt painting, a beer can, or a human being. It is clear that the difference in $S_B$ between a Rembrandt and a similar size canvas covered with the same amount and type of paint by some child is orders of magnitude smaller than the entropy differences we have been talking about earlier. The same is true, I am afraid, for the entropy difference, if at all definable, between a living and a dead person. We therefore need more refined, logically consistent and physically meaningful definitions or organization for a given complex system than those currently available in information or complexity theory.

*Note added in proof:* For a more extensive discussion of some of the points discussed here see, J. L. Lebowitz, *Physica* **A194**, 1 (1993).

## References

[1] R. Feynman, *The Character of Physical Law*, The MIT Press, Cambridge, Mass., 1967. Chapter 5.
[2] O. Penrose, *Foundations of Statistical Mechanics*, Pergamon Press, 1970, Chapter 5.

[3]  R. Penrose, *The Emperor's New Mind*, Oxford University Press, Oxford 199, Chapter 7.

[4]  D. Ruelle, *Chance and Chaos*, Princeton Univ. Press, 1991; chapters 17 and 18.

[5]  E. Schrödinger, *What is Life? and Other Scientific Essays*, Doubleday Anchor Books, Garden City, N.Y., 1965; The Spirit of Science, Sec. 6.

[6]  S. Brush, *The Kind of Motion we Call Heat*, Vol. VI in Studies in Statistical Mechanics, North-Holland, 1976.

[7]  J. Lebowitz and O. Penrose, Modern Ergodic Theory, *Physics Today*, **26**, 155 (1973).

[8]  H. Spohn, *Large Scale Dynamics of Interacting Particles*, Springer-Verlag 1991.

[9]  A. de Masi and E. Presutti, *Mathematical Methods for Hydrodynamic Limits*, Lect. Notes in Math. 1501, Springer-Verlag 1991.

[10] Y. Aharonov has emphasized this point in lectures and private conversations.

[11] A.J. Leggett, *The Arrow of Time in Quantum Mechanics*, The Encyclopedia of Ignorance, ed. R. Duncan and M. Weston-Smith, Pergamon Press (1977), reprinted in *The Enigma of Time*, Adam Hilger Ltd. (1982), ed. P.T. Landsberg.

[12] S.W. Hawking, A Brief History of Time, Bantam Press. (1988).

[13] E. Schrödinger, Irreversibility, *Proc. Roy. Irish Acad.*, **53**, 22 (1950); reprinted in ref. [11].

[14] Y. Aharonov, P.G. Bergmann and J.L. Lebowitz, Time Symmetry in the Quantum Process of measurement, *Phys. Rev.* **B134**, 1410 (1964).

# 9

# Statistical Irreversibility: Classical and Quantum

Robert B. Griffiths

*Physics Department, Carnegie-Mellon University*
*Pittsburgh, PA 15213, USA*

## Abstract

The argument for statistical (thermodynamic) irreversibility in a closed classical system is reviewed, and a parallel argument is constructed for a closed quantum-mechanical system using the consistent-history approach to quantum interpretation.

## 9.1 Introduction

While this manuscript follows the same general course as my talk, it was prepared later, and I have thus replaced some of the things I said with what I think I should have said. A bibliographic note at the end was also not part of the talk, and contains all references to other related work.

A large number of macroscopic phenomena are found to be irreversible in time. Water falls down a waterfall and does not flow back up again. If a hot brick is placed in a cold bucket of water, the water and the brick tend towards the same temperature, and not the reverse. Stars radiate energy into the surrounding space rather than the reverse, although if the reverse were occurring someplace we might simply not be aware of it!

These irreversible effects are connected with increases in entropy, and thus with the second law of thermodynamics, and it is for a time asymmetry of this sort that statistical mechanics can claim to provide some sort of explanation.

There is another type of "irreversibility" which is not connected with the second law. The basic equation of classical mechanics, in the absence of dissipation, are invariant under an operation called time reversal, and the same is true for the simplest equations of quantum mechanics. This symmetry poses a conceptual difficulty: why, if the fundamental microscopic equations are invariant under time reversal, do we observe a macroscopic irreversibility? One is perhaps, tempted to suppose that the *macroscopic* arrow of time is somehow determined by a small violation of *microscopic* time reversal of the sort which particle physicists think may actually be present in nature. But this is not so. In condensed matter physics we

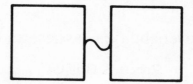

Fig. 9.1. Two metal blocks which can exchange energy through a thin wire.

can remove time reversal invariance by placing a system in an external magnetic field. Experiments show that if two bodies at unequal temperatures but in thermal contact are placed in such a field, they will exchange energy in such a fashion that the temperatures approach one another. This shows that time-reversal invariance (or its absence) is not the key for understanding macroscopic irreversibility. However, the fact that the classical equations of motion preserve the volume element in phase space, corresponding to the unitary time development of quantum systems, is crucial, as we shall see.

The approach which I will use for discussing statistical irreversibility consists in considering a large macroscopic system which is *closed* in the sense that it is completely isolated from all external influences. Real systems in the laboratory are never truly isolated. But the idealization of a closed system is useful in that one knows (in principle) the precise equations of motion (classical or quantum, as the case may be). Once irreversibility is understood in such an idealized system, it is possible to make some estimate of the severity of the approximations involved. Explanations of irreversibility which employ non-closed systems from the outset have the disadvantage of dealing simultaneously with two subtle problems: macroscopic irreversibility, and the influence of unknown perturbation on the microscopic mechanics.

In the case of classical systems the approach I will use is not very new; the basic ideas have been around for about a century. While I do not know of any presentation which is precisely of the form found below, that may simply indicate my ignorance of the literature. In the quantum case, the discussion based on a closed system makes use of a technical development ("consistent histories") which is relatively recent.

## 9.2 Irreversibility in Classical Statistical Mechanics

Imagine two metal blocks placed in an evacuated enclosure and connected by a thin wire, Fig. 9.1. If the blocks are initially at unequal temperatures, it is found that energy in the form of heat flows from one to the other through the thermal link until the temperatures are the same. This is a typical irreversible phenomenon in which the entropy increases. How can this be understood in classical statistical mechanics?

We know that a classical mechanical system of N particles can be described by using a 6N dimensional phase space $\Gamma$ in which the instantaneous microscopic state, or *configuration*, is represented by a point which follows a trajectory in time determined by Hamilton's equations. A *macroscopic* description of the system provides a relatively coarse description compared to that provided by the configuration. In the present example one might think of the total energy of each of the metal blocks as constituting the macroscopic description, and even here we assume the energy is "coarse grained": a precision of one erg, for example, will far exceed the accuracy of most calorimetric experiments. The macroscopic description at time t can be thought of as corresponding to a region or *cell* $E_t$ in the classical phase space, a region containing all the microscopic configurations compatible with the macroscopic information. It is convenient to think of $E_t$ as a function $E_t(\gamma)$ on the phase space which takes the value 1 inside the region and 0 outside. The expression for the statistical mechanical entropy is:

$$S_t = k \, \log \, [\int E_t(\gamma) d\gamma], \qquad (9.1)$$

i.e., a logarithmic measure of the volume in phase space of the region of interest, where k is Boltzmann's constant.

The statistical explanation of entropy increase, or at least *a* statistical explanation, is as follows. At some initial time $t_0$ suppose that a macrostate $E_0 = D$ is given, representing a state away from equilibrium. In terms of our example, the temperatures of the two block are unequal: suppose that one is at $0°C$, the other at $10°C$. The temperatures, of course, tell us the energy in each block, and this enables us to decide which microscopic configurations are possible.

Now, says the statistical mechanician, choose *one* of these *microscopic* configurations at random, and compute its time development. From its time development, calculate the succession of *macroscopic* states which it passes through, and you will find, almost certainly, that the entropy (9.1) increases with time, Fig. 9.2. In terms of the present example this means that energy will flow from the warmer body to the cooler body, in the direction which will eventually bring the temperatures to equilibrium.

Note the probabilistic element which enters here. It is *not* asserted that *all* configurations consistent with the initial macroscopic information, i.e., found in D, will lead to a time development in which the entropy is monotone increasing. In fact there will be some initial configurations for which the entropy decreases with time, and others for which it fluctuates, sometimes increasing and sometimes decreasing. What is being asserted is that the overwhelming majority of configurations will give rise to a behavior consistent with the second law. "Overwhelming majority" means in the sense of Lebesgue measure: the volume of D occupied by "peculiar" configurations which are going to violate the second law is a very tiny fraction of the total.

How small is "tiny"? It is the reciprocal of an enormous number, the ratio of the

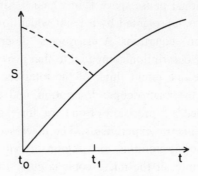

Fig. 9.2. Entropy as a function of time (solid curve) if a microscopic configuration is chosen at time $t_0$. The dashed curve illustrates a point in Section 3.

volumes in phase space,

$$\int D_2(\gamma)\, d\gamma \ / \int D_1(\gamma)\, d\gamma = e^{(S_2 - S_1)/k}, \tag{9.2}$$

of two regions when one of them has a macroscopically larger entropy than the other. For example, a flow of 1 erg of heat from one of our two blocks to the other will produce an entropy increase which makes the ratio (9.2) equal to

$$10^{(10^{13})}. \tag{9.3}$$

This is a very large number. One cannot call it "astronomically large" because the typical big numbers in astronomy, such as the total number of electrons in $10^6$ galaxies, are absolutely trivial in comparison with (9.3). Statistical mechanics never provides absolutely certain predictions, but its practitioners do not lose sleep worrying about violations whose probability is the reciprocal of (9.3)!

To have an intuitive idea as to how these enormous volume ratios can lead to irreversible behavior, consider the following mechanical or "hydrodynamic" analogy. Imagine a series of cups lined up in a row, with volumes of 0.01 cm³, 1 cm³, 100 cm³, $10^4$ cm³, and so forth, each with a volume 100 times larger than its predecessor. Each cup is filled with an incompressible fluid. The discrete time dynamics consists of mixing the fluid from adjacent cups and then pouring it back into the same two cups. As indicated schematically in Fig. 9.3, at one time step one mixes the contents of cups 1 and 2, 3 and 4, 5 and 6, etc., and at the next step the contents of cups 2 and 3, 4 and 5, and so forth. This dynamics in which the fluid is incompressible is supposed to mimic that in a classical phase space in which the volume occupied a set of points representing possible configurations remains constant in time as the points move in accordance with Hamilton's equations.

Even if you don't know the details of the mixing process, you can convince yourself that it is rather likely that a particle of of fluid which at the beginning was in a cup of a given size will, as time advances, tend to move into larger and larger

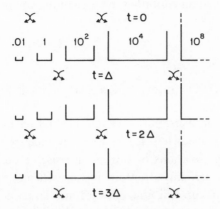

Fig. 9.3. Adjacent cups filled with incompressible fluid are mixed in pairs, as indicated by the double arrows, at successive time intervals.

cups. Thus if it starts off in the cup of volume 1 cm³ and this is mixed with fluid from the 100 cm³ cup, the particle will most likely find itself in the larger cup when the mixed fluid is poured back into the two cups. By the same argument, after the next step, it will most likely be in the $10^4$ cm³ cup. Occasionally, to be sure, it will end up in the smaller of the two cups after mixing. However these events will merely slow down, but will not prevent, the tendency to move to larger cups.

Naturally, we can imagine special forms of mixing dynamics which will prevent the irreversible motion of a typical fluid particle towards larger and larger cups. And in statistical mechanics such situations can arise in the presence of conservation laws, or of some particularly simple model dynamics. A little bit of "dirt" in the microscopic Hamiltonian may be needed to promote the increase of entropy. That dirt should not, however, violate the preservation of volume (Lebesgue measure) by the dynamical flow in phase space. In our mixing model the fact that the fluid is incompressible is crucial, as this prevents, for example, pouring the entire contents of a larger cup into a smaller cup. Of course all analogies are to some extend misleading, and the mixing model of Fig. 9.3 also has its defects. Probably the most severe is that the ratios of the volumes of successive cups is only 100 and not $10^{100}$, or some much larger number.

To summarize, the statistical explanation of irreversibility in a classical system is that if a microscopic state is chosen at random (Lebesgue measure) in the region in the phase space of a closed system corresponding to a macrostate which is away from equilibrium, it is highly likely that the corresponding trajectory, as generated by the laws of classical mechanics, will pass through regions corresponding to larger and larger entropies. The reason that this explanation is plausible is that classical mechanics preserves volumes in phase space, and any macroscopic increase in entropy corresponds to an enormous increase in the volume of the corresponding cell in phase space.

### 9.3 Comments on Classical Irreversibility

(i) The argument for irreversibility does not depend on the details of the microscopic dynamics. Nonetheless, it is not hard to imagine special dynamical laws which will lead to violations of entropy increase. To use the mixing machine analogy, one can certainly draw up mixing schemes such that most of the contents of a small cup will flow outward into larger cups, but then return to the small cup at a later time. One's intuition says that there must be something rather special about the dynamics for this to occur, and that in the "generic" case the result will be a constantly increasing entropy. However, there are clearly some problems lurking here (such as the proper definition of "generic") which deserve further study.

(ii) The argument of Section 2 can also be used to make it plausible that entropy will increase in a subsystem of the main closed system during a time when the subsystem is temporarily isolated from the rest of the main system.

(iii) It is very necessary to make the probability ansatz refer to the beginning of the period of time in which the system's entropy increases, rather than to the end. Thus, suppose that in the situation illustrated in Fig. 9.2 one were to choose at random a microstate within the phase space cell corresponding to the macrostate reached by the original system at time $t_1$, and integrate the classical equations of motion backwards in time. The result (with overwhelming probability) would be an entropy of the type indicated by the dashed curve, in total disagreement with the solid curve. What this tells us is that the probability assumption is only successful in predicting the time development in the direction of increasing entropy. To put it another way, statistical mechanics provides an explanation of an irreversible phenomena by a probabilistic hypothesis which itself singles out a (thermodynamic) direction of time.

(iv) The discussion of Section 2 above requires the introduction of a coarse graining of phase space, and this seems to involve a considerable degree of arbitrariness. In response one can say that thermodynamics employs a certain class of variables to describe systems in which entropy increases, and thus it is natural to make a coarse graining based on these "hydrodynamic" variables. This does not remove all of the arbitrariness, but it should be noted that certain features of the coarse graining, such as whether we use 1/10 erg rather than 1 erg for an energy interval, make no practical difference in the value of the entropy or the discussion of irreversibility.

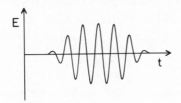

Fig. 9.4. Electric field as a function of time in a radar pulse arriving at a receiver.

## 9.4 Irreversibility in Quantum Statistical Mechanics

The explanation of irreversibility in quantum statistical mechanics can be developed in close parallel to that of the classical case. One begins with the idealization of a closed system of macroscopic size, whose quantum Hilbert space is the analog of the classical phase space. I shall assume the Hilbert space is finite-dimensional, as this seems to represent no limitation that is physically important, and avoids irrelevant mathematical technicalities. The unitary time transformations of this Hilbert space then correspond to the flows of trajectories in the classical phase space. In addition, the regions in the classical phase space corresponding to various macroscopic states of affairs now became subspaces of the Hilbert space, and $E_t$, representing a state at time t, is to be thought of as an orthogonal projection operation, or projector, onto the corresponding subspace. The *trace* (Tr) of this projector corresponds to the volume of a cell in a classical phase space, and thus the entropy corresponding to (9.1) is given by

$$S_t = k \log \mathrm{Tr}\ [E_t].\tag{9.4}$$

There is in quantum mechanics, on the other hand, a conceptual difficulty which is not present in classical mechanics when discussing what goes on in a closed system. In orthodox quantum mechanics one can only speak sensibly about the results of a measurement, and the measurement is always exterior to the system under consideration. Thus the latter cannot be a closed system. The orthodox approach also gives rise to various conceptual and philosophical difficulties about what happens if things are not measured. A colleague once asked me, half-seriously, at a conference, "Is Jupiter really there when no one has his telescope pointed in that direction"?

The relatively recent (1984) technical development which makes it possible to talk sensibly about what is happening in a quantum mechanical closed system (as well as stop worrying about whether Jupiter is still there when we are not looking) is the notion of a "consistent history". While this is not the place for a technical exposition, I think I can make the overall strategy clear using a few analogies.

Consider a radar pulse as it is received at a radar receiver. The electric field as a function of time might have the form shown in Fig. 9.4. The radar pulse can have an approximate frequency $v$ and an approximate arrival time $\tau$, but in the nature of the case (or in the nature of Fourier transforms, if you prefer) the uncertainty $\Delta v$ in the frequency and the uncertaintly $\Delta \tau$ in the arrival time satisfy

$$\Delta \tau \Delta v \geq 1. \tag{9.5}$$

There is nothing very mysterious, and certainly nothing specifically quantum mechanical in such an "uncertainty relationship". The way it entered quantum mechanics was through the decision of our intellectual forbears to describe particles by means of wave functions in which position and momentum, or time and energy, are related through a Fourier transform. Once this decision has been accepted, there is no way within the framework of quantum mechanics to talk about a particle whose position and momentum are more precisely defined than by

$$\Delta p \, \Delta x \geq \hbar, \tag{9.6}$$

because the theory has no way of *describing* such an object, in the same way that a theory of radar receivers has no way of describing a radar pulse violating (9.5). To put it in other words, there are things which make perfectly good sense in classical mechanics, but in quantum mechanics make no sense, are "nonsense" because of the basic assumptions of the latter. It may be that quantum mechanics is not a good theory and needs to be replaced with something else (if you know how to do it, go ahead!). But as long as we want to understand what it is saying we need to pay attention to the rules which separate sense and nonsense within the theory.

Consistent histories provide an extension of this sense-nonsense distinction to *sequences of events occuring at different times* in a closed system. In this context a "history" is a sequence of projectors $D$, $E_1$, $E_2$, ... representing events at a sequence of times

$$t_0 < t_1 < t_2 < \ldots . \tag{9.7}$$

The history is consistent relative to the initial event or projector $D$ provided certain *mathematical conditions*, involving the unitary time transformations of the system, are satisfied. (Usually one speaks of a consistent *family* of histories; I call a history consistent if it can be embedded in such a family.) If the history is consistent it can be assigned a probability. However, inconsistent histories are objects which don't exist, which are "nonsense" (in this way of doing quantum mechanics), the analog of radar pulses violating (9.5). Note that the rules work the same way if all of the inequalties in (9.7) are reversed, in which case the consistency condition is expressed relative to a final rather than an initial event.

While orthodox quantum mechanics is restricted to talking about the results of measurements, consistent histories can involve all sorts of events, microscopic as well as macroscopic, provided the consistency condition is satisfied. This includes the results of measurements, *provided* the measuring apparatus is considered as part of the closed quantum mechanical system. The consistent history approach gives the same measurement statistics as orthodox quantum mechanics properly applied; it is thus to be thought of as an interpretation of standard quantum mechanics, not as a new theory.

For a quantum statistical description of the two interacting blocks, Fig. 9.1, we first introduce a set of projectors $\{A^\alpha\}$, where the superscript is a label, not an exponent, which form a decomposition of the identity on the appropriate Hilbert space,

$$1 = \sum_\alpha A^\alpha \; ; \; A^\alpha A^\beta = \delta_{\alpha\beta} A^\alpha, \tag{9.8}$$

and represent a coarse graining in the sense that each projector specifies within rough limits ($\pm 1$ erg, say) the energy of each of the metal blocks. (As the energy of a block is a function of time, these projectors do *not* commute with the Hamiltonian.)

Suppose that at the initial time $t_0$ our system is away from equilibrium (the blocks have unequal temperatures). We choose an initial state $|\psi_0\rangle$ at random in the appropriate subspace of the Hilbert space corresponding to the specified initial energies. As the dimension of this subspace (the trace of the corresponding projector) is finite, this means choosing a point at random on the corresponding unit sphere. Once such a choice has been made, the corresponding

$$D = |\psi_0\rangle\langle\psi_0| \tag{9.9}$$

serves as the initial condition upon which the events are conditioned, and we can discuss various histories with events at later times drawn from the collection $\{A^\alpha\}$. To actually carry out a calculation of this sort on a macroscopic system is not possible, but we can make a reasonable guess as to what it will yield: most of the probability will be associated with sequences in which the entropy is increasing in time. The reason is that increasing entropy represents an increasing dimensionality of the subspaces, and quantum dynamics is unitary, so that a subspace of a particular dimension is, at later time, mapped into another subspace of precisely the same dimension. That is, just as in our classical phase space analogy, the cups get bigger (by enormous factors) as the entropy increases, so it is easy to imagine little subspaces mapped into bigger ones, but not the reverse.

As this is a quantum mechanical calculation, it is then necessary to check consistency. Again, while a detailed calculation is not possible, we can make a plausible argument that the consistency conditions will be satisfied. The intuitive reason is that violations of consistency are always, in some sense, manifestations of quantum interference phenomena, and in a system of increasing entropy there are lots of mechanisms to inhibit coherent interference.

In summary, a quantum statistical description of irreversibility in a closed system can be constructed which is a close parallel of its classical counterpart. The dimension of a subspace of the Hilbert space corresponds to the volume of a cell in phase space, and the unitarity of the quantum time development is the counterpart of Liouville's theorem. Coarse graining is needed in both cases, and the mixing machine analogy of Fig. 9.3 applies just as well to one as to the other.

## 9.5 Comments on Quantum Irreversibility

Because quantum and classical irreversibility have been discussed in a parallel fashion, all the comments in Section 3 apply (with one or two obvious modifications) to the quantum case. Thus I shall not repeat them, but simply make two points which are specific to quantum irreversibility.

(i) Quantum mechanics (in the consistent histories interpretation) does not *by itself* single out a direction of time, or at least one related to the thermodynamic arrow. One can talk about histories consistent relative to a final event rather than an initial event. The origin of the arrow of time is that it points *away* from some event far from equilibrium *towards* situations of higher entropy, just as in the classical case.

(ii) An increase of entropy probably helps make certain types of histories, in particular those in which the projectors refer to "hydrodynamic" quantities (energy density, matter density, etc.), consistent. At the present time, one of the open questions in the study of quantum mechanics is that of finding good criteria for singling out families of consistent histories in a natural way. It is always unwise to make conjectures while the research is still in progress, but my guess is that statistical irreversibility plays a crucial role in making a long list of everyday (almost) classical events consistent from a quantum perspective.

## 9.6 Conclusion

There are various sorts of time asymmetry present in the universe: the distant galaxies are getting farther away from us, not closer, the laws of particle physics are not completely invariant under time reversal, and the macroscopic irreversibility of the second law of thermodynamics is visible everywhere. I do not know if statistical mechanics can throw any light on the first two, but it has something to say about the last. Namely, if we postulate a beginning far from equilibrium and make an appropriate probability ansatz, then we can argue that entropy is very likely to increase, the way we observe it to do. Both classical and quantum mechanics yield very similar pictures connecting macroscopic irreversibility with the preservation of phase-space volumes, or unitarity in the quantum case.

While I myself find this explanation compelling, I have to admit that it does not answer all questions. Why should there be a beginning? What justifies the assumption of equal probabilities for microscopic starting states consistent with the initial non-equilibrium macrostate? Statistical mechanics has its limitations, and we in the discipline will probably have to leave some of these big unanswered questions to the philosophers, the theologians, and the cosmologists.

## 9.7 Bibliographic Note

Discussions of irreversibility during the nineteenth century were motivated by the kinetic theory of gases. Brush (1966) gives a brief historical overview in the introduction to his collection of reprints (in English translation) of a number of the original papers. Further contributions to the classical theory were made by Gibbs (1902) and the Ehrenfests (1911), and we find quantum generalizations in Tolman (1938). The book by Reichenbach (1956) contains a number of interesting ideas. A detailed discussion of the subject, with copious references, will be found in Davies (1974). I myself have found the work of Jaynes (1957, 1965) very enlightening. The mixing model of Fig. 9.3 was motivated by a somewhat similar idea in Appendix B of Callen (1960).

The consistent history approach to the interpretation of quantum mechanics will be found in Griffiths (1984, 1987). A very similar approach has been employed by Omnés (1990, 1992) and by Gell-Mann and Hartle (1990); see also their contribution to the present volume.

## 9.8 Acknowledgements

My research has been supported in part by the National Science Foundation through grant DMR 9009474.

## Discussion

**Wheeler**  I could not admire more your beautiful account, nor agree more with it. However, you and I deal with different questions. You ask, How does one derive the "No question? No answer!" feature of the elementary quantum phenomenon from quantum mechanics. I ask, How does one derive quantum mechanics from this "No question? No answer!" feature of nature.

**Unruh**  What do you mean by far from equilibrium?

**Griffiths**  In the context of the two metal blocks, it means in a state where the temperatures are different, say one of them is at 0°C and the other at 100°C, or equivalently in a situation where the entropy (as determined by the coarse graining) is significantly less than the maximum value consistent with the given total energy.

**Hartle**  We are used to discussing irreversiblity, both theoretically and experimentally, in terms of a limited, specific class of coarse grainings. To what extent is it known in statistical mechanics (classical or quantum) how these coarse-grainings are distinguished from all others by properties of the Hamiltonian or the initial condition?

**Griffiths**  I don't know a good answer to the question. I hope that you are planning to say something about it in your talk.

**Albrecht**  I have a question about Jupiter. Surely there are consistent histories for which Jupiter has been destroyed by a black hole passing through our solar system. Unless we make an observation we cannot determine if we are on that particular path. It seems that even from the consistent histories point of view "Jupiter is not there unless we observe it."

**Griffiths**  The point I wanted to make was that the consistent histories approach, in contrast to "orthodox" quantum mechanics, permits us to *talk about* whether Jupiter is there without being concerned as to whether we have a telescope pointed at it to make a "measurement". Discourse is not (in consistent histories) limited to outcomes of measurements. Naturally, this discourse does not commit us to asserting that Jupiter has not been perturbed in its orbit by some effect such as the one you mentioned.

**Gell-Mann**  (a) Isn't it a *set* of histories that has to be consistent, a *set* of alternative coarse-grained histories?

(b) The density matrix $\rho$ of the universe *could* be pure ($|\psi\rangle\langle\psi|$) as proposed, for example, by Hartle and Hawking. But I gather you want it to be proportional to a projection operator on a large subspace of Hilbert space. Is that right?

(c) I, too, would like to minimize the role of the "observer", but there is one role that is worth emphasizing. Quantum mechanics assigns probabilities to decohering alternative coarse-grained histories of the universe, but those probabilities are of use only if there is something betting (at least on the quasiclassical near certainties)!

(d) It is very important to understand more about what is an "appropriate" coarse graining for the alternative decohering histories, and the question of what "far from equilibrium" means for the initial $\rho$ is intimately tied to that investigation.

**Griffiths**  (a) You are correct that one normally discusses a *family* of histories; by calling a single history "consistent" I mean that it can be embedded in a consistent family.

(b) I don't object to using, as an initial state, a density matrix or a projector on a large subspace of a Hilbert space corresponding to some macrostate which is far from equilibrium. However, what I suggested in my talk was that one could choose a pure state (one-dimensional subspace) of this large Hilbert space and employ this as the initial state for a family of consistent histories. What one expects is that with high probability (and given histories whose events are chosen from the appropriate coarse graining) the resulting entropy will increase.

(d) I quite agree that the issue of an appropriate coarse graining is an important one, and I have not addressed it in this talk.

**Davies**  (a) You refer to "special dynamics" permitting entropy-reducing evolution, but do you not mean special initial (or boundary) conditions?

(b) Your assertion that, with high probability, the entropy of a far-from-equilibrium state will continuously rise towards its maximum value contains two assumptions of very different degree in terms of rigorous justification. The first is that the entropy will (very probably) rise from its initial value. The second is that, in spite of having risen, it will continue to rise further. Is the latter not much harder to justify?

**Griffiths**  (a) I mean special dynamics, of the sort which could, after some time, bring most of the points initially in some low-entropy region of phase space back into the same region.

(b) I think both assertions can be made very plausible, in terms of the line of thinking I have indicated, provided the time considered is not extraordinarily long; i.e., it should not be vastly greater than the time necessary to allow the system to relax close to equilibrium. Naturally, in order to prove theorems one must be much more precise.

# References

Brush, S.G. (1966) *Kinetic Theory vol. 2: Irreversible Processes*, Pergamon Press, Oxford

Callen, H.B. (1960) *Thermodynamics*, John Wiley, New York.

Davies, P.C.W. (1974) *Physics of Time Asymmetry*, Surrey University Press, London and University of California Press, Berkeley.

Ehrenfest, P. and T. (1911) Begriffliche Grundlagen der statistischen Auffassung in der Mechanik. *Encyklopädie der mathematischen Wissenschaften*, Vol. 4, Part 32, Teubner, Leipzig. English translation by M.J. Moravcsik, *The Conceptual Foundations of the Statistical Approach in Mechanics*, Cornell University Press, Ithaca (1959).

Gell-Mann, M. and Hartle, J.B. (1990). Quantum mechanics in the light of quantum cosmology. In *Complexity, Entropy and the Physics of Information*, Ed. W. H. Zurek, Addison-Wesley. Also in *Proceedings of the Third International Symposium on Quantum Mechanics in the Light of New Technology*, Eds. S. Kobayashi et al., Physical Society of Japan.

Gibbs, J.W. (1902) *Elementary Principles in Statistical Mechanics*, reprinted (1981) by Ox Bow Press, Woodbridge, Connecticut.

Griffiths, R.B. (1984) Consistent Histories and the Interpretation of Quantum Mechanics. *J. Stat. Phys.* **36**, 219-272.

Griffiths, R.B. (1987) Correlations in separated quantum system: a consistent history analysis of the EPR Problem. *Am. J. Phys.* **55**, 11-17.

Jaynes, E.T. (1957) Information Theory and Statistical Mechanics I, II. *Phys. Rev.* **106**, 620-630 and **108**, 171-190.

Jaynes, E.T. (1965) Gibbs vs. Boltzmann entropies. *Am. J. Phys.* **33**, 391-398.

Omnès, R. (1990) From Hilbert space to common sense: a synthesis of recent progress in the interpretation of quantum mechanics. *Ann. Phys. (N.Y.)* **201**, 354-447.

Omnès, R. (1992) Consistent interpretation of quantum mechanics. *Rev. Mod. Phys.* **64**, 339.

Reichenbach, H. (1956) *The Direction of Time*, University of California Press, Berkeley.

Tolman, R. C. (1938) *The Principles of Statistical Mechanics*, Oxford University Press, Oxford.

# 10

## Quantum Fluctuations and Irreversibility[†]

J. Pérez–Mercader

*Laboratorio de Astrofísica Espacial y Física Fundamental*
*Apartado 50727*
*28080 Madrid*
*Spain*

### Abstract

We study how the effects of quantum corrections lead to notions of irreversibility
and clustering in quantum field theory. In particular, we consider the virtual
"charge" distribution generated by quantum corrections and adopt for it a statistical
interpretation. Then, this virtual charge is shown to (*a*) describe a system where
the equilibrium state is at its classical limit ($\hbar \to 0$), (*b*) give rise to spatial diffusion
of the virtual cloud that decays as the classical limit is approached and (*c*) lead to
a scenario where clustering takes place due to quantum dynamics, and a natural
transition from a "fractal" to a homogeneous regime occurs as distances increase.

In quantum field theory, the dynamics stems from both the interactions among the
fields and the quantum fluctuations to which they are subject. Quantum fluctuations
are perhaps the most fundamental feature of quantum field theory, affecting fields,
their sources and the vacuum in which they evolve.

Induced by the fact that quantum fluctuations are to a certain extent random [1],
we will study how this affects irreversibility. Namely, since *true* randomness impairs
one's ability to carry out an exact reconstruction of the past history of the system,
it is legitimate to expect some irreversibility at the microscopic (quantum field) level
because of the presence of quantum fluctuations in this domain.

One possible way to study this irreversibility is to introduce some fundamental
statistical notions into the realm of the quantum field theory. For example, one
can consider the effect that quantum fluctuations have on the "charge" density§ of
the Poisson equation satisfied by the quantum corrected potentials. The quantum

---

† This paper was not presented at the meeting but was accepted for inclusion in these proceedings by
(the other two) editors.

‡ Also at: Instituto de Matemáticas y Física Fundamental, C.S.I.C., Serrano 119–123, 28006 Madrid
and the Theoretical Division, Los Alamos National Laboratory, Los Alamos, New Mexico 87545.

§ We enclose the word charge in quotes because we refer to it in a generic sense, that is, we are not
specifically referring to electric charge. We have in mind the charge for the source of the Poisson
equation satisfied by the effective interaction energy.

fluctuations affect the interaction energy, and this induces a modified "charge" density; this is how the Uehling potential appears in QED. The statistical notions come in when we interpret the induced charge density as a probability density, an action that we are justified in taking due to the mathematical properties of the charge density induced by quantum-fluctuations. We associate with the full interaction energy a "charge" density which (as is done in cosmology and astrophysics [2]) we interpret as a probability density characterizing the spatial distribution of virtual charges surrounding the original charges. By doing this, we are making use of the rich *conceptual* depth of quantum field theory and explicitly taking into account that, in the computation of the quantum corrections at some scale, we are actually including not only simple two body processes with a fixed and precise impact parameter of the order of that scale, but the effects of many body interactions with impact parameter smaller than the distance scale that we are probing. Because of this, divergences appear in the theory which renormalization removes and re–interprets. The physical potential that one measures is thus subject to fluctuations, and it is impossible to specify the *exact* dependence of the potential on the individual components of the virtual "charge" density. We may, however, ask about the probability of occurrence of some particular configuration or similar questions of a statistical nature. This goal is achieved by adopting for the virtual "charge" density a statistical interpretation in terms of a probability distribution whose density is the quantum corrected "charge" density. By studying the *effective* (in the RG [3] sense) form of the interaction energy, one acquires information on how the properties of the virtual cloud change with scale and, through the statistical interpretation, gain insight on how changes of scale affect the dynamics of the cloud in aspects relating to the approach to equilibrium or even structure and form generation properties.

We can obtain the effective interaction energy by solving its RGE. The solution to this equation [4] yields the interaction energy as a function of the scale at which one probes the system, and contains the modifications due to the presence of quantum fluctuations. Unfortunately, the corrections that are included to *all* orders of perturbation theory are only the leading logarithms; but, as is well known, even with this limitation, many interesting physical consequences can still be extracted. Since energy has canonical dimension of inverse time and no anomalous dimension, the RGE has the solution,

$$V(\lambda r_0, g_0, a) = \lambda^{-1} V(r_0, \bar{g}_0(\lambda), a) \tag{10.1}$$

Here $V$ is the interaction energy, $r_0$ is the distance between the interacting sources, and $a$ is a reference distance. The quantity $\bar{g}_0(\lambda)$ is the effective coupling and $\lambda$ the scale parameter; $\bar{g}_0(\lambda)$ satisfies the RGE $\lambda \partial \bar{g}_0 / \partial \lambda = -\beta(\bar{g}_0)$, and $\bar{g}_0(1) \equiv g_0$.

In general, for a massless mediating field (such as photons or gluons), the effective interaction energy for two point particles separated by a distance $a$ is given by

$$V(a, g_0, a) = C_0 \frac{g_0^2}{4\pi a} \tag{10.2}$$

where $C_0$ is an arbitrary constant.

Computing to 1–loop order, where $\beta = -\beta_0 g_0^3$, choosing $\lambda = r/a$, and taking $\sigma$ to be a constant† (for $r$ close to $a$)

$$V(r, g_0, a) = C_0 \frac{g_0^2}{4\pi r} a^{-\sigma} r^{-1+\sigma} \tag{10.3}$$

$\sigma$ is related to the $\beta$–function for the coupling $g_0$, and to one–loop is given by $\sigma = 2\beta_0 g_0^2$. It has three essential properties: (*i*) it is proportional to $\hbar$, (*ii*) it vanishes smoothly in the limit of zero coupling and (*iii*) as defined, it is a function of the momenta which essentially counts the number of degrees of freedom excited from the vacuum at distances less than $r_0$.

For the case of QED one recognizes here the first term in the short distance expansion of the famous Uehling potential; in QCD one recognizes the expression for the interquark potential.

According to potential theory [6], Eq. (10.3) satisfies a Poisson equation whose right hand side is proportional to the "charge" density dictated by quantum corrections. This charge density describes how the phenomenon of vacuum polarization has modified the vacuum and given rise to the formation of virtual pairs that, as is well known, affect the strength and properties of the interaction in a substantial way.

However, we have no information as to the actual distribution of these pairs, but by interpreting the vacuum polarization charge density as a "probability density", we can derive information on the virtual cloud and its physics as a many body (statistical) system. From the mathematical point of view this is possible thanks to the intimate relationship [6] that exists between potential theory and probability theory. The relationship between the potential and the charge density (away from $r = 0$) is through the Poisson equation $\nabla^2 \phi = +4\pi\rho(r)$, which for our isotropic potential gives

$$\rho(r) = A' \ r^{-3+\sigma} \tag{10.4}$$

where $A'$ is a constant. We only need to require from the density $\rho(r)$ that it be a positive and integrable function on its support in order that it can actually be interpreted as a probability density (see below).

We now examine some properties of Eq.(10.4). First of all we see that $\rho(r)$ is the solution to the functional equation

$$\rho(\lambda r) = \lambda^\beta \rho(r) \tag{10.5}$$

with $\beta = -3 + \sigma$; this implies right away that $\rho(r)$ describes a fractal distribution of charge embedded in 3–dimensional configuration space, and with a Hausdorff (or fractal) dimension given by $d_f = +\sigma$. This is not surprising, since $\sigma$ has its origin in

---

† Strictly speaking, $\sigma$ is a function of the distance at which the system is probed. This can be made explicit by computing it in a mass dependent substraction procedure [5]. However, it will be sufficient for our present purposes to take it to be a constant.

the deviation from *canonical* scaling due to the quantum fluctuations. Furthermore, since $\sigma$ changes as we change the size of the domain on which we probe the virtual cloud, it turns out then that the Hausdorff dimension also changes and we are then dealing with a multifractal. For QCD and Quantum Gravity in its asymptotically free regime, $\sigma$ is positive. In QED and other non–asymptotically free theories, $\sigma$ is negative. When the Hausdorff dimension is positive, one sees intuitively that there is a natural tendency to suppress these fluctuations since, contrary to what happens in the case of a negative Hausdorff dimension, the configuration space "cannot accommodate them": it is not big enough!

To reveal some of the physical consequences of the randomness associated with these quantum fluctuations, we will study a few of the features of the "statistical mechanics" of the density in Eq.(10.4). In particular, we will look at the entropy associated with $\rho$ and also will give a glance at the nature of the stochastic processes that it supports.

Requiring normalizability of the probability density, we get

$$\rho = Ar^{-3+\sigma} \tag{10.6}$$

with

$$A = \frac{\sigma}{4\pi} R_0^{-\sigma} \tag{10.7}$$

for $\sigma > 0$, and

$$A = -\frac{\sigma}{4\pi} r_0^{-\sigma} \tag{10.8}$$

when $\sigma < 0$. Here $R_0$ denotes an IR cutoff, necessary in the case of positive $\sigma$, in order to ensure the finiteness of the probability distribution. For non–asymptotically free theories, $r_0$ is an UV cutoff which becomes necessary for the same reasons. The IR cutoff can be identified, e.g., with a typical hadronic size and $r_0$ with Planck's length. The resulting probability density is of the Pareto type [7]; this is a natural consequence of the renormalization group origin for $\rho(r)$, which is ultimately responsible for the functional equation in (10.5) and the associated scaling. Scaling leads to Levy–type distributions, which turn Pareto in some limit. Notice also that both densities go to zero in the classical ($\hbar \to 0$) limit.

To gain information about the "equilibrium configurations" supported by these distributions, we construct and compute an entropy–like quantity. We will assume that, as in thermodynamics, the state of equilibrium corresponds to the maximum for the entropy. In other words, we will assume that it reveals a preferred "direction" for stability in the evolution of the quantum field system. It is possible to discuss two types of entropies: a "coarse grained" entropy based on the probability distribution, and a "fine grained" entropy or "differential" entropy†, based on the probability

---

† In the same sense as in information theory [8].

density $\rho(r)$. We will discuss only the latter, which is defined through

$$S = -k \int d^3\vec{r}\rho(r)\log\left[C\rho(r)\right] \tag{10.9}$$

with the integral extended over the full support of the variable $r$. The constant $C$ needs to be introduced because $\rho(r)$ is a dimensional quantity, and we do not have the equivalent of a Nernst theorem to set a reference valid for *all* physical systems. This constant, as in thermodynamics [9], fixes the minimum entropy of the system. We will set "Boltzmann's constant" $k$ equal to one.

The differential entropy for the case when $\sigma$ is positive gives

$$S^{(\sigma>0)} = 1 - \frac{3}{\sigma} - \log\frac{\sigma}{4\pi}C + 3\log R_0 \tag{10.10}$$

and for negative $\sigma$ one gets,

$$S^{(\sigma<0)} = 1 - \frac{3}{\sigma} - \log\frac{-\sigma}{4\pi}C + 3\log r_0 \tag{10.11}$$

From these two expressions we see that the "entropy constant", $C$, may be related in an interesting and useful way to the "extreme" volume (the volume at the cutoff) of the quantum system. Such a choice has the advantage of eliminating the dependence of the entropy on the cutoff. For asymptotically free theories, if we choose $C$ proportional to the maximum volume, i.e., proportional to the volume of the IR-cutoff to the cube, then the dependence on the cutoff and extreme system size disappears from the expression for the entropy. The equivalent statement is also true for the case of non–asymptotically free theories, where the "Nernst–volume" corresponds to the minimum volume that can be physically reachable.

These entropies are shown in Figures 10.1 and 10.2.

We see that in asymptotically free theories, the equilibrium state corresponds to the *largest* possible value† of $\sigma$. Now, since matter fluctuations contribute negatively to $\sigma$, the most stable state corresponds to a configuration (or system size) where the matter states do not contribute. Because of the decoupling theorem [10], this occurs for the *maximum–possible size* of the system: that is for its *classical* limit!

On the other hand, when the theory is non–asymptotically free, the maximum of the entropy happens when $\sigma$ goes to 0 from the left. This means either the classical limit, in the sense that $\hbar \to 0$, or that one probes the system at distances *so large* that no quantum fluctuation contributes to $\sigma$.

Thus, for both, asymptotically–free and non–asymptotically–free theories one sees that *the maximum of the entropy is attained at the classical limit.*

How does this transition to the equilibrium state take place? Since diffusion is a mechanism for the transition to the equilibrium state for a system where randomness is at work, we will examine diffusion in quantum field theory, and check

---

† For $AF$-theories the entropy has a maximum at $\sigma = 3$. We will not explore this here since, for most theories, this value lies outside the range where we can reasonably trust the approximations we have made.

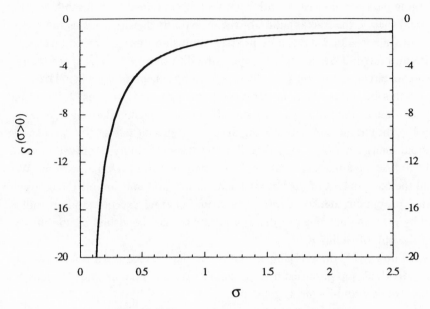

Fig. 10.1. Entropy for assymptotically free theory.

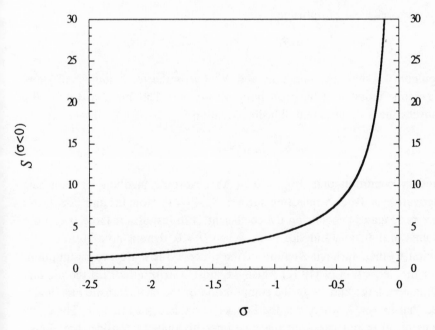

Fig. 10.2. Entropy for non-assymptotically free theory.

if the emerging picture is consistent with what we have obtained from the entropy. In quantum field theory, the fractal nature of the probability density leads us to expect some form of fractal diffusion or, more precisely, diffusion through fractal brownian motion. Because of this "fractality" we *expect* the diffusion to be non–space filling, and to lead to either clustering ($\sigma > 0$) or screening when the system relaxes; it is clear that this behavior ought to depend on the sign of $\sigma$. In what follows we will examine some of the properties of the coefficient of diffusion derived from $\rho(r)$; then we will examine the correlation integral in some specific cases.

The virtual charge density cloud described by the probability density Equation (10.6) will "diffuse" in a manner similar [11] to what occurs in Brownian motion. We expect that the components of the virtual cloud scatter randomly among themselves; the collision probability density is $\rho(r)$; the virtual charges execute random walks controlled by $\rho(r)$; the net effect of these processes is that the virtual cloud diffuses outside of the ball of radius $R$.

As is well known [11] the probability that a "particle" executing a random walk, with each individual step governed by a probability density $\rho(r)$, be after $N$–steps at a position between $\vec{r}$ and $\vec{r} + d\vec{r}$, is given by

$$W(r)d^3\vec{r} = [4\pi Dt]^{-3/2} \exp\left[-|\vec{r}|^2 / (4Dt)\right] d^3\vec{r} \qquad (10.12)$$

where $D$ is the coefficient of diffusion,

$$D = \frac{n}{6} \langle \vec{r}^2 \rangle \ ,$$

$n$ is the number of steps per unit time and $N = nt$ is the total number of steps. $\langle \vec{r}^2 \rangle$ is the second moment of the probability density $\rho(r)$. The function $W(r)$ in Eq. (10.12) satisfies the 3–dimensional diffusion equation

$$\frac{\partial W}{\partial t} = D\nabla^2 W$$

with boundary condition that $W|_s = 0$ at an absorbing medium, or vanishing normal derivative of $W$ at a reflecting surface: $\frac{\partial W}{\partial n}\big|_s = 0$. How far and how fast a disturbance propagates depends on the coefficient of diffusion: for large $D$ one has large amounts of diffusion, and vice versa. When $D = 0$, there is no diffusion.

In addition to the Markovian nature of the process, there are two assumptions involved in the derivation of (10.12) which we must bear in mind: (*a*) the number of steps $N$ is very large and (*b*) in the computation of the characteristic function, it is assumed that $kr \ll 1$, where $k$ is the Fourier variable conjugate to $r$. These two assumptions mean in our case that the time intervals that we consider are "long" compared with virtual time intervals and that the size of the region in which we expect the diffusion to take place, is "large" compared to the Compton wavelength about which we are computing the coefficient $\sigma$. Both of these restrictions are amply met by the validity of our approximations.

In order to compute the coefficient of diffusion one needs to obtain the second moment of the probability density. From a radial distance $s_0$ to a radial distance $s > s_0$, the second moment of $\rho(r)$ is

$$\langle \vec{r}^2 \rangle = \frac{4\pi}{2+\sigma} A \left( s^{2+\sigma} - s_0^{2+\sigma} \right) \tag{10.13}$$

and therefore, in some cases, the coefficient of diffusion *diverges* (naively) as $s$ goes to infinity. This again is not surprising, since our probability distribution is the limit of a Levy–type distribution, and thus its moments are divergent.

As previously, we classify the situation depending on whether we have an asymptotically free theory or not. When $\sigma > 0$, we can take $s_0 = 0$ and set $s = R_0$, the IR cutoff. The coefficient of diffusion is given by

$$D^{(\sigma>0)} = \frac{n}{6} \frac{\sigma}{2+\sigma} R_0^2. \tag{10.14}$$

In non–asymptotically free theories, we take $s_0 = r_0$ (the UV–cutoff) and leave $s$ unspecified. Then,

$$D^{(\sigma<0)} = \frac{n}{6} \frac{-\sigma}{2+\sigma} s^2 \left[ \left( \frac{s}{r_0} \right)^\sigma - \left( \frac{r_0}{s} \right)^2 \right]. \tag{10.15}$$

The expression for $D^{(\sigma>0)}$ reflects the divergence in the second moment of Levy–type distributions. However, quantum corrections again play an important rôle here: physics limits the size of the system ($R_0$ cannot go to infinity) and in the classical limit, when $\sigma \to 0$, diffusion *stops*. Before this happens, i.e., when the typical energies are larger than the inverse Compton wavelength of the IR–cutoff, the system will tend to diffuse itself into clusters or well differentiated pieces. This last statement being a consequence of the positive fractal (Hausdorff) dimension associated with $\sigma > 0$, as was mentioned above.

For $\sigma < 0$ the situation is qualitatively different. We can distinguish two different regimes within non–asymptotically free theories, according as to whether $0 > \sigma > -2$ or $-2 > \sigma$. In the former case, $\langle \vec{r}^2 \rangle$ diverges as diffusion takes place into larger distances, and is only shut–off by the vanishing of $\sigma$ with distance, that is to say, by the transition into the classical regime. When $\sigma < -2$, the coefficient of diffusion goes to zero as $s$ increases; this means that *new* states stationary under diffusion can be created and supported at these very deep values of $\sigma$. When $\sigma > -2$ the diffusion coefficient diverges for large $s$ and quantum corrections will shut the diffusion process off by eventually driving $\sigma$ to zero; what happens is that as $s$ increases, the diffusion coefficient at first increases, to then decrease and cut itself off to zero when $\sigma$ goes to zero or when the classical regime ($\hbar \to 0$) is reached.

We have seen how quantum corrections lead into diffusion, and in some cases, clustered states of the quantum field system. To get more information about clustering and other structure formation properties, it is convenient to look at the correlation function and integral; from an analysis of these objects one can get a better feeling of the structures [12] supported by the quantum field theory. The correlation integral, $C(s)$, gives information on how many correlated pairs there are whose separation is *less* than $s$. This integral is computed from the correlation function $\xi(\vec{r})$ by the following formula [2, 13]

$$C(s) = \int_{V(s)} d^3\vec{r} \, [1 + \xi(\vec{r})]$$

where the region of integration extends over the volume $V(s)$ of radius $s$. The correlation function $\xi(\vec{r})$ is related to $\delta(\vec{k})$, the Fourier transform of the density $\rho(\vec{r})$, via the following Fourier transform

$$1 + \xi(\vec{r}) = (2\pi)^3 \int d^3\vec{k} e^{-i\vec{k}\cdot\vec{r}} \left| \delta(\vec{k}) \right|^2$$

where

$$\delta(\vec{k}) = \int d^3\vec{r} e^{+i\vec{k}\cdot\vec{r}} \rho(\vec{r})$$

The quantity $\left| \delta(\vec{k}) \right|^2$ is called the power spectrum, and measures the mean number of correlated neighbors in *excess* of random over a distance of order $1/k$; because of this, the quantity $1 + \xi(\vec{r})$ (once appropriately normalized) gives information on whether there is "clustering" ($> 1$), "voids" ($< 1$) or a perfectly random distribution ($= 1$). This is a straightforward program that can be carried out for $\rho(r)$. One obtains the following expressions for $1 + \xi(\vec{r})$ and $C(r)$

$$1 + \xi(r) = D(\sigma) \cdot r^{-3+\sigma}$$

$$C(r) = \frac{2\pi}{\sigma} \cdot D(\sigma) \cdot r^{2\sigma}$$

where the coefficient $D(\sigma)$ is

$$D(\sigma) = (4\pi A)^2 \cdot 4\pi \cdot |B(\sigma)|^2 \cdot \Gamma(2 - 2\sigma) \cos \frac{\pi}{2}(2\sigma - 1)$$

and

$$B(\sigma) = \Gamma(\sigma - 1) \cos \frac{\pi}{2}(2 - \sigma).$$

$D(\sigma)$ is shown in Figure 10.3, and in Figure 10.4 we show $D(\sigma)/\sigma$, which are important in determining the behaviors of $1 + \xi(r)$ and the correlation integral.

From Figure 10.3 we see that for positive $\sigma$ and between zero and $3/2$, the quantum field system leads to clustering. For $\sigma$ between $3/2$ and $5/2$ there is a significant qualitative change into a scenario of deep anti-clustering with a tendency for the system to behave near homogeneity when $\sigma \approx 2$. When $\sigma$ increases past $5/2$, we enter a new clustering domain. The rest can be "read off" in the same way from the figure. The plot of Figure 10.4 simply confirms the one for $D(\sigma)$, but for the coefficient of the correlation integral. We also point out that there is a "quantization" of structures, a kind of periodicity, that takes place as $\sigma$ changes and goes through some critical values.

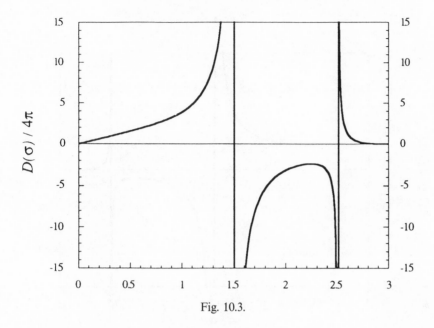

Fig. 10.3.

We end by summarizing our results. Quantum field theory is a many body system par excellence. Intrinsic to it is the randomness associated with quantum fluctuations. The application of a statistical interpretation to one of the most primitive concepts of field theory, that of the induced charge, has interesting and apparently deep consequences which stem from its scaling properties. In particular, fractal behavior in a form related to the Pareto–Levy form of the charge density, leads to systems where the maximum entropy is associated with the classical limit of the quantum system. In other words, the classical regime is the most stable, and the one preferred by the system.

Like in other many body systems, diffusion is a familiar mechanism for the approach to equilibrium (although perhaps not the only one). In concordance to what follows from studying the entropy, the virtual cloud diffuses with a coefficient of diffusion proportional to $\hbar$, and thus diffusion of the cloud stops in the classical limit.

From an analysis of the correlation properties of the virtual cloud system, one learns that the system supports the formation of structures where there is a sharp transition between phases of anticlustering (void formation), clustering and some interspersed quasi–homogeneity. These depend quite clearly on the number of degrees of freedom excited from the vacuum into the system; in other words, they depend on the distance at which the system is probed.

In very general terms, we see that irreversibility is contained in quantum field systems, and that this irreversibility is encountered as the system size grows. The quantum field system tends to "relax" into a classical system or into clustered

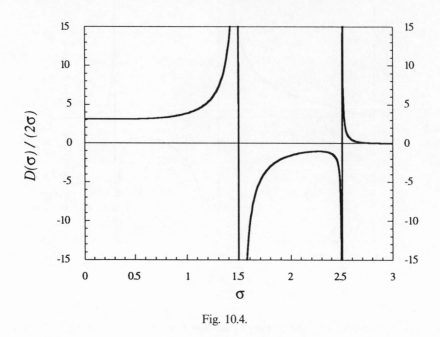

Fig. 10.4.

structures, depending on the scale of the system and the nature of the interactions. The ideas developed here may also find application in the study of intermittency, turbulence, phase transitions, multiparticle physics and other complex phenomena in quantum field theory and the early universe.

## Acknowledgements

This workshop was financed by Fundación Banco de Bilbao Vizcaya and by NATO. The Fundación also provided us with logistic, material and public support. I wish to thank Don José Angel Sánchez Asiaín, Doña Maria Luisa Oyarzábal and Don José Ignacio Oyarzábal for their enthusiastic support. I would also like to thank my co–organizers, Jonathan Halliwell and Wojciech Zurek for the pleasure and fun in working together. I have benefitted from discussions with O. Bertolami, R. Blankebeckler, A. Carro, T. Goldman, A. González–Arroyo, S. Habib, J. Hartle, R. Laflamme, S. Lloyd, C. Morales, M. M. Nieto, E. Trillas, G. Veneziano and G. West.

## References

[1] Bell, J. S., *Speakable and unspeakable in quantum mechanics*, Cambridge University Press, Cambridge, 1991.
[2] See for example, Layzer, D., in *Galaxies and the Universe*, edited by Sandage, A., Sandage, M. and Kristian, J., University of Chicago Press, Chicago, 1975.

[3] Gell–Mann, M. and Low, F. E., *Phys. Rev.* **95** (1954) 1300–1312. For a beautiful presentation full of insight, see West, G. in *Particle Physics: A Los Alamos Primer*, edited by Cooper, N. and West, G., Cambridge University Press, Cambridge, 1988.

[4] Rothe, H. J., *Lattice Gauge Theories*, World Scientific, Singapore, 1992.

[5] Georgi, H. and Politzer, D., *Phys. Rev.* **D14** (1974) 451.

[6] Doob, J., *Classical Potential Theory and Its Probabilistic Counterpart*, Springer–Verlag, New York, 1984.

[7] Schroeder, M., *Fractals, Chaos, Power Laws*, W. H. Freeman and Co., San Francisco, 1991.

[8] Cover, T. M. and Thomas, J. A., *Elements of Information Theory*, J. Wiley and Sons, New York, 1991.

[9] Fermi, E., *Thermodynamics*, Dover Books, New York, 1956.

[10] Appelquist, T. and Carrazone, J., *Phys. Rev.* **D11** (1975) 2896.

[11] Chandrasekhar, S., *Rev. Mod. Phys.* **15** (1943) 1–89.

[12] Thompson, D'Arcy Wentworth, *On Growth and Form: The Complete Revised Edition*, Dover Books, New York, 1992.

[13] Grassberger, P. and Procaccia, I., *Physica* **D9** (1983) 189.

# Part  Three

Decoherence

# 11

# Preferred Sets of States, Predictability, Classicality, and Environment-Induced Decoherence

Wojciech H. Zurek

*Theoretical Astrophysics*
*T-6, MS B288*
*Los Alamos National Laboratory*
*Los Alamos, NM 87545, USA*

## Abstract

Selection of the preferred classical set of states in the process of decoherence – so important for cosmological considerations – is discussed with an emphasis on the role of information loss and entropy. *Persistence of correlations* between the observables of two systems (for instance, a record and a state of a system evolved from the initial conditions described by that record) in the presence of the environment is used to define classical behavior. From the viewpoint of an observer (or any system capable of maintaining records) *predictability* is a measure of such persistence. *Predictability sieve* – a procedure which employs both the statistical and algorithmic entropies to systematicaly explore all of the Hilbert space of an open system in order to eliminate the majority of the unpredictable and non-classical states and to locate the islands of predictability including the preferred *pointer basis* is proposed. Predictably evolving states of decohering systems along with the time-ordered sequences of records of their evolution define the effectively classical branches of the universal wavefunction in the context of the "Many Worlds Interpretation". The relation between the consistent histories approach and the preferred basis is considered. It is demonstrated that histories of sequences of events corresponding to projections onto the states of the pointer basis are consistent.

## 11.1 Introduction

The content of this paper is not a transcript of my presentation given in course of the meeting: Proceedings contributions rarely are. However, in the case of this paper the difference between them may be greater than usual. The reason for the discrepancy is simple. Much of what I said is already contained in papers (see, for example, my recent *Physics Today* article (Zurek 1991) and references therein). While it would be reasonably easy to produce a "conference version" of this contribution, such an exercise would be of little use in the present context. Nevertheless, I have decided to include in the first part of this paper much of the same introductory material

175

already familiar to these interested in decoherence and in the problem of transition from quantum to classical, but with emphasis on issues which are often glossed over in the available literature on the subject, and which were covered in the introductory part of my presentation in a manner more closely resembling Zurek (1991). The second part of the paper ventures into more advanced issues which were considered only briefly during the talk, but are the real focus of attention here. In a sense, the present paper, while it is self-contained, may be also regarded by reader as a "commentary" on the material discussed in the earlier publications. Its union with my 1991 paper would be a more faithful record of my conference presentation, and a familiarity with that publication – while not absolutely essential – could prove helpful.

*Decoherence* is a process which – through the interaction of the system with external degrees of freedom often referred to as the *environment* – singles out a preferred set of states sometimes called *the pointer basis*. This emergence of the preferred basis occurs through the "negative selection" process (*environment-induced superselection*) which, in effect, precludes all but a small subset of the conceivable states in the Hilbert space of the system from behaving in an effectively classical, predictable manner. Here I intend to expand and enhance these past discussions in several directions:

(i) I shall emphasize the relation between the interpretational problems of quantum theory and the existence, within the Universe, of the distinct entities usually referred to as "systems". Their presence, I will argue, is indispensable for the formulation of the demand of classicality. This will justify a similar split of the Universe into the degrees of freedom which are of direct interest to the observer – a "system of interest" – and the remaining degrees of freedom known as the environment. It is therefore not too surprising that the problem (in particular, the measurement problem, but also, more generally, the problem of the emergence of classicality within the quantum Universe) which cannot be posed without recognising systems cannot be solved without acknowledging that they exist.

(ii) I shall discuss the motivation for the concept of the preferred set of "classical" states in the Hilbert space of the system – the pointer basis – and show how it can be objectively implemented by recognising the role of the interaction between the system and the environment. It will be noted that what matters is not a unique definition of the absolutely preferred set of states, but, rather, a criterion for eliminating the vast majority of the "wrong", nonclassical states.

(iii) I shall consider the relation between the states which habitually diagonalise (or nearly diagonalise) the density matrix of an open system – states which can be found on or near the diagonal after a decoherence time has elapsed regardless of what was the initial state – and the preferred states. It will be pointed out that diagonality alone is only a symptom – and not a cause – of the effective classicality of the preferred states. And, as all symptoms, it has to be used with caution in diagnosing its origin: Causes of diagonality may, on occasion, differ from those relevant for the dynamics of the process of decoherence and the quantum to classical transition that decoherence precipitates. For example, "accidental" diagonality in any basis may be a result of complete ignorance of an observer.

(iv) Therefore, I shall regard the ability to preserve correlations between the records maintained by the observer and the evolved observables of open systems as a defining feature of the preferred set of the to-be-classical states, and generalise it to the situations where the observer is not present. This definition, employed already in the early discussions of effective classicality (Zurek, 1981), can be formulated in terms of information-theoretic measures of the persistence of correlations, which leads one to consider the role of *predictability* in the definition of the preferred basis.

(v) I shall analyse the relation of the decoherence approach with the consistent histories interpretation. In particular, I shall show that the histories expressed in terms of the events corresponding to the projections onto the pointer basis states are consistent. Thus, a successful decoherence resulting in a stable pointer basis implies consistency. By contrast, consistency alone does not constrain the histories enough to eliminate even very non-classical branches of the state vector. Moreover, consistency conditions single out not only the initial moment in which all the histories begin, but also the final time instant on a more or less equal footing. Furthermore, in the absence of a fixed set of pointer projectors, consistent histories cannot be expressed in terms of a "global" set of alternative events, which could be then used for all histories at a certain instant of time, but, rather, must be formulated in terms of "local" events which are themselves dependent on the past events in the history of the system, and which do not apply to all of the alternatives coexisting at a certain instant. Unless the records are a part of the physical Universe, this is bound to introduce an additional element of subjectivity into the consistent histories interpretation.

(vi) When decoherence is successful and does yield a preferred basis, the branches of the universal wavefunction can be traced out by investigating sequences of preferred, predictable sets of states singled out by the environment-induced superselection. In this setting the issue of the "collapse of the wavepacket" – formulated as a question of accessibility, through the existing records of the past events and the ability to use such records to predict a future course of events – is posed and discussed.

I shall "iterate" these themes throughout the course of the paper, returning to most of them more than once. Consequently, the paper shall cover the issues listed above only approximately in the order in which they were itemized in this introduction. Sections 2–5 cover the familiar ground but do it with a new emphasis.

Sections 6–11 venture into a new territory and should be of particular interest to these working on the subject.

## 11.2 Motivation for Classicality, or "No Systems – No Problem"

The measurement problem – the most prominent example of the interpretational difficulties one is faced with when dealing with the quantum – disappears when the apparatus-system combination is regarded as a single indivisible quantum object. The same is true not just in course of measurements, but in general – the problems with the correspondence between quantum physics and the familiar everyday classical reality cannot be even posed when we refuse to acknowledge the division of the Universe into separate entities.

The reason for this crucial remark is simple. The Schrödinger equation;

$$i\hbar \frac{d}{dt}|\Phi> = H|\Phi>,\tag{11.1}$$

is deterministic. Therefore, given the initial state, it predicts, with certainty, the subsequent evolution of the state vector of any really isolated entity, including the *joint* state of the apparatus interacting with the measured system. Thus, in accord with Eq. (11.1), the initial state vector;

$$|\Phi_{\mathscr{A}\mathscr{S}}(0)> = |A_0>|\phi_0> = |A_0>\sum_i a_i|\sigma_i>,\tag{11.2}$$

evolves into $|\Phi_{\mathscr{A}\mathscr{S}}(t)>$ in the combined Hilbert space $\mathscr{H}_{\mathscr{A}\mathscr{S}}$. There is no hint of the interpretational problems in this statement until one recognizes that the expected outcome of the measurement corresponds to a state which – while it contains the correlation between $\mathscr{A}$ and $\mathscr{S}$ – also acknowledges the "right" of the apparatus to "a state of its own". By contrast, a typical form of $|\Phi_{\mathscr{A}\mathscr{S}}(t)>$ is;

$$|\Phi_{\mathscr{A}\mathscr{S}}(t)> = \Sigma_i a_i|A_i>|\sigma_i>,\tag{11.3}$$

where more than one $a_i(t)$ are non-zero. Above, $\{|A_i>\}$ and $\{|\sigma_i>\}$ belong to separate Hilbert spaces of the apparatus and of the system, $\mathscr{H}_{\mathscr{A}}$ and $\mathscr{H}_{\mathscr{S}}$, such that

$$\mathscr{H}_{\mathscr{A}\mathscr{S}} = \mathscr{H}_{\mathscr{A}} \otimes \mathscr{H}_{\mathscr{S}}.\tag{11.4}$$

Let me emphasize at this point the distinction between quantum and classical measurements by adopting a somewhat Dirac-like notation to discuss the classical analog of a measurement-like evolution. The initial states of the classical system S and of the classical apparatus A can be denoted by $|s_0\}$ and $|A_0\}$. When these states are "pure" (completely known), and the evolution is classical, the final result of the measurement-like co-evolution is another "pure" state: Reversible evolutions preserve purity in both quantum and classical context, but with a final state which is simpler classically than for a typical quantum case:

$$|\Upsilon_t^{AS}\} = |A_t\}|s_t\}.\tag{11.5}$$

The key difference between Eq. (11.3) and Eq. (11.5) is the absence of the super-position of the multiple alternatives corresponding to the possible outcomes. And even when the classical initial states are known incompletely (as is usually the case in the measurement situations) the initial state of the system is

$$\sum p_i |s_i\} \, , \tag{11.6}$$

where $p_i$ are the probabilities of the various states. The final state of the interacting objects must be also described by a classical probability distribution

$$\sum_i p_i |\Upsilon_i^{AS}\} = \sum_i p_i |A_i\} |s_i\} \, . \tag{11.7}$$

Nevertheless, and in contrast with the quantum case, the range of the final states – pointer positions of the apparatus – will continuously narrow down as the range of the initial states of the system is being restricted. This simultaneous constriction of the uncertainty about the initial state of the system and about the outcome – the final state of the apparatus – shall be referred to below as the *complete information limit*.

Its consequence – the expectation that a complete knowledge of the state of a classical system implies an unlimited ability to predict an outcome of every conceivable measurement – is the cornerstone of our classical intuition. This expectation is violated by quantum theory: Complete information limit in the sense described above does not exist in the quantum case. There, even in the case of the complete information about each of the two objects (i.e., the system and the apparatus) separately, one is still typically forced into a situation represented by Eq. (11.3) in which – following the measurement – neither the apparatus nor the system have "a state of their own". Indeed, for every pure state of the quantum system there exists a corresponding "relative" state of the apparatus (see Everett, 1957). In fact, a correlation between the states of the apparatus and a measured system is, at this stage, analogous to the nonseparable correlation between the two particles in the Einstein – Podolsky – Rosen "paradox" (Zurek, 1981).

A closely related symptom of the quantum nature arises from the superposition principle. That is, pure quantum states can be combined into another pure state. There is no equivalent way of combining pure classical states. The only situation when their combinations are considered is inevitably tied to probability distributions and a loss of information. Thus, formally, classical states denoted above by $|\cdot\}$ correspond not to the vectors in the Hilbert space of the system, but, rather, to the appropriate projection operators, i.e.:

$$|\cdot\} \iff |\cdot ><\cdot | \tag{11.8}$$

Such projection operators can be combined into probability distributions in both quantum and classical cases. A pure classical state – a point in the phase space

– is of course an extreme but convenient idealization. A more realistic and useful concept corresponding to a probability distribution – a patch in the phase space – would be then described by the density matrix with the form given by Eq. (11.6).

In a "hybrid" notation the measurement carried out by a classical apparatus on a quantum system would then involve a transition:

$$|A_0\}|\phi_0 >< \phi_0| \Longrightarrow \sum_i p_i |A_i\}|\sigma_i >< \sigma_i| \qquad (11.9)$$

The difference between Eq. (11.3) and the right hand side of Eq. (11.9) is clear: Classicality of the apparatus prevents the combined object from existing in a superposition of states.

The need for a random "reduction of the state vector" arises as a result of the desire to reconcile consequences of Eq. (11.1) (i.e., the linear superposition, Eq. (11.3)) with the expectation based on the familiar experience (expressed above by Eqs. (11.5), (11.7) and (11.9)) that only one of the outcomes should actually happen (or, at least, appear to happen) in the "real world."

There is, of course, a special set of circumstances in which the above distinction between quantum and classical disappears, and no reduction of the state vector following the interaction is required. This unusual situation arises when – already before the measurement – the to-be-measured quantum system was in an eigenstate (say, $|\sigma_k >$) or in a mixture of the eigenstates of the measured observable. In the case of a single pure state the resulting state of the apparatus-system pair following the interaction between them is given by Eq. (11.3), but with only one non-zero coefficient, $|a_i|^2 = \delta_{ik}$. Hence, the distinction between Eq. (11.3) and Eq. (11.5) disappears. An obvious generalisation of this effectively classical case for a mixture results in an appropriate probability distribution over the possible outcomes of the measurement. Then the correlation between the system and the apparatus has a purely classical character. Consequently, this situation is *operationally* indistinguishable from the case when a classical system in a state initially unknown to the observer is measured by a classical apparatus.

These considerations motivate the decoherence approach – an attempt to resolve the apparent conflict between predictions of the Schrödinger equation and perception of the classical reality. If the operationally significant difference between the quantum and classical characteristics of the observables can be made to continuously disappear, then maybe the macroscopic objects we encounter are "maintained" in the appropriate mixtures by the dynamical evolution of the quantum Universe. Moreover, one can show that purely unitary evolution of an isolated system can never accomplish this goal. Therefore, it is natural to enquire whether one can accomplish it by "opening" the system and allowing it to interact with the environment. The study of this and related questions defines the decoherence approach to the transition between quantum and classical.

## 11.3 Operational Goals of Decoherence

The criterion for success of the decoherence programme must be purely operational. The Schrödinger equation and the superposition principle are at its foundation. One must concede at the outset that they will not be violated in principle. Therefore, the appearance of the violation will have its origin in the fact that the records made and accessed by observers are subject to limitations, since they are governed by quantum laws with specific – rather than arbitrary – interactions and since their records are not isolated from the environment. One important criterion – to be discussed in more detail in the next section – refers to the fact that idealised classical measurements do not change the state of the classical system. That is, for example, a complete information limit – the classically motivated idea that one can keep acquiring information without changing the state of the measured object – should be an excellent approximation in the "everyday" classical context. The recognition of the role of decoherence and the environment allows one to show how this can be the case in a quantum Universe.

An objection to the above programme is sometimes heard that – in essence – the Universe as a whole is still a single entity with no "outside" environment, and, therefore, any resolution involving its division into systems is unacceptable. While I am convinced that much needs to be done in order to understand what constitutes a "system", I have also little doubt that the key aspects of the resolution proposed here are largely independent of the details of the answer to this question. As we have argued in the previous section, without the assumption of a preexisting division of the Universe into individual systems the requirement that they have a right to their own states cannot be even formulated: The state of a perfectly isolated fragment of the Universe – or, for that matter, of the quantum Universe as a whole – would evolve forever deterministically in accord with Eq. (11.1). The issue of the "classicality" of its individual components – systems – cannot be even posed. The effectively stochastic aspect of the quantum evolution (the "reduction of the state vector") is needed only to restore the right of a macroscopic (but ultimately quantum) system to be in its own state.

Quantum theory allows one to consider the wave function of the Universe. However, entities existing within the quantum Universe – in particular us, the observers – are obviously incapable of measuring that wave function "from the outside". Thus, the only sensible subject of considerations aimed at the interpretation of quantum theory – that is, at establishing correspondence between the quantum formalism and the events perceived by us – is the relation between the universal state vector and the states of memory (records) of somewhat special systems – such as observers – which are, of necessity, perceiving that Universe from within. It is the inability to appreciate the consequences of this rather simple but fundamental observation that has led to such desperate measures as the search for an alternative to quantum physics. One of the goals of this paper is to convince the reader that such desperate steps are unwarranted.

One might be concerned that the appeal to systems and a frequent mention of measurement in the above paragraphs heralds considerations with a character which is subjective and may be even inexcusably "anthropocentric", so that the conclusions could not apply to places devoid of observers or to epochs in the prehistory of the Universe. A few remarks on this subject are in order. The rules we shall arrive at to determine the classical preferred basis will refer only to the basic physical ingredients of the model (that is, systems, their states, and the hamiltonians that generate their evolution) and we shall employ nothing but unitary evolution. These rules will apply wherever these basic ingredients are present, regardless of whether an observer is there to reap the benefits of the emerging classicality. Thus, the role of "an observer" will be very limited: It will supply little more than a motivation for considering the question of classicality.

Having said all of the above, one must also admit that there may be circumstances in which some of the ingredients appropriate for the Universe we are familiar with may be difficult or even impossible to find. For example, in the early Universe it might be difficult to separate out time (and, hence, to talk about evolution). While such difficulties have to be acknowledged their importance should not be exaggerated: The state vector of the Universe will evolve in precisely the same way regardless of whether we know what observables could have been considered as "effectively classical" in such circumstances. Therefore, later on in course of its evolution, when one can define time and discern the other ingredients relevant for the definition of classicality, the question of whether they have been present early on or what was or was not classical "at the time when there was no time" will have absolutely no observable consequences.

This last remark deserves one additional caveat: It is conceivable (and has been even suggested, for example by Penrose (1989)) that the quantum theory which must be applied to the Universe as a whole will differ from the quantum mechanics we have become accustomed to in some fundamental manner (that is, for instance, by violating the principle of superposition for phenomena involving gravitation in an essential way). While, at present, there is no way to settle this question, one can nevertheless convincingly argue that the coupling with the gravitational degrees of freedom is simply too weak in many of the situations involving ordinary, everyday quantum to classical transitions (for example, the blackening of a grain of photographic emulsion caused by a single photon) to play a role in the emergence of the classicality of everyday experience.

## 11.4 Insensitivity to Measurements and Classical Reality

The most conspicuous feature of quantum systems is their sensitivity to measurements. Measurement of a quantum system automatically results in a "preparation": It forces the system into one of the eigenstates of the measured observable, which, from that instant on, acts as a new initial condition. The idealized version of this quantum "fact of life" is known as the projection postulate. This feature of quantum systems has probably contributed more than anything else to the perception of a conflict between the *objective reality* of the states of classical objects and the elusive *subjective nature* of quantum states which – it would appear – can be molded into a nearly arbitrary form depending solely on the way in which they are being measured. Moreover, the Schrödinger equation will typically evolve the system from the prepared initial state (which is usually simple) into a complicated superposition of eigenstates of the initially measured observables. This is especially true when an interaction between several systems is involved. Then the initial state prepared by a typical measurement involves certain properties of each of the separately measured systems, but quantum evolution results in quantum correlations – in states which are qualitatively similar to the one given by Eq. (11.3). Thus, when the measurement of the same set of observables is repeated, a new "reduction of the wavepacket" and a consequent restarting of the Schrödinger evolution with the new initial condition is necessary.

By contrast, evolutions of classical systems are – in our experience – independent of the measurements we, the observers, carry out. A measurement can, of course, increase our knowledge of the state of the system – and, thus, increase our ability to predict its future behavior – but it should have no effect on the state of the classical system *per se*. This expectation is based on a simple fact: Predictions made for a classical system on the basis of the past measurements are not (or, at least, need not be) invalidated by an intermediate measurement. In classical physics, there is no need to reinitialize evolution, a measurement does not lead to "preparation", only to an update of the records. This *insensitivity to measurements* appears to be a defining feature of the classical domain.

For a quantum system complete insensitivity to measurements will occur only under special circumstances – when the density matrix of the system commutes with the measured observable already before the measurement was carried out. In general, the state of the system will be influenced to some degree by the measurements. The density matrix is a useful object which can help quantify the degree to which the state of the system is altered by the measurement. This can be done by comparing its density matrix before and after the measurement (which is idealised here in the way introduced by von Neumann (1932)):

$$\rho_{\text{after}} = \sum_i P_i \rho_{\text{before}} P_i \tag{11.10}$$

The projection operators $P_i$ correspond to the various possible outcomes. A generalisation to the situation when a sequence of measurements is carried out is straightforward:

$$\rho_{\text{after}} = \sum_{i,j,\dots,n} P_n \dots P_j P_i \rho_{\text{before}} P_i P_j \dots P_n \tag{11.11}$$

Above, pairs of projection operators will correspond to distinct observables measured at consecutive instants of time. Evolution in between the measurements could be also incorporated into this discussion.

It is assumed above that even though the measurements have happened, their results are not accessible. Therefore, the statistical entropy;

$$h(\rho) = -Tr\rho \ln \rho \qquad (11.12)$$

can only increase;

$$h(\rho_{\text{after}}) \geq h(\rho_{\text{before}}) . \qquad (11.13)$$

The size of this increase can be used as a measure of the sensitivity to a measurement. Unfortunately, it is not an objective measure: It depends on how much was known about the system initially. For example, if the initial state is completely unknown so that $\rho_{\text{before}} \sim 1$, there can be no increase of entropy. It is therefore useful to constrain the problem somewhat by assuming, for example, that the initial entropy $h(\rho_{\text{before}})$ is fixed and smaller than the maximal entropy. We shall pursue this strategy below, in course of the discussion of the predictability in Section 6.

The role of decoherence is to force macroscopic systems into mixtures of states – approximate eigenstates of the same set of effectively classical observables. Density matrices of this form will be then insensitive to measurements of these observables (that is, $\rho_{\text{after}} \approx \rho_{\text{before}}$). Such measurements will be effectively classical in the sense described near the end of Section 2, since the final, correlated state of the system and the apparatus will be faithfuly described by Eqs. (11.7) and (11.9). No additional reduction of the state vector will be needed, as the projection operators $P_i$ defining the eigenspaces of the measured observable commute with the states which invariably (that is, regardless of the initial form of the state of the system) appear on the diagonal of the density matrix describing the system.

This feat is accomplished by the coupling to the environment – which, as is well known, results in noise, and, therefore, in a degraded predictability. However, it also leads to a specification of what are the preferred observables – what set of states constitutes the preferred classical "pointer basis" which can be measured relatively safely, with a minimal loss of previously acquired information. This is because the rate at which the predictability is degraded is crucially dependent on the state of the system. The initial state is in turn determined by what was measured. We shall show that the differences in the rates at which predictability is lost are sufficiently dramatic that only for certain selections of the sets of states – and of the corresponding observables – can one expect to develop the idealised but very useful approximation of classical reality.

Another, complementary way of viewing the process of decoherence is to recognise that the environment acts, in effect, as an observer continuously monitoring certain preferred observables which are selected mainly by the system-environment interaction hamiltonians. Other physical systems which perform observations (such as the "real" observers) will gain little predictive power from a measurement which prepares the system in a state which does not nearly coincide with the states already monitored by the environment. Thus, the demand for predictability forces observers to conform and measure what is already being monitored by the environment. Indeed, in nearly all familiar circumstances, observers gain information by "bleeding off" small amounts of the information which has been already disseminated into the environment (for example by intercepting a fraction of photons which have scattered off the object one is looking at while the rest of the photons act effectively as an environment). In such a case, the observer must be prepared to work with the observables premeasured by the environment. Moreover, even if it was somehow possible to make a measurement of some other very different set of observables with the eigenstates which are given by exotic superpositions of the states monitored by the environment, records of such observables would become useless on a timescale on which the environment is monitoring the preferred states. For, the "measurements" carried out by the environment would invalidate such data on a *decoherence timescale* – that is, well before the predictions about the future measurements made on its basis could be verified.

In what follows we shall therefore search for the sets of observables as well as the corresponding sets of states of open quantum systems which offer optimal predictability of their own future values. This requirement embodies a recognition of the fact that our measuring equipment (including our senses) used in keeping track of the familiar reality is relatively time-independent, and that the only measurements we are capable of refer to small fragments of the Universe – individual systems. We shall be considering sequences of measurements carried out on ensembles of identical open quantum systems. We shall return to the discussion of the practical implementation of this programme in Sections 6 and 7, having introduced – in the next section – a simple and tractable model in which decoherence can be conveniently studied.

## 11.5 Environment-Induced Decoherence in Quantum Brownian Motion

The separation of the Universe into subsystems, indispensable in stating the problem, is also instrumental in pointing out a way to its resolution. Our experience proves that classical systems – such as an apparatus – have a right to individual states. Moreover, these states are not some arbitrary states admissible in $\mathcal{H}_{\mathscr{A}}$ which can be selected on a whim by an external observer (as would be the case for an ordinary quantum system) but, rather, appear to be "stable" in that the state in which $\mathscr{A}$ is eventually found is always one of the "eigenstates" of the same "menu" of options – i.e., the apparatus has a preferred "pointer observable." Thus, even though the correlated system-apparatus state vector $|\Phi_{\mathscr{A}\mathscr{S}}(t)>$, Eq. (11.3), can be in principle rewritten using any (complete) basis in $\mathcal{H}_{\mathscr{A}}$ (Zurek, 1981; 1991), this formal application of the superposition principle and the resulting analogy with nonseparable EPR correlations is not reflected in "the familiar reality". The familiar macroscopic systems tend to be well localized with respect to the usual classical observables – such as position, energy, etc.

Here we shall see how this tendency towards localization, as well as the choice of the preferred observables can be accounted for by the process of decoherence: A classical system (such as an apparatus) will be continuously interacting with its environment. Consequently, its density matrix will obey an irreversible master equation, valid whenever the environment is sufficiently large to have a Poincaré recurrence time much longer than any other relevant timescale in the problem – in particular, much longer than the timescale over which one may attempt to predict its behavior or to verify its classicality.

A useful example of such a master equation can be derived – under certain reasonable, although not quite realistic circumstances – for a particle of mass $m$ moving in a potential $V(x)$, and interacting with the external scalar field (the environment) through the hamiltonian

$$\mathscr{H}_{int} = \epsilon x \dot{\varphi}(0). \tag{11.14}$$

In the high-temperature limit the reduced density matrix of the particle will obey the equation (written below in the position representation):

$$\frac{d}{dt}\rho(x,x') = -\frac{i}{\hbar}[H_R,\rho] - \gamma(x-x')(\frac{\partial}{\partial x} - \frac{\partial}{\partial x'})\rho - \frac{\eta k_B T}{\hbar^2}(x-x')^2\rho. \tag{11.15}$$

Above $H_R$ is the hamiltonian of the particle with $V_R(x)$ renormalized by the interaction with the field, $\eta$ is the viscosity coefficient ($\eta = \frac{\epsilon^2}{2}, \gamma = \frac{\eta}{2m}$), and $T$ is the temperature of the field. Similar equations have been derived under a variety of assumptions by, for example, Caldeira and Leggett (1983; 1985), Joos and Zeh (1985), Unruh and Zurek (1989), and Hu, Paz and Zhang (1992).

The preferred basis emerges through the process of "negative selection": In the "classical" regime associated with $\hbar$ small relative to the effective action the last term of Eq. (11.15) dominates and induces the following evolution of the density matrix:

$$\rho(x,x';t) = \rho(x,x';0)\exp(-\frac{\eta k_B T}{\hbar^2}(x-x')^2 t). \tag{11.16}$$

Thus, superpositions in position will disappear on the decoherence timescale of order:

$$\tau_D = \gamma^{-1}(\lambda_T/\Delta x)^2 \tag{11.17}$$

where $\gamma^{-1}$ is the relaxation rate

$$<\dot{v}> = -\gamma <v> \tag{11.18}$$

This observation† marked an important shift from the idealised models of deco-

---

† See Zurek (1984) for the first discussion of this *decoherence timescale*; Zurek (1991) from the Wigner distribution function perspective; Hartle (1991) for a nice perspective based on effective action; Paz

herence in the context of quantum measurements to the study of more realistic models better suited to the discussion of classicality in the "everyday" context. The conclusion is simple: Only relatively localized states will survive interaction with the environment. Spread-out superpositions of localised wavepackets will be rapidly destroyed as they will quickly evolve into mixtures of localised states.

This special role of the position can be traced to the form of the hamiltonian of interaction between the system and the environment, Eq. (11.14): The dependence on $x$ implies a sensitivity to position, which is, in effect, continuously monitored by the environment. In general, the observable $\Lambda$ which will commute with $H_{int}$;

$$[H_{int}, \Lambda] = 0 , \qquad (11.19)$$

will play a key role in determining the preferred basis: Superpositions of its eigenspaces will be unstable in the presence of the interaction with the environment. Thus, the special role position plays in our description of the physical Universe can be traced (Zurek, 1982, 1986) to the form of the interaction potentials between its subsystems (including the elementary particles and atoms) which is almost invariably position dependent (and, therefore, commutes with the position observable).

The result of such processes is the "effective superselection rule," which destroys superpositions of the eigenspaces of $\Lambda$, so that the density matrix describing the open system will very quickly become approximately co-diagonal with $\Lambda$. Thus, the classical system can be thought of – on a timescale of several $\tau_D$ – as a mixture (and not a superposition) of the approximate eigenstates of position defined with the accuracy of the corresponding $\Delta x$. This approximation becomes better when the self-hamiltonian of the system is less important when compared to $H_{int}$. Thus, for example, overdamped systems have preferred states which are squeezed in position, while weakly damped harmonic oscillator translates position into momentum every quarter of its period and, consequently, favors still localised, but much more symmetric coherent states. Moreover, for macroscopic masses and separations the decoherence timescale is very small compared to the relaxation time (Zurek, 1984, 1991).

The ability to describe the state of the system in terms of the probability distribution of the same few variables is the essence of effective classicality: It corresponds to the assertion that the system already has its own state (one of the states of the preferred basis) which is not necessarily known to the observer prior to the measurement, but is, nevertheless, already quite definite. It is important to emphasize that while this statement is, strictly speaking, incorrect (there is actually a very messy entangled state vector including both the system and the environment) it cannot be falsified by any feasible measurement. Thus, the quantum Universe gives rise to an

---

(1993) for a quick assessment of the role of the spectral density in the environment on the effectiveness of decoherence and Paz, Habib and Zurek (1993) for a more complete assesment of this important issue as well as for a discussion of the accuracy of the high temperature approximation, Eq. (11.15).

effectively classical domain defined through the operational validity of the assertions concerning "reality" of the states of the observables it contains.

## 11.6 Preferred States and the "Predictability Sieve"

In the preceding section we have confirmed that – at least for the model presented above – states localised in the familiar classical observable, position, will be among those selected by the decoherence process and, therefore, are effectively classical. Here I shall reverse the procedure. That is, I shall first formulate a test which will act, in effect, as a sieve, a filter accepting certain states in the Hilbert space of the system and rejecting others. The aim of this procedure will be to arrive at an algorithm capable of singling out preferred sets of states *ab initio*, rather than just confirming the suspicions about classicality of certain observables. The guide in devising the appropriate procedure is a decade-old observation (Zurek, 1981; 1982) that – under some special assumptions – the absolutely preferred states remain completely unaffected by the environment. In the absence of these special conditions it is therefore natural to search for the states which are least affected by the interaction with the external degrees of freedom (Zurek, 1984; Unruh and Zurek, 1989). And a convenient measure of the influence of the environment on the state is – as was already mentioned – the ability to predict its future evolution, to use the results of the past measurements as initial conditions.

Let us consider an infinite ensemble of identical systems immersed in identical environments. The aim of the test will be to find which initial states allow for optimal predictability. In order to settle this question, we shall imagine testing *all* the states in the Hilbert space of the system $\mathscr{H}_{\mathscr{S}}$. To this end, we shall prepare the system in every conceivable pure initial state, let it evolve for a fixed amount of time, and then determine the resulting final state – which, because of the interaction with the environment, will be nearly always given by a mixture.

Entropy of the final density matrix $\rho_k$ which has evolved from the initial state $|k>$;

$$h_k = -Tr\rho_k \log \rho_k \tag{11.20}$$

is a convenient measure of the loss of predictability. It will be different for different initial states. One can now sort the list of all possible pure states in order of the decreasing entropy $h_k$ of the final mixed state: At the head will be the states for which the increase of entropy is the least, followed by the states which are slightly less predictable, and so on, until all of the pure states are assigned a place on the list.

The best candidates for the classical "preferred states" are, clearly, near the top of the list. One may, nevertheless, ask where should one put a cut in the above list to define a border between the preferred classical set of states and the non-classical remainder. The answer to this question is somewhat arbitrary: Where exactly the quantum-classical border is erected is a subjective matter, to be decided by circumstances. What is, perhaps, more important is the nature of the answer which emerges from this procedure aimed at selecting the preferred set of states.

A somewhat different in detail, but similar in the spirit procedure for sorting the Hilbert space would start with the same initial list of the states and evolve them until the same loss of predictability – quantified by the entropy of the final mixture – has ocurred. The list of states would be then sorted according to the length of time involved in the predictability loss, with the states which are predictable to the same degree for the longest time appearing on the top of the list.

It should be pointed out that the qualifying preferred states will generally form (i) an overcomplete set, and (ii) they may be confined to a subspace of $\mathcal{H}_{\mathscr{S}}$. These likely results of the application of the "predictability sieve" outlined above will have to be treated as a "fact of life". In particular, overcompleteness seems to be a feature of many a "nice" basis – for example, coherent states are overcomplete. Moreover, it is entirely conceivable that classical states may not exist for a given system in all of its Hilbert space.

It is reassuring to note that, in the situations in which an exact pointer observable exists (that is, there is a nontrivial observable which commutes with the complete hamiltonian, $H + H_{int}$, of the system of interest; see Zurek, 1981-1983) its eigenstates can be immediately placed on top of either of the predictability lists described above. The predictability sieve selects them because they are completely predictable; they do not evolve at all. Hence, there is no corresponding increase of entropy.

One key ingredient of the first of the predictability sieves outlined above remains arbitrary: We have not specified the interval of time over which the evolution is supposed to take place. This aspect of the selection process will clearly influence the entropy generated in the system. We shall insist that this time interval should remain somewhat arbitrary. That is, the set of the selected preferred states for any time $t$ much longer than the typical decoherence timescale $\tau_D$, but much shorter than the timescale $\tau_{EQ}$ over which the system reaches thermodynamic equilibrium with its environment should be substantially similar. Similarly, for the second proposed sieve we have not specified the loss of predictability (increase of entropy) which must be used to define the timescale characterising the location of the state on the list. Again, we expect that the list should not be too sensitive to this parameter: Indeed, the two versions of the sieve should yield similar sets of preferred states.

The distinction between the preferred set of states and the rest of the Hilbert space – the contrast indicated by the predictability sieve outlined above – should depend on the mass of the system or on the value of the Planck constant. Thus, it should be possible to increase the contrast between the preferred set of states and the other states which decohere much more rapidly by investigating the behavior of the entropy with varying $m$ and $\hbar$ in mathematical models of these systems. It is easy to see how such a procedure will work in the case of quantum Brownian motion described in the preceding section. In particular, reversible Newtonian dynamics can coexist with a very efficient environment-induced superselection process in the appropriate limit (Zurek, 1984; 1991): It is possible to increase mass, decrease the Planck constant, and, at the same time, decrease the relaxation rate to zero, so that the master equation Eq. (11.15) yields reversible Liouville dynamics for the phase space probability distributions constructed from the sets of localised states (such as the coherent states), but eliminates (by degrading them into mixtures of the localised states) all other initial conditions.

A natural generalisation of the predictability sieve can be applied to mixtures. Let us construct a list of all the mixtures which have the same initial entropy $h_m(0)$ using either of the two prescriptions outlined above. We can now consider their evolution and enquire about the entropy of the mixed state after some time $t$. An ordered list of the initial mixtures can be again constructed and the top of the list can be identified. An interesting question can be now posed: Are the mixtures on top of the initial mixture list diagonal in the pure states which are on top of the list of preferred states? When the exact pointer basis in the sense of the early references on this subject (Zurek, 1981; 1982; 1983) exists, the conjecture expressed above can be readily demonstrated: Mixtures of the pointer basis eigenstates do not evolve at all. Therefore, their entropy remains the same. Any other kind of mixture will become even more mixed because superpositions of the pointer basis states decohere. Consequently, at least in this case the most predictable mixtures are diagonal in the basis of the most predictable states.

In this preliminary discussion we have neglected several issues which are likely to play a role in the future developments of the predictability sieve concept. In particular, we have ignored to enforce explicitly the requirement (stated in Section 4) that the measurements which have prepared the set of the candidate states should be suitable for the predictions of their own future outcomes. (For the closed systems this criterion naturally picks out complete sets of commuting observables corresponding to the conserved quantities.) Moreover, we have used only the statistical entropy to assess the cost of the predictability loss.

An addition of an algorithmic component might make for an even more interesting criterion. That is, the place of the pure state $|\ell>$ on the list would be based on the value of the *physical entropy* (Zurek, 1989), given by the sum of the statistical (von Neumann) entropy with the algorithmic information content $K(|\ell>)$:

$$s(|\ell>) = h_\ell + K(|\ell>) . \tag{11.21}$$

Algorithmic contribution would discourage overly complicated states.

It is interesting to consider a possibility of using a similar predictability criterion to justify the division of the Universe into separate systems. The argument (which will be made more precise elsewhere) would start with the list of all of the possible states in some composite Hilbert space, perhaps even the Hilbert space associated with the Universe as a whole. The procedure would then be largely similar to the simpler version of the sieve outlined above, but with one important change:

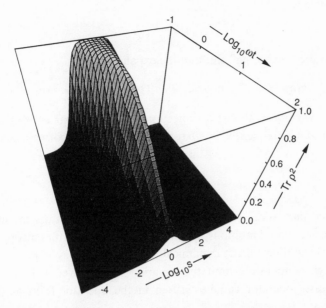

Fig. 11.1. Predictability sieve for an underdamped harmonic oscillator. The plot shows $Tr\rho^2$ (which serves as a measure of purity of the reduced density matrix $\rho$) for mixtures which have evolved from the initial minimum uncertainty wavepackets with different squeeze parameters $s$ in a damped harmonic oscillator with $\gamma/\omega = 10^{-4}$. Coherent states ($s = 1$) which have the same spread in position and momentum, when measured in the units natural for the harmonic oscillator, are clearly favored as a preferred set of states by the predictability sieve. Thus, they are expected to play the crucial role of the pointer basis in the transition from quantum to classical.

One will be allowed to vary the definition of the systems – that is, to test various projection operators, such as the projections corresponding to averaging process – to establish the most predictable collective coordinates, and, simultaneously, to obtain the preferred sets of states corresponding to such observables.

## 11.7 Predictability Sieve: An Example

A simple example of the dependence of the entropy production rate on the set of preferred states is afforded by a model of a single harmonic oscillator interacting with a heat bath we have analysed in Section 5. In this section we shall carry out a restricted version of the more ambitious and general "predictability sieve" programme for this case. In its complete version one would search for the set of states which minimize the entropy over a finite time duration of the order of the dynamical timescale of the system. Below we shall work with an instantaneous *rate* of change of *linear* entropy. A more detailed can be found elsewhere (Zurek, Habib and Paz, 1993).

Linear entropy

$$\varsigma(\rho) \;=\; Tr(\rho - \rho^2) \tag{11.22}$$

is related to the "real thing" through the expansion:

$$h(\rho) \;=\; -Tr\rho\ln\rho \;\approx\; Tr\rho\{(1-\rho) + \cdots \;=\; \varsigma(\rho) + \cdots \tag{11.23}$$

In the limit where $\rho$ is almost a projection operator, linear entropy is an approximation of $h(\rho)$. Consequently, in what follows I shall "cut corners" and consider the quantity:

$$\dot{\varsigma} \;=\; \frac{d\,Tr\rho^2}{dt} \;=\; Tr(\dot{\rho}\rho + \rho\dot{\rho}) \tag{11.24}$$

for the initial pure states (that is, for all the density operators for which $\rho^2 = \rho$). By contrast, the more ambitious programme would deal with entropy increase over finite time of the order of the dynamical timescale of the system.

In the high temperature limit the equation which governs the evolution of the reduced density operator can be written in the operator form as (Caldeira and Leggett, 1983):

$$\dot{\rho} \;=\; \frac{1}{i\hbar}[H_R, \rho] + \frac{\gamma}{2i\hbar}[\{p, x\}, \rho] - \frac{\eta k_B T}{\hbar^2}[x, [x, \rho]] - \frac{i\gamma}{\hbar}([x, \rho p] - [p, \rho x]) \,. \tag{11.25}$$

Above, square brackets and curly brackets indicate commutators and anticommutators, respectively, while $H_R$ is the renormalized hamiltonian.

Only the last two terms can be directly responsible for generation of linear entropy; For, when the evolution of the density matrix is generated by the term of the form $\dot{\rho} = [A, \rho]$, then the linear entropy is unaffected. The proof of the above assertion is simple: It suffices to substitute in the expression for $\dot{\varsigma}$, Eq. (11.24), the above commutator form for the evolution generator of $\rho$. Then it is easy to compute:

$$\dot{\varsigma} \;=\; Tr(A\rho^2 - \rho^2 A)$$

which is obviously zero by the cyclic property of trace.

Consequently, after some algebra one obtains:

$$\dot{\varsigma} \;=\; \frac{4\eta k_B T}{\hbar^2} Tr(\rho^2 x^2 - \rho x \rho x) - 2\gamma Tr\rho^2 \tag{11.26}$$

When the system is nearly reversible ($\gamma \approx 0$) the second term in the above equation is negligible. This is especially true in the high temperature "macroscopic" limit (large mass, small Planck constant). It should be pointed out that the initial "decrease of entropy" predicted by this last term for sufficiently localised wavepackets is unphysical and an artifact of the high temperature approximation (Ambegaokar, private communication; Hu, Paz, and Zhang, 1992). Nevertheless, at late times the existence of that term allowes for the eventual equilibrium with $\dot{\varsigma} = 0$.

For the pure states the first term can be easily evaluated:

$$\dot{\varsigma} = \frac{4\eta k_B T}{\hbar^2} \cdot (<x^2> - <x>^2),$$  (11.27)

where $<x>^2 = |<\varphi|x|\varphi>|^2$ and $<x^2> = <\varphi|x^2|\varphi>$. This is an interesting result. We have just demonstrated that *for pure states the rate of increase of linear entropy in quantum Brownian motion is given by their dispersion in position, which is the preferred observable singled out by the interaction hamiltonian*. More generally, when the system is described by a density matrix so that Eq. (11.26) must be used, it is easy to see that the density matrix which has fewer superpositions of distinct values of $x$ will result in a smaller entropy production. Thus, the special role of the observable responsible for the coupling with the environment is rather transparent in both of the above equations.

In view of the preceding considerations on the role of predictability it is now easy to understand why localized states are indeed favored and form the classical preferred basis. It should be emphasized that this simple result cannot be used to infer that the preferred states are simply eigenstates of position. This is because the self-hamiltonian – while it does not contribute directly to the entropy production, has an important indirect effect. For a harmonic oscillator it rotates the state in the phase space, so that the spread in momentum becomes "translated" into the spread in position and *vice versa*. As a result, linear entropy produced over one oscillator period $\tau$ is given by:

$$\varsigma_\tau = \frac{4\eta k_B T}{\hbar^2} \int_0^\tau dt <\psi|((x-<x>)\cos\omega t + (m\omega)^{-1}(p-<p>)\sin\omega t)^2|\psi>$$

$$= \frac{2\eta k_B T}{\hbar^2}(\Delta x_0^2 + \frac{\Delta p_0^2}{m^2\omega^2}).$$  (11.28)

Above, the dispersions are computed for the pure initial state $|\psi>$ and $\omega$ is the frequency of the oscillator.

It is now not difficult to argue that the initial state which minimizes linear entropy production for a harmonic oscillator must be (i) a minimum uncertainty wavepacket, and (ii) its "squeeze parameter" $s$ – equal to the initial ratio between its spread in position and the spread of the ground state – must be equal to unity. This can be confirmed by more detailed calculations, which yield the value of $Tr\rho^2$ for the reduced density matrices which have evolved from various initial states with different $s$ as a function of time. Coherent states (minimum uncertainty wavepackets with $s = 1$) are clearly selected by the predictability sieve acting on a dynamical timescale of the oscillator. This supplies a physical reason for the role they play in the classical limit of harmonic oscillators, including these encountered in quantum optics.

## 11.8 Insensitivity to Measurements and Consistent Histories

In 1984 Griffiths introduced a notion of history into the vocabulary used in the discussion of quantum systems. A history is a time-ordered sequence of properties represented by means of projection operators:

$$\chi_{i,j,...,l} = P_i(t_i)E(t_i,t_j)P_j(t_j)...P_l(t_l)E(t_l,t_0) \qquad (11.29)$$

where $t_i > t_j > ... > t_l$, $P$'s stand for the projection operators (assumed to be mutually orthogonal at each of the discrete instants at which the corresponding properties are defined) and $E(t,t')$ evolve the system in the time interval $(t,t')$. The set of histories is called *consistent* when they do not interfere, that is, when the probabilities corresponding to different histories can be added. Griffiths (1984) and Omnès (1990, 1992) (who emphasized the potential importance of the consistent histories approach to the interpretation of quantum theory) have demonstrated that a necessary and sufficient condition for the set of histories composed of the initial condition and just two additional events to be consistent is that the projection operators defining the properties at the instants 0, 1, and 2, when expressed in the Heisenberg picture, satisfy the algebraic relation;

$$Re\ Tr P_1 P_0 \bar{P}_1 P_2 = 0 , \qquad (11.30)$$

or, eqiuvalently;

$$Tr[P_1, [P_0, \bar{P}_1]]P_2 = 0 . \qquad (11.31)$$

These equivalent consistency conditions must hold for every of the projection operators in the sets defined at the initial, intermediate, and final instants. Above, $\bar{P}_i = 1 - P_i$ projects on the complement of the Hilbert space. Above, the explicit reference to the time instants at which different projection operators act has been abbreviated to a single index to emphasize that the time ordering is assumed to hold. There is also a stronger consistency condition with a physical motivation which draws on the decoherence process we have discussed above: It is possible to show that a sufficient condition for a family of histories to be consistent is the vanishing of the off-diagonal elements of the *decoherence functional*, introduced by Gell-Mann and Hartle (1990):

$$D(\chi,\chi') = Tr(P_2 P_1 P_0 P_1'). \qquad (11.32)$$

Generalisation to histories with more intermediate properties or with the initial condition given by a density matrix is straightforward. One of the appealing features of the Gell-Mann – Hartle approach is that it supplies a convenient measure of the degree of inconsistency for the proposed sets of histories.

In all of the above consistency conditions above $P_0$ acts as an initial condition and $P_2$ (or, more generally, the set of projection operators corresponding to the "last" instant) acts as a final condition. The two sets of projection operators determine the initial and final sets of alternatives for the system. (It is of course possible to start the system in a mixture of different $P_0$'s.) This special role of the initial and final sets of projection operators was empasized already by Griffiths (1984). It is particularily easy to appreciate in the version of the consistent histories approach due to Gell-Mann and Hartle (1990; 1991; see also their contribution in this volume).

The initial and final properties are set up "from the outside" of the system. Mathematically, the similarity between the role of the initial and final projections in the formula for $D(\chi, \chi')$ follows from the cyclic property of the trace. It is reflected in the special physical function of these two instants, which is defined by the physical context to which the consistent histories approach is applied. In the context originally considered by Griffiths the initial condition and the set of final alternative properties were supplied from the outside. That is, the consistent histories approach is capable of disposing of the "reduction of the wavepacket-like" outside intervention at all the intermediate instants, but not at the initial or final times.

In the cosmological context the initial density matrix could be perhaps supplied by quantum cosmological considerations but the final set of alternatives is more difficult to justify. This distinction between the role of the intermediate and final projections is worth emphasizing especially because it is somewhat obscured by the notation which does not distinguish between the special nature of the "last" property and the associated set of the projection operators and the intermediate properties. (The initial time instant is usually explicitly recognised as different by notation, since density matrices are typically considered for the initial condition in the majority of the papers on the subject (Omnès, 1992; Gell-Mann and Hartle, 1990).) Yet, the final set of options should not be – especially in the cosmological context, where one is indeed dealing with the closed system – determined "from the outside". The question therefore arises: How can one discuss consistency of histories without appealing to this final set of "special" events?

Evolving density matrix can be, at any instant $t$, always written as:

$$\rho(t) = \sum_{\mathcal{H}} \sum_{\mathcal{H}'} \chi(\mathcal{H}) \rho(0) \chi(\mathcal{H}'). \tag{11.33}$$

Above $\chi(\mathcal{H})$ is a collection of the projection operators with the appropriate evolution operators sandwiched in between, and can be thought of as a hypothetical history of the system. A composite index $\mathcal{H}$ designates a specific time-ordered sequence. Projectors corresponding to $\mathcal{H}$ and $\mathcal{H}'$ differ, but the time instants or evolutions do not.

Equation (11.33) is simply an identity, valid for arbitrary sets of orthocomplete projection operators defined at arbitrary intermediate instants $t_i$, $0 < t_i \le t$. There is an obvious formal similarity between the expression for the decoherence functional, Eq. (11.32), and the identity, Eq. (11.33). If one were to perform a trace on the elements of $\rho(t)$ corresponding to pairs of different histories written in the above manner, one would immediately recover the expression for the elements of the decoherence functional. The questions of consistency of the sequences of events should be therefore at least partialy testable at the level of description afforded by the density matrix written in this "historical" notation. In particular, additivity of probabilities of the distinct sequences of events corresponds simply to the *diagonality* of the above expression for $\rho(t)$ in the set of considered histories. That is, the set of histories $\{\chi\}$ will be called *intrinsically consistent* when they satisfy the equality:

$$\rho(t) = \bar{\rho}(t), \tag{11.34}$$

where:

$$\bar{\rho}(t) = \sum_{\mathcal{H}} \chi(\mathcal{H})\rho(0)\chi(\mathcal{H}). \tag{11.35}$$

The physical sense of the equality between the double sum of Eq. (11.33) (which is satisfied for every complete set of choices of the intermediate properties) and the single sum immediately above (which can be satisfied for only very special choices of $\chi$'s) is that the interference terms between the properties defining the intrinsically consistent sets of histories do not contribute to the final result;

$$\sum_{\mathcal{H} \neq \mathcal{H}'} \chi(\mathcal{H})\rho(0)\chi(\mathcal{H}') = 0. \tag{11.36}$$

An equivalent way of expressing the physical content of this equivalence condition, Eqs. (11.34) or (11.36), is to use the terminology of Section 4 and note that the system is insensitive to the measurement of the properties defined by such intrinsically consistent histories. An obvious way in which the above equation can be satisfied, and the only way in which it can be satisfied exactly without demanding cancelations between the off-diagonal terms, is to require that histories should be expressed in terms of the properties which correspond to the operators which commute with the density martix. This simple remark (which, on a formal level, implies the consistency condition, Eq. (11.32)) is, I believe, essential in understanding the connection between the consistent histories approach and the discussions based on the process of decoherence.

An obvious exact solution suggested by the above considerations – a set of histories based on the evolved states which diagonalise the initial density matrix – does supply a formally acceptable set of consistent histories for a closed system, but it is very unattractive in the cosmological, or indeed, any practical context. The resulting histories do not seem to have much in common with the familiar classical evolutions we have grown accustomed to. It is clear that additional constraints will need to be imposed, and the search for such constraints is under way (Gell-Mann and Hartle, 1990; 1991; Omnès, 1992).

In the decoherence approach presented in this paper the additional constraints on the acceptable sets of consistent histories are supplied through insistence on the observables which correspond to separate systems. This assumption significantly constrains sets of the "properties of interest" (see the contribution of Albrecht to these proceedings as well as Albrecht (1992) for a discussion of this point). Indeed, the restriction is so severe that one can no longer hope for the exact probability sum rules to be satisfied at all times. Thus, instead of the ideal additivity of the probabilities (or ideal diagonality of the density matrix in the set of preferred pointer states) one must settle for approximations to the ideal which violate it at a negligible level almost all the time, and which may violate it at a significant level for a negligible fraction of time.

The picture that emerges from our considerations of decoherence in the preceding sections can be then viewed in at least two complementary ways: An evolving quantum system interacting with its quantum environment is nearly always represented by a density matrix which is almost exactly diagonal in the same preferred set of states. This viewpoint is complementary to the "consistent histories" approach providing that the final set of projections is not imposed *ad hoc*, but, rather, constructed from the set of preferred states.

Insensitivity of the system to measurements proposed in Section 4 as a criterion for classicality emerges as a useful concept unifying these two approaches – the environment-induced decoherence and the consistent histories interpretations. The condition of equality between the evolved density matrix, with its double summation over the pairs of histories, and the restricted sum over the diagonal elements only, Eq. (11.35), can be expressed by noting that the measurements of the properties in terms of which the consistent histories are written does not perturb the evolution of the system. This suggests a physically appealing view of the consistent histories approach in the context of the environment-induced decoherence: Measurements of the preferred observable do not alter the evolution of the system which is already monitored by the environment!

### 11.9 Preferred Basis, Pointer Observables, and the Sum Rules

In the last section we have discussed a connection between the consistency of histories and the notion of insensitivity of a quantum system to measurements. Such insensitivity underlies much of our classical intuition. It has served as an important conceptual input in the discussion of the environment-induced superselection, decoherence, and the preferred pointer basis. If the conceptual link between the foundations of the "many histories interpretation" and the preferred basis of an open system does indeed exist, one should be able to exhibit a still more direct relationship between the formalisms of these two approaches.

Let us consider a system interacting with a large environment. In accord with the preceding discussion, I shall focus on its reduced density matrix $\rho(t) \equiv \rho_{\mathscr{S}}(t)$ which can be always computed by evolving the whole, combined system unitarily from the initial condition, and then tracing out the environment. Throughout much of this section we shall also work under a more restrictive assumption: We shall consider only situations in which the reduced density matrix of the system suffices to compute the decoherence functional. It should be emphasized that this is not a trivial assumption, as the consistent histories formalism was introduced for closed quantum systems and its extension to the open quantum systems (required in essentially all practical applications) leads to difficulties some of which have not yet been even pointed out in the literature until now, let alone addressed.

The nature of this demand is best appreciated by noting that (given a split between the system and the environment and the usual tracing over the environment degrees of freedom) we are demanding that the consistency of the histories for the whole Universe with the events defined through $P_k \otimes 1_\mathscr{E}$ must be closely related to the consistency of the histories with the events $P_k$ in the history of the open system. Using the decoherence functional, the strongest form of such equivalence could be expressed by the equality:

$$
\begin{aligned}
D_{\mathscr{S}\mathscr{E}} &\equiv Tr_{\mathscr{S}\mathscr{E}}\{1_\mathscr{E} \otimes P_l \ldots 1_\mathscr{E} \otimes P_i \ldots \rho_{\mathscr{S}\mathscr{E}} \ldots P_i' \otimes 1_{\mathscr{E}\ldots}\} \\
&= Tr_{\mathscr{S}}\{P_l \ldots P_i \ldots \rho \ldots P_i' \ldots\} \equiv D_{\mathscr{S}}
\end{aligned}
\tag{11.37}
$$

Above, "..." stand for the appropriate evolutions inbetween events (which is unitary for $\mathscr{S}\mathscr{E}$ considered jointly and non-unitary for $\mathscr{S}$ alone). While the above equation has the advantage of stating a requirement with brevity, one should keep in mind not only that the diagonality of the decoherence functional is a more restrictive condition than the one necessary to satisfy the sum rules, but also that exact diagonality is not to be expected in the physically interesting situations. Moreover, when the approximate consistency is enough (as would have to be almost always the case) the equality above could be replaced by an approximate equality, or could hold with a very good accuracy almost always, but be violated for a fraction of time which is sufficiently small not to be worrisome. Finally, it is conceivable that the approximate equality could hold only for a certain sets of projection operators, and not for others. We shall not attempt to analyse these possibilities here. For the time being, we only note that whenever the density matrix of the open system obeys a master equation which is local in time (such as Eq. (11.15)) the reduced density matrix will suffice to express the decoherence functional. These issues will be discussed in a separate paper (Paz and Zurek, 1993).

To begin the discussion of the connection between the pointer basis and the consistent histories approach I shall assume – guided by the experience with the environment-induced superselection – that after a decoherence time the density matrix of the system settles into a state which, regardless of the initial form of the state of the system ends up being nearly diagonal in the preferred basis $P_i$. I shall, for the moment, demand exact diagonality and show that it implies exact consistency.

Let us first consider an initial state which is a projection operator $Q_o$, and a historical property which corresponds to a pointer observable:

$$
\Lambda^i = \sum_i \lambda_i P_i \,,
\tag{11.38}
$$

which commutes with the reduced density matrix:

$$\rho_{Q_o}(t^i) = Tr_{\mathscr{E}}\rho_{\mathscr{S}\mathscr{E}}(t^i) , \tag{11.39}$$

at the first historical instant;

$$[\Lambda^i, \rho_{Q_o}(t_i)] = 0 . \tag{11.40}$$

Above, the lower case "*i*" plays a double role: When in subscript, it is a running index which can be summed over (as in Eq. (11.38)). When in the superscript it is used to label the instant of time (i.e. $t^i$, $t^j$, etc.) or a pointer observable ($\Lambda^i$) at that instant.

Given Eq. (11.38), the density matrix can be written as:

$$\rho_{Q_o}(t^i) = \sum_i p(i|o)\rho_{ii|o} = \sum_i P_i\rho_{Q_o}(t^i)P_i . \tag{11.41}$$

In other words, the system is insensitive to the measurement of the pointer observable at that instant. Above, $p(i|o) = Tr_{\mathscr{S}}P_i\rho_{Q_o}(t^i)P_i$ is the conditional probability that a pointer projector $P_i$ would have been measured at a time $t^i$ given that the system started its evolution in the state $Q_o$ at the initial instant. The conditional density matrix;

$$\rho_{ii|o} = P_i\rho_{Q_o}(t^i)P_i/p(i|o) \tag{11.42}$$

which appears in Eq. (11.41) will, in general, contain additional information about the measurements of higher resolution which could have been carried out. It would reduce to unity when $TrP_i = 1$.

Given this form of the reduced density matrix at $t^i$ it is not difficult to evaluate the contribution of that instant to the decoherence functional for histories including other properties $\{P_k\}$ at the corresponding instants $t^k$:

$$D_{Q_o}(\chi_{i,j,...l}, \chi_{i',j',...l}) = Tr\{...(P_i\rho_{Q_o}(t_i)P_i')...\} = p(i|o)D_{\rho_{ii|o}}(\chi_{j,...l}, \chi_{j',...l})\delta_{ii'} \tag{11.43}$$

Above, the projection is followed by the (nonunitary) evolution which continues until the next projection, and so on, until the final instant $t_l$. The off-diagonal elements of the decoherence functional are always zero because of our simplifying assumption of existence of a perfect pointer basis at each "historical" instant.

Such computation can be carried out step by step until finally one can write;

$$D_{Q_o}(\chi, \chi) = p(l|k...j|i|o)...p(j|i|o)p(i|o) , \tag{11.44}$$

and;

$$D_{Q_o}(\chi \neq \chi') = 0 \tag{11.45}$$

for the case when pointer observables existed and were used to define properties at each historical time instant.

When the initial state was mixed and of the form;

$$\rho(t^o) = \sum_o p(Q_o)Q_o \,, \tag{11.46}$$

the elements of the above formula for the decoherence functional would have to be weighted by the probabilities of the initial states, so that;

$$D_\rho = \sum_o p(Q_o)D_{Q_o} \,. \tag{11.47}$$

Conditional probabilities in Eq. (11.44) can be simplified when the "historical process" is Markovian, so that the probability of the "next" event depends only on the preceding historical event. Delineating the exact set of conditions which must be satisfied for this Markovian property to hold is of interest, but, again, beyond the scope of this paper: Markovian property of the underlying dissipative dynamics is not enough. One can, however, see that the process will be Markovian when the projections are onto rays in the Hilbert space. By contrast, conditional probabilities will be determined to a large extent by the properties in the more distant past when such events have had a better resolution – that is, if they projected onto smaller subspaces of the Hilbert space of the system – than the more recent ones.

We have thus seen than when (i) the density matrix of the system alone is enough to study consistency of histories and (ii) when the system evolves into density matrix which is exactly diagonal in the same set of pointer states each time historical properties are established, then the histories expressed in terms of the pointer projectors are exactly consistent. In the more realistic cases the reduced density matrix will be only approximately diagonal in the set of preferred states. It is therefore of obvious interest to enquire how the deviations from diagonality translate into a partial loss of consistency.

To study this case we suppose now that the reduced density matrix of the system at a time $t^i$ does not satisfy Eq. (11.40), but, rather, is given by;

$$\rho_{Q_o}(t^i) = \sum_i p(i|o)\rho_{ii|o} + \sum_{i\neq i'} P_i \rho_{Q_o}(t^i)P_i' \,. \tag{11.48}$$

Below, we shall set $\varrho_{ii'|o} = P_i \rho_{Q_o}(t^i)P_i'$. Following the similar procedure as before one obtains (for the diagonal contribution at the instant $t^i$) the same formula as before. By contrast, an off-diagonal term also contributes;

$$D_{Q_o}(\chi_{i,j,\ldots l}, \chi_{i',j',\ldots l}) = D_{\varrho_{ii'|o}}(\chi_{j,\ldots l}, \chi_{j',\ldots l}) \,. \tag{11.49}$$

The new "initial density matrices" for such off-diagonal terms are non-hermitean. Nevertheless, they can be evolved using – for example – the appropriate master equation and will contribute to the off-diagonal terms of the decoherence functional. Moreover, the size of their contribution to the decoherence functional can be computed by writing them as a product of a hermitean density matrix and an annihilation operator;

$$\varrho_{ii'} = q_{ii'|o} \, \rho_{ii'|o} \, \Pi_{ii'} \, , \tag{11.50}$$

where:

$$P_i = \Pi_{ii'} \, P_i' \, \Pi_{ii'}^\dagger \, , \tag{11.51}$$

and $q(ii'|o)$ is the trace of this component of $\varrho_{ii'}$ which remains after separating out $\Pi_{ii'}$.

The simplest example of this procedure obtains in the case when the projectors are one-dimensional. Then Eq. (11.50) simplifies and the off-diagonal element of the decoherence functional can be evaluated by a formula similar to the one for diagonal elements:

$$D_{Q_o}(\chi_{i,j,\dots l}, \chi_{i',j',\dots l}) = q(ii'|o) D_{\Pi_{ii'}}(\chi_{j,\dots l}, \chi_{j',\dots l}) \, . \tag{11.52}$$

Expressing the decoherence functional in this form allowes one to estimate the decrease of the potential for interference – for the violations of the sum rules – from the relative sizes of the numerical coeffecients such as $p(i|o)$ and $q(ii'|o)$, products of which establish the "relative weight" of the various on-diagonal and off-diagonal terms. Similar procedure applies also more generally and leads to a formula:

$$D_{Q_o}(\chi_{i,j,\dots l}, \chi_{i',j',\dots l}) = q(ii'|o) D_{\rho_{ii'|o}\Pi_{ii'}}(\chi_{j,\dots l}, \chi_{j',\dots l}) \, . \tag{11.53}$$

Equations (11.43), (11.44) and (11.53) – representing individual steps in evaluation of the decoherence functional – can be repeated to obtain "chain formulas" of the kind exhibited before in this section. We shall not go into such details which are relatively straightforward conceptually but rather cumbersome notationally.

These considerations establish a direct link between the process of decoherence with its preferred observables, corresponding sets of states, etc., and the requirement of consistency as it is implemented by the decoherence functional. It is now clear that – given the assumptions stated early on in this section – the environment induced superselection implies consistency of the histories expressed in terms of the pointer basis states. It should be nevertheless emphasized that the issue of the applicability of the sum rules and the consistency criteria address – in spite of that link – questions which differ, in their physical content, from these which are usually posed and considered as a motivation for the environment-induced decoherence and the resulting effective superselection.

The key distinction arises simply from the rather limited goal of the consistent histories programme – the desire to ascribe probabilities to quantum histories – which can be very simply satisfied by letting the histories follow the unitary evolution of the states which were on the diagonal of the density matrix of the Universe at the initial instant. This obvious answer has to be dismissed as irrelevant (Gell-Mann and Hartle, 1990) since it has nothing to do with the "familiar reality". The obvious question to ask is then whether the approximate compliance with the sum rules (which have to be relaxed anyway to relate the histories to the "familiar reality") would not arise as a byproduct of a different set of more restrictive requirements, such as these considered in the discussions of environment-induced decoherence. In the opinion of this author, the division of the Universe into parts – subsystems – some of which are "of interest" while other act as "the environment" is essential. It is required in stating the problem of interpretation of quantum theory. Moreover, it automatically rules out trivial solutions of the kind discussed above, while, at the same time, allowing for the consistency of histories expressed in terms of the preferred basis. In this context the approximate validity of the sum rules which is of course an important prerequisite for classicality, arises naturally as a consequence of the openness of the systems through the dynamics of the process of decoherence.

## 11.10  Preferred Sets of States and Diagonality

Diagonality of the density matrix in some basis alone is not enough to satisfy the physical criteria of classicality. Pointer states should not be defined as the states which just happen to diagonalise the reduced density matrix at a certain instant. What is crucial (and sometimes forgotten in the discussions of these issues) is the requirement of stability of the members of the preferred basis (expressed earlier in this paper in terms of predictability). This was the reason why the preferred observable was defined in the early papers on this subject through its ability to retain correlations with other systems in spite of the contact with the environment (Zurek, 1981). Hence, the early emphasis on the relation of the pointer observable to the interaction hamiltonian and the analogy with the monitoring of the "nondemolition observables" (Caves et al., 1980).

An additional somewhat more technical comment – especially relevant in the context of the many-histories approach – might be in order. One might imagine that the exact diagonality of the density matrix in some basis (rather than existence of the stable pointer basis) might be employed to guarantee consistency. After all, the property of the density matrix we have used earlier in this section to establish consistency of the histories expressed in terms of the pointer basis at a single instant, Eqs. (11.43)–(11.45) – was based on diagonality alone. This might be possible if one were dealing with a single instant of time, but would come with a heavy price when the evolution of the system is considered: The projection operators which diagonalise the reduced density matrices which have evolved from certain "events" in the history of the system will typically depend on these past events! That is, the set of projectors on the diagonal of $\rho_{Q_o}(t_i)$ will be different from the set of projectors on the diagonal of $\rho_{Q'_o}(t_i)$ even when the initial conditions – say, $Q_o$ and $Q'_o$ – initially commute, or even when they are orthogonal. This is because the evolution of the open system does not preserve the value of the scalar products. Thus, for instance, the implication;

$$[\rho(0), \rho'(0)] = 0 \Leftrightarrow [\rho(t), \rho'(t)] = 0 , \tag{11.54}$$

which is valid for unitary evolutions, does not hold for the evolution of open systems (Paz and Zurek, 1993). This can be established, for example, for the dissipative term of the master equation (11.15).

The implications of this phenomenon for both of the approaches considered here remains to be explored. It seems, however, that its consequences for the consistent histories approach could be quite serious: The formalism underlying the many-histories interpretation of quantum mechanics requires an exhaustive set of exclusive properties which can be defined for the system under study as a whole (see papers by Griffiths, Omnès, Gell-Mann, and Hartle cited earlier). This requirement is perfectly reasonable when the system in question is isolated and does not include observers who decide what questions are being asked – what sets of projections operators will be selected – as is surely the case for this Universe (see Wheeler's "game of twenty questions" in his 1979 paper). It could be perhaps satisfied for macroscopic quantum systems of the sort considered in the context of quantum measurements for finite stretches of time. It is, however, difficult to imagine how it could be satisfied for all the branches of the Universe and "for eternity", especially since the effective hamiltonians which decide what are the sensible macroscopic observables are more often than not themselves determined by the past events (including the dramatic symmetry breaking transitions which are thought to have happened early on in the history of the Universe, as well as much less dramatic, but also far less hypothetical transitions in macroscopic systems). A similar comment can be made about the selection of the same "historical" instants of time in a multifarious Universe composed of branches which specifically differ in the instant when a certain event has occured.

The above remark leads one to suspect that such global (or, rather, universal) orthogonal sets of events at reasonably simultaneous instants of time cannot and satisfy the demands of consistency "globally". Only "local" properties defined in a way which explicitly recognizes the events which have occurred in the past history of the branch have a chance of fulfilling this requirement. This suggests the interesting possibility that the consistent history approach can be carried out with the same set of the event-defining properties and for the same sequence of times only for a relatively local neighbourhood of some branch defined by the common set of essential ingredients (which can be in turn traced to the events which contributed to their presence). Such branches would then also share a common set of records – imprints of the past events reflected in the "present" states of physical systems as well as in the effective hamiltonians which in turn determine the division of the Universe into subsystems, the interactions, and the sorts of pointer observables which will aspire to classicality. Indeed, Omnès as well as Gell-Mann and Hartle have mentioned branch-dependence in their papers. Environment-induced noncommutativity (Paz and Zurek, 1993) forces one to consider this possibility even more seriously.

## 11.11  Records, Correlations and Interpretations

The discussion near the end of the last section forces us to shift the focus from the histories defined in terms of "global" projection operators, abstract "events", and equally abstract "records" related to them and existing "outside" of the Universal state vector to far more tangible "local" events which have shaped a specific evolution of "our" branch, and which are imprinted as the real, accessible records in the states of physical systems. It also emphasizes that even an exceedingly "nonperturbative" strategy of trying to identify consistent causally connected sequences of events appears to have inevitable and identifiable physical consequences.

In a sense, we have come a full circle to the essence of the problem of measurement understood here in a very non-anthropocentric manner, as a correlation between the states of two systems, as an appearance of records within – rather than outside – of the state vector of the Everett-like Universe. Moreover, if our conclusions about the need to restrict the analysis of consistent histories to local bundles of branches derived from the records which are in principle accessible from within these branches are correct, and the emphasis on the coarse-grainings which acknowledge a division of the Universe into systems (existence of which will undoubtedly be branch-dependent) is recognised, the difference between the two ways of investigating classicality considered above begins to fade away.

The central ingredient of the analysis is not just a state of the system, but, rather, a correlation between states of two systems – one of which can be regarded as a record of the other. The states of the systems we perceive as an essential part of classical reality are nothing but a "shorthand" for a correlation existing between these states and a corresponding record which exists in our memory (possibly extended by such external devices as notebooks, RAM's, or – on rare occasions – papers in conference proceedings). This "shorthand" is sufficient only because the states of physical system used as memories are quite stable, so that their reliability (and classicality) can be taken for granted at least in this context. Nevertheless, the only evidence for classicality comes not so much from dealing with the bundles of histories which may or may not be consistent, but, rather, from comparisions of these records, and from the remarkable ability to use the old records as initial conditions which – together with the regularities known as "the laws of physics" – allow one to predict the content of the records which will be made in the future.

If our discussion was really about the stability of correlations between the records and the states (rather than just states) it seems appropriate to close this paper with a brief restatement of the problem from such "correlation" point of view. The entity we are concerned with is not just a to-be-classical system $\mathscr{S}$ interacting with the environment, but, rather, a combination of $\mathscr{S}$ with a memory $\mathscr{M}$, each immersed in the environments $\mathscr{E}_\mathscr{S}$ and $\mathscr{E}_\mathscr{M}$. The density matrix describing this combination must have – after the environment has intervened – a form:

$$\rho_{\mathscr{M}\mathscr{S}} = Tr_{\mathscr{E}_\mathscr{S}} Tr_{\mathscr{E}_\mathscr{M}} \rho_{\mathscr{M}\mathscr{S}\mathscr{E}_\mathscr{M}\mathscr{E}_\mathscr{S}} = \sum p(i,j) P_i^\mathscr{M} P_j^\mathscr{S} , \qquad (11.55)$$

where $p(i, j)$ must be nearly a Kronecker delta in its indices;

$$p(i, j) \approx \delta_{ij} , \qquad (11.56)$$

so that the states of memory $\{P_i^\mathscr{M}\}$ can constitute a faithful record of the states of the system.

One might feel that these statements are rather transparent and do not have to be illustrated by equations. Nevertheless, the two equations above clarify some of the aspects of the environment-induced decoherence and the emergence of the preferred sets of states which were emphasized from the very beginning (Zurek, 1981; 1982) but seem to have been missed by some of the readers (and perhaps even an occasional writer) of the papers on the subject of the transition from quantum to classical. The most obvious of these is the need for "consensus" between the preferred sets of states selected by the processes of decoherence occurring simultaneously in the memory and in the system, and the fact that memory is a physical system subjected to decoherence inasmuch as any other macroscopic and open system.

The second point is somewhat more subtle (and, consequently, it has proved to be more confusing): The form of the classical correlation expressed above by the two equations precludes any possibility of defining preferred states in terms of the density matrix of the system alone. The key question in discussing the emergence of classical observables is not the set of states which are on the diagonal after everything else (including the memory $\mathscr{M}$) is traced out, but, rather, the set of states which can be faithfully recorded by the memory – so that the measurement repeated on a timescale long compared to the decoherence time will yield a predictable result. Hence, the focus on predictability, on the form of the interaction hamiltonians, and on the minimum entropy production, and the relative lack of concern with the states which remain on the diagonal "after everything but the system is traced out". The diagonality of the density matrix – emphasized so much in many of the papers – is a useful symptom of the existence of the preferred basis when it occurs (to a good approximation) in the same basis. For, only existence of such stable preferred set of states guarantees the ability of the to be classical systems to preserve the classical amount of correlations. I shall not belabor the discussion of this point any further: The reader may want to go back to the discussion of the role of the pointer basis in preserving some of the correlations in measurements (Zurek, 1981; 1982; 1983), as well as to the earlier sections where some of the aspects of this issue were considered.

Finally, the most subtle point clarified – but only to some extent – by the above two equations is the issue of the perception of a unique reality. This issue of the "collapse of the wave packet" cannot be really avoided: After all, the role of an interpretation is to establish a correspondence between the formalism of a theory describing a system and the results of the experiments – or, rather, the records of these results – accessible to the observer. And we perceive outcomes of measurements and other events originating at the quantum level alternative by the alternative, rather than all of the alternatives at once.

An exhaustive answer to this question would undoubtedly have to involve a model of "consciousness", since what we are really asking concerns our (observers) impression that "we are conscious" of just one of the alternatives. Such a model of consciousness is presently not available. However, we can investigate what sort of consequences about the collapse could be arrived at from a rather modest and plausible assumption that such higher mental processes are based on information processing involving data present in the accessible memory. "To be conscious" must then mean that the information processing unit has accessed the available data, that it has performed logical operations including copying of certain selected records. Such copying results in redundant records: The same information is now contained in several memory cells. Therefore, questions concerning the outcome of a particular observation can be confirmed (by comparing different records) or verified (by comparing predictions based on the past records with the outcomes of new observations) only within the branches.

In a sense, the physical state of the information processing unit will be significantly altered by the information-processing activity and its physical consequences – its very state is to some degree also a record. Information processing which leads to "becoming conscious" of different alternatives will result in computers and other potentialy conscious systems which differ in their physical configuration, and which, because their state is a presently accessible partial record of their past history, exist in different Universes, or rather, in different branches of the same Universal state vector. These differences will be especially obvious when the machinery controlled by the computer reacts to the different outcomes in different ways. A particularly dramatic example of different fates of the machinery is afforded by the example of Schrödingers Cat: The cat in the branches of the Universe in which it is dead cannot be aware of any of the two alternatives because its "machinery" was significantly modified by the outcome of the quantum event – it became modified to the extent which makes it impossible for the cat to process information, and, hence, to be aware of anything at all.

I believe that similar, but (fortunately) much more subtle modification of both the records and the identity – the physical state – of the recording and information processing "observer" (animated or otherwise) are the ultimate cause of the "collapse". This mechanism for the collapse is closely coupled to the phenomenon of decoherence, which makes the changes of the machinery irreversible. The selection of the alternatives arises because of the split into the machinery (including the memory and the information processing unit) and "the rest of the Universe", through the environment induced superselection. The possible set of alternatives of both "what I know" and "what I am" is then fixed by the same process and at the same time. It should be emphasized that these last few paragraphs are considerably more speculative than the rest of the paper, and should be treated as such by the reader.

The interpretation that emerges from these considerations is obviously consistent with the Everett's "Many Worlds" point of view. It is supplemented by a process – decoherence – which arises when the division of the Universe into separate entities is recognised. Its key consequence – emergence of the preferred set of states which can exist for time long compared to the decoherence timescale for a state randomly selected from the Hilbert space of the system of interest – is responsible for the selection of the individual branches. As reported by an observer, whose memory and state becomes modified each time a "splitting" involving him as the system takes place, the apparent collapses of the state vector occur in complete accord with the Bohr's "Copenhagen Interpretation". The role of the decoherence is to establish a boundary between quantum and classical. This boundary is in principle movable, but in practice largely immobilized by the irreversibility of the process of decoherence (see Zeh, 1991) which is in turn closely tied to the number of the degrees of freedom coupled to a macroscopic body. The equivalence between "macroscopic" and "classical" is then validated by the decoherence considerations, but only as a consequence of the practical impossibility of keeping objects which are macroscopic perfectly isolated.

*Added note:* An earlier version of this paper has appeared as an invited contribution in *Prog. Theor. Phys.* **89**, 281–312 (1992).

## Acknowledgements

I would like to thank Robert Griffiths, Salman Habib and Juan Pablo Paz for discussions and comments on the manuscript. I also wish to acknowledge the hospitality of the Santa Fe Institute.

## Discussion

**Halliwell**  You have the nice result that the *off-diagonal* terms of the density matrix (i.e., parts representing interferences) become exponentialy suppressed in comparision with the on-diagonal terms. But why does this mean that you have decoherence? Why is it the peaks of the density matrix that matter? In short, what exactly is your definition of decoherence? In the approach of Gell-Mann and Hartle, decoherence is defined rather precisely by insisting that the probability sum rules are satisfied by the probabilities of histories. These rules are trivially satisfied, always, for histories consisting of one moment of time. You seem to have only one moment of time. How is your notion of decoherence reconciled with theirs?

**Zurek**  Very simply. I prefer to use the word "consistency" employed in the earlier work to describe histories which satisfy probability sum rules (i.e. Griffiths, 1984) and adopted by the others (Omnés, 1991; 1992, and references therein).

I feel that the term *decoherence* is best reserved for the *process* which destroys quantum coherence when a system becomes correlated with the "other" degrees of freedom, its environment (which can include also internal degrees of freedom as would certainly be the case for the environment of a pointer in a quantum apparatus). In spite of this "conservative" attitude to the nomenclature, I am partial to the "decoherence functional" Gell-Mann and Hartle have employed to formulate their version of consistency conditions. This is because much of the motivation for their work goes beyond simple consistency, and in the direction which, when pursued for open systems, will result in a picture quite similar to the one emerging from the recognition of the role of the process of decoherence (see Sections 8 and 9 above).

Now as for your other queries; (i) Decoherence is a process, hence it happens in time. In recognition of this, the discussions of the effects of decoherence have always focused on correlations and on the interaction hamiltonians used, for example, to define preferred basis, rather than on the instantaneous state of the density matrix, which can be always diagonalised. (ii) Consistency rules are "trivially satisfied" at a single moment of time because consistency conditions are always formulated in a manner which effectively results in a "collapse of the state vector" onto the set of alternatives represented by projectors at both the beginning and the end of each history. And applying a formalism which was developed to deal with histories to an object defined at an instant is likely to be a bad idea – and yield results which are either wrong or trivial. (iii) Nevertheless, the effect of decoherence process can be studied at an instant through the *insensitivity* of the system to measurements of the preferred observables. This insensitivity can be related to the existence of classical correlations and seems to be a natural criterion for the *classicality* of the system. It sets in after a decoherence time irregardless of the initial state of the system.

**Griffiths**    The results you have discussed are very interesting, and add to our understanding of quantum mechanics. On the other hand, I think that they could be stated in a much more clearer way if you would use the "grammar" of consistent histories. For example, in the latter interpretation the choice of basis is not left to the "environment"; it is a choice made by the theoretician. In fact, you choose the basis that interests you and, from the consistent histories point of view, what you then did was to show that in this choice of basis, and after a suitably short time, etc., the questions you want to ask form a consistent history. But in order to think clearly about a problem, it is useful to use a clear language, rather than talking about "fairies", which is what one tends to do if one carries over the traditional language of quantum measurement.

**Zurek**    I tend to believe that even though theoreticians have more or less unlimited powers, they cannot settle questions such as the one posed by Einstein in a letter to Born.†
Moreover, I find it difficult to dismiss questions such as this as "fairies". Furthermore, I feel rather strongly that it is the openness of the macroscopic quantum systems which – along with the form of the interactions – allowes one to settle such issues without appealing to anything outside of physics.

---

† The quote from a 1954 letter from Albert Einstein to Max Born "Let $\Psi_1$ and $\Psi_2$ be solutions of the same Schrödinger equation. ... When the system is a macrosystem and when $\Psi_1$ and $\Psi_2$ are 'narrow' with respect to the macrocoordinates, then in by far the greater number of cases this is no longer true for $\Psi = \Psi_1 + \Psi_2$. Narrowness with respect to macrocoordinates is not only *independent* of the principles of quantum mechanics, but, moreover, incompatible with them..." was on one of the transparencies shown during the talk, quoted after the translation of E. Joos from *New Techniques and Ideas in Quantum Measurement Theory*, D. M. Greenberger, ed., (New York Acad. Sci., 1986).

I agree with you on the anticipated relationship between the consistent histories and decoherence, and on the need to establish correspondence between these two formalisms. I believe that they are compatible in more or less the manner you indicate, but I do not believe that they are equivalent. In particular, consistency conditions alone are not as restrictive as the process of decoherence tends to be. They can be, for example, satisfied by violently non-classical histories which obtain from evolving unitarily projection operators which initially diagonalise density matrix. The relation between decoherence and consistency clearly requires further study.

**Lebowitz**  I agree with much of Griffiths's comment although I disagree with him about "fairies". I believe that all questions are legitimate and, as pointed out by Bell and fully analyzed in a recent paper by Durr, Goldstein, and Zanghi (*J. Stat. Phys.* **67**, 843 (1992)) the Bohm theory which assigns complete reality to particle positions gives a completely clear explanation of *non-relativistic* quantum mechanics which is free from both problems of measurement and from "having to talk to ones lawyer" before answering some of the questions. Whether Bohm's theory is right I certainly do not know, but it does have some advantages of clarity, which should be considered when discussing "difficulties" of quantum mechanics.

**Zurek**  I am always concerned with whether the relativistic version of the Bohm–de Broglie theory can even exist. (How could photons follow anything but straight lines? Yet, in order to explain double-slit experiments a lá Bohm–de Broglie one tends to require rather complicated trajectories!) Moreover, having grown accustomed to the probabilities in quantum mechanics, I find the exact causality of particle trajectories unappealing.

**Omnès**  Concerning the basis in which the reduced density operator becomes diagonal: For a macroscopic system this question is linked with another one, which is to define the "macrocoordinates" from first principles. My own guess (relying on the work of the mathematician Charles Fefferman) is that one will find diagonalisation in terms of both position and momentum, as expressed by quasiprojectors representing classical properties.

**Zurek**  I agree with your guess and I can base it on the special role played by coherent states in a quantum harmonic oscillator (as it is shown in Section 7 of the paper in more detail), but I believe that decoherence is sufficient to understand similar spreads in position and momentum. The symmetry between $x$ and $p$ is broken by the interaction hamiltonian (which depends on position) but partialy restored by the rotation of the Wigner function corresponding to the state in the phase space by the self-Hamiltonian. I am, however, not familiar with the work of Fefferman. Therefore, I cannot comment on your motivation for your guess.

**Bennett**  If the interaction is not exactly diagonal in position, do the superpositions of position still decohere?

**Zurek**  This will depend on details of the interaction, self-Hamiltonian, etc., as discussed in the paper and references. It is, however, interesting to note that when the interaction hamiltonian is periodic in position (and, therefore, diagonal in momentum) and the particles are otherwise free (so that the self-Hamiltonian is also diagonal in momentum) diagonalisation in momentum (and delocalisation in position) is expected for the preferred states. Such a situation occurs when electrons are travelling through a regular lattice of a crystal, and the expectation about their preferred states seems to be borne out by their behavior.

**Hartle**   I would like to comment on the connection between the notion of "decoherence" as used by Wojtek in his talk and the concept of "consistent histories" discussed by Bob Griffiths and Roland Omnès that was called "decoherent histories" in the work by Murray Gell-Mann and myself. As Griffiths and Omnès discussed yesterday, probabilities can be assigned to sets of alternative coarse-grained histories if, and only if, there is nearly vanishing interference between the individual members of the set, that is, if they "decohere" or are "consistent".

For those very special types of coarse grainings in which the fundamental variables are divided into a set that is followed and a set that is ignored it is possible to construct a reduced density matrix by tracing the full density matrix over the ignored variables. Then one can show that typical mechanisms that effect the decoherence of histories (as in the Feynman–Vernon, Caldeira–Leggett type of models) also cause the off-diagonal elements of the reduced density matrix to evolve to small values (see, e.g., Hartle 1991). However, the approach of the reduced density matrix to diagonality should not be taken as a fundamental definition of decoherence. There are at least two reasons: First, realistic coarse-grainings describing the quasiclassical domain such as ranges of averaged densities of energy, momentum, etc. do not correspond to a division of the fundamental variables into ones that are distinguished by the coarse-graining and the others that are ignored. Thus, in general and realistic cases there is no precise notion of environment, no basis associated with coarse-graining, and no reduced density matrix.

The second reason is that decoherence for histories consisting of alternatives at a single moment of time is automatic. Decoherence is non-trivial only for histories that involve several moments of time. Decoherence of histories, therefore, cannot be defined at only one time like a reduced density matrix. It is for these reasons that the general discussion of decoherence (consistency) is important. [Note: for further treatment of the terms "decoherence" and "consistent histories", and the concepts occurring in this discussion, please see the note added in proof by Gell-Mann and Hartle on p. 341]

**Zurek**   I have a strong feeling that most of the disagreement between us that you outline in your comment is based on a difference in our vocabularies. You seem to insist on using words "decoherence" and "consistency" interchangeably, as if they were synonymous. I belive that such redundancy is wasteful, and that it is much more profitable to set up a one-to-one correspondence between the words and concepts. I am happy to follow the example of Griffiths and Omnès and use the term "consistency" to refer to the set of histories which satisfy the probability sum rules. I would like, on the other hand, to reserve the term "decoherence" to describe the process which results in the loss of quantum coherence whenever (for example) two systems such as a "system of interest" and an "environment" are becoming correlated as a result of a dynamical interaction. This distinction seems to be well established in the literature (see, in addition to Omnès, also Albrecht, Conradi, DeWitt (in these proceedings) Griffiths (including a comment after DeWitts talk), Fukuyama, Habib, Halliwell, Hu, Kiefer, Laflamme, Morikawa, Padmanabhan, Paz, Unruh, and Zeh, and probably quite a few others).

I see little gain and a tremendous potential for confusion in trying to change this existing usage, especially since the term "decoherence" is well-suited to replace phrases such as "loss of quantum coherence" or "dephasing" which were used to describe similar phenomena (although in more "down to earth" contexts) for a very long time. Thus, I am in almost complete agreement with much of your comment providing that we agree to follow Griffiths and Omnès and continue to use the word "consistent" when referring to histories which satisfy the sum rules. And since you indicate that the two words can be used interchangeably, I assume that you will not object to this proposal. Having done so (and after re-stating your comment with the appropriate substitutions) I find only a few rather minor items which require further clarification.

The procedure required to distinguish the system from the environment is one of them. You seem to insist on such distinction appearing in the *fundamental* variables (which, I assume, would probably take us well beyond electrons and protons, to some version of the string theory). I do not believe that such insistence on fundamental variables is necessary or practical. Indeed, the questions addressed in the context of transition from quantum to classical usually concern variables such as the position and momentum of the center of mass, which are not fundamental, but are nevertheless coupled to the rest of the Hilbert space, with the environment consisting in part of the "internal variables". For the hydrodynamic variables such split is accomplished by defining appropriate projection operators which correspond to the averages (see, e. g., "Equilibrium and Nonequilibrium Statistical Mechanics" by Radu Balescu, as well as papers by Zwanzig, Prigogine and Résibois, and others). Similarly, in Josephson junctions (which motivated the work of Caldeira and Leggett on the influence functional) the environment is "internal" and at least one of the macroscopic quantum observables (the current) is hydrodynamic in nature. Indeed, I am not familiar with the observables for which it is in principle impossible to define a density matrix, and to use projection operator techniques to split the Hilbert space into the system and the environment.

Finally, as to the question whether (1) decoherence and (2) consistency can be studied at a single moment of time, I would like to note that: (i) Decoherence is a process, which occurs in time. Its effects can be seen by comparing the initial density matrix with the final one. The corresponding instantaneous rate of the entropy increase, as well as the discrepancy between the instantaneous density matrix and the preferred set of states can be also assessed at an instant, as discussed in Section 7 above. (ii) I agree with you that it is pointless to study consistency of histories which last an instant, since the first and the last set of the projection operators are "special" and, in effect, enforce a "collapse of the wave packet" (see Sections 8 and 9 for details), and instantaneous histories are rather degenerate. Moreover, (iii) I could not find a better example than the last part of your comment to illustrate why "decoherence" and "consistency" should be allowed to keep their original meanings, so that we can avoid further linguistic misunderstandings and focus on the physics instead.

## References

A. Albrecht (1992); "Investigating Decoherence in a Simple System", *Phys. Rev.* **D46**, 550.

A. O. Caldeira and A. J. Leggett (1983); "Path Integral Approach to Quantum Brownian Motion", *Physica* **121 A**, 587.

A. O. Caldeira and A. J. Leggett (1985); "Influence of Damping on Quantum Intereference: An Exactly Soluble Model", *Phys. Rev.* **A 31**, 1057.

C. M. Caves, K. S. Thorne, R. W. P. Drewer, V. D. Sandberg, M. Zimmermann (1980); "On a Measurement of a Weak Force Coupled to a Quantum Harmonic Oscillator", *Rev. Mod. Phys.* **52**, 341.

M. Gell-Mann and J. B. Hartle (1990); "Quantum Mechanics in the Light of Quantum Cosmology", p. 425 in *Complexity, Entropy, and the Physics of Information*, W. H. Zurek, editor (Addison-Wesley).

M. Gell-Mann and J. B. Hartle (1991); "Alternative Decohering Histories in Quantum Mechanics", in *Proceedings of the 25th International Conference on High-Energy Physics*, K. K. Phua and Y. Yamaguchi, eds. (World Scientific, Singapore).

R. J. Griffiths (1984); "Consistent Histories and the Interpretation of Quantum Mechanics" *J. Stat. Phys.* **36**, 219.

J. B. Hartle (1991); "The Quantum Mechanics of Cosmology", in *Quantum Cosmology and Baby Universes* S. Coleman, J. B. Hartle, T. Piran, and S. Weinberg, eds. (World Scientific, Singapore).

B. L. Hu, J. P. Paz, and Y. Zhang (1992); "Quantum Brownian Motion in a General Environment", *Phys. Rev.* **D45**, 2843.

E. Joos and H. D. Zeh (1985); "The Emergence of Classical Properties Through Interaction with the Environment", *Z. Phys.* **B 59**, 223.

R. Omnès (1991) "From Hilbert Space to Common Sense: A Synthesis of Recent Progress in the Interpretation of Quantum Mechanics" *Ann. Phys. (N. Y.)* **201**, 354.

R. Omnès (1992) "Consistent Interpretations of Quantum Mechanics"; *Rev. Mod. Phys.* **64**, 339.

J. P. Paz, S. Habib, and W. H. Zurek (1993); "Decoherence and the Reduction of the Wavepacket", *Phys. Rev.* **D47**, 488.

J. P. Paz and W. H. Zurek (1993); "Environment-Induced Decoherence, Classicality, and Consistency of Quantum Histories", *Phys. Rev.* **D48**, 2728.

R. Penrose (1989); *The Emperors New Mind* (Oxford University Press, Oxford).

W. G. Unruh and W. H. Zurek (1989); "Reduction of the Wavepacket in Quantum Brownian Motion", *Phys. Rev.* **D 40**, 1071.

J. von Neumann (1932); *Mathematical Foundations of Quantum Mechanics*, English translation by Robert T. Beyer (1955; Princeton University Press, Princeton).

J. A. Wheeler (1979); "Frontiers of Time", in *Problems in the Foundations of Physics, Proceedings of the International School of Physics "Enrico Fermi"* (Course 72), N. Toraldo di Francia, ed. (North Holland, Amsterdam).

H. D. Zeh (1971); "On the Irreversibility of Time and Observation in Quantum Theory", in *Enrico Fermi School of Physics* **IL**, B. d'Espagnat, ed. (Academic Press, New York).

H. D. Zeh (1989) *The Physical Basis for the Direction of Time* (Springer, Heidelberg).

W. H. Zurek (1981); "Pointer Basis of Quantum Apparatus: Into What Mixture Does the Wave Packet Collapse?", *Phys. Rev.* **D 24**, 1516.

W. H. Zurek (1982); "Environment-Induced Superselection Rules", *Phys. Rev.* **D 26**, 1862.

W. H. Zurek (1983); "Information Transfer in Quantum Measurements: Irreversibility and Amplification", p. 87 in *Quantum Optics, Experimental Gravitation, and the Measurement Theory*, P. Meystre and M. O. Scully, eds. (Plenum, New York).

W. H. Zurek (1984); "Reduction of the Wavepacket: How Long Does it Take?", lecture at the NATO ASI *Frontiers of the Nonequilibrium Statistical Mechanics* held in June 3–16 in Santa Fe, New Mexico, published in the proceedings, G. T. Moore and M. O. Scully, eds. (Plenum, New York, 1986).

W. H. Zurek (1986); "Reduction of the Wave Packet and Environment - Induced Superselection", p. 89 in *New Techniques and Ideas in Quantum Measurement Theory*, D. M. Greenberger, ed., (New York Academy of Sciences, New York).

W. H. Zurek (1989); "Algorithmic Randomness and Physical Entropy", *Phys. Rev.* **A 40**, 4731.

W. H. Zurek (1991); "Decoherence and the Transition From Quantum to Classical", *Physics Today*, **44** (10), 36.

W. H. Zurek, S. Habib, and J. P. Paz (1993); "Coherent States Via Decoherence", *Phys. Rev. Lett.* **70**, 1187.

# 12

# Decoherence in Quantum Brownian Motion

Juan Pablo Paz

*T6–Los Alamos National Laboratory*
*Los Alamos, NM 87545, USA*

## Abstract

We examine the dependence of decoherence on the spectral density of the environment as well as on the initial state of the system. We use two simple examples to illustrate some important effects.

## 12.1 Introduction

Decoherence plays a major role in the transition from quantum to classical and has attracted much attention in recent years (see Zurek (1991)). The analysis of this process may allow us to understand in detail the mechanism that prevents observation of some quantum systems in superpositions of macroscopically distinguishable states. In the light of new technologies it can also help us to devise experiments to probe the fuzzy boundary between the quantum and the classical world. The interaction with an external environment is the mechanism responsible for the supression of quantum interference effects. Therefore, there are several questions that arise naturally: How dependent on the environment decoherence is? What are the time scales involved in this process? How are some preferred states of the system dynamically chosen? In this paper we will report on recent work where some of these questions are addressed.

As a first point, let us clarify what we mean here by decoherence. Within the Consistent Histories formulation of quantum mechanics developed by Griffiths, Omnès, Gell–Mann and Hartle (see contributions in this proccedings) the term decoherence is used to refer to the condition that, if satisfied, allows us to assign probabilities to members of sets of coarse grained histories of a closed system. The Decoherence Functional is the basic diagnostic tool used in this framework. On the other hand, in previous works originated in quantum measurement theory, a different notion of decoherence was used. Measurement devices are always open systems that interact with external environments. This interaction dynamically selects a preferred set of states of the apparatus, the so–called pointer basis. This is, in some sense, the set of the most stable states: if the aparatus is prepared in a pointer state, the interaction with the environment has a minimal effect and almost no predictive power is lost. On the contrary, if the initial state is a superposition of pointer states, the interaction with the environment induces correlations and the state of the system tends to evolve into a mixture of pointer states. This process was called decoherence and this is the sense in which we will use this word here. Within this context, there are several important issues that require further attention. The most important one seems to be the definition of an appropriate measure of stability that may be used to determine the pointer states (see Zurek's contribution in this conference). We will study a model describing a particle interacting with an environment formed by a collection of harmonic oscillators. In this case the pointer states of the particle seem to be closely related to coherent states and decoherence is the process that supresses interference between coherent states (our diagnostic tool will be described later). The action of the model is the following:

$$S[x,q] = \int\limits_0^t ds \left[ \frac{1}{2}M\left(\dot{x}^2 - \Omega_0^2 x^2\right) + \sum_n \frac{1}{2}m_n(\dot{q}_n^2 - \omega_n^2 q_n^2) - \sum_n C_n x q_n \right] \qquad (12.1)$$

We will assume that there are no initial correlations between the system and the environment (i.e. the initial density matrix factorizes) and that the initial state of the environment is in thermal equilibrium. In Section 2 we will describe generic features of the evolution of a system interacting with a general environment. We will study in detail the case in which the initial state is a superposition of two coherent states. In Section 3 we will illustrate the fact that decoherence strongly depends on some properties of the environment. We will also illustrate in what sense position is a preferred observable in the model (where position eigenstates are not pointer states). In the Appendix we outline a simple derivation of the master equation for a general environment.

## 12.2 General Properties of the Reduced Dynamics

Due to the interaction with the environment, the evolution of the system is non–unitary since it is affected by a stochastic noise and a "dissipative" force (the word dissipation is used here in a rather vague sense). Noise and dissipation, are entirely determined by two properties of the environment: the spectral density $I(\omega)$ and the initial temperature. The spectral density, defined as $I(\omega) = \sum_n \delta(\omega - \omega_n)C_n^2/2m_n\omega_n$, characterizes the number density of oscillators in the environment and the strength of their coupling with the system. Therefore, in order to analyze how decoherence depends on the environment we can study how this process changes when varying $I(\omega)$ and the temperature since these are the only two environmental properties "seen" by the system.

Luckily enough, the reduced dynamics has some very general features that are entirely independent of the spectral density and the temperature. One of the most striking and important ones is the fact that the reduced density matrix *always* satisfies a master equation that can be written as follows (we use $\hbar = 1$):

$$i\,\partial_t \rho_r(x, x', t) = \ <x|[H_{ren}(t), \rho_r]|x'> -i\gamma(t)(x - x')(\partial_x - \partial_{x'})\,\rho_r(x, x', t)$$
$$- i\,D(t)(x - x')^2\,\rho_r(x, x', t) + f(t)(x - x')(\partial_x + \partial_{x'})\,\rho_r(x, x', t) \quad (12.2)$$

This equation depends on the spectral density and the temperature only through the coefficients appearing in the right hand side: the physical frequency entering in $H_{ren}(t)$, the friction coefficient $\gamma(t)$ and the diffusion coefficients $D(t)$ and $f(t)$ are time-dependent functions that vanish initially and depend on the environment in a fairly complicated way.

The validity of (12.2) for a general environment at arbitrary temperature has been recently demonstrated by Hu et al. (1992) and is an interesting discovery that generalizes previous results concerning the nature of the master equation (see Unruh and Zurek (1989), Caldeira and Leggett (1981), Haake and Reiboldt (1988)). The result is also surprising since a general environment generates a non–Markovian evolution for which one expects highly nonlocal integral kernels in the master equation. However, for this model the non–Markovian effects can be fully encoded in the time dependence of the coefficients. The interested reader can find a simple proof of equation (12.2) (the simplest I could think of) in Appendix 1.

Equation (12.2) is a very useful tool to study generic properties of the evolution and can be exactly solved for some simple initial conditions. To study decoherence we will consider the following initial superposition of coherent states:

$$\Psi(x, t = 0) = N \exp\left(-\frac{(x - L_0)^2}{2\delta^2} + iP_0 x\right) + N \exp\left(-\frac{(x + L_0)^2}{2\delta^2} - iP_0 x\right) \quad (12.3)$$

where $N$ is a constant. In this case it is possible to solve the master equation and show that the Wigner function constructed from the reduced density matrix is:

$$W(x, p, t) = W_1(x, p, t) + W_2(x, p, t) + W_{int}(x, p, t)$$

where

$$W_{\genfrac{}{}{0pt}{}{1}{2}}(x, p, t) = \ \bar{N}^2\,\frac{\delta_2}{\delta_1}\exp\left(-\frac{(x \mp x_c)^2}{\delta_1^2} - \delta_2^2(p \mp p_c - \beta(x \mp x_c))^2\right)$$

$$W_{int}(x, p, t) = \ 2\bar{N}^2\,\frac{\delta_2}{\delta_1}\exp\left(-\frac{x^2}{\delta_1^2} - \delta_2^2(p - \beta x)^2\right)$$
$$\cos\left(\phi_p p + (\phi_x - \beta\phi_p)x\right)\,\exp(-A_{int}) \quad (12.4)$$

The functions $x_c(t)$, $P_c(t)$, $\delta_{\frac{1}{2}}(t)$, $\beta(t)$, $\phi_x(t)$, $\phi_p(t)$ and $A_{int}(t)$ depend on the environment (and on the constants $L_0$, $P_0$ and $\delta$ that appear in (12.3)) in a rather complicated way. For the sake of brevity, we will not discuss here the behavior of all these functions (see Paz et al (1993) for details) but concentrate on $A_{int}$ which is the only relevant for decoherence. Thus, to quantify the importance of interference at a given time we will use the peak to peak ratio between the interference and the direct terms in the Wigner function, a quantity closely related to $A_{int}$:

$$\exp(-A_{int}) = \frac{1}{2} \frac{W_{int}(x,p)|_{peak}}{\left(W_1(x,p)|_{peak} W_2(x,p)|_{peak}\right)^{1/2}} \tag{12.5}$$

As the two initial wave packets have a finite overlap, the above function satisfies $A_{int} \leq \delta^2 P_0^2 + L_0^2/\delta^2$. The system decoheres when $A_{int}$ grows to a value that is large with respect to unity (which can only occur if the initial peaks are well separated, i.e. $\delta^2 P_0^2 + L_0^2/\delta^2 \gg 1$).

To analyze how the evolution of $A_{int}$ is affected by the environment it is convenient to use the master equation (12.2) to show the following identity:

$$\dot{A}_{int} = D(t)\phi_p^2 - 2f(t)\phi_p(\phi_x - \beta\phi_p) \tag{12.6}$$

The first term carries the effect of normal diffusion and always produces decoherence since increases the value of $A_{int}$. On the contrary the sign of the second term in (12.6) may vary in time depending upon the relation between $\phi_p$ and $\phi_x$. Equation (12.6) can be approximately solved if one neglects the anomalous diffusion and considers $D(t)$ as a constant, two conditions met by an ohmic environment in the high tempearture regime (see next section). In this case it can be shown that, for an initial state with $P_0 = 0$, $A_{int}(t) \simeq 4L_0^2 Dt/(1 + 4D\delta^2 t)$ and that the "decoherence rate" is $\Gamma_{dec} = 4L_0^2 D \simeq 8L_0^2 m\gamma_0 k_B T$ (see Paz et al. (1993)). However, this solution is no longer valid when one moves away from the ohmic environment in the high temperature regime or when considers more general initial states. In fact, the behavior of $A_{int}$ strongly depends on the initial conditions (that enter into (12.6) through the functions $\phi_x$ and $\phi_p$ whose initial data are $\phi_x = P_0$, $\phi_p = L_0$) and decoherence will be drastically different in the case $L_0 = 0$, $P_0 \neq 0$ where the two initial gaussian are spatially separated than when $P_0 = 0$ and $L_0 \neq 0$ (where the coherent states are separated in momentum).

## 12.3 Decoherence and the Environment

A wide and interesting class of environments is defined by a spectral density of the form $I(\omega) = \frac{2m\gamma_0}{\pi} \frac{\omega^n}{\Lambda^{n-1}} \exp(-\frac{\omega^2}{\Lambda^2})$ where $\Lambda$ is a high frequency cutoff and $n$ is an index that characterizes different environments. We will consider two examples: $n = 1$ which is the largely studied ohmic environment (Caldeira and Leggett (1985)) and $n = 3$ which is a supra–ohmic environment used to model the interaction between defects and phonons in metals, Grabert et al (1988), and also to mimic the interaction between a charge and its own electromagnetic field, Barone and Caldeira (1991). Using these two environments we want to illustrate how strongly decoherence depends on the spectral density. It can be shown that the process is much more inefficient in the supraohmic than in the ohmic case because the final value of the diffusion coefficient $D(t)$ is much smaller in the former than in the latter environment (as $n = 3$ corresponds to a bath of oscillators with an infrared sector substantially weaker than $n = 1$, the dissipative and diffusive effects are expected to be weaker). The time dependence of the diffusion coefficient for these two environments has been described by Hu et al (1992) and has a rather generic feature: $D(t)$ vanishes initially and develops a very strong peak in a time scale of the order of the collision time $\tau_\Lambda = 1/\Lambda$. Its value after the initial peak is $D(\tau_\Lambda) \simeq m\gamma_0\Lambda$, approximately the same for all environments. After this initial cutoff dominated regime, $D(t)$ approaches (in a dynamical time scale) an asymptotic value that depends on the environment (in the high temperature regime, $D(t) \rightarrow 2m\gamma_0 k_B T$ for $n = 1$ while vanishes for $n = 3$). Thus, there is no generic long time behavior but a quite universal short time regime. One may thus wonder if this general initial behavior produces a rather universal decoherence. We will argue here that this is not the case. The impact of the initial peak has been analyzed in detail (see Unruh and Zurek (1989)) and it was shown that in some cases may completely wipe out the interference effects. However, the physical significance of the decoherence produced by the initial peak is rather questionable since this jolt is certainly related to the initial conditions that do not contain correlations between the system and its environment. In fact, such correlations are likely to wash out the peak, Grabert et al (1988).

To discredit even more the role of the initial peak on decoherence we would like to point out that its effect can be made completely innocuous by appropriately choosing the initial conditions for the system. This is well illustrated by the supra–ohmic environment where the asymptotic value of the diffusion coefficient is too small to produce decoherence and all the effect, if any, should come from the initial peak. In Figure 12.1 we plotted $A_{int}$ for the ohmic and supraohmic environments. We considered an harmonic oscillator with renormalized frequency $\Omega_r$ and fixed $\gamma_0 = 0.3\Omega_r$, $\Lambda = 500\Omega_r$ and $k_B T = 25000\Omega_r$ (high temperature regime).

We can notice that in the ohmic environment decoherence is very fast. For the initial condition I ($L_0 = 3\delta, P_0 = 0$) it takes place in a time of the order of $\tau_\Lambda$ while for condition II ($L_0 = 0, P_0 = 3/\delta$) it requires a time that is also much smaller than $\Omega_r^{-1}$. On the other hand, in the supraohmic environment of Figure 12.1.b decoherence goes as in the ohmic case for condition I while no net decoherence is achieved for condition II. In this case the initial growth of $A_{int}$ is followed by a plateau and a decreasing regime during which coherence is recovered! The reason for the drastic difference between the fate of conditions I and II in the supraohmic environment is clear: decoherence can only be produced by the initial peak but the interaction between the system and the environment is initially effective only if the two coherent states are spatially separated. The non-monotonic behavior of $A_{int}$ seen in curve (II) of Figure 12.1.b is due to the anomalous diffusion that cannot produce any net decoherence since the sign of the second term in the right-hand side of (12.6) changes with time.

The above example not only illustrates the strong dependence of decoherence on the spectral density but also clarifies in what sense position is an observable that

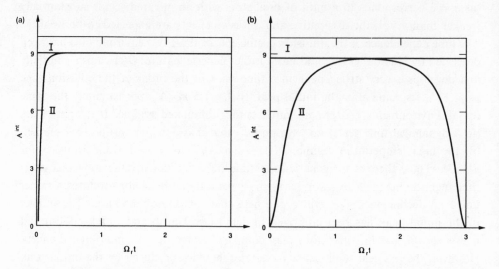

Fig. 12.1. The degree of decoherence $A_{int}$ is plotted as a function of time (in units of $\Omega_r^{-1}$) for the ohmic (a) and supra–ohmic (b) environments in the high temperature regime. Curve (I) corresponds to an initial condition where the initial coherent states are spatially separated ($L_0 = 3\delta$, $P_0 = 0$) while for curve (II) the initial separation is in momentum ($L_0 = 0$, $P_0 = 3/\delta$).

is preferred by the interaction. In fact, in general, coherent states that are spatially separated decohere much faster than those separated only in momentum (see Paz et al (1993) for more details).

## 12.4 Acknowledgements

I would like to thank the organizers for giving me the opportunity to attend such an exciting meeting. I also want to thank B.L. Hu, W. Zurek and S. Habib for many interesting conversations.

## Appendix: Derivation of the Master Equation

I outline here a simple derivation of the master equation (12.2) based on the properties of the evolution operator of the reduced density matrix. This propagator, which we denote as $J(t, t_0)$ and is defined so as to satisfy $\rho_r(t) = J(t, t_0)\rho_r(t_0)$, has a path integral representation of the following form:

$$J(x, x', t \mid x_0, x'_0, t_0) = \int_{x_0}^{x} D\tilde{x} \int_{x'_0}^{x'} D\tilde{x}' \ \exp\frac{i}{\hbar}\left\{S[\tilde{x}] - S[\tilde{x}']\right\} F[\tilde{x}, \tilde{x}'] \qquad (A.1)$$

where $F(x, x')$ is the Feynman–Vernon influence functional that arises due to the integration of the environment variables. For the model we are considering, this functional is well known and can be written as (see Grabert et al (1988)):

$$i \log(F[x, y]) = \int_0^t ds \int_0^s ds'(x - y)(s) \left[ \eta(s - s')(x + y)(s') - iv(s - s')(x - y)(s') \right]$$

where $v(s)$ and $\eta(s)$ are the noise and dissipation kernels defined in terms of the spectral density:

$$\eta(s) = -\int_0^\infty d\omega I(\omega) \sin(\omega s), \qquad v(s) = \int_0^\infty d\omega I(\omega) \coth(\frac{\omega}{2k_B T}) \cos(\omega s)$$

As the integrand of (A.1) is gaussian, the integral can be exactly computed and the result is (written in terms of the variables $X = x + x'$, $\xi = x - x'$)

$$J(X, \xi, t; X_0, \xi_0, t_0) = \frac{b_3}{2\pi} \exp(ib_1 X\xi + ib_2 X_0\xi - ib_3 X\xi_0 - ib_4 X_0\xi_0) \times$$
$$\times \exp(-a_{11}\xi^2 - a_{12}\xi\xi_0 - a_{22}\xi_0^2) \qquad \text{(A.2)}$$

where the functions $b_k(t)$ and $a_{ij}(t)$ depend on the environment and can be constructed in terms of solutions to the equation:

$$\ddot{u}(s) + \Omega_0^2 u(s) + 2 \int_0^s ds' \eta(s - s') \, u(s') = 0 \qquad \text{(A.3)}$$

Thus, if $u_1$ and $u_2$ are two solutions of (A.3) that satisfy the boundary conditions $u_1(0) = u_2(t) = 1$ and $u_1(t) = u_2(0) = 0$ we can write:

$$2 \, b_1(t) = \dot{u}_2(t), \quad 2 \, b_3(t) = \dot{u}_2(0), \quad 2 \, b_2(t) = \dot{u}_1(t), \quad 2 \, b_4(t) = \dot{u}_1(0)$$

$$a_{ij}(t) = \frac{1}{1 + \delta_{ij}} \int_0^t \int_0^t ds \, ds' u_i(s) \, u_j(s') \, v(s - s') \qquad \text{(A.4)}$$

The derivation of the master equation can be done as follows by simply using equations (A.2) and (A.4): Let us take the time derivative of (A.2) and write

$$\dot{J}(t, t_0) = \left[ \frac{\dot{b}_3}{b_3} + i\dot{b}_1 X\xi + i\dot{b}_2 X_0\xi \right.$$
$$\left. - i\dot{b}_3 X\xi_0 - i\dot{b}_4 X_0\xi_0 - \dot{a}_{11}\xi^2 - \dot{a}_{12}\xi\xi_0 - \dot{a}_{22}\xi_0^2 \right] J(t, t_0) \qquad \text{(A.5)}$$

If we multiply (A.5) by an arbitrary initial density matrix and integrate over the initial coordinates $\xi_0$ and $X_0$, we will obtain an equation whose left hand side is $\dot{\rho}_r(x, x', t)$. In the right hand side we will find terms proportional to $\rho(x, x', t)$ that look like some of the ones appearing in the right hand side of equation (12.2). The only potentially problematic terms are the ones that in (A.5) are proportional to the initial coordinates $\xi_0$ and $X_0$. However, their contribution can be easily shown to be local by realizing that the propagator $J(t, t_0)$ satisfies:

$$\xi_0 J(X, \xi, t; X_0, \xi_0, t_0) = \left(\frac{b_1}{b_3}\xi + \frac{i}{b_3}\partial_X\right) J(X, \xi, t; X_0, \xi_0, t_0)$$

$$X_0 J(X, \xi, t; X_0, \xi_0, t_0) = \left(-X\frac{b_1}{b_2} - \frac{i}{b_2}\partial_\xi - i\left(\frac{2a_{11}}{b_2} + \frac{a_{12}b_1}{b_2b_3}\right)\xi\right.$$

$$\left. + \frac{a_{12}}{b_3b_2}\partial_X\right) J(X, \xi, t; X_0, \xi_0, t_0) \tag{A.6}$$

To prove equation (12.2) we just have to show that the coefficients associated to terms like $\partial_X^2$ or $X\partial_X$ cancel and this can be done by exploiting some general properties of the coefficients $b_k$ and $a_{ij}$ that follow directly from their definition in (A.4). In fact, using relations such as $\dot{a}_{22} = -\dot{b}_4 a_{12}/b_2$ (whose proof we omit), it is possible to show that the coefficients of equation (12.2) are:

$$\Omega_{ren}^2(t) = 2\left(\frac{\dot{b}_2 b_1}{b_2} - \dot{b}_1\right); \qquad \gamma(t) = -\left(b_1 + \frac{\dot{b}_2}{2b_2}\right)$$

$$D(t) = \dot{a}_{11} - 4a_{11}b_1 + \dot{a}_{12}\frac{b_1}{b_3} - \frac{\dot{b}_2}{b_2}\left(2a_{11} + a_{12}\frac{b_1}{b_3}\right)$$

$$2f(t) = \frac{\dot{a}_{12}}{b_3} - \frac{\dot{b}_2 a_{12}}{b_2 b_3} - 4a_{11}$$

## Discussion

**Unruh**  Your master equation is local in time. For an arbitrary spectral density I would strongly expect that the equations are strongly non-local in time. Why aren't yours?

**Paz**  There are two observations one can intuitively make for the model described by equation (12.1). On the one hand we expect that a general environment will produce non–Markovian effects and that the master equation will be non local in time. On the other hand, the (reduced) evolution operator must be gaussian since the problem is linear. The crucial observation is that if one admits a gaussian evolution operator, the master equation is always local provided the matrix mixing "old" and "new" coordinates in the propagator can be inverted. Taking this into account, it is surprising to me that the existence of a local master equation has not been noticed until so recently.

**Morikawa**  Why do you get a local coefficient $\gamma(t)$?

**Paz**  I refer to the answer I gave to Prof. Unruh's question.

## References

Barone, P.M.V.B. and Caldeira, A.O. (1991) *Phys. Rev.* **A 43**, 57.
Caldeira, A.O. and Leggett, A.J. (1985) *Phys. Rev.* **A 31**, 1059.
Caldeira, A.O. and Leggett, A.J. (1983) *Physica* **121A**, 587.
Grabert, H., Schramm, P. and Ingold, G. (1988) *Phys. Rep.* **168**, 1 15.
Haake, F and Reibold, R. (1985) *Phys. Rev.* **A 32**, 2462.
Hu B.L., Paz, J.P. and Zhang, Y. (1992) *Phys. Rev.* **D45**, 2843.
Paz, J.P., Habib, S.. and Zurek, W.H., (1993) *Phys. Rev.* **D47**, 488.
Unruh, W.G. and Zurek, W.H. (1989), *Phys. Rev.* **D40**, 1071.
Zurek, W.H. (1991), *Physics Today* **44**, 33.

# 13

# Decoherence Without Complexity and Without an Arrow of Time

Bryce DeWitt

*Physics Department, University of Texas, Austin, Texas 78712, USA*

## 13.1 Introduction

I address this audience with some trepidation for I am neither an expert on decoherence, nor on complexity, nor on the arrow of time. The elementary results that I present constitute a part of my own attempt, with limited knowledge, to understand the emergence of classicality from quantum mechanics. They may well be already known to most of you. But I wish to stress a viewpoint that seems to be not so common: Decoherence can be understood in very simple terms. It is like most concepts in physics; it does not become more transparent by being made more complicated.

In the old Copenhagen days one seldom worried about decoherence. The classical realm existed *a priori* and was needed as a basis for making sense of quantum mechanics. With the emergence of quantum cosmology, and of the need for adopting a viewpoint like Everett's in order to grasp the meaning of a wave function for the whole universe, it became important to understand how the classical realm emerges from quantum mechanics itself. Everett's view leads naturally to the concept of "many worlds," and here I feel compelled to insert a few remarks that are only tangentially related to my main subject. I begin with a question:

*In quantum cosmology, why is Everett's many-world concept not accepted more frankly and openly than it is?* The theologian Paul Tillich once remarked† that among his professional colleagues, of those in the sciences only the physicists seemed capable of using the word "God" without embarrassment. Physicists have known this for a long time. Many examples come to mind, Einstein heading the list. If physicists can handle a concept like God then, except for some imprecision stemming from the use of words rather than equations, a concept like "many worlds" should cause them no pain. Some have suggested replacing the words "many worlds" by "alternative histories." I like "histories," for it suggests a global view of spacetime. But "alternative" is less felicitous, for if one sticks to the original meaning of the Latin *alter* then there can be only *one* alternative, whereas there is

† Talk at the University of North Carolina, *circa* 1960.

truly an astounding number. Also it smacks of elitism by suggesting that there is one "true" history, all the others being mere artifacts of the formalism.

What, in fact, is Everett's conception? It can be summed up on one sentence:

Apart from the incompleteness of our present knowledge of the laws and boundary conditions of Nature, a rigorous *isomorphism* exists between the formalism of physics and reality.

This leads immediately to the following corollary:

The formalism is able to generate its own interpretation.

The demonstration of the corollary has involved the work of many people and includes some of the results being presented at this Workshop. Everett himself focused on the concepts of *apparatus memory, relative states* and *many worlds*. He also tried to show that the formalism of quantum mechanics implies the probability interpretation, but here his reasoning was incomplete. The full proof had to await the convincing arguments of David Deutsch.

## 13.2 Measurements

One of the idealizations used by Everett was the concept of a "good" measurement, first introduced by von Neumann. Thirty years ago I spent some time learning about this concept. Using, as a point of departure, the famous paper by Bohr and Rosenfeld on the measurability of electromagnetic field strengths, I studied the analogous problem in quantum gravity. I was able to show that measurements of spacetime averages of the curvature tensor can in principle be made with a predictive accuracy equal to the limits imposed by quantum mechanics, provided the averaging domains are large compared to the Planck length. Somewhat later, as part of a campaign to publicize Everett, I generalized this analysis to arbitrary "good" measurements.

Much of the earlier work on the measurement problem, influenced no doubt by Bohr's shadow, emphasized the need for permanent information storage and hence for complexity, metastability, ergodicity, etc. This emphasis was misplaced. One does not gain understanding by making a problem more complicated. A measurement is simply the establishment of a correlation between a "system" observable and an "apparatus" observable. It is the function of the apparatus to "observe" the system, not *vice versa*, and hence there is a fundamental asymmetry between them. It turns out that there are two prominent features that characterize a good apparatus: Its "pointer" must be in a localized quantum state, and it must be massive compared to the system. That is all.

## 13.3 Decoherence. A Simple Model.

How does it happen that apparatus (plural) having localized pointer states are readily available in practice? That is the phenomenon of decoherence. Once again it is *massiveness* that provides the answer. Neither complexity of the environment nor an arrow of time are needed. A simple model suffices to show this.

Consider a massive body moving in one dimension in an arbitrary potential $V$. Let it collide with a light body moving freely. Let the collision be mediated by a $\delta$-function interaction. The Hamiltonian of the combined system is

$$H = \frac{1}{2M}P^2 + V(X) + \frac{1}{2m}p^2 + g\delta(x - X), \qquad M \gg m, \tag{13.1}$$

where $M$ and $m$ are the masses of the two bodies, $(X, P)$ and $(x, p)$ are their respective canonical variables, and $g$ is the strength of their interaction. The collision with the light body leaves the motion of the massive body virtually undisturbed. The state of the massive body can be quite arbitrary. We shall assume only that the velocity states into which it can be decomposed correspond to velocities that are small compared to the velocity of the light body.

It is then not difficult to show that the motion of the light body can be described in terms of scattering by a fixed $\delta$-function potential. If the light body is in an "incoming" momentum state at momentum $p(> 0)$ then the transmission and reflection coefficients for the collision are

$$T = \frac{1}{1 + i\frac{gm}{p}}, \qquad R = -\frac{i\frac{gm}{p}}{1 + i\frac{gm}{p}}. \tag{13.2}$$

These coefficients satisfy

$$T = 1 + R, \qquad |T|^2 + |R|^2 = 1, \tag{13.3}$$

the first relation being a property of the $\delta$-function interaction and the second a statement of probability conservation. Let $|\psi\rangle$ be the state vector of the massive body in the absence of interaction with the light body and let $|\Psi\rangle$ the state vector of the combined system. As basis vectors use the eigenvectors $|X, x, t\rangle$ of the position operators of the two bodies at time $t$ in the Heisenberg picture. The wave function of the combined system is then given very accurately by

$$\langle X, x, t \mid \Psi\rangle = L^{-1/2}\Big\{\theta(X - x)\big[e^{ipx} + Re^{ip(2X-x)}\big]$$
$$+ \theta(x - X)Te^{ipx}\Big\}e^{-i(p^2/2m)t}\langle X, t \mid \psi\rangle \tag{13.4}$$

where $\theta$ is the step function and $L$ is the length of an effective "box" controlling the normalization of the momentum wave functions of the light body.

## 13.4 Density Operator

The factorization of the wave function (13.4) into a part referring to the light body and a part $\langle X, t \mid \psi\rangle$ satisfying the Schrödinger equation of the massive body alone is entirely due to the condition $M \gg m$. The reader will note that the first factor does not refer *solely* to the light body. The term involving the reflection coefficient

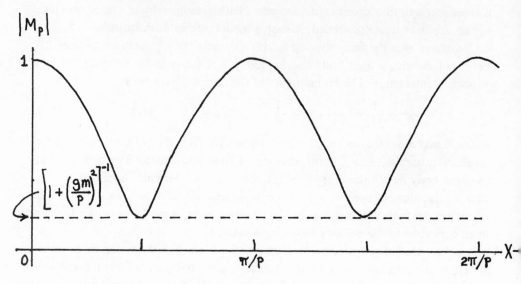

$|M_p|$

$1$

$\left[1+\left(\dfrac{gm}{p}\right)^2\right]^{-1}$

$0$                  $\pi/p$                $2\pi/p$   $X-$

Fig. 13.1.

contains a phase factor $e^{2ipX}$ representing the effect of momentum transfer to the massive body. Although this momentum transfer has no practical effect on the motion of the massive body, which continues to be described by the wave function $\langle X,t \mid \psi \rangle$, its role in decoherence is crucial. In order to see this, construct the density operator $\rho$ of the massive body by "tracing out" the light body:

$$\langle X,t|\, \rho \,|X',t\rangle = \int_{-L/2}^{L/2} \langle X,x,t \mid \Psi \rangle \langle \Psi \mid X',x,t\rangle \, dx$$

$$\xrightarrow[L\to\infty]{} \langle X,t \mid \psi \rangle \langle \psi \mid X',t\rangle \, M_p(X-X'), \qquad (13.5)$$

where

$$M_p(X-X') = \left[1+\left(\frac{gm}{p}\right)^2\right]^{-1} \left[1+\left(\frac{gm}{p}\right)^2 e^{ip(X-X')}\cos p(X-X')\right]. \qquad (13.6)$$

With the light body "traced out" the density operator is no longer that of a pure state. Its matrix representation now includes the *modulation function* $M_p$. The absolute value of this function has the appearance shown in Fig. 13.1. The function $M_p(X-X')$, regarded as a continuous matrix, has the following remarkable properties:

  (i) It is Hermitian and positive definite.

 (ii) If each of its (continuous infinity of) elements is raised to the power $N$, where $N$ is a fixed positive integer, the result is again a positive definite Hermitian matrix. (Note that this is not the same thing as raising the matrix itself to the $N$th power.)

(iii) Properties 1 and 2 are invariant under time reversal.

The last property may be inferred from the fact that the time-reversed wave function (in which the light body is in an "outgoing" momentum state) is simply the complex conjugate of expression (13.4). The time-reversed density matrix and modulation function are also obtained by simple complex conjugation, which leaves properties 1 and 2 intact.

## 13.5 Localization

The density matrix (13.5) does not yet describe the massive body as being in a localized state. However, suppose the massive body is allowed to collide with $N$ identical light bodies, all in the same momentum state and having identical $\delta$-function interactions with the massive body. Then in the wave function of the combined system there will be a factor for each light body, and when all these are "traced out" the density matrix of the massive body will take the form

$$\langle X, t | \rho | X', t \rangle \underset{L \to \infty}{\longrightarrow} \langle X, t | \psi \rangle \langle \psi | X', t \rangle E(X - X') \tag{13.7}$$

where

$$E(X - X') = \left[ M_p(X - X') \right]^N . \tag{13.8}$$

The function $E$ may be called the *environmental modulation function*. Because of property 2 it, like $M_p$, is positive definite and Hermitian when viewed as a continuous matrix. For $N = 20$ its absolute value, based on the previous figure, has the appearance shown in Fig. 13.2.

Here localization is beginning to show itself. But it is a localization *modulo* $\pi/p$. To get true localization it is clear what one must do. One must let the massive body collide with $N_1$ light bodies having momentum $p_1$, $N_2$ having momentum $p_2$, and so on, and choose the $p$'s to be incommensurable. The density matrix takes again the form (13.7) but with an environmental modulation function given by

$$E(X - X') = \left[ M_{p_1}(X - X') \right]^{N_1} \left[ M_{p_2}(X - X') \right]^{N_2} \ldots \tag{13.9}$$

This function, like all the others, defines a positive definite Hermitian continuous matrix, but *its* absolute value has the appearance shown in Fig. 13.3.

As the $N$'s become large $E$ becomes a very narrow function. If the wave function $\langle X, t | \psi \rangle$ varies negligibly over a distance equal to the width of $E$ then, switching to a time-independent basis $\{|X\rangle\}$, one may write

$$\langle X, t | \rho | X', t \rangle = \langle X | \rho(t) | X' \rangle \tag{13.10}$$

where

$$\rho(t) = \int_{-\infty}^{\infty} \left| E_{\bar{X}}^{1/2} \right\rangle \left| \langle \bar{X}, t | \psi \rangle \right|^2 \left\langle E_{\bar{X}}^{1/2} \right| d\bar{X} \tag{13.11}$$

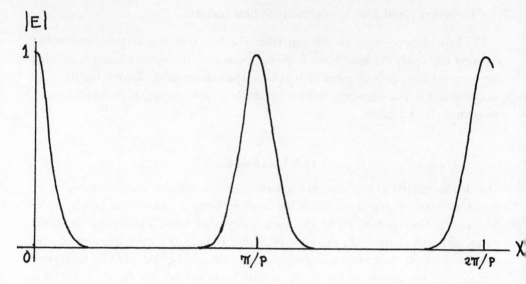

Fig. 13.2.

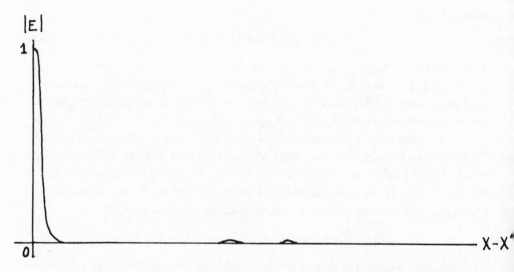

Fig. 13.3.

with

$$\left\langle X \mid E_{\bar{X}}^{1/2} \right\rangle = E^{1/2}(X - \bar{X}), \qquad (13.12)$$

the continuous matrix defined by the function $E^{1/2}$ being the positive definite Hermitian square root of that defined by $E$. We note that

$$\left\langle E_X^{1/2} \mid E_{X'}^{1/2} \right\rangle = E(X - X') \qquad (13.13)$$

$$\left\langle E_X^{1/2} \mid E_X^{1/2} \right\rangle = E(0) = 1. \tag{13.14}$$

The limiting form of the density operator, as the $N$'s become large, evidently describes the massive body as *localized at a point*, the probability that the point is between $X$ and $X + dX$ being $|\langle X, t \mid \psi \rangle|^2 \, dX$.

### 13.6 Discussion

The following remarks are in order:

(i) The localization occurs no matter what $|\psi\rangle$ is.
(ii) The localization is simply a consequence of $M \gg m$.
(iii) Only two incommensurable momenta are needed. One does not need a thermal bath.
(iv) If the light bodies "arrive" in the form of packets, with momentum distribution function $f(p)$, then $M_p$ is modified to $\int M_p |f(p)|^2 \, dp$, which generally narrows $E(X - X')$ further. However, an arrow of time is then introduced, and the density matrix will not take the form (13.7) until "after" the collisions have occurred.
(v) The strict localization achieved here stems from the $\delta$-function interaction potential. Localization normally would not be sharper than the width of the potential.

The above results have the following implications for decoherence in quantum cosmology:

(i) Although complexity (metastability, chaos, thermal baths, wave packets) can only help in driving massive bodies to localized states, it is *massiveness*, not complexity, that is the *key* to decoherence.
(ii) Given the fact that the elementary particles of nature tend, upon cooling, to form stable bound states consisting of massive agglomerations, decoherence at the classical level is a natural phenomenon of the quantum cosmos.
(iii) Given the fact that the interaction described here, between the massive body and the light ones, is a simple scattering interaction and not at all specially designed like a "good" (von Neumann) measurement interaction, the universe is likely to display decoherence in almost all states that it may find itself in.†
(iv) An arrow of time has no basic role to play in decoherence. Let $|\Phi\rangle$ be the state vector of the cosmos. If it has an arrow of time replace it by $\frac{1}{\sqrt{2}}\left(|\Phi\rangle + |\Phi\rangle^T\right)$, "$T$" denoting time reversal. Each Everett world in the latter vector is paired with a time-reversed world. The two worlds are unaware of one another and yet both are overwhelmingly likely to exhibit decoherence.

---

† A good measurement leads to decoherence of apparatus "memory states," which is a special type of decoherence.

## 13.7 Decoherence Function

There is a more sophisticated approach to decoherence, which accounts not only for localization but also for the emergence of classicality.† The formalism used in this approach is easily applied to the present model. One first introduces a set of projection operators that define a *coarse graining* of the possible dynamical histories of the massive body. For example,

$$P_\epsilon(\bar{X}, t) \overset{\text{def}}{=} \int_{\bar{X}-\frac{1}{2}\epsilon}^{\bar{X}+\frac{1}{2}\epsilon} |X, t\rangle \langle X, t| \, dX,\tag{13.15}$$

where $\epsilon$ determines the coarseness of the graining. If $\bar{X}$ is chosen from a discrete set of points, separated by intervals $\epsilon$ from one another, then these projection operators, at a fixed instant of time, are mutually orthogonal. More generally, one introduces projection operators at successive instants of time, and the finest useful graining is controlled by the phenomenon of wave-packet spreading. One chooses

$$\epsilon \gg \sqrt{\Delta t_{\max}/M} \quad (\hbar = 1)\tag{13.16}$$

where $\Delta t_{\max}$ is the largest of the successive time intervals. Note that the larger $M$ is the finer the graining can be.

Using the projection operators one may define the so-called *decoherence function*:

$$\begin{aligned}
D\left(\ldots, \bar{X}_3, \bar{X}_2, \bar{X}_1 \mid \bar{X}_1', \bar{X}_2', \bar{X}_3', \ldots\right)& \\
\overset{\text{def}}{=} \text{tr}\,[\ldots P_\epsilon\left(\bar{X}_3, t_3\right) P_\epsilon\left(\bar{X}_2, t_2\right) P_\epsilon\left(\bar{X}_1, t_1\right) \rho\, P_\epsilon\left(\bar{X}_1', t_1\right)& \\
P_\epsilon\left(\bar{X}_2', t_2\right) P_\epsilon\left(\bar{X}_3', t_3\right) \ldots]& \\
= \int_{-\infty}^{\infty} dX \langle X, t| \ldots P_\epsilon\left(\bar{X}_3, t_3\right) P_\epsilon\left(\bar{X}_2, t_2\right) P_\epsilon\left(\bar{X}_1, t_1\right) \rho\, P_\epsilon\left(\bar{X}_1', t_1\right)& \\
P_\epsilon\left(\bar{X}_2', t_2\right) P_\epsilon\left(\bar{X}_3', t_3\right) \ldots |X, t\rangle.&
\end{aligned}\tag{13.17}$$

The time $t$ in the last line of this equation is arbitrary, but one assumes that the times $t_1$, $t_2$, $t_3$, ... are in either chronological or antichronological order and fixed *a priori*. Obviously one could generalize the above definition by choosing a set of times to the right of the density operator that differs from the set to the left, but the definition (13.17) suffices here.

The function $D$, regarded as a matrix, is positive definite and Hermitian. Its positive real diagonal elements have a ready interpretation. $D\left(\ldots, \bar{X}_3, \bar{X}_2, \bar{X}_1 \mid \bar{X}_1, \bar{X}_2, \bar{X}_3, \ldots\right)$ is the probability that the dynamical trajectory of the massive body will be observed, by measurements at times $t_1$, $t_2$, $t_3$, ..., to pass within a distance $\frac{1}{2}\epsilon$ of the points $\bar{X}_1$, $\bar{X}_2$, $\bar{X}_3$, ... at those times.

---

† My limited reading on this subject has been confined to the work of Gell-Mann and Hartle (see their reports in this volume). Many other references should doubtless be cited.

## 13.8 Emergence of Classicality

Suppose there are just three instants of time, $t_1$, $t_2$ and $t_3$. Then using the explicit form (13.15) for the projection operators, together with the WKB approximation

$$\langle X,t \mid X',t'\rangle \approx \left[-\frac{1}{2\pi i}\frac{\partial^2 S(X,t \mid X',t')}{\partial X \partial X'}\right]^{1/2} e^{iS(X,t \mid X',t')} \tag{13.18}$$

where $S(X,t \mid X',t')$ is the action along the classical trajectory defined by the data $X$, $t$, $X'$, $t'$, and evaluating the resulting integrals in (13.17) by the method of stationary phase, one easily sees that $D\left(\bar{X}_3,\bar{X}_2,\bar{X}_1 \mid \bar{X}_1,\bar{X}_2,\bar{X}_3\right)$ vanishes unless $\bar{X}_2$ lies, at time $t_2$, within a distance of order $\epsilon$ from the classical trajectory defined by the data $\bar{X}_1$, $t_1$, $\bar{X}_3$, $t_3$. The probability that the trajectory of the massive body will be observed to depart from classical, within the limits set by the coarse graining (13.16), is thus vanishingly small. We remark that use of the WKB approximation to derive the emergence of classicality is valid as long as (13.16) is respected, and that the same result will follow if, instead of using the WKB approximation, one appeals directly to the path-integral representation of the decoherence function. We also note that this result is valid only to the extent that one can ignore the buffeting that the massive body gets by the light bodies. An analysis of the effect of the environmental noise that is present in more realistic models would take us into deeper waters.

All the above is just a fancy way of stating Ehrenfest's theorem. It says nothing yet about localization because the density operator has not been taken into account. Consider therefore the full 3-point decoherence function $D\left(\bar{X}_3,\bar{X}_2,\bar{X}_1 \mid \bar{X}'_1,\bar{X}'_2,\bar{X}'_3\right)$. One easily sees from expression (13.17) that, if the width of the environmental modulation function $E(X-X')$ is small compared to $\epsilon$, this function will essentially vanish if $\bar{X}'_1$ differs from $\bar{X}_1$. Because of the cyclic invariance of the trace and the orthogonality of the projection operators the same will be true if $\bar{X}'_3$ differs from $\bar{X}_3$. In order that the decoherence function differ from zero, therefore, $\bar{X}'_1$ must be equal to $\bar{X}_1$ and $\bar{X}'_3$ must be equal to $\bar{X}_3$. But by the stationary phase analysis one sees that this forces both $\bar{X}_2$ and $\bar{X}'_2$ to lie, at time $t_2$, within a distance $\epsilon$ of the classical trajectory between $\bar{X}_1$, $t_1$ and $\bar{X}_3$, $t_3$.

These results are easily generalized. When condition (13.16) is satisfied and when the environmental modulation function is sufficiently narrow, the matrix defined by the decoherence function has the following properties:

(i) It is essentially diagonal.

(ii) Even its diagonal elements will vanish unless the points $\left(\bar{X}_1,t_1\right)$, $\left(\bar{X}_2,t_2\right)$, $\left(\bar{X}_3,t_3\right)$, ..., in spacetime lie within a distance $\epsilon$ of a classical trajectory.

(iii) The diagonal elements will also vanish unless $\left(\bar{X}_1,t_1\right)$ lies in the support of the function $|\langle X,t \mid \psi\rangle|^2$.

Property 1 is a direct consequence of localization and is a signal of true decoherence. Properties 1 and 2 together are consequences of an astute choice of coarse graining of the real world, which, in the present instance, consists of dividing the configuration space of the massive body into intervals of width satisfying condition (13.16), and of ignoring ("tracing out") all the light bodies. The most sophisticated modern investigations are those that turn the problem around and try to discover, in more realistic contexts, the kinds of coarse graining that will lead to decoherence functions having the above properties. Once these are found, the nonvanishing diagonal elements of the associated matrix may be identified with Everett's many worlds or many "histories." These histories are mutually decoherent (no quantum mechanical interference effects), which is another way of saying that Everett's worlds are unaware of one another.

It should finally be remarked that the decoherent histories of a massive system need not always be "classical." Indeed they will not be if the system is a measuring apparatus capable of amplifying and recording the strictly quantum features of a lighter system to which it is coupled. But this is the situation that has already been analyzed by Everett.

## Acknowledgment

I am grateful for the stimulation that I have received from discussions with my student, Donald Marolf, and from exposure to ideas at earlier Workshops in this series. This work was supported by National Science Foundation Grant No. PHY8919177.

## Discussion

**Unruh**   You have an infinite number of particles. You neglect the reaction of the massive particle to the small particles. Your $N \to \infty$ limit does not exist in that the reaction of the large particle becomes large as $N$ goes to $\infty$.

**DeWitt**   You are quite right, of course. My results would be rigorous only in the limit $M \to \infty$ as $N \to \infty$. They are nevertheless sufficient as they stand to illustrate the key role that massiveness plays in decoherence.

**Starobinsky**   Your model is very similar to what arises in the stochastic approach to inflation, which I discussed in my talk. There, a local value of the large-scale part of an inflaton field plays the role of a heavy particle, and small-scale perturbations of the same field, which produce stochastic forces, are analogous to the light particles.

**Griffiths**   I would like to comment on the following points.

(i)   In my opinion, and as I pointed out in my 1984 paper, the Everett interpretation is very different from the consistent histories approach, in particular.

(a)   As you yourself said, Everett did not introduce a satisfactory probabilistic interpretation. In the consistent-history interpretation the probabilistic interpretation is of the essence: it is the new item which allows one to understand quantum mechanics.

(b)   I pointed out in my 1984 paper that the consistent-history approach *rules out* "alternative universes" in a suitably-precise sense, and I was very happy with this. Granted, it is a matter of philosophical predilection whether one does or does not like multiple universes; I don't like them, and I was happy to have one interpretation which ruled them out.

(c) In Deutsch's version of Everett's interpretation it is necessary to introduce a specific basis. The consistent-history approach, on the other hand, is not committed to a particular basis; you can use whatever is convenient. Again it is something of an aesthetic criterion; I myself am happy that the consistent histories approach leaves the question of basis open.

(ii) I would like to point out that there is a considerable difference between "decoherence" as the term is often used, and "consistency" as employed in the consistent history approach. It would be useful for our discussions to keep distinct ideas distinct. [In the context of your talk, "decoherence" is a very different idea from the "decoherence" used by Gell-Mann and Hartle, which is the same thing as "consistency" as I have employed the term.]

**DeWitt**

1.a I disagree that the probability interpretation is a new item that allows one to understand quantum mechanics. As David Deutsch has amply shown, the probability interpretation truly emerges from the mathematical formalism itself.

1.b I also disagree that the consistent-histories approach *needs* to rule out "alternative universes."

1.c An experimental physicist always knows what basis his apparatus is selecting as a "preferred" basis (if he is making an observation of the von Neumann type), so there should be no objection to making use of it when it is there. An "interpretation" is only a guide to understanding. In the end, the formalism must speak for itself. I am happy with any approach that aids in understanding.

2. Perhaps I erred in equating "decoherence" with "localization." I am not an expert in this field, and in using this terminology I only followed what a number of others *seemed* to be doing.

**Gell-Mann**  Do you want to apply the restriction on the meaning of "alter" also to the romance languages derived from latin, to "autre," "otro," and so forth?

**DeWitt**  Alas, it is too late for that. I am fighting a rear guard action!

**Gell-Mann**

(i) I consider myself to be a post-Everett investigator. In fact, I found the modern interpretation independently, in 1963, in conversations with Villars and Feynman, not having heard of Everett at that time. However, the language was closer to our present language. I believe Everett's language is highly misleading. Many distinguished physicists have been confused by the reference to "many worlds" instead of "many histories" and to the "reality" of the "many worlds" instead of to the fact that quantum mechanics does not discriminate among them *a priori* except by their probabilities. Also, Everett denies the importance of probability while discussing "measure" that can really be used in the same way. Apart from the unfortunate language, Everett's physics is okay, although somewhat incomplete. We are trying to fill in the rest of the picture.

(ii)   We can agree with much of what you have said. When some variable (say a quantum variable like a spin projection) comes into full correlation with the quasiclassical domain, we say there is a "measurement situation," whether or not a mongoose or something has come along to see and note the result. The quasiclassical domain is characterized by decoherence and approximate classicality, for which considerable inertia of the distinguished degrees of freedom is important. For decoherence itself, high inertia is not essential. Any correlated photon that has departed can cause the decoherence of whatever it is correlated with.

**Page**   Does the Everett interpretation require a preferred basis? I would have thought the Copenhagen interpretation required a preferred basis for collapse, but that the many-world interpretation does not require such a preferred basis.

**DeWitt**   If a measurement is "good" (in the von Neumann sense) then a preferred basis exists, and the interpretation of the formalism is easy. But the Everett interpretation also permits a straightforward analysis of imperfect measurements. In the end, all that Everett asks is that one regard the total wave function as providing a direct representation of reality and that one let the formalism speak for itself.

**Zeh**   I agree with Murray Gell-Mann's remark that it is ultimately a manner of language whether or not the "other worlds" exist. However, in my opinion it would instead be misleading to verbally *deny* the existence of something that is assumed to be able (in principle) to affect "our world" (by interferring with it).†

**Hartle**   I have two comments. First, I don't know whether he will be too modest to mention it, but Prof. Zeh, in collaboration with E. Joos, analyzed a large number of essentially similar, and indeed more realistic, examples in a paper in Zeitschrift für Physik, about 1986.

My second comment concerns the generality of decohering histories. As I mentioned earlier, if one assumes any pure initial state then an arbitrarily large number of exactly decohering sets of alternative histories can be exhibited. The question is whether among these, or among the approximately, decohering ones, there is one (or several) that constitute a quasiclassical domain. But, I agree that the ubiquity of mechanisms like those modeled by your example, result in the decoherence of familiar quasiclassical variables in universes like ours.

**DeWitt**   Thank you for reminding me of Zeh's work. I have always been an admirer of his, not least because he was one of the first who understood Everett and took him seriously.

Because I have not been active in the field represented by this Workshop I am undoubtedly unaware of much of the relevant work. What I *have* noted (perhaps wrongly) is what I have interpreted as a mistaken emphasis on complexity, metastability, ergodicity, etc. The purpose of my talk has merely been to restore the balance a bit.

---

† Note added in proof: In Everett's interpretation the "other worlds" are considered as dynamically independent components of *one (real) quantum world* that is described by a wave function which always obeys the Schrödinger equation. They are only *perceived* separately because of their dynamical decoupling that has emerged by means of practically irreversible decoherence occurring as a consequence of specific *initial* conditions (the lack of conspirative correlations). Allowing the "other" components to escape from existence would be equivalent to the assumption of a (fundamentally time asymmetric) collapse of the wave function, while the description of "events" (from which to form "histories") by means of concepts introduced *in addition* to the wave function (such as classical ones) would either have to represent a hidden variables theory, or to remain conceptually as vague or inconsistent as the Copenhagen interpretation. Decoherence allows quasi-events to occur, and quasi-classical concepts to emerge, smoothly but on a very short time scale according to the Schrödinger dynamics.

**Zurek**   I would like to agree with you that there is no need for complexity in the state of the environment to achieve decoherence. In an old (1982) paper I have discussed a model which is even simpler than yours or Andy's:  An environment made of spin 1/2 systems with a pure initial state. It does achieve decoherence. The reason why it is convenient to introduce temperature in the discussion is to emphasize the fact that a thermal state is *less* special than a pure initial state.

I would also like to wholeheartedly agree with the remark of Prof. Griffiths that identifying decoherence with consistency is confusing. *Decoherence is a process.* It can help bring about *consistency* (in the sense of Griffiths and Omnès), but the two do not mean the same thing. We should carefully distinguish the goal (consistency in a reasonable preferred basis) from the means (environment-induced decoherence).

# 14

# The Decoherence Functional in the Caldeira-Leggett Model

H. F. Dowker[†]

*NASA/Fermilab Astrophysics Center*
*Fermi National*
*Accelerator Laboratory*
*P.O. Box 500, Batavia, IL 60510-1500, U.S.A.*

J.J. Halliwell[‡]

*Center for Theoretical Physics*
*Laboratory for Nuclear Science*
*Massachusetts Institute of Technology*
*Cambridge, MA 02139, U.S.A.*

## Abstract

We investigate the decoherent histories approach to quantum mechanics in the model of quantum brownian motion due to Caldeira and Leggett. The issue of approximate decoherence is discussed. Quantitative results on decoherence as a function of coarse graining are obtained and the effects of different sorts of coarse graining are shown. It is found that the requirements of decoherence and peaking about classical trajectories compete with each other in the sense that maximizing one tends to decrease the other.

## 14.1 Introduction

In the subject of quantum cosmology one comes face to face with the measurement problem of quantum mechanics. When doing calculations to predict the results of collider experiments, one feels sure that, no matter how vague notions such as *measurement* and *classical apparatus* are in general, one can use the standard Copenhagen interpretation with "good taste and discretion" (Bell 1989) and come up with the right answer. In quantum cosmology, however, where the quantum system under investigation is the universe itself, the vital ingredient of the Copenhagen interpretation of quantum mechanics – the external, classical measuring apparatus – has no meaning at all. In order to make sense of an object such as the "wave function of the universe" and make predictions from it, one must look for an interpretation which does not require such an *a priori* split between quantum and classical regimes.

---

† Present address: Department of Physics, University of California, Santa Barbara, CA 93106, USA.
‡ Present address: Theory Group, Blackett Laboratory, Imperial College, London, SW7 2BZ, UK.

An attempt to do this has been developed by four main workers, variously motivated: R.B. Griffiths (1984), R. Omnès (1990 and references therein), M. Gell-Mann and J.B. Hartle (1990a, 1990b, 1991) (see also Hartle (1991)). This approach is known as the "consistent histories" or "decoherent histories" approach. It specifically concerns closed quantum mechanical systems, and is assumed to apply to microscopic and macroscopic systems alike, up to and including the entire universe. Its most important feature is that it focuses on the possible histories of a system. The case of events at a single moment of time is included as a special case. The formulation makes no reference to external observers, classical apparatus, wave function collapse, or indeed any of the machinery of conventional quantum measurement theory. It is hoped that this approach will provide a framework in which to answer the question: when can a closed quantum mechanical system be said to be behaving classically?

Although application to quantum cosmology is one major aim of this approach, it would be useful first to have a good understanding of how it works in a simple system that we are familiar with. In this article we will report on progress we have made in an investigation of the decoherence functional in non-relativistic quantum mechanics in a model with a single harmonic oscillator interacting with a bath of other oscillators whose state is averaged over. It was proposed by Caldeira and Leggett (1983) as a model of quantum brownian motion. Models of this type have been considered extensively in the reduced density matrix approach to decoherence (Habib and Laflamme 1990, Joos and Zeh 1985, Unruh and Zurek 1989, Zurek 1981, 1982, 1991). Identifying how decoherence and classical characteristics emerge in this simple example, will give us confidence to tackle the (much) more difficult case of even mini-superspace quantum cosmology.

The article is structured as follows. In Section 2 we briefly describe the decoherent histories formalism. Section 3 is a discussion of the issue of approximate decoherence. In Section 4 the Caldeira-Leggett model is introduced and we set up the calculation. We spare the reader pages of gaussian integration and collecting of terms and present our main results in Section 5. Section 6 is a summary. For further details see Dowker and Halliwell (1992).

## 14.2 The Decoherence Functional

We will sketch the decoherent histories formalism. For more details see the references cited above and articles in this volume by Omnès and Griffiths. We assume we start with a quantum mechanical system. This means we have a Hilbert space of states. A history of the system is specified by a sequence of projection operators on that space at times $t_i$, $\{P_{\alpha_1}(t_1), P_{\alpha_2}(t_2), \ldots, P_{\alpha_n}(t_n)\}$, which we write $[\alpha_i]$ for short. Each $P_\alpha(t)$ is a projection operator in the Heisenberg picture. A *set of alternative histories* is given by all such sequences that can be formed from sets of projections at each time $t_i$ which are complete $(\sum_{\alpha_i} P_{\alpha_i} = 1)$ and orthogonal $(P_{\alpha_i} P_{\beta_i} = \delta_{\alpha\beta} P_{\alpha_i})$†. A *fine grained* history is one where projections are specified at all times and each projection is one-dimensional, $P_{\alpha_i} = |\alpha_i\rangle\langle\alpha_i|$. Otherwise the history is *coarse grained*.

Given that the system is described by a density matrix $\rho(t_0)$ at time $t_0$, the *decoherence functional* of two histories in a set of alternative histories is defined to be

$$D([\alpha_i]; [\alpha_i']) = Tr\left(P_{\alpha_n}(t_n)\ldots P_{\alpha_1}(t_1)\rho(t_0)P_{\alpha_1'}(t_1)\ldots P_{\alpha_n'}(t_n)\right). \tag{14.1}$$

We wish to assign $D([\alpha_i]; [\alpha_i])$ as the probability of each history $[\alpha_i]$ in a set of alternatives. In general, however, there are obstructions to doing so. We need the ordinary rules of probability to hold and this imposes consistency conditions on $D$.

Given a set of alternative histories, $\{[\alpha_i]\}$, a coarse graining of the set is another set of alternative histories, $\{[\bar\alpha_i]\}$, constructed using projections which are sums of those used to generate the original set. In that case we write that a particular member of the original set is contained in a particular member of the coarser grained set, $[\alpha_i] \in [\bar\alpha_i]$, if, for each $i$, $P_{\alpha_i} P_{\bar\alpha_i} \neq 0$. Then,

$$D([\bar\alpha_i']; [\bar\alpha_i]) = \sum_{[\alpha_i'] \in [\bar\alpha_i']} \sum_{[\alpha_i] \in [\bar\alpha_i]} D([\alpha_i']; [\alpha_i]). \tag{14.2}$$

For this to be consistent with our desired identification of the diagonal terms of the decoherence functional as probabilities, we require

$$D([\bar\alpha_i]; [\bar\alpha_i]) = \sum_{[\alpha_i] \in [\bar\alpha_i]} D([\alpha_i]; [\alpha_i]). \tag{14.3}$$

This consistency condition must hold for all coarse grainings of the original set $\{[\alpha_i]\}$. In this case the set is said to decohere and we identify the diagonal elements of the decoherence functional as the probabilities for the histories in the set. It can be shown that a necessary and sufficient condition for (14.3) to hold for all coarse grainings is

$$Re\{D([\alpha_i]; [\alpha_i'])\} = 0 \qquad \forall \ [\alpha_i] \neq [\alpha_i'] . \tag{14.4}$$

### 14.3 Approximate Decoherence

In Section 2 we described the formalism of the quantum mechanics of histories and gave the condition, Eq. (14.3), that must be satisfied if probabilities are to be assigned to sets of histories. This condition is for *exact* decoherence, *i.e.* for the probability sum rules for histories to be satisfied exactly. Whilst it is sometimes possible to exhibit histories which decohere exactly, in general, decoherence will

---

† In general, a history is given by a *set* of such sequences and may not be written as a single sequence. Similarly, a set of alternative histories is a partition of the set of all sequences that may be formed from the complete orthogonal sets of projections as described in the text. This is most clearly seen in the path integral formulation of the approach ( see e.g. Hartle 1991). Making this explicit in the text would complicate the notation immensely and obscure the argument. We therefore consider only the simpler, single sequence histories in formulae, which are easily generalized.

only be approximate. This is the case, for example, for the model considered in this article. It therefore becomes an interesting and important question to understand what is meant by approximate decoherence.

From Eq. (14.2) we know that

$$p([\bar{\alpha}_i]) = \sum_{\substack{[\alpha_i] \in [\bar{\alpha}_i]}} p([\alpha_i]) + \sum_{\substack{[\alpha_i] \neq [\alpha_i'] \\ [\alpha_i],[\alpha_i'] \in [\bar{\alpha}_i]}} D([\alpha_i];[\alpha_i']) \qquad (14.5)$$

where

$$p([\alpha_i]) = D([\alpha_i];[\alpha_i]) \qquad (14.6)$$

and $[\alpha_i] \neq [\alpha_i']$ if $P_{\alpha_i}(t_i) \neq P_{\alpha_i'}(t_i)$ for any $i$. The natural generalization of Eq. (14.3) is to say that the probability sum rules are satisfied to order $\epsilon$, where $\epsilon$ is small. By this we mean that the interference terms do not have to be exactly zero, but only suppressed by a factor $\epsilon$; *i.e.*

$$\left| \sum_{\substack{[\alpha_i] \neq [\alpha_i'] \\ [\alpha_i],[\alpha_i'] \in [\bar{\alpha}_i]}} D([\alpha_i];[\alpha_i']) \right| < \epsilon \sum_{\substack{[\alpha_i] \in [\bar{\alpha}_i]}} p([\alpha_i]) \qquad (14.7)$$

for all possible coarser-grainings $[\bar{\alpha}_i]$ of the alternatives $[\alpha_i]$.

In the case of exact decoherence, $\epsilon = 0$, the conditions (14.3) are fully equivalent to the many fewer conditions (14.4). This enormously simplifies the problem of checking the probability sum rules. For the case of approximate decoherence considered here, in the worst possible case, we might have to check the probability sum rules for all possible choices of coarser-grained histories. It could be, for example, that the degree to which the sum rules are satisfied depends on the particular sum rule in question.

Let us investigate the consequence of the most obvious generalizations of the conditions (14.4):

$$| \, 2\text{Re}\, D([\alpha_i];[\alpha_i']) \, | < \epsilon \left( p([\alpha_i]) + p([\alpha_i']) \right) \quad \forall [\alpha_i] \neq [\alpha_i'] \,. \qquad (14.8)$$

Notice that this condition is not simply that the off-diagonal terms are small but that they are small compared with the corresponding diagonal terms. This is a much stronger condition since the diagonal terms themselves may be extremely small†.

Suppose a coarser grained history, $[\bar{\alpha}_i]$, contains $n$ histories in the set of alternatives under scrutiny. Then,

$$\left| \sum_{\substack{[\alpha_i] \neq [\alpha_i'] \\ [\alpha_i],[\alpha_i'] \in [\bar{\alpha}_i]}} D([\alpha_i];[\alpha_i']) \right| < \epsilon(n-1) \sum_{\substack{[\alpha_i] \in [\bar{\alpha}_i]}} p([\alpha_i]) \,. \qquad (14.9)$$

Thus, if there are only a finite number of histories, it would be to possible to enforce

---

† One might argue that such a condition is too stringent and that one needn't worry about the sum rules holding in cases where both diagonal and off-diagonal terms are so small that one might approximate them both by zero. However, Eq. (14.8) is a proposal for the most rigorous expression of consistency.

conditions (14.8) to some order $\epsilon' = \frac{\epsilon}{(n-1)}$ to guarantee that all sum rules held to order $\epsilon$. However, in cases where the number of histories in a set of alternatives is infinite this is no longer possible.

In an attempt to alleviate this problem we propose the alternative sum rule

$$| \operatorname{Re} D([\alpha_i]; [\alpha'_i]) | < \epsilon \left[ p([\alpha_i]) p([\alpha'_i]) \right]^{\frac{1}{2}} \qquad \forall [\alpha_i] \neq [\alpha'_i] \,. \tag{14.10}$$

This condition implies Eq. (14.8). Consider again a coarse grained history, $[\bar{\alpha}_i]$. Using Eq. (14.10) one obtains

$$\left| \sum_{[\alpha_i] \neq [\alpha'_i]} D([\alpha_i]; [\alpha'_i]) \right| < \Delta \epsilon \sum_{[\alpha_i]} p([\alpha_i]) \tag{14.11}$$

where all sums are over $[\bar{\alpha}_i]$ and

$$\Delta = \left( \sum_{[\alpha_i]} p([\alpha_i]) \right)^{-1} \left[ \left( \sum_{[\alpha_i]} p([\alpha_i])^{\frac{1}{2}} \right)^2 - \sum_{[\alpha_i]} p([\alpha_i]) \right] \,. \tag{14.12}$$

$\Delta$ may or may not diverge depending on how the $p([\alpha_i])$ are distributed but at least it is possible that it is convergent, an improvement on the previous situation. Note, however that $\Delta \gg 1$ in general.

This decoherence condition is discussed further in (Dowker and Halliwell 1992). There it is argued that the strict upper bound (14.11) is not representative of the typical case. A statistical analysis based on a random walk analogy suggests that the upper bound for typical course grainings leads to an inequality of the form (14.11) with a $\Delta$ of order 1 (not the same $\Delta$ as in (14.12)). This would bring us close to identifying a subset of the sum rules which, if satisfied to order $\epsilon$, imply that all other sum rules are satisfied to the same order.

## 14.4 The Caldeira-Leggett Model

The Caldeira-Leggett model consists of a distinguished system $A$ with action

$$S_A[x] = \int_0^\tau dt \left[ \frac{1}{2} M \dot{x}^2 - \frac{1}{2} M \omega^2 x^2 \right], \tag{14.13}$$

coupled to a reservoir or environment $B$ consisting of a large number of harmonic oscillators with coordinates $R_k$ and action

$$S_B[\mathbf{R}] = \sum_k \int_0^\tau dt \left[ \frac{1}{2} m \dot{R}_k^2 - \frac{1}{2} m \omega_k^2 R_k^2 \right] \tag{14.14}$$

The coupling is described by the action

$$S_I[x, \mathbf{R}] = - \sum_k \int_0^\tau dt \, C_k R_k x \tag{14.15}$$

where the $C_k$'s are coupling constants.

The histories of the system will be coarse grained by tracing over the environment states and only specifying the position of the special oscillator at a finite number of times $\{t_i\}$ and only to within an accuracy $\sigma_i$ at $t_i$. Projection operators which accomplish this would be

$$P_{\bar{x}_i, \sigma_i} = \int_{x_i \in [\bar{x}_i - \frac{1}{2}\sigma_i, \bar{x}_i + \frac{1}{2}\sigma_i]} dx_i \, | \, x_i \rangle \langle x_i \, | \tag{14.16}$$

for a discrete set of $\bar{x}_i$ separated by a distance $\sigma_i$. For our calculation we approximated these by "gaussian projectors" (which are not exact projectors)

$$P_{\bar{x}_i, \sigma_i} = \int_{-\infty}^{\infty} dx_i \exp\left[-\frac{1}{\sigma_i^2}(\bar{x}_i - x_i)^2\right] | \, x_i \rangle \langle x_i \, | \; . \tag{14.17}$$

It can be shown that, in the so-called Fokker-Planck limit and assuming that the initial density matrix is a product of a density matrix for the special oscillator and a density matrix describing thermal equilibrium for the environment, the decoherence functional is given by:

$$
\begin{aligned}
D[\{\bar{x}_k\}; \{\bar{y}_k\}] &= D[\{\bar{X}_k\}, \{\bar{\xi}_k\}] \\
&= \int dX_{n+1} d\xi_{n+1} dX_n d\xi_n \cdots dX_0 d\xi_0 \; \delta(\xi_{n+1}) \, \rho_A(X_0, \xi_0, t_0) \\
&\quad \times \prod_{k=0}^{n} J(X_{k+1}, \xi_{k+1}, t_{k+1} | X_k, \xi_k, t_k) \\
&\quad \times \exp\left(-\sum_{k=1}^{n} \frac{(X_k - \bar{X}_k)^2}{2\sigma_k^2} - \sum_{k=1}^{n} \frac{(\xi_k - \bar{\xi}_k)^2}{2\sigma_k^2}\right) \tag{14.18}
\end{aligned}
$$

where $X = x + y$, $\xi = x - y$, $\rho_A$ is the initial density matrix of the oscillator and $J$ is the Caldeira-Leggett propagator for the reduced density matrix in the approximation where the influence functional is local in time (see Caldeira and Leggett (1983) and Dowker and Halliwell (1992) for details).

This expression can be evaluated analytically for initial density matrices that are (sums of) exponentials of second order polynomials in $X_0$ and $\xi_0$ since all the integrals are gaussian. We present some of our results in the next section.

## 14.5 Results

### 14.5.1 The Single Wave Packet

We took first the example of a single wave packet initial state, peaked about momentum $p$ and position $x_0$,

$$\rho(X_0, \xi_0, t_0) = exp\left(ip\xi_0 - \frac{1}{\sigma^2}(X_0 - \bar{X}_0)^2 - \frac{1}{\sigma^2}\xi_0^2\right) \tag{14.19}$$

where $\bar{X}_0 = 2\bar{x}_0$. This simple case demonstrated a number of interesting features.

The decoherence functional is calculated to be

$$D[\{\bar{X}_k\}, \{\bar{\xi}_k\}] = \exp\left(-\frac{1}{4}\sum_{kj}\bar{\xi}_k \tilde{M}_{kj}\bar{\xi}_j - \frac{i}{2}U^T M^{-1} V - \sum_{kj}\frac{(\bar{X}_k - Y_k)}{\sigma_k}M_{kj}^{-1}\frac{(\bar{X}_j - Y_j)}{\sigma_j}\right)$$
(14.20)

where $\tilde{M}_{kj}$ and $M_{kj}$ are positive definite matrices depending in a complicated way on the times $\{t_k\}$ and widths $\{\sigma_k\}$ and $\sigma$. $\frac{1}{2}Y_k$ is the position at time $t_k$ along a classical trajectory that starts at $x_0$ with momentum $p$ at time $t_0$. We see that the decoherence functional has the expected qualitative features. The first term in the exponent shows that the decoherence functional is small for large values of $\bar{\xi}_k$, suggesting decoherence. The second term, which is linear in $\bar{\xi}_k$, is purely imaginary. It does not affect the decoherence, and in fact vanishes when $\bar{\xi}_k$ is set to zero. The third term clearly shows that the diagonal part of the decoherence functional is peaked when the slit positions $\bar{x}_k$ lie along the classical trajectory.

An interesting case to look at is that of fine grained position projections at every moment of time. The decoherence functional is then

$$D[x(t); y(t)] = \delta(x_f - y_f)e^{i\tilde{S}[x(t), y(t)]}\exp\left[-2Mk\gamma T\int dt(x - y)^2\right]\rho(x_0, y_0) \quad (14.21)$$

where

$$\tilde{S}[x(t), y(t)] = S_A[x(t)] - S_A[y(t)] + M\gamma\int dt(x - y)(\dot{x} + \dot{y}), \quad (14.22)$$

$k$ is Boltzmann's constant and $\gamma$ is the dissipation. Heuristically, this indicates that there is decoherence on a time scale $\tau = (2Mk\gamma Tl^2)^{-1}$ where $l$ is a typical length scale (Zurek 1986). See, however, remarks in Dowker and Halliwell (1992). Another important observation is that there is no sign of classical peaking, so one needs to coarse grain the position of the special oscillator. As noted by Gell-Mann and Hartle (1991), this is just as expected since one could never know the trajectory of the particle to an accuracy that violated the uncertainty principle.

A tool that has been sometimes used to discuss the peaking of a quantum mechanical state about trajectories in phase space is the Wigner function (Anderson 1990, Calzetta 1989, Habib 1990, Habib and Laflamme 1990, Singh and Padmanabhan 1989, and for extensive reviews see Balazs and Jennings 1984, Hillery et al 1984). In this connection, it is interesting to note here that if an arbitrary initial state gives a decoherence functional that decoheres then the probabilities of the histories are given by

$$p(\{\bar{X}_k\}) = \int dp_0 dX_0 W(p_0, X_0)\exp\left(-\sum_{kj}\frac{(\bar{X}_k - Y_k(p_0, X_0))}{\sigma_k}T_{kj}^{-1}\frac{(\bar{X}_j - Y_j(p_0, X_0))}{\sigma_j}\right)$$
(14.23)

where $W(p_0, X_0)$ is the Wigner transform of the initial density matrix, $T_{kj}$ is a positive definite matrix and $Y_k(p_0, X_0)$ is the classical solution with initial position $X_0$ and initial momentum $p_0$. The appearance of the Wigner function in this context has been noted by Gell-Mann and Hartle (1991).

The form of (14.23) is suggestive of an ensemble of classical paths, with the Wigner transform of the initial density matrix giving the probability distribution of their initial values of position and momentum. This cannot be quite right, however. Firstly, the Wigner function is not always positive, whereas (14.23) is, by construction. Secondly, (14.23) is the probability of a sequence of positions and makes no reference to momenta. The connection with phase space distributions is obtained by considering histories consisting of position samplings at two moments of time. By taking the times very close together, one obtains an approximate position sampling together with a time-of-flight momentum sampling over a short time interval. The resulting probability distribution turns out to be the Wigner function smeared over an $\hbar$-sized region of phase space – just enough to make it positive. These results are described in more detail in a separate paper (Halliwell 1992).

So far we have not analysed Eq. (14.20) quantitatively in terms of the decoherence condition that we proposed in Section 3. We will only note here the general results and refer the reader to Dowker and Halliwell (1992) for details.

In the simplest case of two-time histories without the environment, for short times, decoherence can occur to any desired order by adjusting the widths of the projections. This result is subject to a careful treatment of problems that arise through the approximate nature of the gaussian projections. It was found that the requirements of decoherence and classical peaking compete and one cannot increase one without decreasing the other. In the limit of infinitely long times, there is no decoherence or classical peaking in the case of the free particle. This is an expected result due to the spreading of the wave packet. For the harmonic oscillator, the degrees of decoherence and classical peaking oscillate and do not tend to fixed values in the long time limit.

Next we added the environment. For short times the effect of the environment dropped out and we obtained the no-environment results. For long times, the degrees of decoherence and classical peaking are controlled by a quantity,

$$\sigma_1^2 \alpha_1 = \frac{M(\gamma^2 + \omega^2)}{8kT} \sigma_1^2 \tag{14.24}$$

where $\omega$ is the frequency of the oscillator. This is very roughly the ratio of the energy of the oscillator to the thermal energy of the environment. Classical peaking is increased by increasing $\alpha_1$ and decoherence is increased by decreasing $\alpha_1$. This is in agreement with results of Gell-Mann and Hartle (1991) which say that in order for the system under scrutiny to behave predictably, it must have sufficient inertia to resist the buffeting of the environment. But on the other hand the interaction with the environment must be strong enough to produce the decoherence.

### 14.5.2 *The Double Wave Packet*

In this case we took the initial state to be a superposition of two wave packets. This example shows most clearly how interference is an obstruction to assigning probabilities to histories, and how interference is destroyed by coupling to an environment. It is essentially the double-slit experiment, but paired down to its most basic form.

Consider a particle moving in one-dimension, in a pure state whose wave function at $t = t_0$ is a superposition of wave packets far apart, but moving towards each other. So

$$|\Psi(t_0)\rangle = |\Psi_+(t_0)\rangle + |\Psi_-(t_0)\rangle \tag{14.25}$$

where $|\Psi_+(t_0)\rangle$ is a wave packet at $x = L > 0$, with width $\sigma \ll L$, and with momentum $-p$. Similarly, $|\Psi_-(t_0)\rangle$ is located at $x = -L$, has the same width, but momentum $p$.

First analyse the situation without the environment. The wave packets are approximately orthogonal at $t = t_0$, up to terms of order $\exp(-L^2/\sigma^2)$. Let them meet at the origin at time $t_1$, where they will have substantial overlap. We will assume that the parameters such as the mass of the particle are chosen so that the wave packets do not spread appreciably. In fact, we could consider a harmonic oscillator in which they do not spread at all.

The form of the wave function might tempt one to ascribe definite properties to the history of the particle. In particular, one might wish to say that the particle is in the neighbourhood of either $x = L$ or $x = -L$ at time $t_0$, and then in the neighbourhood of the origin at time $t_1$, with some probability for each of these two histories. We shall show explicitly, however, that this view is not tenable, because this pair of histories do not form a decoherent set.

At time $t_0$, it is sufficient ask whether the particle lies on the positive or negative $x$-axis. This is effected through the projections,

$$P_+ = \int_0^\infty dx\, |x\rangle\langle x|, \qquad P_- = \int_{-\infty}^0 dx\, |x\rangle\langle x|. \tag{14.26}$$

It is easily seen that

$$P_\pm|\Psi_\pm(t_0)\rangle \approx |\Psi_\pm(t_0)\rangle, \qquad P_\pm|\Psi_\mp(t_0)\rangle \approx 0 \tag{14.27}$$

up to terms of order $\exp(-L^2/\sigma^2)$. At time $t_1$, when the wave packets meet, we will ask whether the particle lies in a region of size $\Delta$ around the origin, where $\Delta$ is somewhat less than the wave packet width $\sigma$, but much less than $L$. This proposition is effected by the projection,

$$P_\Delta = \int_{-\Delta/2}^{\Delta/2} dx\, |x\rangle\langle x| \tag{14.28}$$

One has,

$$P_\Delta |\Psi_\pm(t_1)\rangle \approx |x = 0\rangle\langle x = 0|\Psi_\pm(t_1)\rangle \qquad (14.29)$$

An exhaustive set of alternatives at time $t_1$ consists of $P_\Delta$ together with its complement, $1 - P_\Delta$.

The candidate probabilities for the histories in which the particle was either in $x < 0$ or $x > 0$ at $t_0$, and then near the origin at $t_1$ are given by the diagonal elements of the decoherence functional,

$$D(\pm,\pm) = \text{Tr}\left[ P_\Delta\, e^{-iH(t_1-t_0)}\, P_\pm\, |\Psi(t_0)\rangle\langle\Psi(t_0)|\, P_\pm\, e^{iH(t_1-t_0)} \right] \qquad (14.30)$$

The off-diagonal terms, $D(\pm,\mp)$, are given by similar expressions. It is readily shown that the modulus of the off-diagonal terms of the decoherence functional are approximately equal to the diagonal terms

$$|D(\pm,\mp)| \approx D(+,+) \approx D(-,-) \qquad (14.31)$$

and it is *not* possible to satisfy our condition of approximate decoherence Eq. (14.10). The set of histories are therefore not decoherent and the assertion, "the particle was either in $x < 0$ or $x > 0$ at $t_0$, and then near the origin at $t_1$", is meaningless.

Suppose we now couple this system to an environment using the Caldeira-Leggett model. The main difference is that the evolution of the initial density matrix is no longer unitary, but is instead described by the Caldeira-Leggett propagator and one can show that

$$|D(\pm,\mp)| \approx \exp\left(-2\frac{L^2 C}{\sigma^2 \tilde{C}}\right) [D(+,+)D(-,-)]^{\frac{1}{2}} . \qquad (14.32)$$

Here, $\tilde{C} = C + \frac{1}{2\sigma^2}$ and in the short time limit, $C \approx \frac{2}{3}M\gamma kT(t_1 - t_0)$, so

$$\exp\left(-2\frac{L^2 C}{\sigma^2 \tilde{C}}\right) \approx \exp\left(-\frac{8}{3}M\gamma kTL^2(t_1 - t_0)\right) . \qquad (14.33)$$

$M\gamma kTL^2$ can be very large [Zurek 1986].

In the long time limit, $C$ goes to infinity like $e^{2\gamma(t_1-t_0)}$, so

$$\exp\left(-2\frac{L^2 C}{\sigma^2 \tilde{C}}\right) \approx \exp\left(-\frac{2L^2}{\sigma^2}\right) \qquad (14.34)$$

We therefore have very effective decoherence. Probabilities can be assigned to the histories, and it becomes meaningful to say that, "the particle was either in $x < 0$ or $x > 0$ at $t_0$, and then near the origin at $t_1$".

## 14.6 Summary

In this article we have briefly described a calculation of the decoherence functional in the Caldeira-Leggett model of quantum brownian motion. We have demonstrated how decoherence and peaking about classical trajectories, two of the criteria for being able to ascribe "classicality" to a quantum system, emerge quite generally from the formalism. We also saw how the Wigner function of the initial density matrix appears very naturally. In specific cases we have calculated the degrees of decoherence and classical peaking as functions of the widths of the projections and other parameters of the model and shown how the two requirements compete with each other, though in our cases, it seemed there was a parameter range of adequate compromise.

## Acknowledgements

We are grateful to Andreas Albrecht, Murray Gell-Mann, Robert Griffiths, Bei-Lok Hu, Karel Kuchař, Seth Lloyd, Jorma Louko, Miles Blencowe, Roland Omnès, Juan-Pablo Paz, Bob Wald and Wojtek Zurek for useful conversations. We would particularly like to thank Jim Hartle for taking the time to explain his ideas to us, and for many useful conversations over a long period of time. We are grateful to the Aspen Center for Physics for hospitality during the course of this work. H.F.D. warmly thanks Juan Perez-Mercader for organizing the conference in Huelva at which this work was presented. J.J.H. would like to thank Fermilab, and Il Dolce Momento, at which some of this work was carried out.

H.F.D was supported in part by the U.S. Department of Energy (D.O.E.) and by the NASA (grant #2381), at Fermilab. J.J.H. was supported in part by funds provided by the U.S. Department of Energy (D.O.E.) under contract # DE-AC02-76ER03069, at MIT.

## References

Anderson, A. (1990) *Phys. Rev.* **D42**, 585.
Bell, J.S. (1989) Against Measurement, CERN preprint CERN-TH-5611/89.
Balazs, N.L. and Jennings, B.K. (1984) *Phys. Rep.* **104**, 347.
Caldeira, A.O. and Leggett, A.J. (1983) *Physica* **121A**, 587.
Calzetta, E. (19889) *Phys. Rev.* **D40**, 380.
Dowker, H.F. and Halliwell, J.J. (1992) *Phys. Rev.* **D46**, 1580.
Gell-Mann, M. and Hartle, J.B. (1990a) in *Complexity, Entropy and the Physics of Information, SFI Studies in the Sciences of Complexity*, Vol. VIII, Ed. W. Zurek, Addison Wesley, Reading.
Gell-Mann, M and Hartle, J.B. (1990b) in *Proceedings of the Third International Symposium on the Foundations of Quantum Mechanics in the Light of New Technology*, Eds. S. Kobayashi, H. Ezawa, Y. Murayama and S. Nomura, Physical Society of Japan, Tokyo.
Gell-Mann, M. and Hartle, J.B. (1993) *Phys. Rev.* **D47**, 3345.
Griffiths, R.B. (1984) *J. Stat. Phys.* **36**, 219.
Habib, S. (1990) *Phys. Rev.* **D42**, 2566.
Habib, S. and Laflamme, R. (1990) *Phys. Rev.* **D42**, 4056.
Halliwell, J.J. (1992) *Phys. Rev.* **D46**, 1610.
Hartle, J.B. (1991) in *Quantum Cosmology and Baby Universes*, Eds. S. Coleman, J. Hartle, T. Piran and S. Weinberg, World Scientific, Singapore.

Hillery, M., O'Connell, R.F., Scully, M.O. and Wigner, E.P. (1984) *Phys. Rep.* **106**, 121.

Joos, E. and H. D. Zeh, H.D. (1985) *Zeit. Phys.* **B59**, 223.

Omnès, R. (1990) *Ann. Phys.* **201**, 354.

Singh, T.P. and Padmanabhan, T. (1989) *Ann. Phys. (N.Y.)* **196**, 296

Unruh, W.G. and Zurek, W. (1989) *Phys. Rev.* **D40**, 1071.

Zurek, W. (1981) *Phys. Rev.* **D24**, 1516.

Zurek, W. (1982) *Phys. Rev.* **D26**, 1862.

Zurek, W. (1986) in *Frontiers of Nonequilibrium Statistical Physics*, eds. G. T. Moore and Marlan O. Scully, Plenum, NY.

Zurek, W. (1991) *Physics Today* **44**, 36.

# 15

# Two Perspectives on a Decohering Spin

Andreas Albrecht[†]

*NASA/Fermilab Astrophysics Center*
*P.O.B. 500*
*Batavia, IL 60510, USA*

## Abstract

I study the quantum mechanics of a spin interacting with an environment. Although the evolution of the whole system is unitary, the spin evolution is not. The system is chosen so that the spin exhibits loss of quantum coherence, or "wave function collapse", of the sort usually associated with a quantum measurement. The system is analyzed from the point of view of the spin density matrix (or "Schmidt paths"), and also using the consistent histories (or decoherent histories) approach.

## 15.1 Introduction

A cosmologist must face the the issue of utilizing quantum mechanics without the benefit of an outside classical observer. By definition, there is nothing "outside" the universe! The traditional role of an outside classical observer is to cause "wavefunction collapse". This process causes a definite outcome of a quantum measurement to be realized, with the probability for a given outcome determined by the initial wavefunction of the system being measured. It is common to view this process as something that cannot be described by a wavefunction evolving according to a Schrödinger equation, but which instead must be implemented "by hand".

There is a growing understanding that the essential features of wavefunction collapse *can* be present in systems whose evolution is entirely unitary. Pioneering work on this subject has been done by Zeh (1973), Zurek (1981, 1982, 1986), Joos and Zeh (1985), and Unruh and Zurek (1989), building off of ideas of Everett (1957), and von Neumann (1955). The key is the inclusion of an "environment" or "apparatus" within the Hilbert space being studied. A subsystem can exhibit the non-unitary aspects of wavefunction collapse even though the system as a whole evolves unitarily. The wavefunction can then divide up into a number of different terms, each of which reflect a different "outcome". When there is negligible interference among the different terms during subsequent evolution, the "definiteness" of the

---

† Present address: Theory Group, Blackett Laboratory, Imperial College, London, SW7 2BZ, UK

outcome is realized in a restricted sense: Each term evolves as if the others were "not there", so a subsystem state within a given term evolves with "certainty" that its corresponding outcome is the only one. None the less, the total wavefunction describes all possible outcomes, and one is never singled out.

In this work I study a two state "spin" system (subsystem 2) coupled to a 25-state "environment" or "apparatus" system (subsystem 1). The dynamics and the initial state are chosen to give the following behavior: The initial state is

$$|\psi_i\rangle = (a|\uparrow\rangle_2 + b|\downarrow\rangle_2) \otimes |X\rangle_1, \tag{15.1}$$

which evolves into the state

$$|\psi_f\rangle = a|\uparrow\rangle_2 \otimes |Y\rangle_1 + b|\downarrow\rangle_2 \otimes |Z\rangle_1 \tag{15.2}$$

Where $\langle Y|Z\rangle = 0$. This type of evolution is central to the standard way of describing a quantum measurement in the absence of an outside classical observer. Initially the spin subsystem is in a pure state, $(a|\uparrow\rangle_2 + b|\downarrow\rangle_2)$. For the state $|\psi_f\rangle$, the reduced density matrix of the spin ($\rho_2 \equiv \mathrm{tr}_1 (|\psi_f\rangle\langle\psi_f|)$) has two non-zero eigenvalues, and the eigenstates are $|\uparrow\rangle$ and $|\downarrow\rangle$. The spin is no longer in a pure state, but may be said to be in $|\uparrow\rangle$ with probability $a^*a$ and $|\downarrow\rangle$ with probability $b^*b$.

In $|\psi_f\rangle$ each of the spin states ($|\uparrow\rangle$ and $|\downarrow\rangle$) is uniquely correlated with its own state of the environment ($|Y\rangle_1$ and $|Z\rangle_1$ respectively) In this sense the environment has "measured" the spin. The two terms in Eq. (15.2) represent the two possible "outcomes" of the measurement. The fact that initially the probability to find the spin in $|\uparrow\rangle$ or $|\downarrow\rangle$ is also $a^*a$ or $b^*b$ (respectively) illustrates that the measurement is "good": The probabilities of the different outcomes can be predicted from the initial wavefunction of the spin.

Another requirement of a good measurement is that $|\psi_f\rangle$ does not evolve back into the form of $|\psi_i\rangle$. This would amount to the environment "forgetting" the outcome of the measurement. This property is also well exhibited by the toy model studied here. Other features which are needed to match our intuitive notion of a good quantum measurement have to do with interactions of more than two subsystems. For example, one would want agreement among many observers that a particular outcome has been realized. Models can be constructed which exhibit this effect, but that is beyond the scope of this work.

The motivations for this work are twofold. The first goal is to develop some intuition as to what requirements one must place on the dynamics and initial state to produce the behavior just described. Secondly, I wish to explore the links between this approach and the "consistent histories" approach to the study of closed quantum systems (developed by Griffiths 1984, Omnès 1988, and Gell-Mann and Hartle 1990).

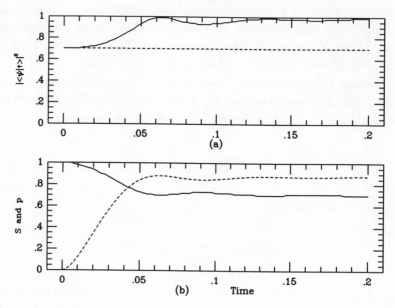

Fig. 15.1. (a): The solid curve is $|\langle\uparrow|1\rangle^S|^2$, and the dashed curve gives $\langle\uparrow|\rho_2|\uparrow\rangle$. (b): The solid curve is the largest eigenvalue of $\rho_2$, the dashed curve is the entropy of the spin.

## 15.2 Results

The analysis presented here follows closely that of Albrecht (1992), and I refer the reader there for a description of the Hamiltonian and other details about the calculation. However, the results presented here are qualitatively different (see especially Section 6.3 of Albrecht 1992).

(For those interested in the technicalities, here are the differences from Albrecht (1992): The environment is larger, with $n_1 = 25$, and the couplings are $E_1 = .1$, $E_2 = .1$ and $E_I = 10$. Most importantly, the initial environment state is a coherent superposition of one eigenstate of $H^\uparrow$ and one eigenstate of $H^\downarrow$, in equal proportions. It is this difference which produces distinctively different behavior. )

Figure 15.1 shows information about the spin as the whole system evolves. Initially, the state is given by Eq. (15.1), with $a = 0.7, b = 0.3$. In the lower plot, the solid curve gives $p_1$, the largest eigenvalue of $\rho_2$. It starts out at unity, as required by the "pure state" form of the initial conditions, and evolves to 0.7, where it holds steady. The dashed curve gives the entropy, $S$, of the spin ($S \equiv -\text{tr}[\rho_2 \log_2(\rho_2)]$), in units where the maximum possible entropy in unity. The entropy starts out zero and increases. This is always the case when a system evolves from a pure to a mixed state. (Note the the combined "spin $\otimes$ environment" system remains in a pure state, so *its* entropy is zero)

In the upper plot, the dashed curve gives the overall probability for the spin to be

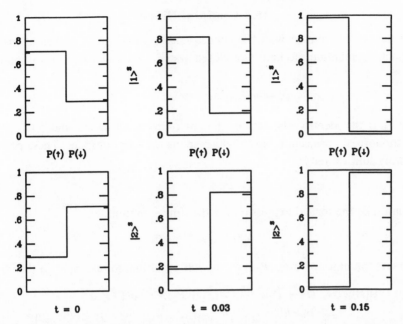

Fig. 15.2. "A collapsing wavefunction." Each plot depicts an eigenstate of $\rho_2$ in terms of $p(\uparrow) \equiv |\langle \uparrow |i\rangle|^2$ and $p(\downarrow) \equiv |\langle \downarrow |i\rangle|^2$. The columns correspond to three different times. The two rows correspond to the two eigenstates.

up, given by $\langle \uparrow |\rho_2|\uparrow \rangle$. This quantity is a "constant of the motion". The solid curve gives $|\langle \uparrow |1\rangle^S|^2$, where $|1\rangle^S$ is the eigenstate of $\rho_2$ (or "Schmidt state") corresponding to the largest eigenvalue. As discussed in Zeh (1973) and Albrecht (1992), the density matrix eigenstates correspond to a "Schmidt decomposition" of $|\psi\rangle$ (Schmidt 1907). When $|\psi\rangle$ is expanded in the eigenstates of $\rho_2$ and $\rho_1$, it always takes the form given by Eq. (15.2), with each eigenstate of $\rho_2$ uniquely correlated with an eigenstate of $\rho_1$. This fact means the eigenstates of $\rho_2$ not only tell about $\rho_2$, but about the correlations with system 1 as well.

Since $|1\rangle^S$ belong to a two state Hilbert space, it is completely specified by $|\langle \uparrow |1\rangle^S|^2$, up to an overall phase. One can see that as the eigenvalue $(p_1)$ approaches 0.7, the eigenvector becomes essentially $|\uparrow\rangle$. Thus the behavior promised in the previous section (Eqs. (15.1) and (15.2)) is realized to a good accuracy.

Figure 15.2 is another representation of the way the eigenstates of $\rho_2$ evolve. The first row represents $|1\rangle^S$, and the second row represents the other eigenvector. The three columns correspond to three times. The histogram in each plot provides two numbers, $p(\uparrow) \equiv |\langle \uparrow |1\rangle|^2$ and $p(\downarrow) \equiv |\langle \downarrow |1\rangle|^2$ for the first row, and similarly for the second eigenvector in the second row. In this way one can visualize a "collapsing wavefunction" by following the eigenstates of $\rho_2$.

## 15.3 Consistent Histories

I will now make contact with the "consistent histories" or "decoherent histories" approach to quantum mechanics of closed systems. Consider two projection operators:

$$\hat{P}_\uparrow \equiv |\uparrow\rangle\langle\uparrow| \otimes I_1; \quad \hat{P}_\downarrow \equiv |\downarrow\rangle\langle\downarrow| \otimes I_1 \tag{15.3}$$

where $I_1$ is the identity operator in the environment subspace, and $\{|\uparrow\rangle, |\downarrow\rangle\}$ form an orthonormal "projection basis" which spans the spin subspace. These projection operators sum to unity:

$$\hat{P}_\uparrow + \hat{P}_\downarrow = I. \tag{15.4}$$

One can take the formal expression for the time evolution:

$$|\psi(t)\rangle = e^{-iHt}|\psi(0)\rangle \tag{15.5}$$

and insert the unit operator $(\hat{P}_\uparrow + \hat{P}_\downarrow)$ at will, resulting, for example, in the identity:

$$\begin{aligned}
|\psi(t)\rangle &= (\hat{P}_\uparrow + \hat{P}_\downarrow)e^{-iH(t-t_1)}(\hat{P}_\uparrow + \hat{P}_\downarrow)e^{-iHt_1}|\psi(0)\rangle \tag{15.6}\\
&= \hat{P}_\uparrow e^{-iH(t-t_1)}\hat{P}_\uparrow e^{-iHt_1}|\psi(0)\rangle + \hat{P}_\uparrow e^{-iH(t-t_1)}\hat{P}_\downarrow e^{-iHt_1}|\psi(0)\rangle \\
&\quad + \hat{P}_\downarrow e^{-iH(t-t_1)}\hat{P}_\uparrow e^{-iHt_1}|\psi(0)\rangle + \hat{P}_\downarrow e^{-iH(t-t_1)}\hat{P}_\downarrow e^{-iHt_1}|\psi(0)\rangle \tag{15.7}\\
&\equiv |[\uparrow,\uparrow]\rangle + |[\uparrow,\downarrow]\rangle + |[\downarrow,\uparrow]\rangle + |[\downarrow,\downarrow]\rangle. \tag{15.8}
\end{aligned}$$

The last line just defines (term by term) a shorthand notation for the previous line. Each term represents a particular choice of projection at each time, and in that sense corresponds to a particular "path". In the path integral formulation of quantum mechanics the time between projections is taken arbitrarily small, and the time evolution is viewed as a sum over paths. For present purposes, the time intervals can remain finite, representing a "coarse graining" in time. Each term in the above expression is called a "path projected state", and the sum is a sum over coarse grained paths.

One attempts to assign the probability "$\langle[i,j]|[i,j]\rangle$" to the path $[i,j]$, but to make sense, the probabilities must obey certain sum rules. For example, one can define

$$|[\uparrow,\cdot]\rangle \equiv |[\uparrow,\uparrow]\rangle + |[\uparrow,\downarrow]\rangle, \tag{15.9}$$

where the "$\cdot$" signifies that *no* projection is made at $t_1$. One would want the probability for the path $[\uparrow,\cdot]$ to be the sum of the probabilities of the two paths of which it is composed:

$$\langle[\uparrow,\cdot]|[\uparrow,\cdot]\rangle = \langle[\uparrow,\uparrow]|[\uparrow,\uparrow]\rangle + \langle[\uparrow,\downarrow]|[\uparrow,\downarrow]\rangle \tag{15.10}$$

However, one can "square" Eq (15.9) to give the general result:

$$\langle[\uparrow,\cdot]|[\uparrow,\cdot]\rangle = \langle[\uparrow,\uparrow]|[\uparrow,\uparrow]\rangle + \langle[\uparrow,\downarrow]|[\uparrow,\downarrow]\rangle + \langle[\uparrow,\uparrow]|[\uparrow,\downarrow]\rangle + \langle[\uparrow,\downarrow]|[\uparrow,\uparrow]\rangle \tag{15.11}$$

Only if the last two terms in Eq (15.11) are small is the sum rule (Eq (15.10)) obeyed.

| Table 1a | | Table 1b | |
|---|---|---|---|
| path | value | path | value |
| $\langle [\uparrow\uparrow] \mid [\uparrow\uparrow] \rangle$ | 0.70 | $\langle [\mathcal{I}\mathcal{I}] \mid [\mathcal{I}\mathcal{I}] \rangle$ | 0.74 |
| $\langle [\uparrow\downarrow] \mid [\uparrow\downarrow] \rangle$ | 0.00 | $\langle [\mathcal{I}\perp] \mid [\mathcal{I}\perp] \rangle$ | 0.03 |
| $\langle [\uparrow \cdot] \mid [\uparrow \cdot] \rangle$ | 0.70 | $\langle [\mathcal{I}\cdot] \mid [\mathcal{I}\cdot] \rangle$ | 0.61 |
| % violation | 0% | % violation | 25% |

Table 15.1. *Testing the probability sum rule (Eq. (15.10)) for different paths. For 1a the sum rules are obeyed for any choice of $t_1$ and $t$. For 1b, $t_1 = .035$ and $t = 0.06$*

When the relevant sum rules are obeyed the paths are said to give "consistent" or "decohering" histories. Advocates of this point of view argue that the only objects in quantum mechanics which make physical sense are sets of consistent histories. For a discussion of how this simple example links up with the (much more general) original work on this subject (Griffiths 1984, Omnès 1988, and Gell-Mann and Hartle 1990), see Albrecht (1992).

## 15.4  Testing for Consistent Histories

Table 15.1a checks the probability sum rule (Eq. (15.10)) for the toy model whose evolution is depicted in Fig 15.1. The projection times are $t_1 = .15, t = .2$, and the projection basis is $\{|\uparrow\rangle, |\downarrow\rangle\}$. The sum rule is obeyed to the accuracy shown. In fact, using the $\{|\uparrow\rangle, |\downarrow\rangle\}$ projection basis, the sum rule is obeyed no matter which projection times are chosen.

This result came as a surprise to me. After all the interesting behavior described in Figs 15.1 and 15.2, the consistent histories approach tells us that "$\uparrow$" and "$\downarrow$" paths are the right way to view the system, right through the period of "wavefunction collapse".

Consider for a moment a static (Hamiltonian = 0) spin, not coupled to any environment. It turns out that as long as the same projection basis is chosen at $t$ and $t_1$, one always gets consistent histories. This is true for any projection basis. One could choose $\{|\uparrow\rangle, |\downarrow\rangle\}$ or one could choose the projection basis $\{|\mathcal{I}\rangle, |\perp\rangle\}$, where $|\mathcal{I}\rangle$ is the initial state of the spin $(a|\uparrow\rangle_2 + b|\downarrow\rangle_2)$, and $|\perp\rangle$ is the state orthogonal to it. A static spin would naturally result in unit probability for the $[\mathcal{I}, \mathcal{I}]$ path, and zero probability for all other paths.

Table 15.1b shows the results for the fully interacting spin, using the $\{|\mathcal{I}\rangle, |\perp\rangle\}$ projection basis, but otherwise the same as Table 15.1a. Clearly the sum rules are not obeyed in this case. When this talk was presented, I felt that these results suggested the following link between the "setting up of correlations" described in Sections 1 and 2, and the consistent histories: The setting up of correlations tends to reduce number of different sets of consistent histories, and help one single out a preferred choice of projection basis.

Since then I have realized that things are not so simple. For one, the space of possible choices of projection basis is extremely large, even for the simple example discussed here. I have found the following new sets of consistent histories which are *not* consistent for the static spin. One chooses the projection basis at $t_1$ to be the eigenstates of $\rho_2$ at that time, and the projection basis at $t$ to be $\{|\uparrow\rangle, |\downarrow\rangle\}$. These histories are consistent to the same accuracy as those shown in Table 15.1a, for any choice of $t_1$ and $t$.

For this article, I will not try to conclude anything about numbers of sets of consistent histories in the static versus interacting cases. I will simply remark that the special behavior of the correlations described by Eqs. (15.1) and (15.2) does manifest itself in the consistent histories approach. It is interesting that this behavior does not show up in all consistent histories. After all, the $\{|\uparrow\rangle, |\downarrow\rangle\}$ projection basis generates consistent histories which are indistinguishable from those of an isolated static spin. With the inclusion of the interactions, however, there are new sets of consistent histories (such as those described in the previous paragraph) which do reflect the special evolution of $\rho_2$.

### 15.4.1 Consistent Histories vs Schmidt Paths

There is a view (well represented at this workshop) that says that consistent histories are the only physically correct objects to discuss in quantum mechanics. I am often asked if following the density matrix (or Schmidt paths) amounts to an alternative "interpretation". My present view is the following: The Schmidt analysis presented in the first part of this paper describes the evolution of correlations among subsystems. It is the point of view I am most familiar with. If I were to ask under what circumstances the evolution of correlations is physically interesting, there would be a number of requirements, such as stability or simplicity of evolution, which would come into play. Under these circumstances I suspect that the two approaches should be essentially equivalent.

For example, one could ask just what it means to follow the wavefunction through the collapse process, as described in Fig 15.1 and 15.2 of this paper. Surely it means nothing unless someone or something makes an observation and "catches it in the act". This would require interactions with a third system, presumably proceeding in a similar manner to the measurement depicted here. Such an interaction would have an impact on the consistent histories. I suspect, for example, that the $\{|\uparrow\rangle, |\downarrow\rangle\}$ would only remain a good projection basis during the collapse process if the new measurement was also made in the $\{|\uparrow\rangle, |\downarrow\rangle\}$ basis.

In particular, I expect the equivalence of the two approaches to emerge from the role correlations among subsystems play in allowing the probability sum rules to be

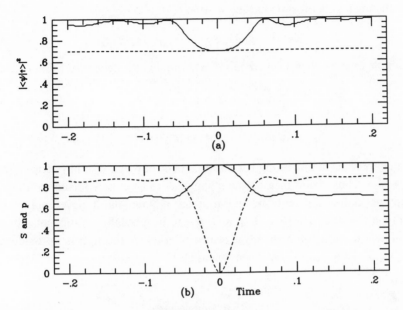

Fig. 15.3. The same plots as Fig 1 extended back to $t = -2$. The "entropy dip" (dashed curve, lower plot) illustrates the special low entropy property of the initial ($t = 0$) state.

obeyed. (This has been elaborated to some degree in Albrecht 1992.) However, I still would like to better understand the relationship between the two approaches.

## 15.5 The Arrow of Time

As has been noted, for example by Zurek (1982) and Zeh (1971, 1990), there is an arrow of time built into the dynamics discussed here. This is dramatized in Fig 15.3, which is identical to Fig 15.1, but the x-axis extended back to $t < 0$. One can seen that the pure "initial" ($t = 0$) state (which has zero entropy for the spin), is a very special state and the "collapse of the wavefunction" proceeds in the direction of increasing spin entropy. The $t < 0$ part of Fig 15.3 illustrates an "un-collapsing" wavefunction, where the correlations present between spin and environment at early times are lost, and the pure state emerges at $t = 0$. Then, for positive values of $t$ correlations are established again. The stability of these correlations (and thus the goodness of the measurement) depend on another such "entropy dip", not occurring for $t > 0$. Even the simple system discussed here is complex enough for such entropy dips to occur very rarely.

Aside from questions of stability, how fundamentally is the arrow of time linked to quantum measurement? The initial state, $|\psi_i\rangle$ has zero entropy for the spin, so it is not surprising that just about anything will cause the entropy to increase. What about starting with a more general initial state? Schmidt tells us that (in a suitable basis) the most general state can be written

$$|\tilde{\psi}_i\rangle = \sqrt{p_1}|1\rangle_2 \otimes |1\rangle_1 + \sqrt{p_2}|2\rangle_2 \otimes |2\rangle_1. \tag{15.12}$$

It is simple to show that if one requires evolution which generalizes Eq (15.2) to give

$$|\tilde{\psi}_i\rangle \quad \rightarrow \quad |\tilde{\psi}_f\rangle \tag{15.13}$$

$$= \quad \sqrt{p_1}\left(\langle\uparrow\,|1\rangle_2|\uparrow\rangle \otimes |A\rangle_1 + \langle\downarrow\,|1\rangle_2|\downarrow\rangle \otimes |B\rangle_1\right) \tag{15.14}$$

$$+\sqrt{p_2}\left(\langle\uparrow\,|2\rangle_2|\uparrow\rangle \otimes |C\rangle_1 + \langle\downarrow\,|2\rangle_2|\downarrow\rangle \otimes |D\rangle_1\right) \tag{15.15}$$

then one must have increasing (or constant) entropy of the spin $(-\text{tr}[\rho_2 \ln(\rho_2)])$ as $|\tilde{\psi}_i\rangle \rightarrow |\tilde{\psi}_f\rangle$. Thus "good measurement" appears to be closely linked with increasing entropy, even for high entropy initial states. (Note that I have chosen all four environment states, $|A\rangle_1$, $|B\rangle_1$, $|C\rangle_1$, and $|D\rangle_1$ to be mutually orthogonal. This means that in $|\tilde{\psi}_f\rangle$ the environment has a record of whether the spin is up or down, *and* which term of Eq. (15.12) has been "measured".)

## 15.6 Conclusions

The ideas put forward by Zeh (1973), Zurek (1982), Joos and Zeh (1985), and Unruh and Zurek (1989), have sufficiently de-mystified the notion of wavefunction collapse that one can actually unitarily follow the evolution of a system right through the collapse process. I have investigated a simple system which exhibits "wavefunction collapse". I find Zeh's idea of watching the evolution of the eigenstates of the reduced density matrix particularly appealing. This approach allows one to follow exactly the evolution of the correlations among subsystems. It also allows one to visualize the collapse process quite explicitly, as illustrated in Fig 15.2.

I also applied the "consistent histories" analysis (of Griffiths 1984, Omnes 1988, and Gell-Mann and Hartle 1990) to the same system. This approach allows one to consider may different sets of histories for the system. In the example studied here, many different sets passed the consistency test. It is intriguing that one set of consistent histories for the spin did not reflect the interesting evolution of the correlations between the spin and the environment. That set of histories would look the same for a static spin, decoupled from the environment. Other consistent histories exhibited quite direct links to the "quantum measurement" process underway.

I have described a system which exhibits interesting behavior, both in terms of evolving correlations, and in terms of consistent histories. All the behavior discussed here lends itself to simple explanation in terms of the nature of the Hamiltonian and the initial state. Understanding this relationship is one of the main goals of this work, and it will be spelled out in a future publication.

**Added note:** The "future-publication" now exists (Albrecht, 1993). It provides a more thorough discussion of the relationship between the consistent histories and the Schmidt paths. The relationship between the behavior of the system, the form of the Hamiltonian and the initial state is discussed. The utility of the Schmidt decomposition in the degenerate case is investigated.

## Acknowledgements

This work was supported in part by the DOE and the NASA (grant NAGW-2381) at Fermilab. Special thanks are due to the organizers for a very stimulating workshop.

## Discussion

**Barbour** Could you just clarify one point? Do you accept the many-worlds interpretation and put quotes around collapse to mean that it is effective collapse brought about by decoherence?

**Albrecht** Yes

**Page** Isn't it true that when there are interactions you don't lose consistent histories, but that the projection operators are no longer products of projection operators on the subsystems?

**Albrecht** It would seem that with complete freedom to choose the projections the total number of consistent histories might not change when the dynamics are changed. (But could approximately consistent histories become *less* consistent?) In any case, under a change of dynamics one can certainly lose (or gain) the ability to describe consistent histories for a particular subsystem, just as you say.

**Lloyd** When are the trajectories identified by Schmidt paths decohering histories? When are they not?

**Albrecht** My views are still evolving on this question, so here is an "anecdotal" answer: (a) The case of the static, isolated spin has but a single Schmidt path. However, there are many sets of consistent histories, each determined by an arbitrary initial choice of projection basis (which then must be used at all other projection times as well). (b) For the coupled spin, there are at most two Schmidt paths, while, with a suitable environment there could be sets with many more than two consistent histories for the spin. However, in that case the consistency of the histories must be due to correlations with the environment. I will bet that if the necessary correlations are there to produce a set of multiple consistent histories, then these correlations can be exposed by further subdividing the environment according to Schmidt (resulting in an increase in the number of Schmidt paths). This type of analysis may bring the two points of view very close together.

**Omnès** Concerning the question of finding the relevant "classical" properties and knowing whether they exist, by starting from first principles, I strongly suspect that the main tools and maybe very significant results are already contained in a paper by Charles Fefferman (*Bull. Am. Math. Soc.* **9**, 129 (1983)). We should urge him to publish his proofs at last so that they might be interpreted from the standpoint of physics.

**Hartle** I want to address the question of the number of different sets of alternative decohering histories. If we restrict attention to *exact* decoherence and a pure initial state, then exhibiting all sets of decohering histories is a purely algebraic process. One picks a time later than the beginning and resolves the initial state vector into some number of mutually orthogonal vectors. One then picks a later moment and resolves those vectors into a further set of mutually orthogonal vectors, and so on. The end result is a resolution of the initial state vector into a number of orthogonal "branches". Each branch corresponds to a history, and the whole set is exactly decoherent. There are thus a very large number of sets of decohering histories. However, all this by itself is only an exercise in Hilbert space. We can distinguish among these different sets when they are described in terms of quantum fields. Then the interesting question is whether among the large number of exactly decohering sets of histories and the approximately decohering ones, there are one or more that correspond to a quasiclassical domain. More precisely there is the question of whether there are decohering sets of histories, as non-trivially refined as possible consistent with decoherence, that have high probabilities for classical patterns of correlation in time.

**Albrecht**  Agreed. Part of the motivation of this work is to see how the process of "quantum measurement" might be reflected in the decohering histories.

## References

Albrecht, A. (1992) *Physical Review* **D46**, 550.

Albrecht, A. (1993) *Physical Review* **D**, in press.

Everett, H. (1957) *Reviews of Modern Physics*, **29**, 454.

Gell-Mann, M, and Hartle, J.B. (1990) Quantum mechanics in light of quantum cosmology, In *Complexity, Entropy, and the Physics of Information*, ed. W. Zurek, Addison Wesley.

Griffiths, R. (1984) *Journal of Statistical Physics* **36**, 219.

Joos, E and Zeh, H.D. (1985) *Zeitschrift für Physik* **B59**, 223.

Omnès R. (1988) *Journal of Statistical Physics* **53**, 893, 933, and 957.

Unruh, W. and Zurek W.H. (1989) *Physical Review* **D40**, 1071.

von Neumann, J. (1955) *Mathematical Foundations of Quantum Mechanics*, Princeton University Press, Princeton.

Zeh, H.D (1971) On the irreversibility of time and observation in quantum theory. In *Enrico Fermi School of Physics IL*, Edited by B. d'Espagnat, Academic Press.

Zeh, H.D. (1973) *Foundations of Physics* **3**, 109.

Zeh, H.D. (1990) Quantum measurements and entropy, in *Complexity, Entropy, and the Physics of Information*, ed. W. Zurek, Addison Wesley.

Zurek, W.H. (1981) *Physical Review* **D24**, 1516.

Zurek, W.H. (1982) *Physical Review* **D26**, 1862.

Zurek, W.H. (1986) Reduction of the wave packet: How long does it take? In *Frontiers of nonequilibrium statistical physics*, eds. G.T. Moore and M. Scully, Plenum, NY.

# Part Four

Time Asymmetry and Quantum Mechanics

# 16

# Is Time Asymmetry Logically Prior to Quantum Mechanics?

William K. Wootters

*Department of Physics, Williams College*
*Williamstown, MA 01267 USA*

## Abstract

In trying to imagine how quantum mechanics might be derived from a more fundamental theory, the author is led to consider a framework in which time-asymmetric events, rather than reversible unitary transformations, are taken as basic. In such a scheme it seems likely that quantum mechanics and the second law of thermodynamics would emerge together, both being consequences of a deeper law.

## 16.1 Introduction

Much of the work prsented at this workshop is aimed at understanding the origin of time asymmetry in quantum mechanical terms, taking the formalism of quantum mechanics as fundamental. In this paper I will pursue a line of thinking that entails precisely the opposite relation between quantum mechanics and time asymmetry, namely, that time asymmetry is fundamental and quantum mechanics is derivative.

To begin, let me recall how it happens that we are faced with a paradox of time asymmetry in quantum mechanics. From a pragmatic point of view, quantum mechanics can be thought of as a theory that predicts probabilities of the outcomes of certain measurements given certain prior measurements. One does not in principle have to talk about what goes on between measurements. However, the theory suggests strongly that something does go on between measurements, namely, a unitary evolution (either of the state vector or of the operators). It is very natural that many of us have come to take this unitary evolution not only as fundamental, but also as inviolable. In the Schrödinger representation, this way of thinking leads to a picture of the universe as an evolving wavefunction with no collapse, that is, the Everett picture. The dynamics of the wavefunction in this picture is reversible. That is, for each possible evolution, there is a corresponding time-reversed evolution which is equally allowed by the laws of physics. And yet our world is manifestly asymmetric in time. Herein lies the paradox of time asymmetry in the quantum context.

259

However, this view is not the only one possible. According to the Copenhagen interpretation, measurements are not like other processes but have a special status. They do not correspond to reversible unitary transformations but rather to irreversible projections. If one adopts this view, then one appears to have gotten around the paradox of time asymmetry, or at least the most blatant form of the paradox. There is no longer any glaring incongruity between the world we experience and the laws of physics: both are time-asymmetric. (Landau and Lifshitz 1958, p. 31. For a recent review, see Zeh 1989.)

In one respect, this approach to a solution is very appealing. The emergence of a definite outcome of a quantum measurement, if one regards it as a fundamental process, is the only thing in fundamental physics that is not reversible. It is therefore a natural place to look for the origin of time asymmetry.

Of course there are problems with this idea. If measurements are treated as special, then we would appear obligated to say exactly what conditions define a measurement. This obligation brings us directly to the ancient Problem Of Measurement in quantum mechanics. In blaming time asymmetry on the measurement process, one seems to be replacing the problem of time asymmetry with the more formidable problem of measurement, while at the same time excluding from consideration the Everett approach, which many regard as the only viable foundation for quantum cosmology. Moreover, in trying to define "measurement," one would certainly be tempted to invoke the irreversibility of thermodynamics, which is precisely what one wanted to explain in the first place (Zeh 1971).

Despite these objections, I wish to explore briefly in this paper a line of thinking based on the view just presented, to see where it might lead. My main reason for doing so is spelled out in the following section.

## 16.2 What's Behind Quantum Mechanics?

We usually treat the principles of quantum theory as though they were absolutely fundamental. We do not normally try to explain quantum theory; rather we use it as the underpinning for the rest of physics. However, it seems to me highly unlikely that nature could have chosen the particular structure of quantum mechanics for no particular reason. One can obviously imagine many other possible worlds with very different frameworks for physics: a probabilistic world without complementarity, for example; or a quantum-like world in which all probability amplitudes are real. In this section I would like to try to imagine what answer one might give to the question, "Why quantum mechanics?"

One of the most striking features of quantum theory is its universality. Consider, for example, the quantum mechanics of systems that have just two orthogonal states. Regardless of whether one is thinking of the polarization of a photon, or the spin of a nonrelativistic electron, or the identity of a mixed muon-electron neutrino, the theory is essentially the same: the allowed pure states can be represented by two-component complex vectors; probabilities are computed by squaring inner products between such vectors; and so on. The essential quantum framework is independent of the actual physical system in which it is instantiated. Therefore, if there is to be a reason for quantum mechanics, it should not be a reason that is at all system-specific. The universality of quantum theory suggests that its explanation will rely on only the most central and universal concepts that one finds in physical theory.

The most obvious such concept to consider is that of probability. The framework of quantum mechanics is, after all, ultimately a framework for computing probabilities. Once one abstracts away from all the specific systems to which we apply quantum theory, one is left with what seems to be a summary description of the way in which nature determines probabilities: in brief, nature determines probabilities by squaring complex amplitudes. It is this prescription that one wants to explain.

Now there are at least two possible ways to proceed. One hypothesis is that the above prescription is a convenient summary of some underlying counting process, through which each probability is determined as the ratio of one large integer to another. A model along these lines was proposed many years ago by Penrose (Penrose 1971). He imagined a large network of points connected by lines, with an integer associated with each line. The lines he interpreted as physical systems, the integers being their total angular momenta in units of $\hbar/2$. The network can be used to model the production of a polarized spin-$\frac{1}{2}$ particle by one apparatus and the subsequent measurement of its spin along a different axis by another apparatus. The probabilities of the two possible outcomes of this measurement are computed according to well defined counting rules. Out of this scheme Penrose was able to derive the familiar set of pure states of a spin-$\frac{1}{2}$ particle, along with the standard inner-product rule for computing probabilities.

I should point out that Penrose did not claim that his model should be taken as *the* explanation of quantum mechanics. Too much of the desired result is built into the counting rules, which are somewhat more baroque than one would expect of a basic law of nature. But I think the model serves as a valuable illustration of how one might derive the quantum framework from a combinatorial argument.

The other approach I have in mind is to try to derive the quantum framework from a simple principle. My own work suggests the following principle: The set of pure states of any system constitutes a Riemannian manifold, in which the distance between states is given by their distinguishability (Wootters 1981). This principle, when spelled out more precisely, strongly suggests a vector space structure for the set of states, with probabilities as the squares of components. It does not, however, favor a complex vector space over a real vector space; if anything it does just the opposite. Nevertheless, I regard the relation between distinguishability and the geometry of states as a genuine clue to the underlying explanation of quantum theory. One can imagine a more far-reaching principle which would encompass the one stated above and would yield quantum theory uniquely.

I now come to the main point of this section. No matter which of the above approaches one takes, the aim of the search is to deduce the structure of quantum mechanics *via* the probabilities. That is, one hopes to obtain, from an argument that assumes nothing about Hilbert space or unitary transformations, a relationship among probabilities that is most conveniently *summarized* in terms of Hilbert space and unitary transformations. Probabilities come logically before, not after, state vectors. In particular, the existence of "measurement events," the probabilities of whose outcomes are to be computed, is taken here to be fundamental. Without such events one has nothing to work with, no starting place on which to base a sub-quantum theory. These events thus play a role similar to that played by molecules in the kinetic theory of gases. (This idea is very much in the spirit of Wheeler 1990.) In what follows I will replace the term "measurement event" with the more generic term "branching event," so as not to suggest that anyone actually has to set up an apparatus or see the outcome in order for an event to be allowed in the description. A branching event is any event that (i) defines a set of possibilities and (ii) selects one of these possibilities, such that the outcome becomes an irrevocable component of the history of the universe. A photon encountering a beam-splitter does not constitute a branching event, but the subsequent detection of the photon by a photomultiplier does. (More on this shortly.)

The above considerations lead me to consider the following picture of the world: The world consists, at bottom, of branching events. They are the building blocks. Moreover, each of these branching events carries with it its own direction of time. It has a "before," when the outcome of the event is not determined, and an "after," when it is. A branching event is thus inherently irreversible; so in such a scheme time asymmetry would be built into the very foundations of physics. This asymmetry would be logically prior to quantum mechanics in that quantum mechanics itself would be derived from a sub-quantum principle or construction that is expressed in terms of branching events.

Filling in the picture a bit more, I imagine the history of the world as a network, or graph, of connected branching events. In order to compute the probabilities of the outcomes of a given event, one looks at the outcomes of other events and uses the Great Underlying Probability Law—this is the principle or construction that I assume will be discovered and that underlies quantum mechanics—to compute the probabilities. Moreover, considering that each individual branching event has a temporal orientation, I assume that this orientation plays a non-trivial role: in the network of events, a connecting link can be attached either to the "before" end or to the "after" end of an event, and the Great Underlying Law distinguishes between these two types of connection. The picture is thus quite asymmetric, but a very natural one to consider if one is looking for a sub-quantum theory based on probabilities.

One could object that there is an alternative, "time-free" way to introduce probabilities, namely, to start with a static joint probability distribution over the outcomes of all possible events. Ordinary everyday probabilities would then emerge as conditional probabilities derived from this global distribution. I appreciate the appeal of this approach (cf. Page and Wootters 1983). Note, however, that in order for these conditional probabilities to be usefully applicable to any given event, we must assume that there are observers whose total experience will include both a state of not knowing the outcome and a state of know outcome; that is, there must implicitly be a "before" and an "after." I admit that it is a non-trivial step to transfer this "before" and "after" from the *observers* to the *event*—what this step does in effect is to enforce a certain kind of agreement among the observers—but everything in our experience supports this step, and frankly, if one is trying to derive quantum mechanics, one wants to take advantage of whatever solid ground one can find to stand on. Hence my use of branching events.

Before addressing some of the questions raised by the notion of a branching event, I would like to say a few more words about the business of deriving quantum mechanics. One may well ask: how would we know if we had found the correct sub-quantum theory? Is it not always going to be a matter of aesthetic judgment? I would say that the answer to this question is the same as in any case of finding a deeper theory in physics: one believes the theory to be correct if it brings together parts of physics that were previously disjoint. In the nineteenth century, the atomic theory of matter not only explained the ideal gas law; it also made sense out of the ratios one observed in chemical reactions. Similarly, when we finally understand quantum mechanics in more primitive terms, who knows what other parts of physics will be explained in the same terms? Possibly the dimensionality of spacetime. Possibly even the value of the fine-structure constant. (It has been argued recently that the fine-structure constant, being the low-energy limit of a running coupling constant, appears to be very far removed from the basic laws of physics and is therefore not likely to have a simple explanation (Gross 1989). However, if measurement-like events are taken as fundamental, then the fine-structure constant may be quite close to the foundations of physics.) The uncovering of a such a connection in physics should be the mark of a valid sub-quantum theory.

## 16.3 What is a Branching Event?

Invoking the notion of a branching event brings us back to the measurement problem. What is it exactly that separates a given branching event from the rest of the history of the universe? How does one know when such an event starts and when it is completed? These questions are difficult enough when the event is a standard quantum measurement, such as the detection of a photon. They are even more difficult when one is talking about, say, the nuclear burning of the sun. Without being completely arbitrary, how could one possibly break up the burning of the sun into elementary branching events?

The only answer that makes any sense is that the representation of the history of the universe as a set of branching events is not unique. The physicist must be free to choose his or her own representation, but within certain limits. One can divide up the world's history rather finely into events, but not arbitrarily finely. In the case of the sun's burning, every measurement we make or can imagine making on the sun without disturbing it allows us to divide the sun's history into more detailed events. But we cannot go so far as to say that an elastic collision between two of its particles constitutes a branching event, because such a collision by itself leaves no definite trace in the history of the universe. It is obvious from this example that there can be no sharp line between what counts as a branching event and what does not. Therefore if one is to take such events as the basic building blocks, one must forsake the goal of mirroring perfectly in one's theory an independently existing reality. The statement, "the world consists of branching events," may have the appearance of an objective description, but in fact there is an unremovable subjective element in the notion of a branching event.

Many people object to such an approach on principle, preferring a framework that claims to be based on a description of objective reality (e.g., the wavefunction of the universe). Without justifying it, let me simply record here my own view that it will probably not be possible to maintain a rigorous separation between the objective and the subjective, but that this limitation need not keep our physical theories from being rigorous.

### 16.4 The Second Law of Thermodynamics

I conclude by addressing one of the main questions of this conference: what is the origin of the second law of thermodynamics?

In a theory of the sort I am trying to imagine, time asymmetry would be built into the foundations of physics, in the sense that every elementary event would have its own direction of time. One might think that this degree of asymmetry would be sufficient to guarantee the second law of thermodynamics. But I do not think this is the case. Imagine, for example, a beaker of water in which ink initially diffused throughout the water spontaneously collects itself into a localized blob. We have never seen such a process, but we can imagine it. We can also imagine watching this process and taking photographs of it, while the ink continues the whole time to coalesce into a blob, oblivious to the many branching events that we create all around it by our measurements (and to the many other branching events that we do not create). The very fact that we can imagine this scenario suggests that the second law does not follow logically from the asymmetry of the branching events.

It is standard practice to base the second law on the initial state of the universe, which is either assumed or argued to be a very special low-entropy state. But the concept of an initial state of the universe is quite foreign to the framework I am proposing. There is no initial state of the universe. The only things that exist are branching events, their outcomes, and the connections among them. So I cannot invoke the initial state as the origin of the second law.

But remember that there would be in this theory a deep Law, the Great Underlying Probability Law, that would allow one to compute probabilities and would give rise to quantum mechanics. Now, the second law of thermodynamics also concerns the computation of probabilities, e.g., the probability that the ink will spontaneously coalesce. Moreover, there should be only one correct way of computing probabilities, so the same Law that gives rise to quantum mechanics must also be used to compute the probability that the ink coalesces. In other words, I imagine that both quantum mechanics and the second law of thermodynamics emerge at the same logical level, from the same underlying principle.

At this point it would be a great help to my argument if I could produce the Great Underlying Law. Unfortunately I cannot, but the above considerations suggest the following clue: given that the Law has to replace the concept of an initial state of the universe, one should expect to find built into the Law certain regularities of the universe that might otherwise be attributed in part to the initial state. One particular regularity I have in mind is the following fact: our universe is such that we can make all sorts of useful predictions by (i) generalizing from past experience (this is related to Gell-Mann and Hartle (1990), pp. 426-7) and (ii) avoiding unjustified bias (cf. Jaynes 1957). These are very general principles worthy of being incorporated in a fundamental law. Whether such principles, when spelled out precisely, could form the basis of a derivation of quantum mechanics is thoroughly a matter of speculation at this point, but it is nevertheless intriguing to try to imagine such a derivation.

Interestingly, Peres has recently demonstrated a connection between the structure of quantum mechanics and the second law: it turns out that many changes that one can imagine making in the structure of quantum mechanics would lead to violations of the second law (Peres 1990). In a very different context, Gell-Mann and Hartle, building on work by others, have shown that imposing a final condition on the universe would affect not only thermodynamics but also quantum mechanics (Gell-Mann and Hartle 1994, Aharonov et al 1964, Griffiths 1984). These results reinforce the idea that quantum mechanics and the second law of thermodynamics may prove to be two aspects of a single theory.

To repeat the main point of this paper, the attempt to envision a sub-quantum theory suggests a picture of the world in which time-asymmetric events are taken as fundamental. Unitary evolution would appear as a derived concept and would of course not be universal. Many questions remain unanswered, of course, but I am convinced that in the end we will learn much by not taking quantum mechanics for granted.

## 16.5 Acknowledgments

I would like to thank Ben Schumacher and David Park for very helpful discussions.

## Discussion

**Lloyd**   A comment: The entire physical rationale for studying time asymmetry comes from the historical fact that during the nineteenth century, physicists deduced that the underlying equations of motion for physical systems were Hamiltonian and time symmetric. The observed asymmetry in time then became a paradox, rather than the simple fact of dynamics postulated by Aristotle.

**Wootters**   I am going back to thinking of time asymmetry as a simple fact of the world. The challenge for the view I propose will not be to explain time asymmetry but to explain the time symmetric equations of motion.

**Griffiths**   Is there anything in your presentation which does not apply equally well to classical as to quantum mechanics?

**Wootters**   I understand the question: Even if the world ran according to classical mechanics, we would be able to access it only through irreversible measurements, and one could argue that these measurements should be taken as the basic building blocks of the universe. But the motivation for doing this is much greater in quantum mechanics. In classical mechanics, measurements on a system allow one to follow arbitrarily closely the system's evolution, with negligible disturbance of that evolution. But in quantum mechanics this is not the case, and the evolution of a system between measurements takes on much more the character of something that we have constructed.

There is also the question of the initial state of the universe. In classical mechanics it was necessary to assume a complex initial state, in order to produce all the variety we see around us. But in quantum mechanics the initial state can be so simple that one can imagine doing without it altogether.

**Davies**   On a point of information, I believe that a theoretical scheme along the lines you are suggesting has been developed by Alastair Rae of the University of Birmingham.

**Wootters**   I am familiar with Rae's popular book on quantum mechanics, in which he endorses Prigogine's view of irreversibility and applies it to the problem of measurement. Rae does indeed take irreversible measurement processes as fundamental, as I do. One difference between my view and Rae's is that he claims that one can make an objective distinction between reversible and irreversible processes (based on the concept of strong mixing), whereas I do not think such an objective distinction is possible. I would say simply that in order to do physics at all we must have some concrete measurement results—this is the one item we cannot do without—and I assume that we have them.

## References

Aharonov, Y., Bergmann, P., and Lebowitz, J. (1964) Time Symmetry in the Quantum Process of Measurement. *Physical Review* **B134**, 1410.

Gell-Mann, M. and Hartle, J. B. (1990) Quantum Mechanics in the Light of Quantum Cosmology. In *Complexity, Entropy, and the Physics of Information*, Ed W. H. Zurek, Addison-Wesley, Redwood City, California.

Gell-Mann, M. and Hartle, J. B. (1994) Time Symmetry and Asymmetry in Quantum Mechanics and Quantum Cosmology. In this volume.

Griffiths, R. B. (1984) Consistent histories and the interpretation of quantum mechanics. *Journal of Statistical Physics* **36**, 219.

Gross, D. J. (1989) On the calculation of the fine-structure constant. *Physics Today* **42(12)**, 9–11.

Jaynes, E. T. (1957) Information Theory and Statistical Mechanics. *Physical Review* **106**, 620.

Landau, L. D. and Lifshitz, E. M. (1958) *Statistical Physics*, Pergamon Press, London.

Page, D. N. and Wootters, W. K. (1983) Evolution without evolution: Dynamics described by stationary observables. *Physical Review* **D27**, 2885.

Penrose, R. (1971) Angular momentum: an approach to combinatorial space-time. In *Quantum Theory and Beyond*, Ed T. Bastin, Cambridge University Press, Cambridge.

Peres, A. (1990) Thermodymanic Constraints on Quantum Axioms. In *Complexity, Entropy, and the Physics of Information*, Ed W. H. Zurek, Addison-Wesley, Redwood City, California.

Wheeler, J. A. (1990) Information, Physics, Quantum: The Search for Links. In *Complexity, Entropy, and the Physics of Information*, Ed W. H. Zurek, Addison-Wesley, Redwood City, California.

Wootters, W. K. (1981) Statistical distance and Hilbert space. *Physical Review* **D23**, 357–362.

Zeh, H. D. (1971) On the Irreversibility of Time and Observation in Quantum Theory. In *Proceedings of the International School of Physics "Enrico Fermi," Course IL: Foundations of Quantum Mechanics*, Ed B. d'Espagnat, Academic Press, New York.

Zeh, H. D. (1989) *The Physical Basis of the Direction of Time*, Springer, Heidelberg.

# 17

## Time and Quantum Mechanics

Roland Omnès

*Laboratoire de Physique Théorique et Hautes Energies*
*Université Paris Sud, Bâtiment 211, 91405 Orsay, France*

### 17.1  General Background

Several attempts at a consistent interpretation of quantum mechanics (i.e., a theory explicitly free of any possible self-contradiction or paradox) have been made recently [1-3]. Such an interpretation must rely upon the notions of consistent histories (meaning consistency with probability calculus), decoherence, a clearcut logical background and a well-suited formulation of semi-classical physics. These last two features characterize the difference between my own approach and the ones followed by Robert Griffiths [1] or by Murray Gell-Mann and James Hartle [2], although the three of them have much in common and may be roughly considered as variants of the same basic theory. The power of a precise logical foundation is essential in order to show that the overall theory is completely consistent so that it might replace the Copenhagen interpretation while explaining why and when the older theory is safe from pitfalls or not, and also how it must be modified.

The background concerning histories has been given by Robert Griffiths in this Meeting so that one can be very brief: A property of a quantum system asserts the value of an observable as being in some real domain at a given time t and it can be associated with a specific time-dependent projector $E(t)$. A history is a sequence of properties holding at successive times (like some sort of a motion picture) and it can be given a probability when it belongs to a family of different histories satisfying well-defined algebraic consistency conditions.

As well known (or perhaps, as should be well known), one can use plain logic confidently once a field of propositions has been defined (mentioning all what could be said within some domain of thinking) and also when one knows what precise meaning is given to a few logical links (i.e., how to use not, and, or; ... is logically the same as ...; if ..., then ...) obeying codified rules. In quantum mechanics, the propositions can be built from any consistent family of histories and the corresponding logical treatment is straightforward [3].

This framework has been used to provide a clearcut foundation for the interpreta-

† Laboratoire associé au Centre National de la Recherche Scientifique

268

tion of quantum mechanics, which relies upon a unique rule replacing the customary Copenhagen rules of measurement theory. This axiom states that any description of a physical system (respectively: any valid reasoning concerning it) should be made of properties belonging to an overall consistent family of histories (respectively: the reasoning consists in valid logical inferences). A valid logical inference $a \Rightarrow b$ between two properties $(a, b)$ amounts to their conditional probability $p(b|a)$ being equal to 1, or practically equal to 1. This "universal" rule provides a complete control over logical consistency and it turns the theory of interpretation into a completely deductive process [3].

A key step in the construction consists in recovering classical physics as a good description of a macroscopic system. One assumes that good collective coordinates can be defined and they are the only ones of importance in this respect. The correspondance between classical dynamical variables $a(q, p)$ and quantum observables $A$ acting in the collective Hilbert space is obtained as usual by the Wigner-Weyl formalism, the best-known example being the correspondance between a density operator and a Wigner probability distribution in phase space.

A typical classical property asserts simultaneously the values of the position and momentum coordinates up to some given errors $(\Delta q, \Delta p)$ with $\Delta q \Delta p \gg h$. This property can be given a meaning according to the universal rule by associating it with a quantum projector $E$. As a matter of fact, $E$ is not strictly a projector but a so-called quasi-projector such that [4]

$$Tr(E - E^2) = (\Delta q \Delta p / h) 0(\epsilon)$$

in the simplest case of a unique degree of freedom. The first factor in the right-hand side is the number of semiclassical states in the given classical cell and the parameter $\epsilon$ ("classicity parameter"), which is equal to $(h/\Delta q \Delta p)^{1/2}$, provides an estimate for the unavoidable relative error to be met when thinking in classical terms, i.e., when relying upon common sense logic.

What has been said up to this point is all that is needed for what we have to say concerning time. We shall therefore leave aside the next steps in the construction, giving the conditions for applicability of classical determinism and finally yielding a tightproof measurement theory, this last step relying strongly upon decoherence.

### 17.2 An Objection Against Decoherence and an Answer

I shall again rely on another talk, Zurek's one, to assume that decoherence is known to the reader and I shall concentrate upon an important conceptual problem concerning it. It was called to attention by John Bell [5], following an attempt towards a consistent interpretation by Klaus Hepp, who already made use of decoherence [6,7]. The difficulty was also acknowledged by Zurek, who dismissed it for all practical purposes [8], but it remains nevertheless present at a deeper level.

Bell's point consists in noticing that, if one starts from a quantum superposition for a macroscopic system, the reduced density operator (obtained after tracing out the environment) becomes diagonal by decoherence but the full density operator is still a quantum superposition. Accordingly, "as a matter of principle", there exists at least one observable $A$ for the environment that is able to exhibit this superposition. Bell concluded that decoherence is a nice physical result but of no value as far as the foundations of quantum mechanics are concerned.

I have recently investigated this problem. The idea is to analyze how a measurement of the observable $A$ can be performed, at least theoretically. The present theory of decoherence, though it is still very rough, strongly suggests that the observable $A$ must be able to distinguish between two neighbouring energy eigenstates of the environment and to select essentially only one such state. In view of the tremendously large number of these states, a measurement of $A$ will most often give no indication about the value of $A$, the probability $p$ for a positive result being very small. The idea is to compare $p$ with the classicity parameter $\epsilon$ for the measuring device, which can be shown to give in the present case the probability for obtaining a misleading positive measurement coming only from a misfunctioning (a quantum fluctuation) in the measuring device. Only when $p$ is larger than $\epsilon$ can one believe in the result, either logically or according to its most probable cause.

As an example, if the macroscopic object $O$ to be tested contains $N$ particles and the measuring device $M$ contains $N'$ particles, both $O$ and $M$ being made of metallic copper, one finds that the condition $p > \epsilon$ amounts to

$$N' > \exp(0.1N^{2/3}),$$

the exponential keeping memory of the exponentially small energy difference among elementary eigenenergies. For all but very small objects, the apparatus would have to be much bigger than the universe. Even from the standpoint of pure theory, its working is inconsistent with special relativity and, according to general relativity, it should collapse into a black hole.

Although this argument has not yet reached the level of a rigorous proof, it strongly indicates that the conventional axiom according to which every observable is measurable is highly questionable. This axiom contradicts other basic laws of physics, namely the particle structure of matter in a finite universe and relativity. As a consequence, Bell's objection is groundless and decoherence is absolutely valid for a large enough system. It may be mentioned that this conclusion rests upon the logical foundations of the theory and decoherence alone cannot justify by itself its own fundamental character.

## 17.3 About the Direction of Time

A simple but presumably important consequence of the previous result has to do with thermodynamical irreversibility: Consider for instance a gas initially prepared in a quantum state where all the atoms are located in a small region within a vessel.

Its density operator at a later time $t$ represents a gas filling up the vessel. If one could prepare the time-reversed state, its entropy would decrease and it would tend to a more ordered state. This is essentially the basic problem of statistical mechanics going back to Boltzmann and never yet solved in a completely satisfactory way.

One can offer now a new answer: In order to prepare the time-reversed state, one would need a preparing device much more elaborate than the measuring device necessary to perform Bell's ideal measurement. Its existence would contradict the particle nature of matter in a finite universe and also relativity. It cannot exist, even in principle.

So, one can understand better the thermodynamical direction of time from the standpoint of a consistent interpretation of quantum mechanics. Decoherence has the same direction of time since it goes along with thermodynamical relaxation. It should be recalled at this point that the formalism of consistent histories, when provided with a logical setup, must also select a specific (logical) direction of time. The deductive theory of quantum measurements can then be used to prove that *the two directions of time, in thermodynamics and in logic, must be the same.*

Finally, one can use quantum logic and its definite direction of time to get a framework of quantum mechanics differing from the one introduced by Everett [9]. Rather than starting from the state of the universe at an initial singular time zero, one starts from the present time so as to reconstruct logically the past by a consistent reasoning from what exists by now.

This is best understood by using a model of a toy universe containing three measuring devices $M$, $M'$, $M''$ and a unique quantum object $Q$ (e.g., an atom) going from $M$ to $M'$ and then to $M''$. The device $M$, initially in a neutral state, measures a non-degenerate observable $A$ relative to $Q$ at a time $t$. It registers a data as well as the value of $t$. This behaviour is perfectly consistent with quantum mechanics ($M$ behaving according to quantum mechannics), as I have shown some time ago [3]. Let it be assumed that the present time $T$ is later than the two times $t$ and $t'$ when the measurements by $M$ and $M'$ occured, but previous to the time $t''$ where $M''$ is expected to act.

Let us now assume that there is at the present time T a unique *state of facts*, namely that the (classically and quantum mechanically meaningful) set of records shown by $M$ and $M'$ together with the still neutral situation of $M''$ is unique. It should be stressed that this assumption does not follow directly from quantum mechanics. It is taken from the observation of reality and added to the theory from outside.

One can then use logical retrodiction to infer the "classical" past properties of $M$, $M'$ and $M''$ as well as the quantum state of $Q$ at any previous time (except during a measurement) in a unique way. As for the outcome of the third measurement, it remains in the realm of probabilities. Accordingly, *present facts determine uniquely past phenomena* and the essential difference between past and future is found to be a necessary feature (and even in some sense a consequence) of logical quantum mechanics. It is furthermore possible to fit quantum mechanics with a unique reality (consisting of facts), without an akward many-worlds representation.

There is however a price to pay: "Present" time $T$ evolves so that some potentialities become realities and other ones fade away. In other words, Reality remains unique, at least as far as facts (i.e., classically meaningfull properties) are concerned. Quantum mechanics cannot give a mechanism and not even an account for that. This is the essence of wave packet reduction, which goes much deeper than its description by decoherence [8] from which it appears as a logical theorem. At the deepest level, it has to do with an encounter between theory and reality in their most essential features: the probabilistic character of the theory, which is a building block of its logic before acquiring an empirical meaning, hurts an essential character of Reality, namely the uniqueness of facts. I personally believe that this conflict is not a failure of quantum mechanics but its greatest triumph, suggesting that it has reached the ultimate limit of a theory: the point of no-return beyond which one would have to identify Reality with its mathematical representation, the ever-evolving with the timeless abstraction.


## Discussion

**Lloyd**   Could you clarify what you mean by the logical arrow of time?

**Omnès**   There are two possible orderings of the properties (projectors) in the probabilities of histories: logic is not time-reversal invariant. Therefore, a specific time ordering must be chosen in the histories.

**Unruh**   What about the spin echo effect?

**Omnès**   There is no decoherence in that case (i.e., no diagonalization of the reduced density operator, though the average magnetization tends to vanish) so that this effect lies outside the results I mentioned.

**Unruh**   Does not the dynamical arrow of time (decoherence) just arise from the logical one (i.e., initial conditions)?

**Omnès**   Essentially yes.

**Zeh**   Is not at least a part of what you call "logic" usually called "dynamics"?

**Omnès**   Dynamics (i.e., the Schrödinger equation) is used by logic. Inversely, some "understanding" of the dynamical effects is necessary and therefore always present in usual formulations. I try to do it consistently by clearly separating the two structures of dynamics and logic in the theory.

**Zurek**   In your discussion of an apparatus measuring another apparatus, are they classical or quantum systems?

**Omnès**   Quantum systems.

**Albrecht**   Is the past unique, as you say? I will probably never know which shirt my son wore yesterday: there are many consistent histories on which we may reside.

**Omnès**   Presently, I prove that it is unique in a simplified model and I assume it is also unique in reality.

**Schulman**   Do your results apply to a superfluid system?

**Omnès**   No. Decoherence must be occurring and this needs dissipation.

## References

[1]  Griffiths R. (1984) *J. Stat. Phys.* **36**, 219.

[2]  Gell-Mann, M., J.B. Hartle (1990) In *Complexity, Entropy and the Physics of Information*, W.H. Zurek (ed.), Santa Fe Institute, Studies in the science of complexity, Addison-Wesley.

[3]  Omnès, R. (1990) *Ann. Phys.* **201**, 354.

[4]  Omnès, R. (1991) *J. Stat. Phys.* **62**, 841.

[5]  Bell, J.S. (1975) *Helv. Phys. Acta* **48**, 93.

[6]  Hepp, K. (1974) *Comm. Math. Phys.* **35**, 265.

[7]  Hepp, K., and E.H. Lieb (1973) *Helv. Phys. Acta* **46**, 573.

[8]  Zurek, W.H. (1982) *Phys. Rev.* **D26**, 1862.

[9]  Everett III, H. (1957) *Rev. Mod. Phys.* **29**, 454.

# 18

## Is Time Sharp or Diffused?

Iwo Bialynicki-Birula

*Centre for Theoretical Physics*
*Lotników 32/46, 02-668 Warsaw, Poland*

### Abstract

It is argued that in quantum field theory the time parameter used to label the state vectors in the Schrödinger picture cannot be interpreted as the time of observation because the states cannot be characterized in terms of observables that refer to one instance of time only. This argument is directly related to a more technical observation made by Stueckelberg in 1951 that the time evolution operator for finite time intervals has additional divergences which are not removed by the standard mass, charge, and wave function renormalization.

### 18.1  Introduction

Classical space-time is made of events — structure-less, pointlike abstractions representing elementary physical happenings. On the other hand, in the real world even the most elementary happenings, like for example acts of emission or absorption of photons by charged particles, have finite spacetime extension. As has been pointed out by Schwinger (Schwinger 1958), this discrepancy makes the "observational basis of quantum field theory self-contradictory" and it is most likely the cause of divergences that plague this theory. It is not possible to shrink the real events to a point because, again using Schwinger's words, "The localization of charge with infinite precision requires for its realization a coupling with the electromagnetic field that can attain arbitrarily large magnitudes. The resulting appearance of divergences, and contradictions, serves to deny the basic measurement hypothesis. We conclude that a convergent theory cannot be formulated consistently within the framework of present space-time concepts."

The fact that real physical events are fuzzy must, in my opinion, influence our notions of time and also of time's arrow. Time enters the description of the physical world in several different ways. There is cosmological time, there is thermodynamical time, there is time of the laboratory clock, and there is also time that enters the definition of the state of the system. I will deal here exclusively with this last case. In quantum theory this will be the time parameter labelling the state vectors.

## 18.2 Time Label of the State Vector

The fundamental notion of any quantum theory, be it simple nonrelativistic quantum mechanics or relativistic quantum field theory, is the notion of the state of the system *at a given time*, described by a wave function or a state vector representing this state. In the Schrödinger picture the state vector carries the time mark explicitly, while in the Heisenberg picture the time mark is implicit. In both cases different state vectors are distinguished by labelling them with the eigenvalues of various measurable physical quantities. In the Schrödinger picture these labels refer to time-independent operators, while in the Heisenberg picture the assigned labels are the eigenvalues of operators taken *at a given time*. Unfortunately, in relativistic quantum theory such a description can be realized only for free particles. For interacting particles, and that means for all real particles observed in actual experiments, the notion of the state vector at a given time and also the notion of the time evolution from time $t_1$ to time $t_2$ cannot be given precise meaning within the mathematical framework of relativistic quantum field theory. It is not that one cannot introduce unitary operators that evolve state vectors for a finite length of time — the operators $\exp(-iH(t_1 - t_2))$ clearly have this property — but rather that one cannot label the state vectors and the matrix elements of the Hamiltonian $H$ using the eigenvalues that refer only to instantaneous properties of the system.

## 18.3 S-matrix Theory of Interacting Particles

How does one avoid the problems with instantaneous description of the states? In comparing theory with experiment, one always makes a tacit assumption that all particles are mutually noninteracting when they are being observed, even though in reality one is always dealing with interacting particles. This abstraction, embodied in the S-matrix theory, has so far apparently not resulted in significant discrepancies between theory and experiment, but such an oversimplified treatment of reality may lead to serious consequences when it comes to the study of fundamental problems. The interaction of particles is described in the S-matrix theory by a unitary transformation that maps one free motion (the *in* motion) into another free motion (the *out* motion). These two free motions are assumed to be separated by an infinitely long time interval. In other words, we neglect the interaction time in comparison with the observation time. This mode of description will clearly become untenable from the practical point of view in the future, when one will learn how to make measurements on the time scale comparable to the interaction time. From the theoretical point of view this mode of description is highly questionable even today. If one cannot give meaning to the notion of the state of the system at a given time, what is to become of causality? It is true that causality has already suffered a severe setback due to the probabilistic nature of the predictions of quantum theory, but at least at the level of the state vector (or the wave function) the time evolution is believed to be completely causal. In relativistic quantum theory even such a restricted form of microcausality cannot be retained and it must be replaced by macrocausality. A simple physical argument supporting this statement is based on the time-energy uncertainty principle. Precise localization in time requires infinite energies and in relativistic theory this, in turn, leads to infinite fluctuations in the number of particle which makes the state of the system ill-defined. One way out of this dilemma is to completely abandon the notion of the state vector at a given time and rely exclusively on the notion of quantum states defined over finite time intervals. I believe that quantum field theory almost forces us to introduce these changes because, as has been already argued by Bohr and Rosenfeld (Bohr and Rosenfeld 1933, 1950) only smeared in time (and in space) field operators are mathematically meaningful.

## 18.4  Smearing in Time

The smearing of the state of the system in time will always have profound consequences for the interpretation of the theory. It might even be related to the problem of hidden parameters. When the specification of the state vector at a given time is not adequate one has to introduce additional characteristics. Whether they can be identified with hidden parameters is not clear, but they certainly change our notion of the state of the system.

A necessity to modify the naïve notion of a state of the system and to introduce a smearing in time is already encountered in the classical relativistic theory of charged particles. The Lorentz-Dirac equation of motion for a charged particle that includes the radiation reaction contains the third time derivative. This by itself requires the specification of acceleration (as a "hidden" parameter) in addition to position and velocity, but the existence of unphysical, self-accelerating, the so-called runaway solutions leads to the replacement of the differential equation of motion by an integral equation of the form (see eg. Rohrlich 1965)

$$m\dot{v}^{\mu} = \int_0^{\infty} da\, K^{\mu}(t + at)e^{-a}, \tag{18.1}$$

where the parameter $\tau_0$ is related to the classical radius of the particle by the formula

$$\tau_0 = r_0/6\pi c \approx 0.6 \times 10^{-23} \text{ sec.} \tag{18.2}$$

The equation (18.1) has only physically acceptable solutions. Since the force term $K^{\mu}$ under the integral contains the velocity of the particle, the specification of "the initial data" requires in principle the knowledge of the motion of particles for all future times, although in practice, owing to the exponential damping of the force term, this knowledge is restricted only to the times of the order of $\tau_0$.

The values of the velocity at all future times are an analog of hidden variables and the smearing in time of the initial data is quite similar to the smearing in time of the state of the quantum system. Such smearing is needed in relativistic quantum theory in order to eliminate the so called "surface infinities" which are known to occur if one wants to follow the evolution of the system for finite time intervals. This problem was first analyzed by Stueckelberg, who immediately after the formulation of the renormalization theory published a paper (Stueckelberg 1951) on this subject. In the abstract of this paper he wrote: "If transition probabilities are evaluated for transitions occurring during a finite time interval, additional divergences occur different from those commonly encountered for infinite time intervals. The expressions obtained can however be made convergent, if an indeterminacy of time is attributed to each epoch of observation." The problem discovered by Stueckelberg has been studied further by Bogoliubov and Shirkov (Bogoliubov and Shirkov 1959) and more recently by Symanzik (Symanzik 1981).

## 18.5 States at a Given Time Versus States of Motion

The idea that the instantaneous specification of the states of relativistic quantum systems may be untenable has been already put forward by von Neumann in 1932 in his influential treatise on the foundations of quantum mechanics (von Neumann 1955), where he pointed out that "the chief weakness of quantum mechanics is: its nonrelativistic character, which distinguishes the time $t$ from the three space coordinates $x, y, z$, and presupposes an objective simultaneity concept."

Similar ideas have been discussed in the works of the three virtuosi of quantum electrodynamics: Dirac, Feynman and Schwinger. Dirac in a paper (Dirac 1965) that has surprisingly attracted very little attention questioned the equivalence of the Heisenberg and the Schrödinger pictures, arguing that only the Heisenberg picture makes sense because "the state vector of the Schrödinger picture does not remain in the Hilbert space".

Feynman in his Nobel lecture said in his characteristic Feynmanese: "Instead of wave functions we could talk about this; that if a source of a certain kind emits a particle, and a detector is there to receive it, we can give the amplitude that the source will emit and the detector receive. We do it without specifying the exact instant that the source emits or the detector receives, without trying to specify the state of anything at any particular time in between, but by just finding the amplitude for the complete experiment." I took this quote from a recent letter by Schwinger (Schwinger 1989), who wrote in it that he might have started to work on his source theory after subconsciously absorbing this idea of Feynman. In Schwinger's source theory, the sources of particles that play the role of the fundamental building blocks for all processes are extended in space and in time.

What still remains to be done is to formulate a consistent and simple framework that would eliminate completely the unphysical notion of instantaneous states of the system and replace it by the states of motion whose specifications extend over finite time intervals. Such states of motion can be introduced in a natural way in quantum electrodynamics where we have classical, external electromagnetic fields at our disposal. This mode of description cannot only be made consistent with the mathematical concepts of relativistic quantum field theory, but also seems to be more appropriate when it comes to comparing it with realistic experimental arrangements. State description in terms of electromagnetic fields is not restricted to typical quantum electrodynamic processes since in *every* real experiment the preparation of the state can ultimately be expressed in terms of configurations of electromagnetic fields and the final measurement can be always reduced to detection of charged particles. Therefore, it seems justified to define the states in terms of classical electromagnetic field configurations in a wide range of situations and treat the states vectors as functionals of the field potential, $\Psi[\mathscr{A}]$, in which $\mathscr{A}$ extends over a *finite region* of spacetime. Not only will this remove the sharp-time infinities noticed by Stueckelberg, but as a bonus we would also obtain expressions free of infrared divergences. As usual, there is a price to be paid for these highly desirable features; the theory becomes quite cumbersome. Since the description of even the simplest experiment requires complicated field arrangements, the completion of the calculations beyond the lowest approximation requires the development of new calculational tools; standard Feynman diagram techniques are no longer sufficient because the initial state cannot be described in terms of free particles.

## 18.6  Final Notes

When the state of the system at a given instant becomes ill-defined, time itself becomes diffused. After all, what is the meaning of an instant if there is no element of reality associated with it? One may say that the idea of a diffused time offers the ultimate resolution of the Zeno paradox. The paradox disappears since there is no instantaneous, "static" state; at all times Zeno's arrow carries with it the notion of flight — the internal notion of change. The notion of state is not static; the static state gives way to a dynamic state of motion. Strange as these ideas may sound at first, they do find their natural place in the world of wave phenomena. There is no music, for example, without the time duration. Does it make sense to talk about the instantaneous state of Beethoven's IX-th Symphony during the performance? And in quantum theory we are dealing all the time with wave phenomena.

The concept of diffused time has also consequences for the time arrow. Once we accept the idea that the notion of a state at the given time should be replaced by the concept of a dynamic state of motion, we can include the sense of direction of that motion in time. There are no nontrivial (i.e. truly evolving) states of motion that would be time symmetric, as opposed to the states at a given time which never carry any information about the time arrow. We may easily restrict the states of motion to only those whose time dependence agrees with the boundary conditions appropriate for our expanding Universe; only outgoing radiation is allowed.

## References

Bogoliubov, N.N. and D.V. Shirkov (1959) *Introduction to the Theory of Quantized Fields*, Interscience, New York.

Bohr, N. and L. Rosenfeld (1933) Zur Frage der Messbarkeit der elektromagnetischen Feldgrössen, Mat.-fys. Medd. Dan. Vid. Selsk. no.8 (English translation in: *Selected Papers of Léon Rosenfeld*, Eds. R.S.Cohen and J.J.Stachel, Reidel, Dordrecht 1979).

Bohr, N. and L. Rosenfeld (1950) Field and charge measurements in quantum electrodynamics. *Phys. Rev.* **78**, 794.

Dirac, P. A. M. (1965) Quantum electrodynamics without dead wood. *Phys. Rev.* **139B**, 684.

Rohrlich, F. (1965) *Classical Charged Particles*, Addison-Wesley, Reading, Ch.6.

Schwinger, J. (1958) Preface to *Quantum Electrodynamics*, Dover, New York.

Schwinger, J. (1989) Letter to the Editor, *Physics Today* **42**, No.5, 13.

Stueckelberg, E.C.G. (1951) Relativistic quantum theory for finite time intervals. *Phys. Rev.* **81**, 130.

Symanzik, K. (1981) Schrödinger representation and the Casimir effect in renormalizable quantum field theory. *Nucl. Physics* **B190**, 1.

von Neumann, J. (1955) *Mathematical Foundations of Quantum Mechanics*, Princeton University Press, Princeton 1955. Ch.5.

# 19

# Time Asymmetry and the Interpretation of Quantum Theory

V. Mukhanov†

*Institute for Theoretical Physics*
*ETH Hönggerberg, 8093 Zürich*

## 19.1 Introduction

The purpose of these notes is to show that there is a fundamental difference in understanding of time irreversability in two different approaches to the interpretation of quantum theory. (I mean "conventional" or Copenhagen interpretation and Many-Worlds interpretation [1,2]). This difference is due to the fact that Copenhagen and Many-Worlds interpretations are not only different interpretation of the same physical theory but rather different physical theories [3].

According to Copenhagen approach to quantum theory there exist some special interactions (measurements) for which the reduction ("collapse") of the state vector of quantum system takes place [4]. These interactions cannot be described by Schrödinger's equation and it puts the bounds on applicability of Schrödinger's equation. So the measurement breaks down the unitary evolution of the state vector.

If we assume that the Schrödinger equation describes also the measurements and there is only unitary evolution of the state vector of the closed system, it is then quite difficult to avoid the conclusion about the existence of other Universes [1]. In this case quantum theory itself says us a lot about it's own interpretation [3].

To express the difference between Copenhagen and Many-Worlds versions of quantum theory in an explicit manner let us describe the measurements in terms of density matrices.

## 19.2 Actual vs. "Effective" Reduction

We shall consider the ensemble of microscopic quantum systems which are prepared at the begining in the same initial state $| \Psi^S \rangle$. The purpose is to measure the observable $\mathscr{A}$ to which the operator $\hat{A}$ corresponds. Let a quantum system $S$ interact with a device $M$. Then according to different interpretations of quantum theory the description of this interaction should be different. We shall consider here three possible types of interactions.

† on leave from: Institute for Nuclear Research, 117312 Moscow

First, we assume that the interaction between the quantum system $S$ and the device $M$ can not be described by Schrödinger's equation and as a result of measurement the reduction of the state vector takes place. (Proccess I according to von Neumann [4]). Then the measurement can be described as

$$
| \Psi^S \rangle \longrightarrow
\begin{aligned}
& \xrightarrow{|c_1|^2} | \Psi_1^S \rangle \\
& \xrightarrow{|c_2|^2} | \Psi_2^S \rangle \\
& \longrightarrow \cdots \\
& \xrightarrow{|c_n|^2} | \Psi_n^S \rangle \\
& \longrightarrow \cdots
\end{aligned}
\tag{19.1}
$$

where

$$
| \Psi^S \rangle = \sum c_n | \Psi_n^S \rangle
\tag{19.2}
$$

and $| \Psi_n^S \rangle$ are the eigenvectors of operator $\hat{A}$ $\left( \hat{A} | \Psi_n^S \rangle = A_n | \Psi_n^S \rangle \right)$. The coefficient $| c_n |^2$ gives the probability to find the system $S$ in corresponding pure state $| \Psi_n^S \rangle$ after the measurement. If there are $N \gg 1$ quantum systems in the same state $| \Psi^S \rangle$ before the measurements, then after the measurement we will find $\sim N | c_1 |^2$ systems in the state $| \Psi_1^S \rangle$, ..., $\sim N | c_n |^2$ systems in the state $| \Psi_n^S \rangle$ etc.

Thus as a result of interaction the ensemble of $N$ systems, the density matrix of which was

$$
\hat{\rho}^S = | \Psi^S \rangle \langle \Psi^S | ,
\tag{19.3}
$$

evolves to the ensemble with density matrix

$$
\hat{\rho}^I = \sum | c_n |^2 | \Psi_n^S \rangle \langle \Psi_n^S | = \sum | c_n |^2 \hat{\rho}_n^S
\tag{19.4}
$$

where $\hat{\rho}_n^S = | \Psi_n^S \rangle \langle \Psi_n^S |$. I will refer to this ensemble as Ensemble I.

Now, let us describe the measurements in terms of Schrödinger's equation. In this case the device $M$ should also be described by the state vector $| \Psi^M \rangle$. Let $| \Psi_0^M \rangle$ to be initial state of the device before measurement. To simplify the consideration we consider only so-called nondemolition or good measurements of the variable $\mathscr{A} \leftrightarrow \hat{A}$. It means that the state of the quantum system is not changed as a result of interaction of this system with the device $M$ if it was the eigenstate of operator $\hat{A}$. So, if the initial state of the system $S$ is an eigenvector of the operator $\hat{A}$, which corresponds to the eigenvalue $A_k$, $| \Psi_k^S \rangle$, then the result of interaction between system $S$ and device $M$ should take the form

$$
| \Psi_k^S \rangle | \Psi_0^M \rangle \rightarrow | \Psi_k^S \rangle | \Psi_{A_k}^M \rangle ,
\tag{19.5}
$$

where $| \Psi_{A_k}^M \rangle$ is the state of device after the measurement. The condition (19.5) is necessary for realizing a good measurement of the quantity $\mathscr{A}$ by the device $M$.

If

$$| \Psi^S \rangle \; = \; \sum_k c_k \, | \Psi^S_k \rangle \; , \tag{19.6}$$

then (19.5) and the linearity of the Schrödinger's equation imply

$$| \Psi^S \rangle \, | \Psi^M_0 \rangle \; \rightarrow \; \sum_k c_k \, | \Psi^S_k \rangle \, | \Psi^M_{A_k} \rangle \; . \tag{19.7}$$

The quantum system $S$ is not in pure state after the measurement and the final state (19.7) is the superposition of different macroscopical (macro) states of the device $M$ (analog of "Schrödinger cat" in the case of measurements). If we want to refer only to the state of $S$, we should introduce the density matrix $\hat{\rho}^S$:

$$\hat{\rho}^S \;\; = \;\; Tr_M \, | \Psi^{S+M} \rangle \langle \Psi^{S+M} | \tag{19.8}$$

$$= \;\; Tr_M \left( \sum_{\ell, m} c_\ell c^*_m \, | \Psi^S_\ell \rangle \Psi^M_{A_\ell} \rangle \langle \Psi^M_{A_m} | \, \langle \Psi^S_m | \right) . \tag{19.9}$$

If the vectors $| \Psi^M_{A_\ell} \rangle$ form a complete set, then

$$\hat{\rho}^S \; = \; \sum_{\ell, m, k} c_\ell c^*_m \langle \Psi^M_{A_k} | \, \Psi^M_{A_\ell} \rangle \langle \Psi^M_{A_m} | \, \Psi^M_{A_k} \rangle \, | \Psi^S_\ell \rangle \langle \Psi^S_m | \; . \tag{19.10}$$

When the state vectors of the device $| \Psi^M_{A_k} \rangle$ are orthogonal, that is $\langle \Psi^M_{A_k} | \, \Psi^M_{A_\ell} \rangle = 0$ for $A_k \neq A_\ell$ (a necessary condition to distinguish the different results ($A_k$ and $A_\ell$) of the measurements) we obtain

$$\hat{\rho}^S \; = \; \sum | c_n |^2 | \, \Psi^S_n \rangle \langle \Psi^S_n | \; . \tag{19.11}$$

If we have the ensemble of $N$-systems, then the density matrix of this ensemble (to which we will refer as Ensemble II) after the measurements is

$$\hat{\rho}^{II} \; = \; \hat{\rho}^S \; = \; \sum | c_n |^2 | \, \Psi^S_n \rangle \langle \Psi^S_n | \; . \tag{19.12}$$

Thus we found that the density matrices of Ensembles I and II are the same and correspondingly these two ensembles have the same statistical properties (expectation values of any operator $\hat{B}$ : $\langle \tilde{B} \rangle = Tr \, \hat{\rho} \hat{B}$ are the same).

However the internal structure of these ensembles is different (see (19.4), (19.10), (19.11)) and the question is: could we distinguish them experimentally?

To clarify this question let us consider also the gedanken Ensemble III with the same density matrix as (19.4), (19.11). To obtain it we assume that as a result of interaction of the system $S$ with device $M$, the quantum system $S$ with the initial state vector (19.6) will stay in the pure state, but the coefficients in expansion (19.6) of $| \Psi^S \rangle$ in terms of eigenstates $| \Psi^S_m \rangle$ will take the random phases:

$$| \Psi^S \rangle \; \equiv \; \sum c_m \, | \Psi^S_m \rangle \; \rightarrow | \Psi^S_L \rangle \; = \; \sum c_m \, e^{i \alpha^L_m} \, | \Psi^S_m \rangle \tag{19.13}$$

where we use the index $L = 1, 2, ...N$ to distinguish the different systems of the ensemble which were initially in the same state but as a result of interaction took the different phases $\alpha_L$. The resulting density matrix of the Ensemble III is

$$\hat{\rho}^{III} \approx \frac{1}{N} \sum_L \hat{\rho}_L^S = \frac{1}{N} \sum_L | \Psi_L^S \rangle \langle \Psi_L^S | \tag{19.14}$$

$$= \sum | c_n |^2 | \Psi_n^S \rangle \langle \Psi_n^S | + \frac{1}{N} \sum_{\substack{k \neq j \\ L}} c_k c_j^* \, e^{i(\alpha_k^L - \alpha_j^L)} \, | \Psi_k^S \rangle \langle \Psi_j^S | \tag{19.15}$$

The last term in this equation vanishes if the phases $\alpha_K^L$ are random and $N \to \infty$.

Thus the density matrices of Ensembles I, II, III are equal

$$\hat{\rho}^I = \hat{\rho}^{II} = \hat{\rho}^{III} \tag{19.16}$$

Nevertheless the Ensembles I and III can be distinguished experimentally using *repeatablilty*. Actually we can separate the Ensemble I into a set of subensembles basing on the information obtained in the experiment. Let us refer to the same subensemble, the quantum systems for which the results of the measurement are the same. Then all systems in the subensemble will be in the same pure state $| \Psi_\ell^S \rangle$. The information about the measurement is enclosed in the state vector $| \Psi_\ell^S \rangle$ of every system, so when the experiment is over we don't need to refer to the state of the device to characterize the quantum system. For every system from each subensemble we can predict the result of the next same type measurement with probability 100% (repeatablilty).

It is evident that it is not the case for the Ensemble III, since here the state vector of every system after the measurement is, in general, the superposition (19.12) of eigenvectors of the operator $\hat{A}$. Thus, the result of the next experiment is not predictable and the *repeatibility* doesn't take place in this case.

We conclude that as a rule the explanation of the actual reduction cannot be reduced to diagonality of the density matrix in the corresponding basis. The diagonality of the density matrix is insufficient to explain the repeatibility. We have seen above that two ensembles (I and III) can have the same density matrixes, but for one of them (Ensemble I) the property of repeatability takes place, while for the other one it doesn't. Thus, we should distinguish the actual reduction (Ensemble I) from the "effective" reduction (diagonality of density matrix; Ensemble III) (compare with [5]).

The comparison of the Ensembles I and II is more complicated. These two ensembles also have the same statistical properties. Concerning the repeatability there is no problem with the explanation of this phenomenon in the case of Ensemble I: if, for example, the system $S$ went to the state $| \Psi_1^S \rangle$ from the initial state $| \Psi^S \rangle = \sum c_n | \Psi_n^S \rangle$ as a result of the reduction, then the next experiment to measure $\mathscr{A} \leftrightarrow \hat{A}$ will give us certainly the $A_1$-result. The information which permits us to predict the result of the next experiment is enclosed in the state vector of the system $S$.

In the case of Ensemble II every of the systems is not in a pure state and all systems have the same density matrices: $\hat{\rho}^S = \hat{\rho}^{II}$. There is no place to enclose information about the result of first experiment only in the density matrix of the system $S$. To explain repeatability we need in information (even if the first measurement is already over) which only the density matrix (or state vector) of the whole system (quantum system $S$ + device $M$) contains:

$$\hat{\rho}_{Ens.}^{A+M} = \hat{\rho}^{(A+M)} = |\Psi^{(A+M)}\rangle\langle\Psi^{(A+M)}| \tag{19.17}$$

The quantum theory describes not only the ensembles of the systems, but also a single quantum system. (It is clear since the quantum theory permits us to predict the result of repeatable experiments for a single system). Then repeating the Everett's analysis of repeatability and keeping one-to-one correspondence between the mathematical description of the state of the physical system and reality (this correspondence takes place in any physical theory with the exception of Bohr's version of quantum theory) we arrive at the conclusion: the quantum theory tells us that there exist other Universes which are real in the same extent as our Universe. This is Everett's or Many-Worlds interpretation of quantum theory. From my point of view, we would get to the same conclusion in "consistent histories" interpretation of quantum theory [6,7] if we would analyze the repeatability and correspondence between mathematics and reality in this theory.

To interpret the state vector in Many-Worlds theory we need the preferred basis $|\Psi_{pref.}^i\rangle$ in terms of which the state vector should be expanded:

$$|\Psi^{(S+M)}\rangle = \sum_i c_i |\Psi_{pref.}^i\rangle . \tag{19.18}$$

The preferred basis defines the ensemble of different Universes which makes sense only if the different (quasi) classical behavior of some macroscopic variables corresponds to different $|\Psi_{pref.}^i\rangle$ (compare with "consistent histories" approach [6,7]). It can be shown [8] that this preferred basis can be fixed unambiguously.

## 19.3 The Arrow of Time

The fact that according to Many-Worlds interpretation the evolution is always unitary, while in the Copenhagen version of quantum theory the unitary evolution is broken down under the measurements leads to rather surprising conclusion: these two interpretations (or rather theories) can be distingushed experimentally [3].

Actually, according to the Many-World theory the evolution of the state vector is time reversable at a fundamental level (the irreversibility is the question of practice). Thus, there is the possibility for the future influence (usual interference) of other branches (Universes), which were separated, for example, as a result of some measurement, on our branch (Universe) under some special circumstances.

In the Copenhagen version of quantum theory the reduction leads to the disappearence of all branches besides one and there will be no even in principle possibility for the future interference of these branches.

It leads to the conclusion that there should be a fundamental difference in understanding of time irreversability in two versions of quantum theory.

To follow it in detail let us consider how the thermodynamical arrow of time arises in the case of measurement. This arrow of time is defined by the direction in which the entropy increases.

If the reduction of the state vector of the quantum system $S$ takes place (as in Copenhagen interpretaion) the total entropy after the measurement can be represented as

$$S^{M+S} = S^M + S^S \tag{19.19}$$

as well as before the measurement, since there is no correlation between the system $S$ and the device $M$. The ensemble of quantum systems with zero entropy evolves into the Ensemble I with density matrix $\hat{\rho}^I$ under the measurement, the entropy of which is

$$S^I = -Tr\hat{\rho}^I \ln\hat{\rho}^I \tag{19.20}$$

This entropy always increases irreversibly and we could use it to define completely an irreversible thermodynamical arrow of time.

In the case when both the system $S$ and the device $M$ are described quantum mechanically (Many-Worlds interpretation), the total entropy defined as

$$S^{M+S} = -Tr\hat{\rho}^{M+S} \ln\hat{\rho}^{M+S} \tag{19.21}$$

is conserved. So, if $S^{M+S} = 0$ (the pure state) before the measurement, then $S^{M+S} = 0$ also after the measurement. We cannot use the total entropy $S^{M+S}$ to define the thermodynamical arrow of time in the case under consideration. However, this entropy can be represented in form

$$0 = S^{M+S} = -Tr\hat{\rho}^M \ln\hat{\rho}^M - Tr\hat{\rho}^S \ln\hat{\rho}^S + S_{corr.}^{M+S} \tag{19.22}$$
$$= S^M + S^S + S_{corr.}^{M+S} \tag{19.23}$$

where $\hat{\rho}^M = Tr_S\,\hat{\rho}^{M+S}$ and $\hat{\rho}^S = Tr_M\,\hat{\rho}^{M+S}$ are the density matrices of $M$ and $S$, and $S_{corr.}^{M+S}$ is the negative correlation entropy (see Everett's paper in [2]), which reflects the presence of correlation between the states of the quantum system and the device after the measurement (nonseparability). As a rule $S^S$ and $S^M$ are increasing and we could use the entropy $S^M + S^S$ to define for a finite time interval the thermodynamical arrow of time (compare with coarse grained entropy [9]). Nevertheless the unitary evolution is time reversable and, for a example, for the system with finite motion, it is quasiperiodic. Hence in some (as a rule, very long) time the system will return to the initial state ($S^M + S^S$ should decrease). Thus here the thermodynamic arrow of time is reversable at a fundamental level.

## Acknowledgements

I would like to thank the organizers of the conference "Physical Origins of Time Asymmetry" for a very nice meeting. This work was supported in part by the Swiss National Science Foundation.

## Discussion

**Unruh**  If you want to make a correspondence between the theory and reality you have two options. You can change the theory to correspond with reality, or you can change your notion of reality to correspond with the theory. Your view is the second option.

**Mukhanov**  Nobody knows in advance what is reality. We study it experimentally and then try to describe the experimental results by a theory. Usually, a good theory tells us much more about reality than we assumed at the begining. This is namely the case of quantum mechanics and the many-worlds interpretation of reality is the simplest way to understand the interference phenomena.

**Griffiths**  It seems to me that while you have properly pointed out the problems with wave-function collapse, you have not taken account of the consistent histories possibilities which eliminates the "multiple worlds" with a single logical extension of standard quantum mechanics, namely an extended probability postulate.

**Mukhanov**  From my point of view the consistent histories approach to quantum theory is still underdeveloped and logically incomplete in some respects. In particular, if we will analyze the *repeatability* and correspondence between reality and the mathematical description of the state of physical system in this theory then this analysis will naturally lead us to Everett's conclusion about "multiple worlds".

## References

[1] Everett H. III (1957) *Rev. Mod. Phys.*, **29**, 454.
[2] DeWitt B. and Graham N. (1973) *The many-worlds interpretation of quantum mechanics*, Princeton University Press, Princeton.
[3] Deutsch D. (1985) *Int. J. Theor. Phys.*, **24**, 1.
[4] von Neumann (1955) *Mathematical foundations of quantum mechanics*, Princeton University Press, Princeton.
[5] Zurek W. (1986) in: *Frontiers of Nonequilibrium Statistical Physics*, eds. Moore G.T. and Scully M.O. (Plenum).
[6] Griffiths R.B. (1984) *J. Stat. Phys.*, **36**, 219.
[7] Gell-Mann M. and Hartle J.B. (1990) in: *Complexity, Entropy and the Physics of Information, SFI Studies in the Sciences of Complexity*, ed. Zurek W., (Addison Wesley, Reading)
[8] Mukhanov V. (1988) *Phys. Lett.*, **127A**, 251.
[9] Zaslavsky O. (1985) *Chaos in dynamical systems.*

# 20

## Clock Time and Entropy

### Don N. Page

*CIAR Cosmology Program*
*Institute for Theoretical Physics*
*Department of Physics*
*University of Alberta*
*Edmonton, Alberta*
*Canada T6G 2J1*

### 20.1 What is Fundamental in a Quantum Description of the World?

It is a matter of debate what the basic entities of the universe are in quantum theory. Expanding the metaphor used by Griffiths at this conference, one might imagine that the conversations of the three meals of the day could be restricted to three possible answers:

(a) Amplitudes (subject of lunch conversation). One might imagine formulating quantum theory in the morning and concluding at lunch that complex amplitudes were the fundamental entities. Each fine-grained history $h_i$ might be assigned an amplitude

$$A[h_i] = e^{iS[h_i]}, \tag{20.1}$$

where $S[h_i]$ is the action (in units of $\hbar$) of that history (Feynman and Hibbs 1965). The fine-grained histories could then be combined into a weighted group of histories $\alpha$ (called a coarse-grained history) with amplitude

$$A[\alpha] = \sum_i w_\alpha [h_i] e^{iS[h_i]}, \tag{20.2}$$

where $w_\alpha$ is an $\alpha$-dependent weight functional of the fine-grained histories $h_i$. For example, $A[\alpha]$ could be given by a path integral over a set of histories, so that $w_\alpha$ could be 1 if $h_i$ is in the set and 0 if $h_i$ is not. More generally, $w_\alpha[h_i]$ might be weighted by the complex amplitude for the initial configuration of the history $h_i$. If $\alpha$ consisted of all of the histories that reached a certain final configuration, and no others, $A[\alpha]$ would be what is called the wave function evaluated at that final configuration. A problem with these amplitudes as basic entities, however, is that it is not clear what direct interpretation they have.

(b) Probabilities for decohering sets of histories (subject of dinner conversation) (Griffiths 1984; Omnès 1988a, 1988b, 1988c, 1989; Gell-Mann and Hartle 1990; Hartle 1990a, 1990b, 1991). One might imagine interpreting the amplitudes in the

---

† E-mail address: don@phys.ualberta.ca

afternoon so that the (unnormalized) probability associated with each coarse-grained history $\alpha$ is

$$p(\alpha) = |A(\alpha)|^2. \tag{20.3}$$

In order to get a set of probabilities obeying the usual axioms of probability, one then needs a set of $\alpha$'s such that for any pair ($\alpha \neq \alpha'$),

$$p(\alpha + \alpha') \equiv |A(\alpha) + A(\alpha')|^2 = p(\alpha) + p(\alpha') \equiv |A(\alpha)|^2 + |A(\alpha')|^2. \tag{20.4}$$

This means that the real part of the interference terms must vanish,

$$2\,\mathrm{Re}\,A(\alpha)A^*(\alpha') = A(\alpha)A^*(\alpha') + A^*(\alpha)A(\alpha') = 0. \tag{20.5}$$

More generally, one may have a decoherence functional $D(\alpha, \alpha')$ with

$$\mathrm{Re}\,D(\alpha, \alpha') = 0, \tag{20.6}$$

and then

$$p(\alpha) = D(\alpha, \alpha). \tag{20.7}$$

One then calls this set of $\alpha$'s a decohering set of (coarse-grained) histories. One says that probabilities are defined only for each member of such a decohering set. Then the viewpoint is that such probabilities are the fundamental entities of the quantum theory. Feeling that his task is basically completed, the theorist retires for the night.

(c) Testable conditional probabilities (subject of breakfast conversation) (Page and Wootters 1983; Page 1986a, 1986b, 1987, 1988, 1989, 1991a, 1991b, 1992a).

After a good night's sleep, the theorist, able to think more clearly, realizes that even decohering sets of histories are too broad to be directly interpreted. Instead, it is a much narrower set of conditional probabilities that we, living within the universe, can test. So far as we know, each of these occurs on a single hypersurface (at a single "time"), and perhaps is also highly localized on that hypersurface.

That is, we cannot directly compare things at different times, but only different records at the same time. We cannot know the past except through its records in the present, so it is only present records that we can really test. For example, we cannot directly test the conditional probability that the electron has spin up at $t = t_f$, given that it had spin up at $t = t_i < t_f$, but only given that there are records at $t = t_f$ that we interpret as indicating the electron had spin up at $t = t_i$. Wheeler (1978) states this even more strongly: "the past has no existence except as it is recorded in the present."

This principle should perhaps also be extended to say that we cannot directly compare things at different locations either, so that all observations are really localized in space as well as in time. Certainly each of my observations seems to be localized within the spatial extent of my brain and the temporal extent of a single conscious moment, though it appears to be an exaggeration to say it is localized to a single spacetime point. Whether it is actually localized on a single hypersurface of a certain spatial extent is admittedly an open question, but since the duration of a single conscious moment is so much shorter than the apparent age of the universe, it seems reasonable and consistent with all we do know to idealize each observation as occurring at a single time, on a single spatial hypersurface.

If the quantum state of the universe on a spatial hypersurface is given by a density matrix $\rho$, the conditional probability of the result $A$, given a testable condition $B$, is

$$p(A|B) = \frac{\mathrm{Tr}\,(P_A P_B\,\rho\,P_B)}{\mathrm{Tr}\,(P_B\,\rho\,P_B)} = \frac{\mathrm{Tr}\,(P_B P_A P_B\,\rho)}{\mathrm{Tr}\,(P_B\,\rho)}, \tag{20.8}$$

where $P_A = P_A^\dagger = P_A^2$ and $P_B = P_B^\dagger = P_B^2$ are the corresponding projection operators. The testable conditional probabilities are then the subject of rational breakfast conversation.

Of course, the theorist's job is not completed with the mere formulation of Eq. (20.8): one must discover the density matrix of the universe, formulate the projection operators, and calculate the quantities in Eq. (20.8). Then there is the question of which results and conditions are testable. One could avoid this problem by simply postulating that the conditional probabilities associated with all projection operators $P_A$ and $P_B$ are meaningful fundamental entities. However, most of these would not be readily interpretable, so there would seem to be little motivation not to go back to the complex amplitudes (20.1) and (20.2) as more fundamental.

A conclusion of these various meal discussions is that amplitudes seem to be the most basic entities, but that if one wishes a more restricted set that can be readily interpreted and tested, it does not seem to be going far enough to say they are the probabilities for decohering sets of histories. Testable conditional probabilities appear in reality to be restricted to a single hypersurface, not to histories, though even this restriction alone is almost certainly not enough.

## 20.2 Inaccessibility of Coordinate Time

As mentioned above, testable conditional probabilities (20.8) seem to be confined to a single "time" (e.g., a spatial hypersurface in canonical gravity). Yet they cannot depend on the value of the coordinate time labeling the hypersurface, which is completely unobservable. One can give three arguments for this fact:

(a) For a closed universe, the Wheeler-DeWitt equation, $H\psi = 0$, implies that $\psi$ is independent of $t$.

(b) For an asymptotically-flat open universe, the long-range gravitational field provides a superselection rule for the total energy, just as the long-range Coulomb electric field provides a superselection rule for the total charge (Strocchi and Wightman 1974). This means that phases between states of different energy are unmeasurable, so the coordinate-time dependence of the density matrix $\rho$ is not detectable (Page and Wootters 1983).

(c) In any case, one has no access to the coordinate time $t$, so one should average over this inaccessible variable to get a density matrix for the other variables (Page 1989). This time-averaged density matrix in the Schrödinger picture is

$$\bar{\rho} = \langle \rho(t) \rangle \equiv \lim_{T \to \infty} \frac{1}{2T} \int_{-T}^{T} dt \, \rho(t) \tag{20.9}$$

and in the Heisenberg picture is

$$\bar{\rho} = \langle e^{-iHt} \rho e^{iHt} \rangle \equiv \lim_{T \to \infty} \frac{1}{2T} \int_{-T}^{T} dt \, e^{-iHt} \rho \, e^{iHt}. \tag{20.10}$$

Without access to coordinate time $t$, we can only test conditional probabilities of the form

$$
\begin{aligned}
p(A|B) &= \frac{\langle \text{Tr}\,[P_B P_A P_B \, \rho(t)] \rangle}{\langle \text{Tr}\,(P_B \, \rho(t)) \rangle} = \frac{\text{Tr}\,(P_B P_A P_B \, \bar{\rho})}{\text{Tr}\,(P_B \, \bar{\rho})} \\
&= \frac{\langle \text{Tr}\,(P_B P_A P_B e^{-iHt} \rho(0)\, e^{iHt}) \rangle}{\langle \text{Tr}\,(P_B e^{-iHt} \rho(0)\, e^{iHt}) \rangle} \\
&= \frac{\text{Tr}\,[\langle e^{iHt} P_B P_A P_B \, e^{-iHt} \rangle \rho(0)]}{\text{Tr}\,[\langle e^{iHt} P_B \, e^{-iHt} \rangle \rho(0)]} = \frac{\text{Tr}\,[\overline{P_B P_A P_B} \, \rho(0)]}{\text{Tr}\,[\overline{P}_B \, \rho(0)]},
\end{aligned} \tag{20.11}
$$

where in the Schrödinger picture

$$\overline{O} \equiv \langle e^{iHt} O e^{-iHt} \rangle \tag{20.12}$$

for an observable $O$ without explicit time dependence, and in the Heisenberg picture

$$\overline{O} \equiv \langle O(t) \rangle = \langle e^{iHt} O(0) e^{-iHt} \rangle. \tag{20.13}$$

$$[H, \overline{O}] = 0, \tag{20.14}$$

so the observable $\overline{O}$ is stationary.

Thus testable conditional probabilities depend only on the stationary (time-averaged) density matrix $\bar{\rho}$, or alternatively, only on $\rho(0)$ and stationary observables like $\overline{P}_B$ and $\overline{P_B P_A P_B}$ ($\neq \overline{P}_B \, \overline{P}_A \, \overline{P}_B$). In a constrained Hamiltonian system like canonical quantum gravity, these observables, but not the individual projection operators, would commute with the constraints (e.g., the Wheeler-DeWitt operator).

## 20.3 Observable Evolution as Dependence on Clock Time

Although unobservable coordinate time cannot be part of the condition B of $p(A|B)$ given by Eq. (20.8) or (20.11), the reading of a physical clock can be. The dependence of $p(A|B)$ upon the clock reading is then the observable time evolution of the system.

For simplicity, consider the case where the condition is entirely the reading of a clock subsystem ($C$) with states $|\psi_C(T)\rangle$. Let

$$P_B \to P_T = |\psi_C(T)\rangle\langle\psi_C(T)| \otimes I_R, \tag{20.15}$$

where $I_R$ is the unit operator for the rest $(R)$ of the closed system (e.g., the rest of the universe). Then

$$p(A|T) = \frac{\text{Tr}(P_T P_A P_T \bar{\rho})}{\text{Tr}(P_T \bar{\rho})} \tag{20.16}$$

can vary with the clock time $T$.

In the special case that the clock does not interact with the rest of the system, so

$$H = H_C \otimes I_R + I_C \otimes H_R, \tag{20.17}$$

and in the case that the result $A$ does not concern the clock, so

$$P_A = I_C \otimes P_{AR} \tag{20.18}$$

with $P_{AR}$ acting only on $R$, then the choice of clock states of reading $T$ as

$$|\psi_C(T)\rangle \equiv e^{-iH_C T} |\psi_C(0)\rangle \tag{20.19}$$

leads to the following familiar-looking equation for the conditional probability of $A$ given the clock reading $T$ :

$$p(A|T) = \underset{R}{\text{tr}} \, [P_{AR} \, \rho_R(T)]. \tag{20.20}$$

Here, the conditional density matrix for $R$ at clock reading $T$ is

$$\begin{aligned}
\rho_R(T) &= \underset{C}{\text{tr}} (P_T \, \bar{\rho} \, P_T) / \text{Tr}(P_T \, \bar{\rho}) \\
&= e^{-iH_R T} \rho_R(0) e^{iH_R T}, \tag{20.21}
\end{aligned}$$

thus evolving (in the Schrödinger picture) according to the von Neumann equation with respect to the clock time $T$ (Page and Wootters 1983).

In the more general case that the clock does interact with the rest of the system and/or if the clock states are not defined by Eq. (20.19) (which gives

$$P_T = e^{iHT} P_{T=0} e^{-iHT} \tag{20.22}$$

and has the somewhat undesirable consequence that $P_T P_{T'}$ is not necessarily zero for $T \neq T'$), then generically $p(A|T)$ can be obtained from Eq. (20.16), but it will not have the $\text{tr}(P_{AR} \rho_R)$ form of Eq. (20.20) with a unitarily evolved density matrix $\rho_R(T)$. In this more general case clock time will not have been defined so that motion is so simple as Eqs. (20.20) and (20.21), in contrast to the criterion of Misner, Thorne, and Wheeler (1973).

Karel Kuchař (1992, and at lunch during this conference) has given two objections to the conditional probability interpretation outlined here:

(a) The application of the condition violates the constraints. Eq. (20.8) or (20.11) may be written as

$$p(A|B) = \text{Tr}(P_A \, \rho_B), \tag{20.23}$$

where the conditional density matrix

$$\rho_B = P_B \rho P_B / \text{Tr}(P_B \rho) \tag{20.24}$$

generally does not satisfy the constraints, e.g.,

$$[H, \rho_B] \neq 0, \tag{20.25}$$

if

$$[H, P_B] \neq 0. \tag{20.26}$$

However, $\rho_B$ is merely a calculational device to simplify Eq. (20.8) or (20.11) to Eq. (20.23). One could instead rewrite Eq. (20.8) or (20.11) as

$$p(A|B) = \text{Tr}\, \rho_{AB}, \tag{20.27}$$

where

$$\rho_{AB} = \langle P_A P_B \rho P_B P_A \rangle / \text{Tr}\langle P_B \rho P_B \rangle \tag{20.28}$$

does obey the constraints.

(b) The conditional probability formula (20.8) does not give the right propagators. For example, if

$$P_{T_1} P_{T_2} = P_{T_2} P_{T_1} = 0, \tag{20.29}$$

$$P_{A_1} P_{T_1} = P_{T_1} P_{A_1} \neq 0, \tag{20.30}$$

$$P_{A_2} P_{T_2} = P_{T_2} P_{A_2} \neq 0, \tag{20.31}$$

then Kuchař wants $p(A_2, T_2 | A_1, T_1)$ to be the absolute square of the propagator (i.e., the transition probability) to go from $A_1$ at $T_1$ to $A_2$ at $T_2$, whereas the formula

$$p(A_2, T_2 | A_1, T_1) = \frac{\text{Tr}(P_{A_2} P_{T_2} P_{A_1} P_{T_1} \rho P_{T_1} P_{A_1} P_{T_2} P_{A_2})}{\text{Tr}(P_{A_1} P_{T_1} \rho P_{T_1} P_{A_1})} \tag{20.32}$$

gives zero.

In response, I would say that the absolute square of the propagator is

$$|\langle A_2 \text{ at } T_2 | A_1 \text{ at } T_1 \rangle|^2 = p(A_2 \text{ at } T_2 | A_1 \text{ at } T_1), \tag{20.33}$$

that is, the probability to have $A_2$ at $T_2$ if one had $A_1$ at $T_1$, whereas Eq. (20.32) gives

$$p(A_2, T_2 | A_1, T_1) = p(A_2 \text{ and } T_2 | A_1 \text{ and } T_1), \tag{20.34}$$

the probability that one has $A_2$ *and* $T_2$ if one has $A_1$ *and* $T_1$ on the same hypersurface. Since Eq. (20.29) says that $T_1$ and $T_2$ are mutually exclusive, this probability is zero: one cannot simultaneously have two distinct clock readings.

The conditional probability interpretation is dynamical in the sense that it gives the $T$ dependence of $p(A|T)$ by Eq. (20.16). However, it is not so dynamical as Kuchař would like to give the transition probability of Eq. (20.33) directly. The reason it does not is that this quantity is not directly testable, since $T_1$ and $T_2$ are different conditions that cannot be imposed together. At $T = T_2$, one has no direct access to what happened at $T = T_1$. One must instead rely upon records, which can be checked at $T_2$.

One way that one could try to calculate the transition probability (20.33) theoretically is to try to construct a projection operator $P_{A_1 T_1}$ ($\neq P_{A_1} P_{T_1}$) which, when applied to $\rho$, gives a density matrix

$$\rho_{A_1 T_1} = P_{A_1 T_1} \rho P_{A_1 T_1} / \mathrm{Tr}\,(P_{A_1 T_1}\, \rho) \qquad (20.35)$$

which satisfies the constraints and which gives

$$p(A_1 | T_1 ; \rho_{A_1 T_1}) = \frac{\mathrm{Tr}\,(P_{T_1} P_{A_1} P_{T_1}\, \rho_{A_1 T_1})}{\mathrm{Tr}\,(P_{T_1}\, \rho_{A_1 T_1})} = 1. \qquad (20.36)$$

Then one could say

$$p(A_2 \text{ at } T_2 \mid A_1 \text{ at } T_1) = p(A_2 | T_2 ; \rho_{A_1 T_1}). \qquad (20.37)$$

However, it is not clear whether the answer is unique and whether it has certain desirable properties. More importantly, it is not directly testable, and therefore it is rather *ad hoc* which definition to propose for the transition probability.

## 20.4 The Variation of Entropy with Clock Time

The time asymmetry of the second law of thermodynamics should of course be expressed in terms of physical clock time rather than in terms of unobservable coordinate time (Page 1992b).

How to express entropy is more problematic. One mathematical expression is

$$S_T \quad = \quad -\mathrm{Tr}\,\rho_T\, \ln \rho_T, \qquad (20.38)$$

$$\rho_T \quad = \quad P_T\, \bar{\rho}\, P_T / \mathrm{Tr}\,(P_T\, \bar{\rho}\, P_T). \qquad (20.39)$$

If

$$\bar{\rho} = |\psi\rangle\langle\psi|, \qquad (20.40)$$

a pure state, then

$$\rho_T = |\psi_T\rangle\langle\psi_T| \qquad (20.41)$$

with

$$|\psi_T\rangle = P_T |\psi\rangle / \langle\psi|P_T|\psi\rangle^{1/2}, \qquad (20.42)$$

another pure state, so $S_T = 0$.

But if $\bar{\rho}$ is not pure (e.g., is obtained from

$$\rho = |\psi\rangle\langle\psi| \qquad (20.43)$$

pure but time dependent), then $S_T$ can exceed zero and vary with the clock time $T$.

*Example*: Consider two coupled spin-$\frac{1}{2}$ systems in a vertical magnetic field, with Hamiltonian

$$H = \frac{1}{4}\vec{\sigma}_C \cdot \vec{\sigma}_R + \frac{1}{2}\sigma_{C_z} \otimes I_R + \frac{1}{2}I_C \otimes \sigma_{R_z} + \frac{3}{4}I$$

$$= 2|\uparrow\uparrow\rangle\langle\uparrow\uparrow| + \frac{1}{2}(|\uparrow\downarrow\rangle + |\downarrow\uparrow\rangle)(\langle\uparrow\downarrow| + \langle\downarrow\uparrow|), \tag{20.44}$$

giving a triplet and a singlet degenerate with the lowest state of the triplet. Take the time-dependent pure state

$$|\psi\rangle = 0.1\sqrt{5}\left[(e^{-it} + 1)|\uparrow\downarrow\rangle + (e^{-it} - 1)|\downarrow\uparrow\rangle + 4|\downarrow\downarrow\rangle\right], \tag{20.45}$$

so the coordinate-time-averaged density matrix is

$$\begin{aligned}\bar{\rho} = \ & 0.1(|\uparrow\downarrow\rangle\langle\uparrow\downarrow| + 2|\uparrow\downarrow\rangle\langle\downarrow\downarrow| + 2|\downarrow\downarrow\rangle\langle\uparrow\downarrow| + |\downarrow\uparrow\rangle\langle\downarrow\uparrow| \\ & -2|\downarrow\uparrow\rangle\langle\downarrow\downarrow| - 2|\downarrow\downarrow\rangle\langle\downarrow\uparrow| + 8|\downarrow\downarrow\rangle\langle\downarrow\downarrow|).\end{aligned} \tag{20.46}$$

Let

$$P_T = \frac{1}{2}(|\uparrow\rangle + e^{-iT}|\downarrow\rangle)(\langle\uparrow| + e^{iT}\langle\downarrow|)_C \otimes I_R. \tag{20.47}$$

This leads to

$$\rho_T = \rho_{TC} \otimes \rho_{TR} \tag{20.48}$$

with

$$\rho_{TC} = \frac{1}{2}(|\uparrow\rangle + e^{-iT}|\downarrow\rangle)(\langle\uparrow| + e^{iT}\langle\downarrow|)_C, \tag{20.49}$$

$$\begin{aligned}\rho_{TR} = \ & (10 + 4\cos T)^{-1}\big[|\uparrow\rangle\langle\uparrow| - 2|\uparrow\rangle\langle\downarrow| - 2|\downarrow\rangle\langle\uparrow| \\ & + (9 + 4\cos T)|\downarrow\rangle\langle\downarrow|\big].\end{aligned} \tag{20.50}$$

Therefore,

$$S(\rho(t)) = 0, \quad S(\bar{\rho}) = \ln 10 - 1.8\ln 3 \approx 0.3251, \tag{20.51}$$

$$\begin{aligned}S_T = S(\rho_T) = \ & \ln\left(\frac{10 + 4\cos T}{\sqrt{5 + 4\cos T}}\right) - \frac{2\sqrt{1 + (2 + \cos T)^2}}{5 + 2\cos T} \\ & \times \ln\left(\frac{5 + 2\cos T + 2\sqrt{1 + (2 + \cos T)^2}}{\sqrt{5 + 4\cos T}}\right),\end{aligned} \tag{20.52}$$

which varies in a rather complicated way with clock time $T$.

Even when $\rho_T$ is pure and gives $S_T = 0$, we may divide the system into subsystems and add the entropy of each subsystem density matrix (which ignores the information or negentropy of the correlations between the subsystems):

$$S_{T,\text{coarse}} = -\sum_{i=1}^{n} \text{tr}\,\rho_{T_i} \ln \rho_{T_i}, \tag{20.53}$$

$$\rho_{T_i} = \underset{j\neq i}{\text{tr}}\, P_T\,\bar{\rho}P_T / \text{Tr}\,(P_T\,\bar{\rho}). \tag{20.54}$$

*Example*: Consider three spin-$\frac{1}{2}$ systems in a vertical magnetic field, with Hamiltonian

$$H = \frac{1}{2}\sigma_{C_z} \otimes I_{R1} \otimes I_{R2} + \frac{1}{2}I_C \otimes \sigma_{R1_z} \otimes I_{R2}$$

$$+ \frac{1}{2}I_C \otimes I_{R1} \otimes \sigma_{R2_z} + \frac{1}{4}I_C \otimes \vec{\sigma}_{R1} \cdot \vec{\sigma}_{R2} + \frac{1}{4}I$$

$$= 2|\uparrow\uparrow\uparrow\rangle\langle\uparrow\uparrow\uparrow|$$

$$+ \frac{1}{2}(|\uparrow\uparrow\downarrow\rangle + |\uparrow\downarrow\uparrow\rangle)(\langle\uparrow\uparrow\downarrow| + \langle\uparrow\downarrow\uparrow|) + |\downarrow\uparrow\uparrow\rangle\langle\downarrow\uparrow\uparrow|$$

$$- \frac{1}{2}(|\downarrow\uparrow\downarrow\rangle - |\downarrow\downarrow\uparrow\rangle)(\langle\downarrow\uparrow\downarrow| - \langle\downarrow\downarrow\uparrow|) - |\downarrow\downarrow\downarrow\rangle\langle\downarrow\downarrow\downarrow|, \quad (20.55)$$

so the first spin acts as a noninteracting clock and the remaining two have the same spin-spin coupling as the previous example. Take a pure-state linear combination of the zero-energy eigenvectors,

$$|\psi\rangle = \frac{1}{2}(|\uparrow\uparrow\downarrow\rangle - |\uparrow\downarrow\uparrow\rangle + |\downarrow\uparrow\downarrow\rangle + |\downarrow\downarrow\uparrow\rangle), \quad (20.56)$$

so Eq. (20.40) holds. Let $P_T$ be given by Eq. (20.47) again, except that now $I_R$ is the $4 \times 4$ unit matrix for the last two spins. This gives $\rho_T$ of the pure form (20.41) with

$$|\psi_T\rangle = 2^{-1/2}(|\uparrow\rangle + e^{-iT}|\downarrow\rangle)_C \otimes \frac{1}{2}[(e^{iT} + 1)|\uparrow\downarrow\rangle + (e^{iT} - 1)|\downarrow\uparrow\rangle]_R, \quad (20.57)$$

so $\rho_T$ again factorizes as in Eq. (20.48) between the clock and the rest, with $\rho_{TC}$ given by Eq. (20.49) and

$$\rho_{TR} = \frac{1}{2}[(1 + \cos T)|\uparrow\downarrow\rangle\langle\uparrow\downarrow| - i\sin T|\uparrow\downarrow\rangle\langle\downarrow\uparrow|$$

$$+ i\sin T|\downarrow\uparrow\rangle\langle\uparrow\downarrow| + (1 - \cos T)|\downarrow\uparrow\rangle\langle\downarrow\uparrow|], \quad (20.58)$$

a pure state which does not factorize between the last two spins. The reduced density matrices of the two spins are

$$\rho_{TR1} = \frac{1}{2}I_{R1} + \frac{1}{2}\cos T \; (|\uparrow\rangle\langle\uparrow| - |\downarrow\rangle\langle\downarrow|)_{R1}, \quad (20.59)$$

$$\rho_{TR2} = \frac{1}{2}I_{R2} - \frac{1}{2}\cos T \; (|\uparrow\rangle\langle\uparrow| - |\downarrow\rangle\langle\downarrow|)_{R2}, \quad (20.60)$$

each with eigenvalues

$$p_1 = \cos^2\frac{1}{2}T, \qquad p_2 = \sin^2\frac{1}{2}T. \quad (20.61)$$

Hence, the coarse-grained entropy given by Eq. (20.53) is

$$S_{T,\text{coarse}} = S(\rho_{TC}) + S(\rho_{TR1}) + S(\rho_{TR2})$$

$$= -2\cos^2\frac{1}{2}T \ln\cos^2\frac{1}{2}T - 2\sin^2\frac{1}{2}T \ln\sin^2\frac{1}{2}T, \quad (20.62)$$

oscillating between 0 at $T = n\pi$ and $2\ln 2$ at $T = (n + 1/2)\pi$.

Although they can vary with clock time, the mathematical and coarse-grained entropies defined by Eqs. (20.38) and (20.53) above are not really observables. One would like a truly observable entropy operator $\widehat{S}$ and then define the clock-time-dependent entropy as its conditional expectation value:

$$S_{T,\text{observable}} = E(\widehat{S}|T) = \text{Tr}\,(P_T \widehat{S} P_T\, \bar{\rho})/\text{Tr}\,(P_T\, \bar{\rho}). \tag{20.63}$$

Can one find suitable definitions of $\widehat{S}$ and $T$ so that $S_{T,\text{observable}}$ increases fairly monotonically with $T$ for the actual density matrix $\rho$ of the universe?

I have benefited from discussions on these points at the conference with Bryce DeWitt, Murray Gell-Mann, James Hartle, Karel Kuchař, Emil Mottola and William Unruh. Financial assistance has been provided by the Canadian Institute for Advanced Research, the Fundación Banco Bilbao Vizcaya, the National Science Foundation, and the Natural Sciences and Engineering Research Council.


## Discussion

**Kuchař**  You always apply the conditional probability formula to calculate the conditional probability of a projector at a single instant of an internal clock time. You never apply it to answering the fundamental DYNAMICAL question of the internal Schrödinger interpretation, namely, "If one finds the particle at $Q'$ at the time $T'$, what is the probability of finding it at $Q''$ at the time $T''$?" By your formula, that conditional probability differs from zero only if $T' = T''$ and $Q' = Q''$. In brief, your interpretation prohibits the time to flow and the system to move!

For me, this virtually amounts to a reductio ad absurdum of the conditional probability proposal. One can trace this feature back to the fact that the conventional derivation of the conditional probability formula amounts to the violation of the super-Hamiltonian constraint.

**Page**  At this time I cannot give much more of an answer than I did in my lecture (which is slightly expanded for the Proceedings). In my viewpoint, only quantities at a single instant of time are directly accessible, and so one cannot directly test the two-time probability you discuss. One could instead at one time test the conditional probability that the particle is at $Q''$, given that the time is $T''$ and that at this time there is a *record* indicating that the particle was at $Q'$ at the time $T'$. However, there is no direct way to test whether the record is accurate, though one can check whether different records show consistency. After all, that is the only way we have to increase our confidence in the existence of historical events.

**Kuchař**  Don, you are the first person I met who simultaneously believes in the existence of many worlds and is a solipsist of an instant.

**Page**  I believe that different instants, i.e., different clock times, are actually examples of the different worlds. They all exist, but each observation, and its associated conditional probability, occurs at one single time (assuming that the condition includes or implies a precise value of the clock time in question). We can only directly observe and be aware of the world, and the time, in and at which we exist, though the correlations in memories and other structures we observe in one world give indirect evidence of other worlds, and other clock times, in the full quantum state of the universe. I do not deny the existence of historical events at different instants of clock time, but I do not believe that they have conditional probabilities (given our conditions here and now) exactly equal to unity, or even that they can be precisely assigned any probabilities at all, say in the sense of Eq. (20.7) when Eq. (20.6) holds. In any case, any imputed probabilities for events in the past or future cannot be directly tested.

## References

Feynman, R.P. and Hibbs, A.R. (1965) *Quantum Mechanics and Path Integrals,* McGraw-Hill, New York.

Gell-Mann, M. and Hartle, J.B. (1990) In *Complexity, Entropy, and the Physics of Information, SFI Studies in the Sciences of Complexity,* Vol. VIII, Ed. W. Zurek, Addison Wesley, Reading, and in *Proceedings of the 3rd International Symposium on the Foundations of Quantum Mechanics in the Light of New Technology,* Eds. S. Kobayashi, H. Ezawa, Y. Murayama, and S. Nomura, Physical Society of Japan, Tokyo.

Griffiths, R. (1984) *J. Stat. Phys.,* **36**, 219.

Hartle, J.B. (1990a) In *Gravitation and Relativity 1989: Proceedings of the 12th International Conference on General Relativity and Gravitation,* Eds. N. Ashby, D. F. Bartlett, and W. Wyss, Cambridge University Press, Cambridge.

Hartle, J.B. (1990b) In *Proceedings of the 60th Birthday Celebration of M. Gell-Mann,* Eds. J. Schwarz and F. Zachariasen, Cambridge University Press, Cambridge.

Hartle, J.B. (1991) In *Quantum Cosmology and Baby Universes,* Eds. S. Coleman, J. B. Hartle, T. Piran, and S. Weinberg, World Scientific, Singapore.

Kuchař, K. (1992) In *Proceedings of the 4th Canadian Conference on General Relativity and Relativistic Astrophysics,* Eds. G. Kunstatter, D. Vincent, and J. Williams, World Scientific, Singapore.

Misner, C.W., Thorne, K. S., and Wheeler, J.A. (1973) *Gravitation,* Freeman, San Francisco.

Omnès, R. (1988a) *J. Stat. Phys.,* **53**, 893.

Omnès, R. (1988b) *J. Stat. Phys.,* **53**, 933.

Omnès, R. (1988c) *J. Stat. Phys.,* **53**, 957.

Omnès, R. (1989) *J. Stat. Phys.,* **57**, 357.

Page, D.N. (1986a) In *Quantum Concepts in Space and Time,* Eds. R. Penrose and C. J. Isham, Clarendon Press, Oxford.

Page, D.N. (1986b) *Phys. Rev.,* **D34**, 2267.

Page, D.N. (1987) In *String Theory – Quantum Gravity and Quantum Cosmology (Integrable and Conformal Invariant Theories): Proceedings of the Paris-Meudon Colloquium,* Eds. H. DeVega and N. Sanchez, World Scientific, Singapore.

Page, D.N. (1988) In *Proceedings of the Fourth Seminar on Quantum Gravity,* Eds. M. A. Markov, V. A. Berezin, and V. P. Frolov, World Scientific, Singapore.

Page, D.N. (1989) Time as an inaccessible observable. University of California at Santa Barbara report NSF-ITP-89-18.

Page, D.N. (1991a) In *Conceptual Problems in Quantum Gravity,* Eds. A. Ashtekar and J. Stachel, Birkhäuser, Boston.

Page, D.N. (1991b) In *Gravitation: A Banff Summer Institute,* Eds. R. Mann and P. Wesson, World Scientific, Singapore.

Page, D.N. (1992a) A physical model of the universe. Submitted for publication in *The Origin of the Universe: Philosophical and Scientific Perspectives,* Eds. R. F. Kitchener and K. T. Freeman.

Page, D.N. (1992b) The arrow of time. In, The Sakharov Memorial Lectures in Physics, Eds. L.V. Keldysh and V.Ya. Feinberg, Nova Science Publishers, Commack, New York.

Page, D.N. and Wootters, W.K. (1983) *Phys. Rev.*, **D27**, 2885.

Strocchi, F. and Wightman, A.S. (1974) *J. Math. Phys.*, **15**, 2198.

Wheeler, J.A. (1978) In *Mathematical Foundations of Quantum Theory*, Ed. A. R. Marlow, Academic Press, New York.

# 21

# Time-Symmetric Cosmology and Definite Quantum Measurements

L. S. Schulman

*Physics Department, Clarkson University*
*Potsdam, NY 13699-5820 USA*

### Abstract

The implications of a time-symmetric cosmology are considered, in particular the reversal of the thermodynamic arrow of time and the possibility of deducing an impending collapse. Within such a cosmology there is a rationale for obtaining definite quantum measurements by means of the "special" states proposed by the author. Other issues related to the author's measurement theory are discussed.

## 21.1 Introduction

In this article I will discuss time asymmetry and quantum measurement theory. My points will be that things may not be as asymmetric as they appear and that this may have a lot to do with the quantum measurement problem. In the context of the present conference such a position is not extreme.

To develop the first theme, that entropy need not increase monotonically with time, I will go back to an idea of Gold (1962), who suggested that coffee cools because the universe is expanding. His reasoning is essentially, COSMOLOGICAL ARROW $\Rightarrow$ RADIATION ARROW $\Rightarrow$ THERMODYNAMIC ARROW. I have criticized his arguments (Schulman 1973), but I accept his thesis. However, there have been other criticisms (Davies 1977, Weinberg 1972, Penrose 1979), based on what has been called the "switch-over" (if the universe enters a contracting phase, and if that implies a reversed thermodynamic arrow, what happens at the moment of reversal?), in which the thesis itself is rejected. For the purposes of my quantum theoretical ideas, this reversed arrow is not an inconvenience but an essential feature of the world. Therefore I will devote the first part of this article to examining, indeed exorcising, the "switch-over" bogeyman.

Then I will present my solution to the quantum measurement problem. Since the present conference appears to be a hotbed of (post-) Everettists, there will be agreement with two of the ideas I find fundamental. First, nothing ever happens except the unitary time evolution of quantum mechanics. After a time $t$, $\psi$ becomes

$\exp(-iHt/\hbar)\psi$. That's it: no nonlinear dynamics, no collapse, no extra set of dynamical rules. Secondly, the entire universe has a single wave function. You don't draw a line between system and apparatus, between apparatus and observer. However, lest you think that you are going to hear only the old and familiar, let me add the feature I maintain despite the previous assertions: There is only one version of macroscopic reality. You evolve by pure quantum dynamics, but the cat does not split. This will be accomplished with something I call "special" states, as I will outline below. Other issues that I will touch on are the recovery of standard probabilities (despite the determinism) and the existence of experimental tests to confirm or deny the theory I propose.

## 21.2 InterMediate Time Dynamics with Two-Time Boundary Conditions: Exorcising the "Switch-Over" Bogeyman

To say that the thermodynamic arrow follows the cosmological arrow, is to suggest that if the universe were to enter a contracting phase the thermodynamic arrow would reverse. Entropy would *decrease*, and one could have a big crunch roughly resembling a big bang played backwards. An image that arises in contemplating this scenario is that of a clock suddenly having its second hand switch round. The implausibility of such a "switch-over" has been raised as an objection to the correlation of thermodynamic and cosmological arrows. In Schulman (1991a) will be found specific quotations that make this point.

I will show here that the switch-over, to the extent that there is one, looks nothing like what was just suggested. The first step for understanding this issue is to realize that the natural and automatic use of "arbitrary" or "controllable" *initial* conditions is already equivalent to the setting of a thermodynamic arrow of time. For this reason, in discussing the arrow of time the specification of system states should not be made at one particular time ("naturally" taken as the initial time) but rather should be nonprejudicial, for example the giving of boundary conditions at two times. This approach is one that I and others have taken (Schulman 1973, 1974, 1976, 1977, Schulman and Shtokhamer 1977, Cocke 1967, Wheeler 1975) and it is helpful for examining the "switch-over." We therefore posit low entropy states at early and late times in the history of the universe. In practice we give incomplete information at each time. Under reasonable evolution, the entropy in between will be greater than at the endpoints. We want to know what happens at intermediate times and with stochastic time evolution it is possible to get comprehensive analytical results. Here is the precise framework: Let $X_t$ be the random position (or state) of a system at time $t$. In the absence of future conditioning this is a Markov process with the transition probability

$$W(x \leftarrow y) \equiv \Pr(X_{t+1} = x \mid X_t = y)$$

Define the propagator to be the conditional probability $G(x, t; y) = \Pr(X_t = x \mid X_0 =$

$y$). Using matrix notation $(W)_{xy} = W(x \leftarrow y)$, it follows that $G(x, t; y) = (W^t)_{xy}$. We impose the condition that the system be in state $\alpha$ at $t = 0$ and state $\beta$ at $t = T$. To deduce the intermediate dynamics, all we need are standard identities for conditional probabilities. The reasoning is summarized in our calculation of the modified propagator $G^{(\alpha,\beta)}(x, t)$

$$
\begin{aligned}
G^{(\alpha,\beta)}(x, t) &\equiv \Pr(X_t = x \mid X_0 = \alpha \text{ and } X_T = \beta) \\
&=^a \frac{\Pr(X_t = x \text{ and } X_T = \beta \mid X_0 = \alpha)}{\Pr(X_T = \beta \mid X_0 = \alpha)} \\
&=^b \frac{\Pr(X_T = \beta \mid X_t = x)\Pr(X_t = x \mid X_0 = \alpha)}{\Pr(X_T = \beta \mid X_0 = \alpha)} \\
&=^c \frac{G(\beta, T - t; x)G(x, t; \alpha)}{G(\beta, T; \alpha)}
\end{aligned}
\tag{21.1}
$$

Equality $a$ is the conditional probability identity. Equality $b$, uses the Markov property. Step $c$ uses time-translation invariance and invokes the definition of $G$.

The consequences of these identities follow from the properties of $W$. Its spectrum $\{\lambda\}$ generally has the following form: $1 = \lambda_0 > \lambda_1 \geq \lambda_2, \dots$. Since it is powers of $W$ that determine the properties of $G^{(\alpha,\beta)}(x, t)$, the issue will be how far $\lambda_1$ is from 1. One writes $\lambda_1 = e^{-1/\tau_1}$, and recognizes that $\tau_1$ is the longest relaxation time in the problem. In (Schulman 1991a), I examine various cases in detail. Here I summarize the conclusions and present graphical illustration. If $\tau_1 \ll T$, then long before $t = T/2$ the system is fully equilibrated and not only do clocks not reverse, but there are no clocks. If $\tau_1$ is comparable to $T$, then the system does not relax in its normal fashion at any time during the interval $[0, T]$. This may be explained with a casino analogy. Imagine a game in which one bets a stake and gets back 90% or 109% with equal probability. After about 500 plays, if you entered the casino with \$100, you would, on the average, be unable to meet the \$1 minimum ante. Suppose I tell you though that one evening you played 500 times and walked out with \$100. The most likely history is not that everything proceeded normally for most of the evening (so that after 300 plays you would have $\$100 \times (0.9 \times 1.09)^{150}$), but rather that all along you would be lucky. In fact, Eq. (21.1) is a precise description of the expected state at any given time.

There is no change in the structure if there are several time scales, for example $\tau_1 > T \gg \tau_2$ (with $\tau_2 = -1/\log \lambda_2$). In Figures 21.1 and 21.2, I illustrate the predictions of Eq. (21.1) for a system with two significant time scales, satisfying the inequality just written. You can think of it as a three state system with the state $x$ a highly excited metastable state that can decay quickly to another metastable state $y$. The state $y$, however, decays slowly. With initial conditions (only), the state $x$ empties quickly (heading for an asymptotic, nonzero, value). State $y$ fills, and then itself begins to lose probability to the state $z$. However, if we require that the systems begin *and* end in the state $x$, then for $y$ there is not much change in the

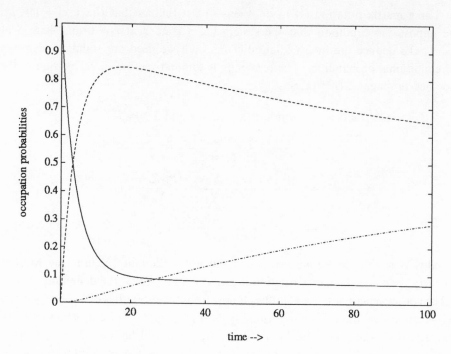

Fig. 21.1. History of a three state system with transition rates $x \to y$: 0.2, $y \to x$: 0.02, $y \to z$: 0.05, $z \to y$: 0.005. Initially $\Pr(X_1 = x) = 1$. Solid, dashed and dot-dashed lines are respectively the occupation probabilities of $x$, $y$, and $z$.

initial behavior. But when it comes to filling $z$ the final condition imposes a serious constraint and the system simply does not equilibrate "normally" (i.e., as in the initial condition case).

Several conclusions can be drawn from the preceding discussion. First about the switch-over: Well, there isn't any. Processes with short relaxation times are in equilibrium by $T/2$. Processes with long relaxation times never behave "normally" and at $T/2$ their behavior is no more peculiar than at other times. Of course there may be fluctuations in which short relaxation time processes are out of equilibrium at $T/2$, but for them the distant boundary conditions have no effect anyway, so that again there is no switch-over.

A second observation is that the initial emptying of $x$ is essentially the same both with and without final conditions. Since the future condition is many relaxation times away, the early time behavior of the fast processes is indistinguishable from initial-condition-only behavior. This means that although there may be a future condition on our universe, we might not notice it today.

A third lesson is the possibility of finding contemporary portents of a future collapse. Wheeler (1975) has suggested that samples of long-lived nuclides might show departures from exponential decay. Based on a model two-time boundary

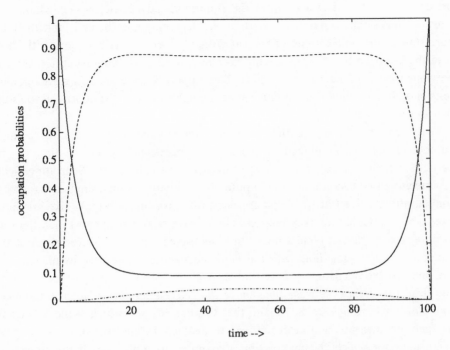

Fig. 21.2. As in Fig.1, except that we also require $\Pr(X_{100} = x) = 1$.

condition calculation (Schulman 1977), I found that laboratory samples would not exhibit this effect and that one would need a handle on the cosmological distribution of the nuclide, information that does not seem to be available with sufficient precision. However, I suggested another possibility. The dynamics of galaxies are slow, and the time scales one might assign are surely cosmological. Here, for example, is a way to note the effect of future boundary conditions: Look at the galaxy distribution in several old (distant) regions of the universe and use this inevitably imperfect information to predict the current (geometrical (Schulman and Seiden 1978)) entropy of this galaxy gas. Now measure the entropy of corresponding contemporary (nearby) regions. If a future low entropy condition is influencing the evolution, then the current entropy would not be as high as expected from the (inevitably) coarse grained information from the early measurements. As in Fig. 21.2, the lower entropy would reflect the future boundary condition.

### 21.3 Definite Quantum Measurements: "Special" States for a Single Exemplar of the Universe

As indicated, I believe that the only time evolution is that given by letting $e^{-iHt/\hbar}$ act on the wave function and that there is no reason not to let that wave function be the wave function of the entire universe. The theory (Schulman 1991b) I will present can be stated easily within the framework advocated by Gell-Mann and Hartle, namely take a pure state both at the beginning and the end (and the later pure state is the time evolute of the early one). Where I part company with them is in the assertion that despite having nothing but pure quantum evolution there is only one version of macroscopic reality. The support of the universe's wave function rests entirely on a single macroscopic state. I emphasize that it is nontrivial to assert that this is possible.

To be specific, let $\psi_k$ be the wave function at time $t_k$, $k = 1, 2$, of a large number of particles contained in a particular space-time region. At time $t_1$, the particles constitute a cat, radioactive source, observer, etc., in the Schrödinger-cat-experiment initial condition. So $\psi_1$, a pure state, has support on a single macroscopic configuration, at least at $t_1$. If you ask what the wave function looks like one hour later (and you forbid monkey business about wave packet reduction) then the way to answer is to take a typical microstate associated with the macrostate (not the unknown $\psi_1$, but something with the same macrostate) and evolve by microscopic dynamics. If the microstate was typical, $\psi_2$ will consist of a superposition of living and dead cat. *Hence the wave function of the universe no longer has all its support on a single macroscopic state.* I claim that this conclusion—which is the essence of the quantum measurement problem—can be avoided. I claim that there are atypical microstates for which the microscopic dynamics give only one of the outcomes. These microstates are designated "special." If $\psi_1$ is one of these "special" states, then $\psi_2$ will itself have support on only a single macroscopic state. But that is not all that is asserted. I further claim that for things that actually happen (as opposed to theoretical calculations) the initial conditions are always "special" in the sense indicated.

Before taking up further questions, let me say a little about the existence of "special" states. A classical analogy can be made with the states of an ice cube in an isolated glass of water. If asked to predict, one takes a typical microstate associated with the macrostate, evolves it forward and describes the final macrostate. The prediction will be: smaller ice cube, colder water. But we know that there are specific microscopic states of this system for which the ice cube grows. We know this by time reversal of the final state of a melting ice cube. My expectation is that the vast number of microscopic states in cat experiments, or any other macroscopic measurement situation, will have among them those for which the typical wave-function-splitting behavior does not occur, and a final microscopic state is reached for which the support is entirely on a single macroscopic state. Time reversal arguments do not suffice to show the existence of such states. For this reason I have made detailed microscopic models of (large) systems that can serve as measuring apparatus. In these models I have shown that certain particular microscopic initial conditions can lead to single macroscopic final states. This is a long way from showing that *all* apparatus have such "special" states, but it is a start.

However, even granting the existence of "special" states, several questions remain:

- Why should Nature choose them as the initial conditions in all measurement situations?
- If everything is deterministic, how do we recover the usual probabilities of quantum mechanics?
- Can this theory be tested experimentally?

## 21.4 Short Answers to Difficult Questions

- *Why "special" initial conditions?* The possibility of unusual initial conditions was the subject of Section 2. We saw that the imposition of far future final conditions selected a small subset of sample space points. This selection need not be apparent, as in the mutual resemblance of the initial portions of Figures 21.1 and 21.2. However, the selection ultimately dominates, in that the approach to the final state is determined by just this precise selection.

I will give an example of a final condition that under quantum evolution could select the "special" states. It differs from the stochastic dynamics example in that the effect of the future constraint does *not* get washed out at intermediate times. What is different about quantum mechanics is wave packet spreading. One can see this by looking at the sum of the squares of wave packet spreads at two times. We give a Hamiltonian $H$ and a pair of positions and times, $(x_1, t_1, x_2, t_2)$. For a particular $\psi_0$, let $\psi_t \equiv \exp(-iHt/\hbar)\psi_0$. Consider

$$W(\psi_0; x_1, t_1, x_2, t_2) \equiv \int dx |\psi_{t_1}(x)|^2 (x - x_1)^2 + \int dx |\psi_{t_2}(x)|^2 (x - x_2)^2$$

There are lower bounds on $W$ (Schulman 1991b, 1992). For example, for a nonrelativistic harmonic oscillator of mass $m$ and frequency $\omega$, no matter how clever you are in the selection of $\psi_0$ you cannot violate

$$W \geq \frac{\hbar}{m} \frac{|\sin \omega T|}{\omega} \qquad \left( W \geq \frac{\hbar T}{m} \quad \text{for the free particle} \right) \qquad (21.2)$$

(with $T = t_2 - t_1$). In contrast to classical mechanics, for quantum mechanics the satisfying of remote two-time boundary conditions may become *more difficult* the more remote those times, not easier.

The proposal I have made for the source of "special" states (in effect for the existence of macroscopic features in coordinate space) relates to a reasonable guess about the way hadrons were formed from the quark sea. The nucleation process for hadrons involves interactions that are local in coordinate space, and as such the wave function of the newly born hadron is highly localized (in the appropriate relative coordinate). In a loosely time symmetric universe there will be a similar localization on the way to the big crunch. From Eq. (21.2) it is clear that localizing a wave packet at two separate times may be impossible, for example if the particle is free throughout a long interval. One could reduce spreading by keeping particles in clumps, but even with this there would still be substantial spreading each time a "measurement" was made. For example, our unfortunate cat, split into two exemplars in respective Everett worlds, would have considerable spreading for its many constituent particles. My view of the need for "special" states is that these are the states that avoid the measurement-induced spread. They are the initial conditions that are consistent with Nature's wave packet-localized two-time boundary conditions.

- *Recovery of standard probabilities.* Although "special" states are rare, I expect this to be a *relative* statement; in absolute terms there should be many. Note that the "special" states for each possible outcome of an experiment form a Hilbert subspace. Now when one performs a measurement one knows a priori that the initial state had to be "special," but not which "special" state actually occurred. I postulate that one deals with this in the same way that one deals with ignorance in classical mechanics. The probability of an experiment having a given outcome is the phase space measure (dimension of the Hilbert subspace) of the set of "special" states for that outcome. To show that I recover the standard probabilities of quantum mechanics I must then show that the appropriate dimension of apparatus "special" states is correct. This is the most difficult technical feature of my theory. It demands a universality in apparatus. Recently I made progress on this and found that a condition for satisfying this requirement is that apparatus possess a set of Cauchy distributed modes. For details see (Schulman 1991b). A point of interest is that this works only if probability goes like $|\psi|^2$, not any other power of $\psi$. There is thus a theoretical justification for the wave function *squared* rule.

- *Experimental tests.* There are two aspects. First, *in principle* is an experimental test possible? And secondly, is there a feasible test? Satisfying the first requirement establishes that there is physical content to the theory.

A straightforward test consists of modifying an apparatus to prevent the occurrence of some of its "special" states without affecting the Copenhagen predictions. I have designed a detection situation in which the detector is small compared to the wave packet of the particle to be detected. The difficult part is to be sure the wave packet is well spread. In dealing with "special" states one encounters extreme coherence (the classical analog would be coordinating water molecules to make an ice cube grow) and in proposing an experiment one must be sure that no coherent interactions of the environment could possibly keep the wave packet from spreading. In the experiment I have proposed one does this by using a small source and high vacuum. I believe this experiment is at the edge of feasibility.

A second class of tests relies on the Cauchy distributed modes that I found to be universally necessary in apparatus. First, one can ask whether such modes could be recognized in phenomena other than quantum measurements. Secondly, I recover standard probability predictions only in the limit that the Cauchy parameters are negligible. If not, there would be differences and these might be measurable.

A third class of experiment involves seeing whether the interconnectedness of the wave function of the universe could imply causal anomalies. According to me, the "special" states of an apparatus must be coordinated with the measurement that actually will occur, even if that measurement is a delayed choice measurement. Is there any way that an experimental system can reveal the fact that it possesses such information (but not the actual microscopic information)? I do not have a specific proposal along these lines, but it seems a promising direction because of the developed technology of EPR/delayed choice experiments.

## Acknowledgements

I am grateful to the University of Paris VI (Université Pierre et Marie Curie) and to the University of Antwerp for the kind hospitality shown me. This work was supported by NSF grants PHY 88-11106 and PHY 90-15858.

## Discussion

**Lloyd**   Why can't you explain the lack of success of experiments to predict proton decay from the idea of final boundary conditions?

**Schulman**   Well, maybe I can. It's just that to preserve credibility one limits the number of outrageous claims per seminar. The idea is that if proton decay has a lifetime long compared to the big bang-big crunch time, then future constraints might act on it as they act on the state $z$ in my figure, and, in particular, give it an effective decay rate below the dynamical (calculated) rate. This recalls Wheeler's idea about long lived nuclides. As I indicated, using a two-time boundary condition model, I found that that this would work only for the cosmological distribution of the nuclide. One could view our information on proton decay as a cosmological absence, and, as such, place bounds on the lifetime of the universe.

**Wootters**   It seems that you need a Cauchy distribution of kicks in order to get the right probabilities. Is there any other theoretical reason for expecting a Cauchy distribution?

**Schulman**   Mathematically, I need a distribution without second moment to avoid the central limit theorem. Among such distributions, Cauchy stands out in technical ways. The more interesting version of your question is whether there are physical processes that select Cauchy. With Bernard Gaveau I have been exploring the idea that there is Cauchy distributed collection of photons available because of the Lorentz line shape (which has the same functional form). When an atom decays from a metastable state such photons are emitted. Another colleague, Roger Bidaux, has been looking at ways to generate Cauchy as fluctuations in statistical mechanics models. So far none of us has anything firm.

**DeWitt**   A year ago the examples you gave led only approximately to the boundary conditions you desired. Have you now found examples that give exactly what you want?

**Schulman**   "Exactly" is a strong word, but practically the answer to your question is "yes." The model I described last year, while pedagogically useful, did not become arbitrarily accurate as system size $\rightarrow \infty$. For my present models there is no such limitation. Secondly, in the year since you asked me a similar question at Sante Fe, I have thought about whether even extremely small errors would mess things up. I believe they don't. My reasoning is given in Appendix B of (Schulman 1991b).

**Gell-Mann**  Jim Hartle mentioned on our behalf a time-symmetric universe with $\rho_i$ and $\rho_f$ taken as pure states. Then, for a given coarse-grained set of histories, only one (or possibly two) coarse-grained histories can have non-zero probability, given the requirement of decoherence. Now Jim did not wax very enthusiastic about this idea, but it is certainly worth investigating whether our experience (say the success of statistical predictions of quantum mechanics in reproducible situations) really contradicts this picture. Aren't your proposal and this picture very similar?

**Schulman**  Yes. I believe that we would also agree that $\rho_f$ is $\exp(-iHt/\hbar)\rho_i\exp(iHt/\hbar)$. What is difficult about this proposal is showing that the pure quantum evolution and the "macroscopicness" of $\rho_f$ (that it have support on a single macrostate) are compatible. This is where I use "special" states. One point I would like to understand, that could perhaps connect our ideas, is whether your way of recovering probabilities (under the pure $\rho_i$ and $\rho_f$ hypotheses) and mine are the same.

## References

Cocke, W.J. (1967) *Phys. Rev.* **160**, 1165 (1967).

Davies, P.C.W. (1977) *The Physics of Time Asymmetry*, University of California Press, Berkeley, p. 194.

Gold, T. (1962) *Am. J. Phys.*, **30**, 403.

Penrose, R. (1979). In *General Relativity*, Eds. S. Hawking and W. Israel, Cambridge University Press, Cambridge, p. 597.

Schulman, L.S. (1973) *Phys. Rev.* **D7**, 2868.

Schulman, L.S. (1974) *J. Math. Phys.* **15**, 295.

Schulman, L.S. (1976) *Phys. Lett.* **57A**, 305.

Schulman, L.S. (1977) *J. Stat. Phys.* **16**, 217.

Schulman, L.S. (1991a) *Physica* **A177**, 373.

Schulman, L.S. (1991b) *Ann. Phys.* **212**, 315.

Schulman, L.S. (1992) *J. Phys.* **A25**, 3007.

Schulman, L.S. and Shtokhamer, R. (1977) *Int. J. Theor. Phys.* **16**, 287.

Schulman, L.S. and Seiden, P.E. (1978) *J. Stat. Phys.* **19**, 293.

Weinberg, S. (1972) *Gravitation and Cosmology*, Wiley, New York, p. 597.

Wheeler, J.A. (1975) In *General Relativity and Gravitation* (GR7) Eds. G. Shaviv and J. Rosen, Wiley, New York, and Israel University Press, Jerusalem.

# Part Five

Quantum Cosmology and Initial Conditions

# 22

# Time Symmetry and Asymmetry in Quantum Mechanics and Quantum Cosmology

Murray Gell-Mann

*Lauritsen Laboratory*
*California Institute of Technology*
*Pasadena, CA 91125, USA*

James B. Hartle

*Department of Physics*
*University of California*
*Santa Barbara, CA 93106-9530, USA*

## 22.1 Introduction

The disparity between the time symmetry of the fundamental laws of physics and the time asymmetries of the observed universe has been a subject of fascination for physicists since the late 19th century.† The following general time asymmetries are observed in this universe:

- The thermodynamic arrow of time – the fact that approximately isolated systems are now almost all evolving towards equilibrium in the same direction of time.
- The psychological arrow of time – we remember the past, we predict the future.
- The arrow of time of retarded electromagnetic radiation.
- The arrow of time supplied by the $CP$ non-invariance of the weak interactions and the $CPT$ invariance of field theory.
- The arrow of time of the approximately uniform expansion of the universe.
- The arrow of time supplied by the growth of inhomogeneity in the expanding universe.

All of these time asymmetries could arise from time-symmetric dynamical laws solved with time-asymmetric boundary conditions. The thermodynamic arrow of time, for example, is implied by an initial condition in which the progenitors of today's approximately isolated systems were all far from equilibrium at an initial time. The $CP$ arrow of time could arise as a spontaneously broken symmetry of the Hamiltonian. The approximate uniform expansion of the universe and the growth of inhomogeneity follow from an initial "big bang" of sufficient spatial

---

† For clear reviews, see [1], [2], [3].

311

homogeneity and isotropy, given the attractive nature of gravity. Characteristically such arrows of time can be reversed temporarily, locally, in isolated subsystems, although typically at an expense so great that the experiment can be carried out only in our imaginations. If we could, in the classical example of Loschmidt [4], reverse the momenta of all particles and fields of an isolated subsystem, it would "run backwards" with thermodynamic and electromagnetic arrows of time reversed.

Quantum cosmology is that part of physics concerned with the theory of the boundary conditions of our universe. It is, therefore, the natural and most general context in which to investigate the origin of observed time asymmetries. In the context of contemporary quantum cosmology, several such investigations have been carried out [2, 5–11], starting with those of Penrose [2] on classical time-asymmetric initial and final conditions and those of Page [5] and Hawking [6] on the emergence of the thermodynamic arrow of time from the "no-boundary" theory of the initial condition of the universe. It is not our purpose to review these results or the status of our understanding of the time asymmetries mentioned above. Rather, we shall discuss in this essay, from the perspective of quantum cosmology, a time asymmetry not specifically mentioned above. That is the arrow of time of familiar quantum mechanics.

Conventional formulations of quantum mechanics incorporate a fundamental distinction between the future and the past, as we shall review in Section 2. This quantum-mechanical arrow of time has, in a way, a distinct status in the theory from the time asymmetries discussed above. It does not arise, as they do, from a time-asymmetric choice of boundary conditions for time-neutral dynamical laws. Rather, it can be regarded as a time asymmetry of the laws themselves. However, the quantum mechanics of cosmology does not have to be formulated in this time-asymmetric way. In Section 3, extending discussions of Aharonov, Bergman, and Lebowitz [12] and of Griffiths [13], we consider a generalized quantum mechanics for cosmology that utilizes both initial and final conditions to give a time-neutral, two-boundary formulation that does not necessarily have an arrow of time [14]. In such a formulation all time asymmetries arise from properties of the initial and final conditions, in particular differences between them, or, at particular epochs, from nearness to the beginning or end. A theory of both initial and final conditions would be the objective of quantum cosmology.

In the context of a time-neutral formulation, the usual quantum mechanics results from utilizing a special initial condition, together with what amounts to a final condition representing complete indifference with respect to the future states, thus yielding the quantum-mechanical arrow of time, which is sufficient to explain the observed time asymmetries of this universe. However, a time-neutral formulation of quantum mechanics allows us to investigate to what extent the familiar final condition of indifference with respect to future states is mandated by our observations. In particular, it allows us to investigate whether quantum cosmologies with less blatantly asymmetric initial and final conditions might also be consistent with the observed general time asymmetries. As a step in this direction we discuss a quantum cosmology that would be, in a sense, the opposite extreme – a cosmology with a time-symmetric pair of initial and final conditions leading to a universe that is statistically symmetric about a moment of time. Such boundary conditions imply deviations from the thermodynamic arrow of time and the arrow of time supplied by the $CP$ non-invariance of the weak interactions. We investigate such deviations to see if they are inconsistent with observations. The classical statistical models reviewed in Section 4 and the models of $CP$ symmetry breaking discussed in Section 5 suggest that the predicted deviations may be insufficient to exclude time-symmetric boundary conditions if the interval between initial and final conditions is much longer than our distance in time from the initial condition. Next, we review and augment the arguments of Davies and Twamley that electromagnetic radiation may supply a probe of the final condition that *is* sufficiently accurate to rule out time-symmetric boundary conditions.

We should emphasize that we are not advocating a time-symmetric cosmology but only using it as a foil to test the extent to which observation now requires the usual asymmetric boundary conditions and to search for more refined experimental tests. The important result of this paper is a quantum framework for examining cosmologies with less asymmetric boundary conditions than the usual ones, so that the quantum-mechanical arrow of time (with its consequent time asymmetries) can be treated, or derived, as one possibility out of many, to be confronted with observation, rather than as an axiom of theory.

Relations between the initial and final conditions of a quantum-mechanical universe *sufficient* for both $CPT$-symmetric cosmologies and time-symmetric cosmologies are discussed in Section 5. Ways in which the $T$-violation exhibited by the weak interaction could arise in such universes are described there as well. In Section 6 we discuss the limitations on time-symmetric quantum boundary conditions following from the requirements of decoherence and classicality. Specifically, we show that for a set of alternative histories to have the negligible interference between its individual members that is necessary for them to be assigned probabilities at all, there must be some impurity in the initial or final density matrices or both, except in the highly unorthodox case in which there are only one or two coarse-grained histories with non-negligible probability.

We should make clear that our discussion of time-symmetric cosmologies, based on speculative generalizations of quantum mechanics and causality, with separate initial and final density matrices that are related by time symmetry, is essentially different from the conjecture that has sometimes been made that *ordinary* causal quantum or classical mechanics, with just a single boundary condition or a single prescribed wave function, $CPT$-invariant about some time in the distant future, might lead to a $T$-symmetric or $CPT$-symmetric cosmology with a contracting phase in which the arrows of time are reversed. [15–17, 6, 18] It is the latter notion, by the way, that Hawking refers to as his "greatest mistake"[19]. We shall return to this topic in Section 5.

## 22.2  The Arrow of Time in Quantum Mechanics

As usually formulated, the laws of quantum mechanics are not time-neutral but incorporate an arrow of time. This can be seen clearly from the expression for the probabilities of histories consisting of alternatives at definite moments of time $t_1 < t_2 < \cdots < t_n$. Let $\{\alpha_k\}$ be an exhaustive set of alternatives at time $t_k$ represented by $\{P^k_{\alpha_k}(t_k)\}$, a set of projection operators in the Heisenberg picture. For example, the alternatives $\{\alpha_k\}$ might be defined by an exhaustive set of ranges for the center-of-mass position of a massive body. A particular history corresponds to a specific sequence of alternatives $(\alpha_1, \cdots, \alpha_n)$. The probability for a particular history in the exhaustive set of histories is

$$p(\alpha_n, \cdots, \alpha_1) = Tr\left[ P^n_{\alpha_n}(t_n) \cdots P^1_{\alpha_1}(t_1) \rho P^1_{\alpha_1}(t_1) \cdots P^n_{\alpha_n}(t_n) \right], \qquad (22.1)$$

where $\rho$ is the density matrix describing the initial state of the system and the projection operators are time-ordered from the density matrix to the trace.†

The expression for the probabilities (22.1) is not time-neutral. This is not because of the time ordering of the projection operators. Field theory is invariant under $CPT$ and the ordering of the operators could be reversed by a $CPT$ transformation of the projection operators and density matrix, leaving the probabilities unchanged. (See e.g. [20] or [14]). Either time ordering may therefore be used; it is by convention that we usually use the one with the condition represented by the density matrix $\rho$ in the past.

Rather, (22.1) is not time-neutral because there is a density matrix on one end of the chain of projections representing a history while at the other end there is the trace [12, 13, 14]. Whatever conventions are used for time ordering, there is thus an asymmetry between future and past exhibited in the formula for probabilities (22.1). That asymmetry is the arrow of time in quantum mechanics.

The asymmetry between past and future exhibited by quantum mechanics implies the familiar notion of causality. From an effective density matrix describing the present *alone* it is possible to predict the probabilities for the future. More precisely, given that alternatives $\alpha_1, \cdots, \alpha_k$ have "happened" at times $t_1 < \cdots < t_k$ before time $t$, the conditional probability for alternatives $\alpha_{k+1}, \cdots, \alpha_n$ to occur in the future at times $t_{k+1}, \cdots, t_n$ may be determined from an effective density matrix $\rho_{\mathrm{eff}}(t)$ at time $t$. Specifically, the conditional probabilities for future prediction are

$$p(\alpha_n, \cdots, \alpha_{k+1} | \alpha_k, \cdots, \alpha_1) = \frac{p(\alpha_n, \cdots, \alpha_1)}{p(\alpha_k, \cdots, \alpha_1)} . \qquad (22.2)$$

† This compact expression of the probabilities of ordinary quantum mechanics has been noted by many authors. For more details of this and other aspects of the quantum-mechanical formalism we shall employ the reader is referred to [20] and [14] where references to earlier literature may be found.

These can be expressed as

$$p(\alpha_n, \cdots, \alpha_{k+1} | \alpha_k, \cdots, \alpha_1) = Tr\left[P_{\alpha_n}^n(t_n) \cdots P_{\alpha_{k+1}}^{k+1}(t_{k+1}) \rho_{\text{eff}}(t_k) P_{\alpha_{k+1}}^{k+1}(t_{k+1}) \cdots P_{\alpha_n}^n(t_n)\right],$$
(22.3)

where the effective density matrix $\rho_{\text{eff}}$ is

$$\rho_{\text{eff}}(t_k) = \frac{P_{\alpha_k}^k(t_k) \cdots P_{\alpha_1}^1(t_1) \rho P_{\alpha_1}^1(t_1) \cdots P_{\alpha_k}^k(t_k)}{Tr\left[P_{\alpha_k}^k(t_k) \cdots P_{\alpha_1}^1(t_1) \rho P_{\alpha_1}^1(t_1) \cdots P_{\alpha_k}^k(t_k)\right]}.$$
(22.4)

The density matrix $\rho_{\text{eff}}(t_k)$ can be said to define the effective state of the universe at time $t_k$, given the history $(\alpha_1, \cdots, \alpha_k)$.

What is the physical origin of the time asymmetry in the basic laws of quantum mechanics and what is its connection with the other observed time asymmetries of our universe? The rest of this Section addresses that question.

The reader may be most familiar with the expression for probabilities (22.1) in the context of the approximate "Copenhagen" quantum mechanics of measured subsystems. In that case operators, the density matrix, etc. all refer to the Hilbert space of the subsystem. The sets of projection operators $\{P_{\alpha_k}^k(t_k)\}$ describe alternative outcomes of measurements of the subsystem.

Formula (22.1) for the probabilities of a sequence of measured outcomes is then a unified expression of the "two forms of evolution" usually discussed in the quantum mechanics of subsystems — unitary evolution in between measurements and the "reduction of the state vector" on measurement. The time asymmetry of (22.1) does not arise from the unitary evolution of the projection operators representing the measured quantities in the Heisenberg picture; that is time-reversible. Rather, it can be said to arise from the successive reductions represented by the projections in (22.4) that occur on measurement. The common explanation for the origin of the arrow of time in the quantum mechanics of measured subsystems is that measurement is an irreversible process and that quantum mechanics inherits its arrow of time from the arrow of time of thermodynamics.† If that is the case, then the origin of the quantum-mechanical arrow of time must ultimately be cosmological, for the straightforward explanation of the thermodynamic arrow of time is a special initial condition for the universe implying that its constituents were far from equilibrium across a spacelike surface. Let us, therefore, investigate more fundamentally the quantum-mechanical arrow of time, not in an approximate quantum mechanics of

---

† This connection between the thermodynamic arrow of time and the quantum-mechanical arrow of time can be ambiguous. Suppose, for example, a measuring apparatus is constructed in which the local approach to equilibrium is in the opposite direction of time from that generally prevailing in the larger universe. If that apparatus interacts with a subsystem (perhaps previously measured by other apparatus adhering to the general thermodynamic arrow of time) should the operators representing those measurements be ordered according to the thermodynamic arrow of the apparatus or of the larger universe with respect to which it is running backwards? Such puzzles are resolvable in the more general quantum mechanics of closed systems to be discussed below, where "measurements", the "thermodynamic arrow of time", and any connection between the two are all approximate notions holding in only special situations.

measured subsystems, but in the quantum mechanics of a closed system — most realistically and generally the universe as a whole.

The formula (22.1) for the probabilities of histories also holds in the quantum mechanics of a closed system such as the universe as a whole, at least in an approximation in which gross fluctuations in the geometry of spacetime are neglected. The sets of projection operators describe alternatives for the whole system, say the universe, and the density matrix can be thought of as describing its initial condition.† Not every set of histories that may be described can be assigned probabilities according to (22.1). In the quantum mechanics of closed systems consistent probabilities given by (22.1) are predicted only for those sets of histories for which there is negligible interference between the individual members of the set [14] as a consequence of the particular initial $\rho$. Such sets of histories are said to "decohere". We shall defer until Section 5 a discussion of the precise measure of the coherence between histories and the implications of decoherence for time symmetry in quantum mechanics. We concentrate now on the theoretical status of the arrow of time exhibited by (22.1) in the quantum mechanics of cosmology.

An arrow of time built into a basic quantum mechanics of cosmology may not (as in the approximate "Copenhagen" quantum mechanics of measured subsystems) be attributed to the thermodynamic arrow of an external measuring apparatus or larger universe. In general, these external objects are not there. An arrow of time in the quantum mechanics of cosmology would be a fundamental time asymmetry in the basic laws of physics. Indeed, given that, as we mentioned in the Introduction, the other observed time asymmetries could all arise from time-symmetric dynamical laws solved with time-asymmetric boundary conditions, a fundamental arrow of time in the laws of quantum mechanics could be the only fundamental source of time asymmetry in all of physics.

There is no inconsistency between known data and a fundamental arrow of time in quantum mechanics. General time asymmetries *are* exhibited by our universe and there is no evidence suggesting any violation of causality. The observed time asymmetries such as the thermodynamic arrow of time, the arrow of retarded electromagnetic radiation, the absence of white holes, etc. *could* all be seen to follow from a fundamental quantum-mechanical distinction between the past and future. That is, they could all be seen to arise from a special initial $\rho$ in a quantum-mechanical framework based on (22.1).

But might it not be instructive to generalize quantum mechanics so that it does not so blatantly distinguish past from future? One could then investigate a more general class of quantum cosmologies and identify those that are compatible with the observed time asymmetries. Even if it is highly unlikely that ordinary quantum mechanics needs to be replaced by such a generalization, the generalization can still provide an instructive way of viewing the origin of time asymmetry in the

---

† For a more detailed exposition of this quantum mechanics of cosmology, the reader is referred to our previous work [20], [21], and [14], where references to the earlier literature may also be found.

universe and provide a framework for discussing tests of the usual assumptions. We shall discuss in the next section a quantum mechanics that employs two boundary conditions, one for the past and one for the future, to give a time-neutral formulation. Each condition is represented by a density matrix and the usual theory is recovered when the future density matrix is proportional to the unit matrix while the one for the past is much more special.

## 22.3  A Time-Neutral Formulation of Quantum Mechanics for Cosmology

Nearly thirty years ago, Aharonov, Bergmann, and Lebowitz [12] showed how to cast the quantum mechanics of measured subsystems into time-neutral form by considering final conditions as well as initial ones. The same type of framework for the quantum mechanics of closed systems has been discussed by Griffiths [13] and ourselves [14]. In this formulation the probabilities for the individual members of a set of alternative histories is given by

$$p(\alpha_n, \cdots, \alpha_1) = NTr\left[\rho_f P_{\alpha_n}^n(t_n) \cdots P_{\alpha_1}^1(t_1)\, \rho_i\, P_{\alpha_1}^1(t_1) \cdots P_{\alpha_n}^n(t_n)\right], \qquad (22.5)$$

where

$$N^{-1} = Tr\left(\rho_f \rho_i\right) . \qquad (22.6)$$

Here, $\rho_i$ and $\rho_f$ are Hermitian, positive operators that we may conventionally call Heisenberg operators representing initial and final conditions for the universe respectively. They need not be normalized as density matrices with $Tr(\rho) = 1$ because (22.5) is invariant under changes of normalization.

The expression (22.5) for the probabilities of histories is time-neutral. There is a density matrix at both ends of each history. Initial and final conditions may be interchanged by making use of the cyclic property of the trace. Therefore, the quantum mechanics of closed systems based on (22.5) need not have a fundamental arrow of time.

Different quantum-mechanical theories of cosmology are specified by different choices for the initial and final conditions $\rho_i$ and $\rho_f$. For those cases with $\rho_f \propto I$, where $I$ is the unit matrix, this formulation reduces to that discussed in the previous Section because then (22.5) coincides with (22.1).

Of course, the condition for decoherence must also be extended to incorporate initial and final conditions. That extension, however, is straightforward [13, 14] and will be reviewed briefly in Section 5. The result is a generalized quantum mechanics in the sense of Refs. [14] and [21].

Lost in this generalization is a built-in notion of causality in quantum mechanics. Lost also, when $\rho_f$ is not proportional to $I$, is any notion of a unitarily evolving "state of the system at a moment of time". There is generally no effective density matrix like $\rho_{\text{eff}}(t)$ in (22.4) from which *alone* probabilities for either the future or past could be computed. What is gained is a quantum mechanics without a fundamental arrow of time in which all time asymmetries may arise in particular cosmologies because of differences between $\rho_i$ and $\rho_f$ or at particular epochs from their being near to the beginning or the end. That generalized quantum mechanics embraces a richer variety of possible universes, allowing for the possibility of violations of causality and advanced as well as retarded effects. These, therefore, become testable features of the universe rather than axioms of the fundamental quantum framework.

From the perspective of this generalized quantum mechanics the task of quantum cosmology is to find a theory of *both* the initial and final conditions that is theoretically compelling and fits our existing data as well as possible. Certainly a final condition of indifference, $\rho_f \propto I$, and a special initial condition, $\rho_i$, seem to fit our data well, and there is no known reason for modifying them. But how accurately is $\rho_f \propto I$ mandated by the data? What would be the observable consequences of a completely time-symmetric boundary condition that is, in a sense, the opposite extreme?

Our ability to detect the presence of a final condition differing from $\rho_f \propto I$ depends on our experimental access to systems whose behavior today predicted with $\rho_f \not\propto I$ would be measurably different from the predictions of that behavior with $\rho_f \propto I$. Loosely speaking, it depends on our finding physical systems which can "see" the final condition of the universe today. In the following we examine several candidates for such systems, beginning with simple classical analyses in Section 4 and proceeding to more quantum-mechanical ones in Section 5.

## 22.4 Classical Two-Time Boundary Problems

### 22.4.1 A Simple Statistical Model

The simplest explanation of the observed thermodynamic arrow of time is the asymmetry between a special, low-entropy,† initial condition and a maximal-entropy final condition describable as indifference with respect to final state (or no condition at all!). Studying deviations of the entropy increase predicted by statistical mechanics with these boundary conditions from that predicted with time-symmetric boundary conditions is a natural way to try to discriminate between the two. Such studies were carried out in classical statistical models by by Cocke [22], Schulman [23], Wheeler [24], and others in the late '60s and early '70s. Schulman, in particular, has written extensively on these problems both in classical and quantum mechanics [25]. We briefly review such statistical models here.

Relaxation to equilibrium is a time-symmetric process in a universe with an underlying dynamics that is time reversal invariant. Without boundary conditions, a system out of equilibrium is just as likely to have evolved from a state of higher entropy as it is to evolve to a state of higher entropy. The characteristic relaxation

---

† For quantitative estimates of how low the initial entropy is, see [2].

time for a system to evolve to equilibrium depends on the size of the system and the strength of the interactions that equilibrate it. Other factors being equal, the larger the system the longer the relaxation time.

There is no simpler instructive model to illustrate the approach to equilibrium than the Ehrenfest urn model [26]. For this reason, it and related models have been much studied in connection with statistical two-time boundary problems [22], [23]. The model consists of two boxes and a numbered set of $n$ balls that are distributed between them. The system evolves in time according to the following dynamical rule: At each time step a random number between 1 and $n$ is produced, and the ball with that number is moved from the box containing it to the other box. This dynamical rule is time-symmetric.

The fine-grained description of this system specifies which ball is in which box (a "microstate"). An interesting coarse-grained description involves following just the total number of balls in each box (a "macrostate") irrespective of *which* balls are in which box. Let us use this coarse graining to consider an initial condition in which all the balls are in one box and follow the approach to equilibrium as a function of the number of time steps, with no further conditions. Figure 22.1 shows a numerical calculation of how the entropy averaged over many realizations of this evolution grows with time to approach its maximum, equilibrium value. The relaxation time, obtained either analytically or numerically, is approximately the total number of balls, $t_{relax} \sim n$. If there are no further constraints, the system tends to relax to equilibrium and remain there.

Consider evolution in the Ehrenfest urn model when a final condition identical to the initial one is imposed at a later time $T$. Specifically, construct an ensemble of evolutions consistent with these boundary conditions by evolving forward from an initial condition where all the balls are in one box but accepting only those evolutions where all the balls are back in this box at time $T$. Figure 22.1 shows the results of two such calculations, one for a system with a small number of balls (where the relaxation time is significantly smaller than $T$) and the other for a system with a larger number of balls (where it is significantly larger than $T$.)

For both systems the time-symmetric boundary conditions imply a behavior of the average entropy that is time-symmetric about the midpoint, $T/2$. For the system with a relaxation time short compared to the time at which the final condition is imposed, the initial approach to equilibrium is nearly indistinguishable from that in the case with no final condition. That is because, in equilibrium, the system's coarse-grained dynamics is essentially independent of its initial *or* final condition. It, in effect, "forgets" both from whence it started and whither it is going.

By contrast, if the relaxation time is comparable to or greater than the time interval between initial and final condition, then there will be significant deviations from the unconstrained approach to equilibrium. Such systems typically do not reach equilibrium before the effect of the final condition forces their entropy to decrease.

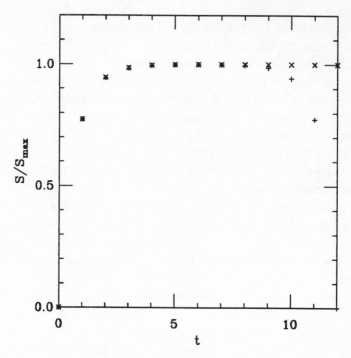

Fig. 22.1. The entropy of a coarse-grained state, in which only the total number of balls in each box is followed, is the logarithm of the number of different ways of distributing the balls between the boxes consistent with a given total number in each. This figure shows this entropy averaged over a large number of different evolutions of the system for several situations. These simulations were carried out by the authors but are no different in spirit from those discussed by Cocke [22].

The left figure shows the evolution of a system of four balls. In each case the system starts with all balls in one box – a configuration of zero entropy as far from equilibrium as it is possible to get. The $\times$'s show how the average entropy of 12,556 cases approaches equilibrium when there are no further constraints. The entropy approaches its equilibrium value in a relaxation time given approximately by its size, $t_{\text{relax}} \sim 4$, and remains there. The curve of $+$'s shows the evolution when a time-symmetric final condition is imposed at $T = 12$ that all balls have returned to the one box from whence they started at $t = 0$. A total of 100,000 evolutions were tried. (Figure 22.1 continued on the next page).

The evolution of the entropy in the presence of time-symmetric initial and final conditions must itself be time-symmetric when averaged over many evolutions, as the simulations in Figure 22.1 show. However, in a statistical theory with time-symmetric boundary conditions the individual histories need not be time-symmetric. Figure 22.2 shows an example of a single history from an urn model calculation for which the average behavior of the entropy is shown in Figure 22.1. The ensemble of histories is time-symmetric by construction; the individual histories need not be. Since, by definition, we experience only one history of the universe, this leaves open the possibility that the time-asymmetries that we see could be the result of

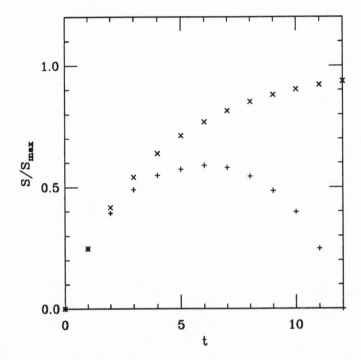

Fig. 22.1. *Continued* – The average entropy of the 12,556 cases that met the final condition is shown. It is time symmetric about the midpoint, $T/2 = 6$. The initial approach to equilibrium is virtually indistinguishable from the approach without a final condition because $t_{\text{relax}} \sim 4$ is significantly less than the time $T = 12$ at which the final condition is imposed. Only within one relaxation time of the final condition is the deviation of the evolution from the unconstrained case apparent.

The right figure shows the same two cases for a larger system of twenty balls. The unconstrained approach to equilibrium shown by the ×'s was calculated from the average of 1010 evolutions and exhibits a relaxation time, $t_{\text{relax}} \sim 20$. The average entropy when time-symmetric boundary conditions are imposed at $T = 12$ is shown in the curve of +'s. 10,000,000 evolutions were tried of which 1010 met the time-symmetric final condition. (This demonstrates vividly that it is very improbable for the entropy of even a modest size system to fluctuate far from equilibrium as measured by the *CPU* time needed to find such fluctuations.) The deviation from the unconstrained approach to equilibrium caused by the imposition of a time-symmetric final condition is significant from an early time of about $t = 3$ as the differences between the +'s and the ×'s show. These models suggest that to detect the effects of a time-symmetric final condition for the universe we must have access to systems for which the relaxation time is at least comparable to the time difference between initial and final conditions.

a statistical fluctuation in a universe with time-symmetric boundary conditions. In quantum cosmology, we would not count such a theory of the initial and final conditions as successful if the fluctuations required were very improbable. However, in some examples the magnitude of the fluctuation need not be so very large. For instance, consider a classical statistical theory in which the boundary conditions

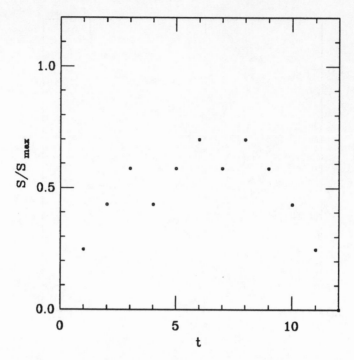

Fig. 22.2. An individual evolution in the urn model with time-symmetric initial and final conditions. The history of the entropy averaged over many evolutions must clearly be time-symmetric in a statistical theory with time-symmetric initial and final conditions as Figure 22.1 shows. However, the individual evolutions need not be separately time-symmetric. This figure shows the entropy for the case of twenty balls in the first evolution among the 10,000,000 described in Figure 22.1 that met the time-symmetric final condition. It is not time-symmetric. For systems such as this with relaxation time, $t_{\text{relax}}$, larger than the time between initial and final conditions, significant deviations from exact time symmetry may be expected.

allow an ensemble of classical histories each one of which displays an arrow of time and also, with equal probability, the time-reverse of that history displaying the opposite arrow of time. The boundary conditions and the resulting ensemble are time-symmetric, but predict an observed time-asymmetry with probability one. Of course, such a theory might be indistinguishable from one that posited boundary conditions allowing just one arrow of time. However, other theoretical considerations may make it reasonable to consider such proposals, for example, the "no boundary" initial condition which is believed to have this character [5]. In the subsequent discussion of time-symmetric cosmological boundary conditions we shall assume that they predict with high probability *some* observable differences from the usual special initial conditions and final condition of indifference and investigate what these are.

### 22.4.2 *Classical Dynamical Systems with Two-Time Statistical Boundary Conditions*

The analysis of such simple classical statistical models with two-time boundary conditions suggests the behavior of a classical cosmology with time-symmetric initial and final conditions. A classical dynamical system is described by a phase space and a Hamiltonian, $H$. We write $z = (q, p)$ for a set of canonical phase-space coördinates. The histories of the classical dynamical system are the phase-space curves $z(t)$ that satisfy Hamilton's equations of motion.

A statistical classical dynamical system is described by a distribution function on phase space $\rho^{cl}(z)$ that gives the probability of finding it in a phase-space volume. For analogy with quantum mechanics it is simplest to use a "Heisenberg picture" description in which the distribution function does not depend explicitly on time but time-dependent coördinates on phase space are used to describe the dynamics. That is, if $z_0$ is a set of canonical coördinates at time $t = 0$, a set $z_t$ appropriate to time $t$ may be defined by $z_t = z_0(t)$ where $z_0(t)$ is the classical evolution of $z_0$ generated in a time $t$ by the Hamiltonian $H$. The statistical system at time $t$ is then distributed according to the function $\rho^{cl}$ expressed in terms of the coördinates $z_t$, viz. $\rho^{cl}(z_t)$. The distributions $\rho^{cl}(z_t)$ and $\rho^{cl}(z_{t'})$ will therefore typically have *different* functional forms.

An ensemble of histories distributed according to the probabilities of a statistical classical dynamical system with boundary conditions at two times $t_i$ and $t_f$ might be constructed as follows: Evolve a large number of systems distributed according to the initial distribution function $\rho_i^{cl}(z_i)$ forward from time $t_i$ to time $t_f$. If a particular system arrives at time $t_f$ in the phase space volume $\Delta v$ centered about point $z_f$, select it for inclusion in the ensemble with probability $\rho_f^{cl}(z_f)\Delta v$ where $\rho_f^{cl}$ is the distribution function representing the final boundary condition. Thus, if $\rho_i^{cl}$ and $\rho_f^{cl}$ are referred to a common set of phase-space coördinates, say $z_t$, the time-symmetric ensemble of systems will be distributed according to the function

$$\bar{\rho}^{cl}(z_t) = N\rho_f^{cl}(z_t)\rho_i^{cl}(z_t) , \tag{22.7}$$

where

$$N^{-1} = \int dz_t \; \rho_f^{cl}(z_t)\rho_i^{cl}(z_t) . \tag{22.8}$$

Referred to the initial time, (22.7) has a simple interpretation: Since classical evolution is unique and deterministic the selection at the final time could equally well be carried out at the initial time with $\rho_f^{cl}$ evolved back to the initial time. The distribution $\bar{\rho}^{cl}$ is the result.

We now discuss the relation between $\rho_i^{cl}$ and $\rho_f^{cl}$ that is necessary for the probabilities of this classical cosmology to be symmetric about a moment of time. Take this time to be $t = 0$ and introduce the operation, $\mathcal{T}$, of time-reversal about it.

$$\mathcal{T}\rho^{cl}(q_0, p_0) \equiv \rho^{cl}(q_0, -p_0) . \tag{22.9}$$

If we assume that the Hamiltonian is itself time-reversal invariant

$$H(q_0, p_0, t) = H(q_0, -p_0, -t) , \qquad (22.10)$$

this implies

$$\mathscr{T} \rho^{cl}(q_t, p_t) = \rho^{cl}(q_{-t}, -p_{-t}) . \qquad (22.11)$$

The distribution function (22.7) may then be conveniently rewritten

$$\bar{\rho}^{cl}(q_t, p_t) = N \rho_i^{cl}(q_t, p_t) \mathscr{T} \rho_f^{cl}(q_{-t}, -p_{-t}) . \qquad (22.12)$$

A relation between $\rho_i^{cl}$ and $\rho_f^{cl}$ sufficient to imply the time-symmetry of the distribution $\bar{\rho}^{cl}$ is now evident, namely

$$\rho_f^{cl}(q_t, p_t) = \mathscr{T}^{-1} \rho_i^{cl}(q_t, p_t) . \qquad (22.13)$$

The final condition is just the time-reversed version of the initial one.

The imposition of time-symmetric statistical boundary conditions on a classical cosmology means in particular that the entropy must behave time-symmetrically if it is computed utilizing a coarse graining that is itself time-symmetric. The entropy of the final distribution must be the same as the initial one. The thermodynamic arrow of time will run backwards on one side of the moment of time symmetry as compared to the other side. This does not mean, of course, that the histories of the ensemble need be individually time-symmetric, as the example in Figure 22.2 shows. In particular, subsystems with relaxation times long compared to the interval between initial and final conditions might have non-negligible probabilities for fluctuations from exactly time-symmetric behavior. There would appear to be no principle, for example, forbidding us to live on into the recontracting phase of the universe and see it as recontracting. It is just that as time progressed events judged to be unexpected on the basis of just an initial condition would happen with increasing frequency. It is by the frequency of such unexpected events that we could detect the existence of a final condition.

Could we infer the existence of a time-symmetric final condition for the universe from the deviations that it would imply for the approach to equilibrium that would be expected were there no such final condition? The statistical models reviewed in above suggest that to do so we would need to study systems with relaxation times comparable to or longer than the lifetime of the universe between the "big bang" and the "big crunch". If the lifetime of the universe is comparable to the present age of the universe from the "big bang", then we certainly know such systems. Systems of stars such as galaxies and clusters provide ready examples. Any single star, with the ambient radiation, provides another as long as the star's temperature is above that of the cosmic background radiation. Black holes with lifetime to decay by the Hawking radiation longer than the Hubble time are further examples. Indeed, from the point of view of global cosmological geometry, the singularities contained within black holes can be considered to be parts of the final singularity of the universe, where a final condition would naturally be imposed [27]. The singularities of detectable black holes may be the parts of this final singularity closest to us. On smaller scales, samples of radioactive material with very long half-lives may be other such examples and Wheeler [24] has discussed experiments utilizing them to search for a time-symmetric final condition. We may hope, as mentioned above, that the evolving collective complex adaptive system of which we are a part could be such a long-lasting phenomenon!

However, if the lifetime of the universe is much *longer* than the present age from the "big bang", then it might be much more difficult to find systems that remain out of equilibrium long enough for their initial approach to equilibrium to be significantly affected by a time-symmetric final condition. That could be the case with the $\Omega$-near-one universe that would result from a rapid initial inflation. If its lifetime were long enough, we might never be able to detect the existence of a time-symmetric final condition for the universe.

The lifetime of our classical universe obeying the Einstein equation is, of course, in principle determinable from present observations (for example, of the Hubble constant, mean mass density, and deceleration parameter). Unfortunately we do not have enough data to distinguish observationally between a lifetime of about twenty-five billion years and an infinite lifetime. Very long lifetimes are not only consistent with observations, but also, as we now describe, are suggested theoretically by quantum cosmology as a consequence of inflation.

The quasiclassical cosmological evolution that we observe should be, on a more fundamental level, a likely prediction of a quantum theory of the universe and its boundary conditions. We shall discuss time symmetry in the context of quantum cosmology in later sections, but for the present discussion we note that, in quantum cosmology, the probabilities for different lifetimes of the universe are predictable from a theory of its initial and final conditions. That is because, in a quantum theory that includes gravitation, the geometry of spacetime, including such features as the time between the "big bang" and a "big crunch", if any, is quantum-probabilistic.

A quantum state that predicts quasiclassical behavior does not typically predict a unique classical history but rather an ensemble of possible classical histories with different probabilities. This is familiar from wave functions of WKB form, which do not predict single classical trajectories but only the classical connection between momentum and classical action. Similarly, in the quantum mechanics of closed cosmologies, we expect a theory of quantum boundary conditions to determine an ensemble of different classical cosmological geometries with different probabilities.†
The geometries in the ensemble will have *different times* between the "big bang" and the "big crunch" because in quantum gravity that time is a dynamical variable and not a matter of our choice. In this way, the probability distribution of lifetimes of the universe becomes predictable in quantum cosmology.

---

† For further discussion see e.g. [14].

Cosmological theories that predict inflation lead to very large expected values for the lifetime of the universe; and inflation seems to be implied by some currently interesting theories of the boundary conditions of the universe. The question has been analyzed only for theories with a special initial condition, such as the "no-boundary proposal" and the "tunneling-from-nothing proposal". Analyses by Hawking and Page [10], Grishchuk and Rozhansky [28], and Barvinsky and Kamenshchik [29] suggest expected lifetimes for the "no-boundary proposal" that are at least large enough to give sufficient inflation to explain the present size of the universe. Analyses by Vilenkin [30] and by Mijić, Morris, and Suen [31] also suggest very large expected lifetimes for the "tunneling-from-nothing" case.

### 22.4.3 Electromagnetic Radiation

The above discussion suggests that in order to probe the nature of a non-trivial final condition, one should study processes today that are sensitive to that final condition no matter how far in the future it is imposed. At the conference, P.C.W. Davies suggested that electromagnetic radiation might provide such a mechanism for "seeing" a final condition in the arbitrarily far future in realistic cosmologies. In an approximately static and homogeneous cosmology, radiation must travel through ever more material the longer the time separation between initial and final conditions. For sufficiently large separations, the universe becomes opaque to the electromagnetic radiation necessary to probe the details of the final condition directly. However, in an expanding universe the dilution of matter caused by the expansion competes with the longer path length as the separation between initial big bang and final big crunch becomes longer and longer. Davies and Twamley [32] show that, under reasonable conditions, the expansion wins and that the future light cone is transparent to photons all the way to a distance from the final singularity comparable to ours from the big bang.

Transparency of the forward light cone raises the possibility of constraining the final condition by present observations of electromagnetic radiation and perhaps ruling out time-symmetric boundary conditions. Partridge [33] has actually carried out experiments which could be interpreted in this way and Davies and Twamley discuss others. The following is an example of a further argument of a very direct kind.

Suppose the universe to have initial and final classical distributions that are time-symmetric in the sense of (22.13). Suppose further that these boundary conditions imply with high probability an initial epoch with stars in galaxies distributed approximately homogeneously and a similar final epoch of stars in galaxies at the symmetric time. Consider the radiation emitted from a particular star in the present epoch. If the universe is transparent, it is likely to reach the final epoch without being absorbed or scattered. There it may either be absorbed in the stars or proceed past them towards the final singularity. If a significant fraction of the radiation proceeds past, then by time-symmetry we should expect a corresponding amount of radiation to have been emitted from the big bang. Observations of the brightness of the night sky could therefore constrain the possibility of a final boundary condition time-symmetrically related to the initial one. The alternative that the radiation is completely absorbed in future stars implies constraints on present emission that are probably inconsistent with observation because the total cross section of future stars is only a small fraction of the whole sky, as it is today.†

By such arguments, made quantitative, and extended to neutrinos, gravitational and other forms of radiation, we may hope to constrain the final condition of the universe no matter how long the separation between the big bang and the big crunch.

### 22.5 Hypothetical Quantum Cosmologies with Time Symmetries

#### 22.5.1 *CPT- and T-Symmetric Boundary Conditions*

The time-neutral generalized quantum mechanics with initial and final conditions developed in Section 3 permits the construction of model quantum cosmologies that exhibit symmetries with respect to reflection about a moment of time. By this we mean that the probabilities given by (22.5) for a set of alternative histories are identical to those of the symmetrically related set. This section explores the relations between $\rho_f$ and $\rho_i$ and the conditions on the Hamiltonian under which such symmetries exist.

$CPT$-symmetric universes are the most straightforward to implement because local field theory in flat spacetime is invariant under $CPT$. We expect $CPT$ invariance as well for field theories in curved cosmological spacetimes such as closed Friedmann universes that are symmetric under a space inversion and symmetric about a moment of time.

To construct a $CPT$-invariant quantum cosmology, choose the origin of time so that the time reflection symmetry is about $t = 0$. Let $\Theta$ denote the antiunitary $CPT$ transformation and for simplicity consider alternatives $\{P_{\alpha_k}^k(t_k)\}$ such that their $CPT$ transforms, $\{\widetilde{P}_{\alpha_k}^k(-t_k)\}$, are given by

$$\widetilde{P}_{\alpha_k}^k(-t_k) = \Theta^{-1} P_{\alpha_k}^k(t_k)\Theta \ . \tag{22.14}$$

A $CPT$-symmetric universe would be one in which the probabilities of histories of alternatives at times $t_1 < t_2 < \cdots < t_n$ would be identical to the probabilities of the $CPT$-transformed histories of alternatives at times $-t_n < \cdots < -t_2 < -t_1$. Denote by $C_\alpha$ the string of projection operators representing one history

$$C_\alpha = P_{\alpha_n}^n(t_n) \cdots P_{\alpha_1}^1(t_1) \ , \tag{22.15}$$

and by $\widetilde{C}_\alpha$ the corresponding string of $CPT$-transformed alternatives written in

† Thanks are due to D. Craig for discussions of this example.

standard time order with the earliest alternatives to the right

$$\widetilde{C}_\alpha \equiv \widetilde{P}^1_{\alpha_1}(-t_1) \cdots \widetilde{P}^n_{\alpha_n}(-t_n) \,. \tag{22.16}$$

Thus,

$$\widetilde{C}_\alpha = \Theta^{-1} C^\dagger_\alpha \Theta \,. \tag{22.17}$$

The requirement of $CPT$ symmetry is then, from (22.5),

$$Tr\left(\rho_f C_\alpha \rho_i C^\dagger_\alpha\right) = Tr\left(\rho_f \widetilde{C}_\alpha \rho_i \widetilde{C}^\dagger_\alpha\right) \,. \tag{22.18}$$

Using (22.14), (22.16), the cyclic property of the trace, and the identity $Tr[A\Theta^{-1}B\Theta] = Tr[B^\dagger \Theta A^\dagger \Theta^{-1}]$, following from the antiunitarity of $\Theta$, the right hand side of (22.18) may be rewritten to yield the following form of the requirement of $CPT$ symmetry

$$Tr\left(\rho_f C_\alpha \rho_i C^\dagger_\alpha\right) = Tr\left(\Theta \rho_i \Theta^{-1} C_\alpha \Theta \rho_f \Theta^{-1} C^\dagger_\alpha\right) \,. \tag{22.19}$$

Evidently a sufficient condition for a $CPT$-symmetric universe is that the initial and final conditions be $CPT$ transforms of each other:

$$\rho_f = \Theta \rho_i \Theta^{-1} \tag{22.20}$$

because, acting on Bose operators, $\Theta^2$ is effectively unity, and as a consequence of (22.20), $\rho_i = \Theta \rho_f \Theta^{-1}$ .

As stressed by Page [34], a $CPT$-symmetric universe can also be realized with within the usual formulation of quantum mechanics with an initial $\rho_i$ and a final $\rho_f = I$, provided the $\rho_i$ representing the condition at the initial instant is $CPT$-*invariant* about some time in the future. Thus, initial and final conditions that are not related by (22.20) do not *necessarily* imply differing probabilities for sets of histories connected by $CPT$. Further, as discussed in the previous section, both ways of realizing a $CPT$-symmetric universe can, with appropriate kinds of initial and final conditions and coarse-graining, lead to sets of histories in which each individual member is $CPT$-asymmetric about the moment of symmetry. Thus, neither are $CPT$-symmetric boundary conditions *necessarily* inconsistent with arrows of time that extend consistently over the whole of the universe's evolution.

A universe is time-symmetric about a moment of time if the probabilities of any set of alternative histories are identical to those of the time-inverted set. The relation between initial and final conditions necessary for a purely time-symmetric universe is analogous to that for a $CPT$-symmetric one and derived in the same way. However, we cannot expect boundary conditions to impose time symmetry if the Hamiltonian itself distinguishes past from future. We must assume that the Hamiltonian is symmetric under time inversion, $\mathscr{T}$,

$$\mathscr{T}^{-1} H(t) \mathscr{T} = H(-t) \,. \tag{22.21}$$

Given (22.21), a time-symmetric universe will result if the initial and final conditions are related by time inversion:

$$\rho_f = \mathcal{T} \rho_i \mathcal{T}^{-1} \,. \tag{22.22}$$

For realistic quantum cosmologies, the time-neutral quantum mechanics of universes in a box, described in Section 3, must be generalized to allow for significant quantum fluctuations in spacetime geometry, and notions of space and time inversion must be similarly generalized. A sketch of a generalized quantum mechanics for spacetime can be found in [14,42] and discussions of time inversion in the quantum mechanics of cosmology in [34], [5], and [6].

### 22.5.2  *T  Violation in the Weak Interactions*

The effective Hamiltonian describing the weak interaction on accessible energy scales is not $CP$-invariant. As a consequence of the $CPT$ invariance of field theory it is also not $T$-invariant. $T$ violation of this kind is a small effect in laboratory experiments but is thought to be of central importance in the evolution of the matter content of the universe. It is believed to be responsible, together with the non-conservation of baryons, for the emergence of a matter-dominated universe from an initial equality of matter and antimatter, as originally pointed out by Sakharov.† Can the symmetric universes just discussed be consistent with this effective $T$ violation in the weak interaction?

The violation of time-inversion symmetry that we observe in the effective weak interaction Hamiltonian could arise in three ways: First, it could be the result of $T$ violation in the fundamental Hamiltonian. Second, it could arise throughout the universe, even if the fundamental Hamiltonian were time-inversion-symmetric, from asymmetries in the cosmological boundary conditions of the universe. Third, it could be an asymmetry of our particular epoch and spatial location arising dynamically in extended domains from a time-inversion symmetric Hamiltonian and boundary conditions. We shall now offer a few comments on each of these possibilities.

If the fundamental Hamiltonian is time-inversion asymmetric, then we cannot expect a time-symmetric universe, as we have already discussed. One could investigate whether such a fundamental time asymmetry is the source of the other observed time asymmetries. So far such an approach has neither been much studied nor shown much promise.

Even though a $T$-symmetric universe is inconsistent with a $T$-asymmetric fundamental Hamiltonian, a $CPT$-symmetric universe could be realized if the initial and final density matrices were related by (22.20). That is because a field-theoretic Hamiltonian is always $CPT$-symmetric even if it is not $T$-symmetric. But $CPT$ symmetry needs to be reconciled with the observed matter-antimatter asymmetry over large

† Ref [36]. For an accessible recent review of these ideas see [37].

domains‡ of the universe and the classical behavior of their matter content. If the universe is *homogeneously* matter-dominated now, then $CPT$ symmetry would imply that it will be homogeneously antimatter-dominated at the time-inverted epoch in the future. What evolution of the present universe could lead to such an inversion? One possibility is a universe that lasts much longer than the proton lifetime.§

There is no evidence for $CP$ violation in the basic dynamics of superstring theory. If it is the correct theory, the effective $CP$ violation in the weak interaction in four dimensions has to arise in the course of compactification or from some other form of spontaneous symmetry breaking. From the four-dimensional point of view, which we are taking for convenience in this article, this would correspond to having a non-zero expected value of a $CP$-odd quantity. Then, as discussed above, it is possible to investigate time-symmetric universes with initial and final conditions related by (22.22). An effective $CP$ violation could arise from $CP$ asymmetries of the initial or final states or both. Typical theories of these boundary conditions relate them to the Hamiltonian or an equivalent action. Each density matrix, $\rho_i$ or $\rho_f$, may either inherit the symmetries of the fundamental Hamiltonian or be an asymmetrical member of a symmetrical family of density matrices determined by it. This is the case, for example, with "spontaneous symmetry breaking" of familiar field theory where there are degenerate candidates for the ground state not individually symmetrical under the symmetries of the Hamiltonian. Before discussing the possibility of effective $CP$ violation in time-symmetric universes, let us review how an effective $CP$ violation can arise in familiar field theory and in usual quantum cosmology with just an initial condition.

Effective $CP$ violation can arise in field theory even when the fundamental Hamiltonian is $CP$-invariant, provided there is a non-vanishing vacuum expected value of a $CP$-odd field $\phi(\vec{x}, t)$ [39], i.e. one such that

$$\phi(-\vec{x}, t) = -(\mathscr{C}\mathscr{P})^{-1}\phi(\vec{x}, t)(\mathscr{C}\mathscr{P}) . \qquad (22.23)$$

Usually the vacuum state $|\Psi_0\rangle$ inherits the symmetry of the Hamiltonian that determines it and the vacuum expected value of a $CP$-odd field would vanish if the Hamiltonian is $CP$-invariant. However, if there is a symmetrical family of degenerate candidates for the ground state that are individually not $CP$-invariant, then the expected value

$$\langle\phi(\vec{x}, t)\rangle = Tr\left[\phi(\vec{x}, t)|\Psi_0\rangle\langle\Psi_0|\right] \qquad (22.24)$$

may be non-zero for the physical vacuum.

Similarly, in usual quantum cosmology with just an initial condition $\rho_i$, a non-zero value of

$$\langle\phi(\vec{x}, t)\rangle = Tr\left[\phi(\vec{x}, t)\rho_i\right] \qquad (22.25)$$

‡ For a classic review of the observational evidence that there is a matter-antimatter asymmetry over a
   domain at least the size of the local group of galaxies see [38].
§ We owe this suggestion to W. Unruh.

can lead to effective $CP$ violation. The "no-boundary" wave function of the universe [34] is the generalization of the flat space notion of ground state, i.e. vacuum, to the quantum mechanics of closed cosmological spacetimes. The "no-boundary" prescription with matter theories that would lead to spontaneous $CP$ violation in flat space thereby becomes an interesting topic for investigation. In such situations, we expect the "no-boundary" construction to yield a $CP$-symmetric set of possible wave functions for the universe that are individually $CP$-asymmetric.

We now turn to effective $CP$ violation in time-symmetric universes with initial and final states related by (22.22). An expected value for a field is defined when probabilities are assignable to its alternative values — that is, when there is decoherence among the alternatives. The requirements of decoherence will be discussed in the next Section. They are automatically satisfied for alternatives at a single moment of time when $\rho_f \propto I$ but they are non-trivial when $\rho_f$ is non-trivial. We have not analyzed the circumstances in which the values of the field decohere but we assume those circumstances to obtain here so that the expectation value of the field may be defined.

A consequence of decoherence and the probability formula (22.5) is the validity of two equivalent expressions for the expected value of the field that are analogous to (22.24) and (22.25):

$$\langle \phi(\vec{x}, t) \rangle = NTr \left[ \rho_f \phi(\vec{x}, t) \rho_i \right] = NTr \left[ \rho_i \phi(\vec{x}, t) \rho_f \right] . \tag{22.26}$$

These are demonstrated in the Appendix. The symmetry between the initial and final conditions in (22.26) can be understood from the fact that it is not probabilities at one moment of time that distinguish the future from the past. We shall now show that for a $CP$-odd field this expected value is odd under time inversion for a time-symmetric universe.

We carry over from flat space field theory the assumption that we are dealing with a $CPT$-even field $\phi(\vec{x}, t)$. In flat space that is necessary if the field is to have a non-vanishing vacuum expected value. The $CPT$ invariance of field theory then means that it is possible to choose a (real) representation of $\phi(\vec{x}, t)$ such that

$$\phi(-\vec{x}, -t) = (\mathscr{CPT})^{-1} \phi(\vec{x}, t)(\mathscr{CPT}) . \tag{22.27}$$

Therefore, since $\phi(\vec{x}, t)$ is $CP$-odd it must be $T$-odd and then

$$\langle \phi(\vec{x}, -t) \rangle = -Tr \left[ \rho_f \mathscr{T}^{-1} \phi(\vec{x}, t) \mathscr{T} \rho_i \right] . \tag{22.28}$$

But if $\rho_i$ and $\rho_f$ are related by (22.22) this relation may be written

$$\langle \phi(\vec{x}, -t) \rangle = -Tr \left[ \mathscr{T}^{-1} \rho_i \mathscr{T} \mathscr{T}^{-1} \phi(\vec{x}, t) \mathscr{T} \mathscr{T}^{-1} \rho_f \mathscr{T} \right] \tag{22.29}$$

$$= -Tr \left[ \rho_i \phi(\vec{x}, t) \rho_f \right] = -\langle \phi(\vec{x}, t) \rangle . \tag{22.30}$$

The conclusion is that it is possible to choose initial and final conditions so that a universe is time-symmetric and has a non-vanishing expected value of a $CP$-odd field. That expected value is odd in time (the correct time-*symmetric* behavior for a $T$-odd field.) As a consequence the sign of $CP$ violation would be opposite on opposite sides of the moment of time symmetry and the magnitude of $CP$ violation would decrease on cosmological time scales as the moment of time symmetry is approached. The $CP$ violation in the early universe might well be larger than generally supposed and Sakharov's mechanism for the generation of the baryons more effective. However, if the moment of time symmetry is far in our future, then such variation in the strength of $CP$ violation would be small and it would be difficult to distinguish this time-symmetric situation from the kind of $CP$ violation that arises from just an initial condition as discussed above.

In the class of time-symmetric universes just discussed, $CP$ violation arises from initial and final conditions that are not $CP$-symmetric. However, an effective $CP$ violation could also exist in our epoch, in local spatial domains, even if both Hamiltonian and initial and final states were $CP$-symmetric:

$$H = (\mathscr{CP})^{-1} H (\mathscr{CP}) \,, \quad \rho_i = (\mathscr{CP})^{-1} \rho_i (\mathscr{CP}) \,, \quad \rho_f = (\mathscr{CP})^{-1} \rho_f (\mathscr{CP}) \,. \qquad (22.31)$$

Dynamical mechanisms would need to exist that make likely the existence of large spacetime domains in which $CP$ is effectively broken, say by the expected value of a $CP$-odd field that grows to be homogeneous over such a domain. In such a picture the set of histories of the universe would be overall $CP$-symmetric and $T$-symmetric, as follows from (22.31). Individual histories would display effective $CP$ violation in domains with sizes and durations that are quantum-probabilistic. If very large sizes and durations were probable it would be difficult to distinguish this kind of mechanism from any of those discussed above.

Overall matter-antimatter symmetry would be expected for such universes with matter or anti-matter predominant only in local domains. Their size must therefore be larger than the known scales on which matter is dominant [38]. The calculation of the probabilities for these sizes and durations thus becomes an important question in such pictures. An extreme example occurs in the proposal of Linde [40], in which such domains are far larger than the present Hubble radius.

## 22.6 The Limitations of Decoherence and Classicality

As we mentioned in Section 2, the quantum mechanics of a closed system such as the universe as a whole predicts probabilities only for those sets of alternative histories for which there is negligible interference between the individual members in the set. Sets of histories that exhibit such negligible interference as a consequence of the Hamiltonian and boundary conditions are said to decohere. A minimal requirement on any theory of the boundary conditions for cosmology is that the universe exhibit a decoherent set of histories that corresponds to the quasiclassical domain of everyday experience. This requirement places significant restrictions on the relation between $\rho_i$ and $\rho_f$ in the generalized quantum mechanics for cosmology, as we shall now show.

### *22.6.1 Decoherence*

Coherence between individual histories in an exhaustive set of coarse-grained histories is measured by the decoherence functional [20]. This is a complex-valued functional on each pair of histories in the set. If the cosmos is replaced by a box, so that possible complications from quantum gravity disappear, then individual coarse-grained histories are specified by sequences of alternatives $\alpha = (\alpha_1, \cdots, \alpha_n)$ at discrete moments of time, $t_1, \cdots, t_n$. The decoherence functional for the case of two-time boundary conditions is given by [14]

$$D(\alpha', \alpha) = N Tr \left[ \rho_f C_{\alpha'} \rho_i C_\alpha^\dagger \right] . \qquad (22.32)$$

A set of histories decoheres when the real parts of the "off-diagonal" elements of the decoherence functional — those between two histories with any $\alpha_k \neq \alpha'_k$ — vanish to sufficient accuracy. As first shown by Griffiths [13], this is the necessary and sufficient condition that the probabilities (22.5), which are the "diagonal" elements of $D$, satisfy the sum rules defining probability theory.

The possibility of decoherence is limited by the choice of initial and final density matrices $\rho_i$ and $\rho_f$. To see an example of this, consider the case in which both are pure, $\rho_i = |\Psi_i ><\Psi_i|$ and $\rho_f = |\Psi_f ><\Psi_f|$. The decoherence functional would then factor:

$$D(\alpha', \alpha) = N < \Psi_f|C_{\alpha'}|\Psi_i ><\Psi_i|C_\alpha^\dagger|\Psi_f > , \qquad (22.33)$$

where $N$ now is $| < \Psi_i|\Psi_f > |^{-2}$. In this circumstance the requirement that the real part of $D$ vanish for $\alpha' \neq \alpha$ can be satisfied only if there are at most two non-vanishing quantities $< \Psi_i|C_\alpha|\Psi_f >$, with phases differing by 90°, giving at most two histories with non-vanishing probabilities! Thus initial and final states that are both pure, such as those corresponding to a "wave function of the universe", leads to a highly unorthodox quantum mechanics in which there are only one or two coarse-grained histories. All the apparent accidents of quantum mechanics would be determined† by the boundary conditions $\rho_i$ and $\rho_f$. The usual idea of a simple $\rho_i$ (or $\rho_i$ and $\rho_f$), with the algorithmic complexity of the universe contained almost entirely in the throws of the quantum dice, would here be replaced by a picture in which the algorithmic complexity is transferred to the state vectors $|\Psi_i\rangle$ and $|\Psi_f\rangle$. Presumably these would be described by a simple set of rules plus a huge amount of specific information, unknowable except by experiment and described in practice by a huge set of parameters with random values.

This bizarre situation refers to the use of a pure $\rho_i$ and a pure $\rho_f$, whether or not there is any kind of time symmetry relating them.

---

† This situation is closely related to the one described by L. Schulman [41].

### 22.6.2 *Impossibility of a Universe with $\rho_f = \rho_i$.*

We shall now give a very special example of a relation between $\rho_i$ and $\rho_f$, stronger than time symmetry, that is inconsistent with the existence of a quasiclassical domain. More precisely, we shall show that in the extreme case

$$\rho_f = \rho_i \equiv \rho \tag{22.34}$$

only sets of histories exhibiting trivial dynamics can exactly decohere. This condition means that $\rho_f$ has the same form when expressed in terms of the initial fields $\phi(\check{x}, t_0)$ as $\rho_i$ does. Such a situation could arise if, in addition to time symmetry, we had $\rho_i$ and $\rho_f$ separately, individually time-symmetric and with effectively no time difference between the initial and final conditions. We know of no theoretical reason to expect such a situation, but it does supply an example that leads to a contradiction with experience.

Given the artificial condition, (22.34), we can write the decoherence condition as

$$(Tr\rho^2)^{-1}ReTr(\rho C_{\alpha'}\rho C_{\alpha}^{\dagger}) = \delta_{\alpha'\alpha}p(\alpha) , \tag{22.35}$$

where $p(\alpha)$ is the probability of the history $\alpha$. Summing over all the $\{\alpha_n\}$ and $\{\alpha'_n\}$ except $\alpha_k$ and $\alpha'_k$, we have

$$(Tr\rho^2)^{-1}ReTr[\rho P^k_{\alpha'_k}(t_k)\rho P^k_{\alpha_k}(t_k)] = \delta_{\alpha'_k\alpha_k}p(\alpha_k) . \tag{22.36}$$

We note that $P^k_{\alpha_k}(t_k)$ and $P^k_{\alpha'_k}(t_k)$ are just projection operators and thus of the form

$$\sum_n |n><n| \quad \text{and} \quad \sum_{n'} |n'><n'|$$

respectively, where the $|n>$ and $|n'>$ are mutually orthogonal for $\alpha_k \neq \alpha'_k$. Eq. (22.36) then tells us that

$$(Tr\rho^2)^{-1} \sum_{n,n'} |<n|\rho|n'>|^2 = 0, \quad \text{for } \alpha_k \neq \alpha'_k . \tag{22.37}$$

Thus $\rho$ has no matrix elements between any $|n>$ and any $|n'>$ for $\alpha_k \neq \alpha'_k$. In other words, $\rho$ commutes with all the $P$'s and therefore with all the chains $C_{\alpha}$ of $P$'s:

$$[C_{\alpha}, \rho] = 0 \quad \text{for all } \alpha . \tag{22.38}$$

This consequence of perfect decoherence for the special case (22.34) has some important implications. For one thing, the decoherence formula can now be written

$$(Tr\rho^2)^{-1}Tr(C_{\alpha'}\rho^2 C_{\alpha}^{\dagger}) = \delta_{\alpha'\alpha}p(\alpha) , \tag{22.39}$$

so that we are back to ordinary quantum mechanics with only an initial density matrix $\bar{\rho} \equiv (Tr\rho^2)^{-1}\rho^2$ [cf. (22.11) in the classical case] but with the very restrictive condition

$$[C_{\alpha}, \bar{\rho}] = 0 \quad \text{for all } \alpha . \tag{22.40}$$

The cosmology with the symmetry (22.34) was supposed to be in contrast to the usual one with only an initial density matrix, and yet it turns out to be only a special case of the usual one with the stringent set of conditions (22.40) imposed in addition. The resolution of this apparent paradox is that Eq. (22.40) permits essentially no dynamics and thus achieves symmetry between $\rho_i$ and $\rho_f$ in a rather trivial way. That is not surprising in view of the nature of this condition discussed above.

We have seen that any $P_{\alpha_k}^k(t_k)$ has to commute with $\bar{\rho}$ if it is to be permitted in a chain of $P$'s constituting a member of a decohering set of alternative coarse-grained histories. Now it is unreasonable that for a given projection operator $P$ there should be only a discrete set of times at which it is permissible to use it in a history (e.g., for a measurement). Thus we would expect that there should be a continuous range of such times, which means that $\dot{P} = -i[P, H]$ must commute with $\bar{\rho}$. But $\bar{\rho}$ and $P$, since they commute, are simultaneously diagonalizable, with eigenvalues $\pi_i$ and $q_i$ respectively. The time derivative of the probability $Tr(\bar{\rho}P)$ is

$$Tr(\bar{\rho}\dot{P}) = -iTr(\bar{\rho}[P, H]) = -i\sum_i \pi_i(q_i - q_i)H_{ii} = 0 . \qquad (22.41)$$

The probabilities of the different projections $P$ remain constant in time, so that there is essentially no dynamics and certainly no second law of thermodynamics.

### 22.6.3 *Classicality*

A theory of the boundary conditions of the universe must imply the quasiclassical domain of familiar experience. A set of histories describing a quasiclassical domain must, of course, decohere. That is the prerequisite for assigning probabilities in quantum mechanics. But further, the probabilities must be high that these histories are approximately correlated by classical dynamical laws, except for the intervention of occasional amplified quantum fluctuations.

There are, of course, limitations on classical two-time boundary conditions. We cannot, for example, specify both coördinates and their conjugate momenta at *both* an initial and a final time. There would, in general, be no corresponding solutions of the classical equations of motion. Even if initial and final conditions in quantum cosmology allow for decoherence as discussed above, they could still be too restrictive to allow for classical correlations. One would expect this to be the case, for example, if they required a narrow distribution of both coördinates and momenta both initially and finally. Quantum cosmologies with two boundary conditions are therefore limited by both decoherence and classicality.

## 22.7 Conclusions

Time-symmetric quantum cosmologies can be constructed utilizing a time-neutral generalized quantum mechanics of closed systems with initial and final conditions related by time-inversion symmetry. From the point of view of familiar quantum mechanics such time-symmetric cosmologies are highly unusual. If we think of Hilbert space as finite-dimensional, we could introduce a normalization $Tr(\rho_f)\, Tr(\rho_i) = Tr(I)$, which would agree with the usual case $Tr(\rho_i) = 1$, $Tr(\rho_f) = Tr(I)$. (Note that both $N = Tr(\rho_f \rho_i)$ and the quantity $Tr(\rho_f)\, Tr(\rho_i)$ are invariant under multiplication of $\rho_i$ by a factor and $\rho_f$ by the inverse factor.) With this normalization we may think of $N^{-1} = Tr(\rho_f \rho_i)$ as a measure of the likelihood of the final condition given the initial one. The similarly defined quantity $N^{-1}$ in the analogous classical time-symmetric cosmologies is just that. It is the fraction of trajectories meeting the initial condition that also meet the final one [cf. (22.7)]. The measure $N^{-1}$ is unity for the usual cases where $\rho_f = I$. It can be expected to be very small for large systems with time-symmetric boundary conditions, as the simple model described in Figure 22.1 suggests. The measure $N^{-1}$ is likely to be *extraordinarily* small in the case of the universe itself. Were it exactly zero the initial and final boundary condition construction would become doubtful. We are unsure how much of that doubt survives if it is merely extraordinarily small.

As a prerequisite for a time-symmetric quantum cosmology, the fundamental Hamiltonian must be time-inversion symmetric to give a meaningful notion of time-symmetry and this restricts the mechanisms by which the effective $CP$ violation in the weak interactions can arise. There must be some impurity in the initial or final density matrices or in both for any non-trivial probabilities to be predicted at all. If we wish to exclude the highly unorthodox quantum mechanics in which $|\Psi_i\rangle$ and $|\Psi_f\rangle$ determine all the throws of the quantum dice, then we could not have, for example, a time-symmetric quantum cosmology with both the initial and final conditions resembling something like the "no-boundary" proposal. These results have been obtained by assuming unrealistic exact decoherence and by neglecting gross quantum variations in the structure of spacetime, which may be important in the early universe. It would be desirable to extend the discussion to remove these special restrictions.

Even if these purely theoretical requirements for time-symmetry were met, observations might rule out such boundary conditions. Deviations from the usual thermodynamic or $CP$ arrows of time may be undetectably small if the time between initial and final conditions is long enough. But, as suggested by Davies and Twamley, an expanding and contracting time-symmetric cosmology may be transparent enough to electromagnetic and other forms of radiation that the effects of certain time-symmetric initial and final conditions would be inconsistent with observations today. In the absence of some compelling theoretical principle mandating time symmetry, the simplest possibility seems to be the usually postulated universe where there is a fundamental distinction between past and future — a universe with a special initial state and a final condition of indifference with respect to state. Nevertheless, the notion of complete $T$ symmetry or $CPT$ symmetry remains sufficiently intriguing to warrant further investigation of how such a symmetry could occur or what observations could rule it out. In this paper we have provided a quantum-mechanical framework for such investigations.

## Acknowledgments

An earlier version of this paper appeared in the *Proceedings of the 1st International Sakharov Conference on Physics*, Moscow, USSR, May 27–31, 1991 as a tribute to the memory of A.D. Sakharov.

Part of this research was carried out at the Aspen Center for Physics. The work of MG-M was supported by DOE contract DE-AC-03-81ER40050 and by the Alfred P. Sloan Foundation. That of JBH was supported by NSF grant PHY90-08502.

## Appendix

We derive the expression (22.26) for the expected value of a scalar field in the time-neutral quantum mechanics of cosmology with an initial condition represented by a density matrix $\rho_i$ and a final condition represented by a density matrix $\rho_f$. Consider alternatives such that the value of the field $\phi$ at $(\vec{x}, t)$ lies in one of an exhaustive set of infinitesimal exclusive intervals $\{\Delta_\alpha\}$ with central values $\{\phi_\alpha\}$. Let $P_\alpha(\vec{x}, t)$ denote the corresponding projection operators. The decoherence functional for this set of alternatives is, according to (22.32)

$$D(\alpha', \alpha) = N Tr \left[ \rho_f P_{\alpha'}(\vec{x}, t) \rho_i P_\alpha(\vec{x}, t) \right] . \tag{A.1}$$

We assume that this is diagonal, that is, proportional to $\delta_{\alpha\alpha'}$. The diagonal elements give the probabilities of the alternative values of the field according to (22.5). Thus the expected value of $\phi(\vec{x}, t)$ is

$$\langle \phi(\vec{x}, t) \rangle = \sum_\alpha \phi_\alpha D(\alpha, \alpha) . \tag{A.2}$$

Because the alternatives decohere, this can be written in two equivalent forms

$$\langle \phi(\vec{x}, t) \rangle = \sum_{\alpha'\alpha} \phi_{\alpha'} D(\alpha', \alpha) = \sum_{\alpha'\alpha} \phi_\alpha D(\alpha', \alpha) . \tag{A.3}$$

But, utilizing $\sum_\alpha P_\alpha(\vec{x}, t) = 1$ and $\sum_\alpha \phi_\alpha P_\alpha(\vec{x}, t) = \phi(\vec{x}, t)$, as well as (A.1) and the cyclic property of the trace, we get

$$\langle \phi(\vec{x}, t) \rangle = N Tr \left[ \rho_f \phi(\vec{x}, t) \rho_i \right] = N Tr \left[ \rho_i \phi(\vec{x}, t) \rho_f \right] \tag{A.4}$$

as in (22.26).

## Discussion

**DeWitt**    If you propose that the universe is in a particular quantum state, determined by particular initial conditions, why do you bother with a complete Hilbert-space framework for discussion?

**Hartle**    I interpret your question as mainly referring to the status of the superposition principle in quantum cosmology. It is true that if the initial condition of our universe is described by a single wave function then it is never necessary to superpose it with another to make predictions. However, the principle of superposition enters centrally elsewhere in the predictive framework. Specifically, it enters into the construction of the probabilities for the coarse-grained sets of histories that we observe. If $P_A$ and $P_B$ are projection operators representing exclusive alternatives $A$ and $B$ at one time, then the alternative $A$ *and* $B$ is represented by the *sum* of the projections, $P_A + P_B$. That is a specific instance of the principle of superposition. More generally, the decoherence functionals for sets of histories related by an operation of coarse graining must be connected by the superposition principle. That is one reason we assume the full apparatus of Hilbert space when discussing the quantum cosmology of matter fields in a fixed background spacetimes or generalizations of that formalism consistent with the superposition principle when quantum gravity is taken into account. Even in the most general cosmological context it should still be possible to test these aspects of the principle of superposition.

**Kuchař**    Both Murray's and your talk were based on the assumption that there is a true Hamiltonian and that there is a single time parameter which orders the projection operators. How does one formulate the difference between time-symmetric initial and final conditions if the dynamics is driven by constraints and there is no privileged time parameter?

**Hartle**    To keep the discussion in the talk manageable, we assumed a fixed, background spacetime. That, of course, is an excellent approximation any time much more than a Planck time after the initial singularity and a Planck time before the final singularity if there is one. That fixed spacetime geometry supplies the notion of time used to order the operators and define the Hamiltonian. However, in regimes near the singularity, where quantum gravity is important, and the geometry of spacetime fluctuates quantum mechanically, there will be no fixed spacetime geometry to supply a notion of time. A further generalization of quantum mechanics is thus required. I have described in several places the basic elements of one such generalization based on sum-over-histories quantum mechanics.† In that generalization, the histories are four-dimensional cosmological spacetimes with boundaries where the analogs of "initial" and "final" conditions represented by wave functions are imposed. The decoherence functional for coarse-grained sets of alternative histories of the universe, including diffeomorphism-invariant coarse grainings of spacetime geometry, is represented in a sum-over-histories form that does not single out a privileged time parameter. Decoherence is thus defined and probabilities for the individual members of decoherent sets of coarse-grained histories can be calculated. In order not to be manifestly inconsistent with observations, the specific initial and final conditions of our universe had better predict the approximately classical behavior of spacetime geometry on accessible

---

† See, for example, my lectures "The Quantum Mechanics of Cosmology" in *Quantum Mechanics and Baby Universes: Proceedings of the 1989 Jerusalem Winter School*, edited by S. Coleman, J. Hartle, T. Piran, and S. Weinberg, World Scientific, Singapore, 1990, or in more complete detail, in my lectures "Spacetime Quantum Mechanics and the Quantum Mechanics of Spacetime" in *Gravitation and Quantizations*, Proceedings of the 1992 Les Houches Summer School, edited by B. Julia and J. Zinn-Justin, North Holland, Amsterdam, 1993.

scales in our epoch. That is, semiclassically, realistic boundary conditions predict an ensemble of possible classical spacetimes of which we live in one. The probabilities of suitably coarse-grained matter field histories in each spacetime in the ensemble would be approximately given by the kind of quantum mechanics we have limited ourselves to in this talk with a notion of time given by the particular classical spacetime geometry. The discussion we gave is thus both a model for the more general case of quantum gravity and an approximation to it in all directly accessible circumstances.

If the initial and final conditions are suitably related, I would expect the ensemble of possible spacetimes predicted semiclassically by such a theory to exhibit statistical time symmetry. Further, for the probable spacetimes that are time symmetric I would expect the quantum mechanics of matter fields and small fluctuations of geometry to be time-symmetric in the sense have described in this talk. Put briefly, I expect the present discussion that assumes a fixed spacetime is a good approximation in interesting circumstances to the more general case where it is allowed to fluctuate. It is fair to say, however, that detailed calculations have not been done to check on these expectations.

**Omnès** Your probability formula, in a universe with a destiny, violates Gleason's theorem and therefore one of its assumptions at least. There are two possibilities: (i) Not all properties are possible, which is what you are aiming at. (ii) Maybe, the Hilbert space is highly non-separable.

**Hartle** I think it's the former.

**Halliwell** You argue that for initial and final density matrices satisfying a certain condition (and in particular, for pure initial and final states) the decoherence functional factors, and therefore, will not decohere except for certain trivial histories. You suggested that there will therefore be problems for the no-boundary state, which is a pure state. It seems to me that this result may depend rather crucially on the existence of a Hilbert space structure etc., and in particular, on the possibility of folding in initial and final states using the usual inner product. My point is that all of this structure is not known to exist for quantum cosmology. The decoherence functional for quantum cosmology is yet to be constructed, and is likely to have a structure rather different to the quantum mechanical one. You may therefore be premature in your conclusions about the no-boundary proposal.

**Hartle** In the proposals for the decoherence functional for quantum cosmology that I have put forward, the result that pure initial and final states permit the decoherence of only trivial sets of histories continues to hold in much the same way it does for the quantum cosmologies in a box described in the text. Initial and final conditions are represented by density matrices and pure conditions by single wave functions. The principle of superposition of amplitudes is maintained as is the relation between amplitudes and probabilities. When the initial and final states are pure, the decoherence functional factors into a term for one history times a term for the other as in (22.33) and the rest of the argument goes through. In these generalizations imposing the "no boundary" proposal for both initial and final conditions does not lead to interesting sets of decohering histories. However, there is much to be investigated here, and there could be other generalizations of quantum mechanics for which the result does not hold.

**Bialynicki-Birula** I would like to make a comment on the possible role of soft photons in the time-symmetric quantum theory. In the presence of massless particles, and that is a typical case, even for time-symmetric Hamiltonians there is a difficulty in implementing the time-symmetry condition,

$$\rho_f = \mathcal{T}^{-1} \rho_i \mathcal{T} \, ,$$

due to the existence of infrared radiation. There is no $\rho_f$ that will give a nonvanishing transition probability whenever charged particles are being accelerated. In order to obtain a finite result we must perform an integration over the momenta of final, unobservable soft photons. The necessity to perform this integration implies that there is an asymmetry between the initial state described by $\rho_i$ and the final state for which a density operator does not exist.

**Hartle**   Soft photons in the universe certainly provide an important and widespread mechanism for decoherence. However, I haven't thought through the effect soft photons might have on the size of the normalizing factor $N^{-1} = Tr(\rho_f \rho_i)$ that occurs in the expression (22.5) for probabilities. I would be surprised if, in a proper formulation of quantum electrodynamics, $N^{-1}$ *necessarily* vanished identically for the finite (although cosmologically long) time interval between time-symmetric initial and final conditions that we have been discussing. However, even if $N^{-1}$ does not vanish identically, but is only very small, that would signal a significant difference between the statistics of histories in a time-symmetric universe and the usual case. One guesses that $N^{-1}$ is likely to be small in any realistic time-symmetric universe even in the absence of electrodynamics but if soft photons play a significant role in determining its size that would be very interesting.

**Page**   What are the consequences if you require a $CPT$-invariant universe instead of a T-invariant universe?

**Hartle**   I didn't get to $CPT$-symmetric cosmologies in my talk, but they are discussed in the written contribution that Murray Gell-Mann and I have submitted to the proceedings. In the generalization of quantum mechanics that we discuss, a $CPT$-symmetric universe will result if the initial and final density matrices are related by a $CPT$ transformation. Since all local field theories are $CPT$-invariant, $CPT$-symmetric cosmologies are possible even if the $CP$ violation observed in the weak interactions arises from a fundamental Hamiltonian that is $CP$-non-invariant. That is in contrast to the case of $T$ symmetry which can only be achieved with a $CP$-symmetric Hamiltonian so that the observed $CP$ violation must arise from one of the symmetry breaking mechanisms we discussed. However, also as discussed in the text, $CPT$ symmetry may be difficult to reconcile with universes that are homogeneously dominated by matter near one singularity (and therefore antimatter dominated near the other) unless the lifetime is very long.

**DeWitt**   So if the no boundary condition leads to a $CP$ invariant family of $CP$ violating states, you must pick one member of this family out by hand?

**Hartle**   If the "no boundary" condition leads to a $CP$ invariant family of $CP$ non-invariant states then I would prefer to say that the initial condition is a density matrix with probabilities distributed uniformly among these possibilities. But for predictive purposes that amounts to what you said.

**Albrecht**   If one discusses the thermodynamic arrow of time in a time symmetric universe, one has a region near the "beginning" where the arrow runs toward the middle, and a region near the "end" where the arrow runs towards the middle. In the middle there is no particular arrow of time. The probability that we survive into the future epoch where the arrow is reversed is no greater than the probability that some IGUS is present right now, evolving in the opposite direction of time.

**Hartle**   I think that's essentially right with a few qualifications. It follows from the assumed time symmetry that the statistics of IGUS's at the present age from the big bang must be the same as that at a comparable time from the final singularity if the only input to estimating those statistics is the initial and final conditions. If the lifetime of the universe is long compared to those times, both the probability that an IGUS survives into the far future and the probability that there are IGUSes in the present living backward in time may be very low. If we are a typical IGUS then those probabilities apply to us. However, we have more information about our particular history with which the conditional probabilities of our surviving into the far future could, in principle, be calculated and compared with the probability that there are IGUSes that have evolved backward from the far future around today. I'll leave it to you to make the estimate of whether our particular history makes it more or less probable that we survive farther into the future than the typical IGUS!

**Davies** Your model is most plausible if all asymmetric physical processes relax to equilibrium before the time reversal occurs. But long ago it was found that the future light cone in Robertson-Walker cosmological models is transparent to photons. (In the recontracting case it is transparent at least until the turnover point.) Thus, retarded radiation cannot equilibrate before time reversal, so that the imposing of time-symmetric boundary conditions would surely show up experimentally in the emission of radiation.

**Hartle** As mentioned in the talk, there are several different examples of physical systems that will not come to equilibrium in the Hubble time, and the system of matter and radiation is one of them. I take your comment to be a suggestion that observation in the electromagnetic system could supply the best lower bound on the time between initial and final conditions beyond which these are indistinguishable. That may well be the case and is an interesting subject for further research. It's an important question. [For further discussion see Section 4.3, added after the conference.]

**Starobinsky** (in response to P. Davies' comment)
This is not always the case. If the width of a domain wall in flat spacetime is larger than the radius of curvatures "gravitational radius" corresponding to the field energy density in a metastable state at the top of the field potential, then regions with the different sign of $CP$ violation are always beyond an observer's particle horizon and they are never accessible to him for any future evolution. A good example can be constructed using the "new" inflationary scenario and relating the sign of an inflation field to the sign of $CP$ violation.

*Note added by M. Gell-Mann and J. B. Hartle, November 1993.* We would like to clear up any possible confusion over the relationship among several ideas in the quantum mechanics of closed systems as well as the history of the subject and the terminology employed. In this note, added two years after the discussion, we attempt to clarify these matters as we see them, benefiting from research in the intervening time.

As far as we are aware, in the context of the modern interpretation of quantum mechanics, the term "decoherence" was first used by us in lectures and discussion after 1986 (summarized in our paper published in 1990) to describe a property of a set of alternative time-*histories* of a closed system. Specifically a set of histories is said to (medium-) decohere when there is negligible or vanishing quantum mechanical interference between the individual histories in the set as measured by the "off-diagonal" elements of the decoherence functional $D(\alpha', \alpha)$. However, the term "decoherence" has subsequently come to be used to refer also to the decay over time of the off-diagonal elements of a reduced density matrix defined by a division of a complete set of variables into ones that are traced over in constructing the reduced density matrix and the rest. The dynamics of the remaining variables is described by the reduced density matrix. This decay was discussed in connection with the interpretation of quantum mechanics in the '70s and early '80s by Zeh, Zurek, and others (although not referred to as decoherence) and by many others since. These two notions of decoherence – of histories and of density matrices – are not the same but are not unconnected either. The vanishing, at a sequence of times, of the off-diagonal elements of a reduced density matrix in certain variables is neither mathematically or physically equivalent to the decoherence of the corresponding set of alternative histories. However, the ideas are connected in certain idealized models, where it can be shown that the physical mechanisms causing the decoherence of histories coarse-grained by ranges of values of certain sets of coordinates suitably spaced in time *also* lead to the diagonalisation of the reduced density matrix in these variables over similar intervals of time [14]. It has also been shown that in similar models, under restrictive conditions, the diagonalization of the reduced density matrix implies the decoherence of the histories associated with the "Schmidt basis" for that density matrix at suitable times [42,43]. Finally, a certain interpretation of "mechanism of decoherence" can be defined [44,45] that generalizes the reduced density matrix concept of decoherence *in the context of a stronger form of decoherence of histories*. A precise connection is thereby established between the two kinds of decoherence. The reader should therefore keep in mind that in these and other discussions and papers at this conference, the word "decoherence" is used for two distinct but connected ideas – the decoherence of reduced density matrices and the decoherence of histories.

We now try to clarify the connection of our ideas with the work of Bob Griffiths and Roland Omnès, another topic that was raised in the discussion. In any quantum-mechanical theory a rule is needed to discriminate between those sets of histories that can be assigned probabilities and those that cannot because of quantum-mechanical interference. Griffiths was the first to propose a quantum mechanics of closed systems with a rule that did not involve a fundamental notion of measurement. Instead, probabilities are assigned to just those sets of histories that are "consistent" in the sense that their probabilities obey correct sum rules – essentially the consistency requirement that the theory not offer two different results for the same probability. (In this connection it may be helpful to note that what is sometimes referred to as the quantum mechanics of an "open system" means a set of effective rules for describing the quantum-mechanical behavior of *part* of a closed system.) Griffiths' ideas were extended by Omnès and a similar formulation was arrived at independently but later by ourselves. As far as we are aware, *all* formulations of the quantum mechanics of closed systems under serious consideration, including ours, are "consistent history formulations" in the sense of requiring the consistency of a set of probability sum rules for those histories that are assigned probabilities. However, there are different formulations depending on just what probability sum rules are required and the strength of the conditions used to ensure their consistency.

The consistency conditions of Griffiths are not the same as the decoherence conditions used in our work. To explain the difference a small amout of notation is useful. Let $\{C_\alpha\}$ denote a set of chains of projections representing a set of alternative histories, and $D(\alpha', \alpha) = Tr(C_{\alpha'} \rho C_\alpha)$ the associated decoherence functional. The consistency condition of Griffiths is that $RcD(\alpha', \alpha) \approx 0$, $\alpha' \neq \alpha$, if and only if $C_{\alpha'} + C_\alpha$ is *another chain of projections*. This is the necessary condition for the probability sum rules if only histories corresponding to independent sets of alternatives at different moments of time (histories which are represented by chains of projections) are allowed. What we called the weak-decoherence condition, $ReD(\alpha', \alpha) \approx 0$, $\alpha' \neq \alpha$, is *stronger* than that of Griffiths. It is the necessary condition if histories that are *sums* of chains of projections are allowed (corresponding to a rule for coarse-graining that allows arbitrary unions of histories as new histories). Our medium-decoherence condition, $D(\alpha', \alpha) \approx 0$, $\alpha' \neq \alpha$, is *still stronger*, implying both of the above conditions but not being implied by either of them. Thus there are at least three different notions of decoherence of histories, *all* of which imply the consistency of (sometimes different) sets of probability sum rules.

In our work we have sought a notion of decoherence of histories that would capture generally the idea that, in physically interesting situations, relevant for quasiclassical behavior, quantum-mechanical interference between histories vanishes *for a reason*. That is, we sought to provide a general characterization of the mechanisms of dissipation of phases described by Zeh, Zurek and others, but in a way that would not require an artificial or poorly defined division of the closed system into subsystem and environment. For this reason we were led to notions of the decoherence of histories that imply consistency, of course, but are stronger and should not be confused with it.

The differences we have described should not obscure the fact that our work lies within the class of consistent histories formulations of the quantum mechanics of closed systems, but should also not obscure the fact that decoherence of histories, as we have defined it, is not the same as just consistency.

## References

[1] P.C.W. Davies (1976) *The Physics of Time Asymmetry*, University of California Press, Berkeley.
[2] R. Penrose (1979) in *General Relativity: An Einstein Centenary Survey* ed. by S.W. Hawking and W. Israel, Cambridge University Press, Cambridge.
[3] H.D. Zeh (1989) *The Physical Basis of the Direction of Time*, Springer, Berlin.
[4] J. J. Loschmidt (1876) *Wiener Ber.*, **73**, 128; *ibid.* (1877) **75**, 67.
[5] D. Page (1985) *Phys. Rev.*, **D32**, 2496.
[6] S.W. Hawking (1985) *Phys. Rev.*, **D32**, 2989.
[7] J. Halliwell and S.W. Hawking (1985) *Phys. Rev.*, **D31**, 1777.
[8] R. Laflamme (Unpublished) "Wave Function of an $S^1 \times S^2$ Universe".

[9] S.W. Hawking (1987) *New Scientist*, **115**, 46.

[10] S.W. Hawking and D. Page (1988) *Nucl. Phys.*, **B298**, 789.

[11] S. Wada (1990) in *Proceedings of the 3rd International Symposium on the Foundations of Quantum Mechanics in the Light of New Technology*, ed. by S. Kobayashi, H. Ezawa, Y. Murayama, and S. Nomura, Physical Society of Japan, Tokyo.

[12] Y. Aharonov, P. Bergmann, and J. Lebovitz (1964) *Phys. Rev.*, **B134**, 1410.

[13] R. B. Griffiths (1984) *J. Stat. Phys.*, **36**, 219.

[14] J.B. Hartle (1991) in *Quantum Cosmology and Baby Universes, Proceedings of the 1989 Jerusalem Winter School*, ed. by S. Coleman, J. Hartle, T. Piran, and S. Weinberg, World Scientific, Singapore.

[15] T. Gold (1958) in *La structure et l'evolution de l'universe, Proceedings of the 11th Solvay Congress*, Editions Stoops, Brussels; (1962) *Amer. J. Phys.*, **30**, 403.

[16] M. Gell-Mann (unpublished) comments at the 1967 Temple University Panel on Elementary Particles and Relativistic Astrophysics.

[17] Y. Ne'eman (1970) *Int. J. Theor. Phys.*, **3**, 1.

[18] H.-D. Zeh (1994) This volume.

[19] S.W. Hawking (1994) This volume.

[20] M. Gell-Mann and J.B. Hartle (1990) in *Complexity, Entropy, and the Physics of Information, SFI Studies in the Sciences of Complexity*, Vol. VIII, ed. by W. Zurek, Addison Wesley, Reading or in *Proceedings of the 3rd International Symposium on the Foundations of Quantum Mechanics in the Light of New Technology* ed. by S. Kobayashi, H. Ezawa, Y. Murayama, and S. Nomura, Physical Society of Japan, Tokyo.

[21] M. Gell-Mann and J.B. Hartle (1990) in the *Proceedings of the 25th International Conference on High Energy Physics, Singapore, August, 2-8, 1990*, ed. by K.K. Phua and Y. Yamaguchi, South East Asia Theoretical Physics Association and Physical Society of Japan, distributed by World Scientific, Singapore.

[22] W.J. Cocke (1967) *Phys. Rev.*, **160**, 1165.

[23] L.S. Schulman (1973) *Phys. Rev.*, **D7**, 2868; L.S. Schulman (1977) *J. Stat. Phys.*, **16**, 217; L.S. Schulman and R. Shtokhamer (1977) *Int. J. Theor. Phys.*, **16**, 287.

[24] J.A. Wheeler (1979) in *Problemi dei fondamenti della fisica*, Scuola internazionale di fisica "Enrico Fermi", Corso 52, ed. by G. Toraldo di Francia, North-Holland, Amsterdam.

[25] L.S. Schulman (1991) *Physica A*, **177**, 373.

[26] M. Kac (1959) *Probability and Related Topics in Physical Sciences*, Interscience, New York.

[27] R. Penrose (1978) in *Confrontation of Cosmological Theories with Observational Data* (IAU Symp. 63) ed. by M. Longair, Reidel, Boston (1974) and in *Theoretical Principles in Astrophysics and Relativity*, ed. by N.R. Lebovitz, W.H. Reid, and P.O. Vandervoort, University of Chicago Press, Chicago.

[28] L. Grishchuk and L.V. Rozhansky (1988) *Phys. Lett.*, **B208**, 369; (1990) *ibid.*, **B234**, 9.

[29] A. Barvinsky and A. Kamenshchik (1990) *Class. Quant. Grav.*, **7**, L181.

[30] A. Vilenkin (1988) *Phys. Rev.*, **D37**, 888.

[31] M. Mijić, M. Morris and W.-M. Suen (1989) *Phys. Rev.*, **D39**, 1486.

[32] P.C.W. Davies and J. Twamley (1993) *Class. Quant. Grav.*, **10**, 931.

[33] R.B. Partridge (1973) *Nature*, **244**, 263.

[34] D. Page (1993), *No Time Asymmetry from Quantum Mechanics*, University of Alberta preprint.

[35] J.B. Hartle and S.W. Hawking (1983) *Phys. Rev.*, **D28**, 2960.

[36] A.D. Sakharov (1967) *ZhETF Pis'ma*, **5**, 32; [(1967) *Sov. Phys. JETP Lett.*, **5**, 24]; (1979) *ZhETF*, **76**, 1172, 1979; [(1979) *Sov. Phys. JETP*, **49**, 594].

[37] E.W. Kolb and M.S. Turner (1990) *The Early Universe*, Addison-Wesley, Redwood City, Ca.

[38] G. Steigman (1976) *Ann. Rev. Astron. Astrophys.*, **14**, 339.

[39] T.D. Lee (1974), *Physics Reports*, **9C**, 144.

[40] A. Linde (1986) *Mod. Phys. Lett. A*, **1**, 81; (1987) *Physica Scripta*, **T15**, 169.

[41] L. Schulman (1986) *J. Stat. Phys.*, **42**, 689.

[42] J.P. Paz and W.H. Zurek (1993) *Phys. Rev.* **D48**, 2728.

[43] M. Gell-Mann and J.B. Hartle (1993) *Phys. Rev.* **D47**, 3345.

[44] J. Finkelstein (1993) *Phys. Rev.* **D47**, 5430.

[45] M. Gell-Mann and J.B. Hartle (to be published).

# 23

## The No Boundary Condition And The Arrow Of Time

Stephen W. Hawking

*Department of Applied Mathematics and Theoretical Physics*
*Silver Street, Cambridge, CB3 9EW, UK*

When I began research, nearly 30 years ago, my supervisor, Dennis Sciama, set me to work on the arrow of time in cosmology. I remember going to the university library in Cambridge, to look for a book called *The Direction of Time*, by the German philosopher, Reichenbach (Reichenbach, 1956). However, I found the book had been taken out by the author, J B Priestly, who was writing a play about time, called Time and the Conways. Thinking that this book would answer all my questions, I filled in a form, to force Priestly to return the book to the library, so I could consult it. However, when I eventually got hold of the book, I was very disappointed. It was rather obscure, and the logic seemed to be circular. It laid great stress on causation, in distinguishing the forward direction of time from the backward direction. But in physics, we believe there are laws that determine the evolution of the universe uniquely. So if state A evolved into state B, one could say that A caused B . But one could equally well look at it in the other direction of time, and say that B caused A. So causality does not define a direction of time.

My supervisor suggested I look at a paper by a Canadian, called Hogarth (Hogarth, 1962). This applied to cosmology a direct action formulation of electro dynamics. It claimed to derive a connection between the expansion of the universe and the electro-magnetic arrow of time. That is, whether one got retarded or advanced solutions of Maxwell's equations. The paper said that one would obtain retarded solutions in a steady state universe, but advanced solutions in a Big Bang universe. This was seized on by Hoyle and Narlikar (Hoyle and Narlikar, 1964), as further evidence, if any were needed, that the steady state theory was correct. However, now that no one except Hoyle believes that the universe is in a steady state, one must conclude that the basic premise of the paper was incorrect.

Shortly after this, there was a meeting on the direction of time, at Cornell in 1963 (Gold, 1967). Among the participants, there was a Mr X, who felt the proceedings were so worthless that he didn't want his name associated with them. It was an open secret that Mr X was Feynman.

Mr X said that the electro-magnetic arrow of time didn't come from an action at a distance formulation of electro-dynamics, but from ordinary statistical mechanics.

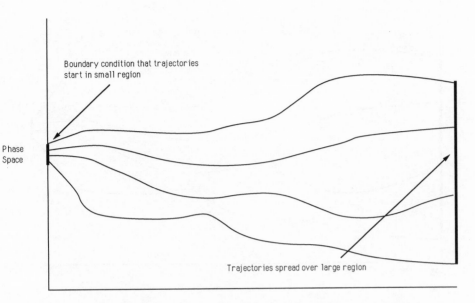

Fig. 23.1. Evolution of a system with an initial boundary condition.

Guided by his comments, I came to the following understanding of the arrow of time. The important point is that the trajectories of a system should have the boundary condition that they are in a small region of phase space at a certain time. In general, the evolution equations of physics will then imply that at other times, the trajectories will be spread out over a much larger region of phase space. Suppose the boundary condition of being in a small region is an initial condition. Then this will mean that the system will begin in an ordered state, and will evolve to a more disordered state (see Figure 23.1). Entropy will increase with time, and the second law of thermodynamics will be satisfied.

On the other hand, suppose the boundary condition of being in a small region of phase space was a final condition, instead of an initial condition. Then at early times, the trajectories would be spread out over a large region, and they would narrow down to a small region as time increased (see Figure 23.2). Thus disorder and enropy would decrease with time, rather than increase. However, any intelligent beings who observed this behavior would also be living in a universe in which entropy decreased with time. We don't know exactly how the human brain works in detail, but we can describe the operation of a computer. One can consider all possible trajectories of a computer interacting with its surroundings. If one imposes a final boundary condition on these trajectories, one can show that the correlation between the computer memory and the surroundings is greater at early times, than at late times. In other words, the computer remembers the future, but not the past.

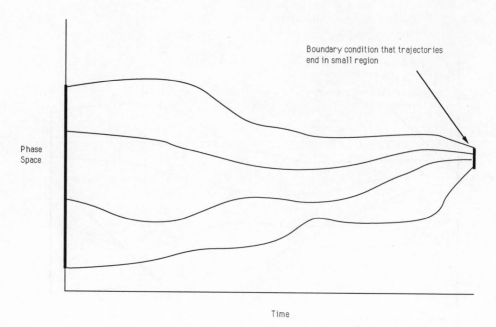

Fig. 23.2. Evolution of a system with a final boundary condition.

Another way of seeing this, is to note that when a computer records something in memory, the total entropy increases. Thus computers remember things in the direction of time in which entropy increases. In a universe in which entropy is decreasing in time, computer memories will work backward. They will remember the future and forget the past.

Although we don't really understand the workings of the brain, it seems reasonable to assume that we remember in the same direction of time that computers do. If it were the opposite direction, one could make a fortune, with a computer that remembered who won tomorrows horse races. This means that the psychological arrow of time, our subjective sense of time, is the same as the thermodynamic arrow of time, the direction in which entropy increases. Thus, in a universe in which entropy was decreasing with time, any intelligent beings would also have a subjective sense of time that was backward. So the second law of thermodynamics is really a tautology. Entropy increases with time, because we define the direction of time to be that in which entropy increases.

There are, however, two non-trivial questions one can ask about the arrow of time. The first is, why should there be a boundary condition at one end of time, but not the other. It might seem more natural to have a boundary condition at both ends of time, or at neither. As I will discuss, the former possibility would mean that the arrow of time would reverse, while in the latter case, there would be no well defined arrow of time. The second question is, given that there is a boundary condition at one end of time, and hence a well defined arrow of time, why should this arrow point in the direction of time in which the universe is expanding. Is there a deep connection, or is it just an accident.

I realized that the problem of the arrow of time, should be formulated in the manner I have described. But at that time, in 1964, I could think of no good reason why there should be a boundary condition at one end of time. I also needed something more definite, and less airy fairy than the arrow of time, for my PhD. I therefore switched to singularities and black holes. They were a lot easier. But I retained an interest in the problem of the direction of time. This surfaced again when Jim Hartle and I (Hartle and Hawking, 1983), formulated the no boundary proposal for the universe. This was the suggestion that the quantum state of the universe was determined by a path integral over positive definite metrics on closed spacetime manifolds. In other words, the boundary condition of the universe was that it had no boundary.

The no boundary condition, determined the quantum state of the universe, and thus what happened in it. It should therefore determine whether there was an arrow of time, and which way it pointed. In the paper that Hartle and I wrote, we applied the no boundary condition to models with a cosmological constant and a conformally invariant scalar field. Neither of these gave a universe like we live in. However, a minisuperspace model with a minimally coupled scalar field gave an inflationary period that could be arbitrarily long (Hawking, 1984). This would be followed by radiation and matter dominated phases, like in the chaotic inflationary model. Thus it seemed that the no boundary condition would account for the observed expansion of the universe. But would it explain the observed arrow of time? In other words, would departures from a homogeneous and isotropic expansion, be small when the universe is small, and grow larger as the universe got bigger? Or would the no boundary condition, predict the opposite behavior? Would the departures be small when the universe was large, and large when the universe was small? In this latter case, disorder would decrease as the universe expanded. This would mean that the thermodynamic arrow, pointed in the opposite way to the cosmological arrow. In other words, people living in such a universe, would say that the universe was contracting, rather than expanding.

To answer the question of what the no boundary proposal predicted for the arrow of time, one needed to understand how perturbations of a Friedmann model would behave. Jonathan Halliwell and I (Halliwell and Hawking, 1984), studied this problem. We expanded perturbations of a minisuperspace model in spherical harmonics, and expanded the Hamiltonian to second order. This gave us a Wheeler-Dewitt equation,

$$[m_p^{-2} G_{ijkl} \frac{\partial^2}{\partial h_{ij} \partial h_{kl}} - m_p^2 h^{\frac{1}{2}3} R] \Psi(h_{ij}) = 0,$$

$$h_{ij} = a^2(\Omega_{ij} + \epsilon_{ij}),$$

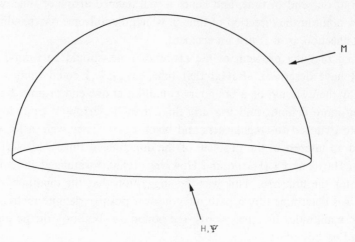

Fig. 23.3. The no boundary condition.

for the wave function of the universe. We solved this, as a background mini-superspace wave function, times wave functions for the perturbation modes. These perturbation mode wave functions obeyed Schrödinger equations, which we could solve approximtely. To obtain the boundary conditions for these Schrödinger equations, we used a semiclassical approximation to the no boundary condition

$$\Psi(h) = \int d[g]e^{-I},$$

(see Figure 23.3).

Consider a three geometry and scalar field that are a small perturbation of a three sphere, and constant field. The wave function at this point in superspace will be given by a path integral over all Euclidean four geometries and scalar fields, that have only that boundary. One would expect the dominant contribution to this path integral to come from a saddle point. That is, a complex solution of the field equations, which has the given geometry and field on one boundary, and which has no other boundary. The wave function for the perturbation mode, will then be exponential of minus the action of the complex solution for the perturbation.

In this way, Halliwell and I calculated the spectrum of pertubations predicted by the no boundary condition. The exact shape of this spectrum doesn't matter for the arrow of time. What is important is that, when the radius of the universe is small, and the saddle point is a complex solution that expands monotonically, the amplitudes of the perturbations are small. This means that the trajectories, corresponding to different probable histories of the universe, are in a small region of phase space when the universe is small. As the universe gets larger, the amplitudes of some of these perturbations will go up. Because the evolution of the universe, is governed by a Hamiltonian, the volume of phase space remains unchanged. Thus while the perturbations are linear, the region of phase space that the trajectories are in will change shape only by some matrix of determinant one. In other words, an initially spherical region, will evolve to an ellipsoidal region of the same volume. Eventually however, some of the perturbations can grow so large, that they become non-linear. The volume of phase space is still left unchanged by the evolution, but in general, the initially spherical region will be deformed into long thin filaments. These can spread out, and occupy a large region of phase space. Thus one gets an arrow of time. The universe is nearly homogeneous and isotropic when it is small. But it is more irregular when it is large. In other words, disorder increases as the universe expands. So the thermodynamic and cosmological arrows of time agree, and people living in the universe will say it is expanding, rather than contracting.

I wrote a paper (Hawking, 1985), in which I pointed out that these results about perturbations would explain both why there was a thermodynamic arrow, and why it should agree with the cosmological arrow. But I made what I now realize was a great mistake. I thought that the no boundary condition, would imply that the perturbations would be small whenever the radius of the universe was small. That is, the perturbations would be small, not only in the early stages of the expansion, but also in the late stages of a universe that collapsed again. This would mean that the trajectories of the system would be that subset that lies in a small region of phase space, at both the beginning, and the end of time. But they would spread out over a much larger region at times in between. This would mean that disorder would increase during the expansion, but decrease again during the contraction (see Figure 23.4). So the thermodynamic arrow would point forward in the expansion phase, and backward in the contracting phase. In other words, the thermodynamic and cosmological arrows would agree in both expanding and contracting phases. Near the time of maximum expansion, the entropy of the universe would be a maximum. This would mean that an intelligent being, who continued from the expanding to the contracting phase, would not observe the arrow of time pointing backward. Instead, his subjective sense of time would be in the opposite direction in the contracting phase. So he would not remember that he had come from the expanding phase, because that would be in his subjective future.

If the thermodynamic arrow of time were to reverse in a contracting phase of the universe, one might also expect it to reverse in gravitational collapse to form a black hole. This would raise the possibility of an experimental test of the no boundary condition. If the reversal took place only inside the horizon, it would not be much use, because someone that observed it could not tell the rest of us. But one might hope that there would be slight effects, that could be detected outside the horizon.

The idea that the arrow of time would reverse in the contracting phase had a satisfying ring to it. But shortly after having my paper accepted by the Physical review, discussions with Raymond Laflamme and Don Page, convinced me that the prediction of reversal was wrong. I added a note to the proofs, saying that entropy would continue to increase, during the contraction, but I fell ill with pneumonia,

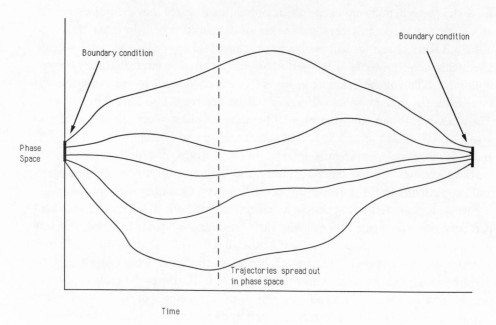

Fig. 23.4. Evolution of a system with initial and final boundary conditions.

before I could write a paper to explain it properly. So I want to take this opportunity to show how I went wrong, and what the correct result is.

One reason I made my mistake, was that I was misled by computer solutions, of the Wheeler Dewitt equation, for a minisuperspace model of the universe (Hawking and Wu, 1985). In these solutions, the wave function didn't oscillate in a so-called "forbidden region", at very small radius. I now realize that these computer solutions (see Figure 23.5), had the wrong boundary conditions. But at the time, I interpreted them as indicating that the Lorentzian four geometries, that corresponded to the WKB approximation, didn't collapse to zero radius. Instead, I thought they would bounce and expand again (see Figure 23.6). My feelings were strengthened when I found that there was a class of classical solutions that oscillated. The computer calculations of the wave function seemed to correspond to a superposition of these solutions. The oscillating solutions were quasi periodic. So it seemed natural to suppose that the boundary conditions on the perturbations should be that they were small whenever the radius was small. This would have led to an arrow of time, that pointed forward in the expanding phase, and backward in the contracting phase, as I have explained.

I set my research student, Raymond Laflamme, to work on the arrow of time in more general situations than a homogeneous and isotropic Friedmann background. He soon found a major objection to my ideas. Only a few solutions, like the spherically symmetric Friedmann models, can bounce when they collapse. Thus the

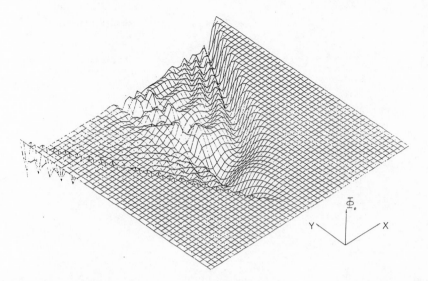

Fig. 23.5. The wavefunctions for a homogeneous, isotropic universe with a scalar field. The wavefunction does not oscillate the lines $y = |x|$.

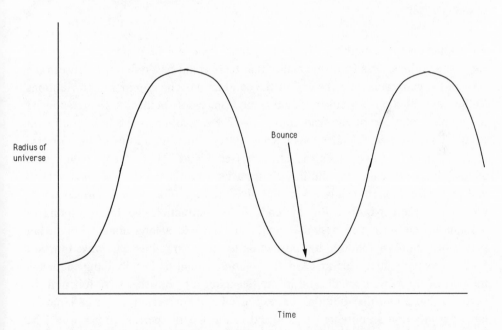

Fig. 23.6. A quasi-periodic solution for a Friedmann universe filled with a massive scalar field.

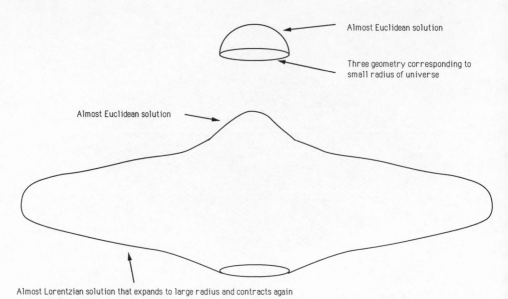

Almost Euclidean solution

Three geometry corresponding to
small radius of universe

Almost Euclidean solution

Almost Lorentzian solution that expands to large radius and contracts again

Fig. 23.7. Two possible saddle points in the path integral for the wavefunction of a given radius.

wave function for something like a black hole could not be concentrated on non singular solutions. This made me realize that there could be a difference between the start of the expansion, and the end of the contraction. The dominant contributions to the wave functions for either, would come from saddle points, that corresponded to complex solutions of the field equations. These solutions have been studied in detail by my student, Glenn Lyons (Lyons, 1992). When the radius of the universe is small, there are two kinds of solutions (see Figure 23.7). One would be an almost Euclidean complex solution, that started like the north pole of a sphere, and expanded monotonically up to the given radius. This would correspond to the start of the expansion. But the end of the contraction, would correspond to a solution that started in a similar way, but then had a long almost Lorentzian period of expansion, followed by contraction to the given radius. The wave function for perturbations about the first kind of solution, would be heavily damped, unless the perturbations were small, and in the linear regime. But the wave function for perturbations about the solution, that expanded and contracted, could be large for large perturbation amplitudes. This would mean that the perturbations, would be small at one end of time, but could be large and non-linear, at the other end. So disorder and irregularity would increase during the expansion, and would continue to increase during the contraction. There would be no reversal of the arrow of time, at the point of maximum expansion.

Hawking, Laflamme and Lyons (1993) have studied how the arrow of time manifests itself in the various perturbation modes. It makes sense to talk about the arrow of time only for modes that are shorter than the horizon scale, at the time concerned. Modes that are longer than the horizon just appear as a homogeneous background. There are two kinds of behavior for perturbation modes within the horizon. They can oscillate, or they can have power law growth or decay. Modes that oscillate are the tensor modes, that correspond to gravitational waves, and scalar modes, that correspond to density perturbations, of wavelength less than the Jeans length. On the other hand, density perturbations longer than the Jeans length, have power law growth and decay.

Perturbation modes that oscillate, will have an amplitude that varies adiabatically, as an inverse power of the radius of the universe,

$$Aa^p = constant,$$

where $A$ is the amplitude of the oscillating perturbation, $a$ is the radius of the universe and $p$ is some positive number. This means they will be essentially time symmetric, about the time of maximum expansion. In other words, the amplitude of the perturbation, will be the same at a given radius during the expansion, as at the same radius during the contracting phase. So if they are small when they come within the horizon during expansion, which is what the no boundary condition predicts, they will remain small at all times. They will not become non-linear, and they will not show an arrow of time.

By contrast, density perturbations on scales longer than the Jeans length, will grow in amplitude in general,

$$A = Ba^p + Ca^{-q},$$

where $p$ and $q$ are positive. They will be small, when they come within the horizon during the expansion. But they will grow during the expansion, and continue to grow during the contraction. Eventually, they will become non-linear. At this stage, the trajectories will spread out over a large region of phase space.

So the no boundary condition, predicts that the universe is in a smooth and ordered state, at one end of time. But irregularities increase, while the universe expands and contracts again. These irregularities lead to the formation of stars and galaxies, and hence to the development of intelligent life. This life will have a subjective sense of time, or psychological arrow, that points in the direction of increasing disorder. The one remaining question is why this psychological arrow should agree with the cosmological arrow. In other words, why do we say the universe is expanding, rather than contracting. The answer to this, comes from inflation, combined with the weak anthropic principle. If the universe had started to contract a few billion years ago, we would indeed observe it to be contracting. But inflation implies that the universe should be so near the critical density, that it will not stop expanding, for much longer than the present age. By that time, all the stars will have burnt out. The universe will be a cold dark place, and any life will have died out long before. Thus the fact that we are around to observe the universe, means that we must be in the expanding, rather than the contracting phase. This is the explanation why the psychological arrow agrees with the cosmological arrow.

So far I have been talking about the arrow of time, on a macroscopic, fluid dynamical scale. But the inflationary model, depends on the existence of an arrow of time on a much smaller microscopic scale. During the inflationary phase, practically the entire energy content of the universe is in the single homogeneous mode of a scalar field. The amplitude of this mode changes only slowly with time, and its energy momentum tensor causes the universe to expand in an accelerating, exponential way. At the end of the inflationary period, the amplitude of the homogeneous mode begins to oscillate. The idea is that these coherent homogeneous oscillations of the scalar field cause the creation of short wavelength particles, of other fields, with a roughly thermal spectrum. The universe expands thereafter, like the hot big bang model. This inflationary scenario implicitly assumes the existence of a thermodynamic arrow of time that points in the direction of the expansion. It wouldn't work if the arrow of time had been in the opposite direction. Normally, people brush the assumption of an arrow of time under the carpet. But in this case, one can show that this microscopic arrow also seems to follow from the no boundary condition. One can introduce extra matter fields, coupled to the scalar field. If one expands them in spherical harmonics, one obtains a set of Schrödinger equations, with oscillating coefficients. The no boundary condition tells you that the matter fields start in their ground state. One then finds that the matter fields become excited, when the scalar field begins to oscillate. Presumably, the back reaction will damp the oscillations of the scalar field, and the universe will go over to a radiation dominated phase. Thus, the no boundary proposal seems to explain the arrow of time, on microscopic, as well as on macroscopic scales.

I have told you how I came to the wrong conclusion, and what I now think is the correct result, about what the no boundary condition predicts for the arrow of time. This was my greatest mistake, or at least my greatest mistake in science. I once thought there ought to be a journal of recantations, in which scientists could admit their mistakes. But it might not have many contributors.

## Discussion

**Bennett**  When entropy began increasing rapidly again, near the end of the contraction, wouldn't there be a possibility for a new origin of life, even though it had died out from the expansion phase?

**Page**  As you know, both you and I [Phys Rev D32, 2496(1985)] independently proposed this anthropic explanation of why we see the universe expanding in the same direction as the thermodynamic arrow. But since then I have had second thoughts and wondered whether, as Dyson and Tipler have proposed, that life might take other forms and survive long after the death of stars. Do you believe life can exist in the collapsing phase of the universe?

**Hawking**  You and I wrote a paper, in which we claimed the no boundary condition would imply the universe expanded for an almost infinite time, before contracting. I don't think even Dyson could keep life going that long.

I don't think life would develop in the contracting phase, because all the baryons would have decayed, and the universe would be filled with photons and neutrinos. It is an interesting question whether you can have a life form made out of them. But if it were possible, why didn't it develop in the expanding phase?

**Gell-Mann**  There was an amusing incident which I once recounted to you, Stephen, in Pasadena.

In 1963-1964, I thought about the derivation of the second law of themodynamics from an initial density matrix of the universe that would lead to expansion. In 1966, at a meeting in Philadelphia organised by Melvin, I asked the assembled physicists and astronomers whether, in the contracting phase of the universe (if there is one), the same condition could apply, in that time would be reversed during the contraction. Wigner got up and said, "You are a – what do you call it? – a heretic!"

I don't think my remarks or his appear in the proceedings.

## References

Gold, T. (1967) *The Nature of Time*, Cornell University Press, New York.

Halliwell, J.J. and Hawking, S.W. (1984) The Origin of Structure in the Universe. *Physical Review* **D31**, 1777.

Hartle, J.B. and Hawking, S.W. (1983) Wave Function Of The Universe. *Physical Review* **D28**, 2960-2975.

Hawking, S.W. (1984) The Quantum State Of The Universe. *Nuclear Physics,* **B239**, 257.

Hawking, S.W. (1985) The Arrow Of Time In Cosmology. *Physical Review* **D32**, 2489.

Hawking, S.W., Laflamme, R. and Lyons, G.W. (1993) The origin of Time Assymetry. *Physical Review* **D47**, 5342.

Hawking, S.W. and Wu, Z.C. (1985) Numerical Calculations Of Minisuperspace Cosmological Models. *Physics Letters* **B151**, 15.

Hogarth, J.E. (1962) Cosmological Considerations of the Absorber Theory of Radiation. *Proceedings of the Royal Society* **A267**, 365.

Hoyle, F. and Narlikar, J.V. (1964) Time Symmetric Electrodynamics and the Arrow of Time in Cosmology. *Proceedings Of The Royal Society* **A273**, 1.

Lyons, G.W. (1992) Complex Solutions for the Scalar Field Model of the Universe. *Phys. Rev.* **D46**, 1546.

Page, D.N. (1985) Will Entropy Decrease if the Universe Collapses? *Physical Review* **D32**, 2496-2499.

Reichenbach, H.(1956) *The Direction Of Time*, Berkeley: University Of California Press.

# 24

# The Arrow of Time and the No-Boundary Proposal

R. Laflamme†

*Department of Applied Mathematics and Theoretical Physics*
*University of Cambridge*
*Silver Street, Cambridge*
*CB3 9EW, U.K.*

*Peterhouse, Cambridge*
*CB2 1RD, U.K.*

**Abstract**

The consequences of the no-boundary proposal for the arrow of time are studied. A model corresponding to the interior of a black hole and a FRW model with small inhomogeneities are investigated. We show that in both cases there exists a well-defined arrow of time associated with the Weyl tensor or the evolution of inhomogeneities does but that it does not always point in the same direction as the cosmological arrow of time.

## 24.1 Introduction

A physical correlate for our subjective experience of directed temporality can be found in the second law of thermodynamics, the observed increase of entropy. Since this correlate apparently cannot be derived from known physical laws, because they are time-reversal invariant, some have sought to explain its origin in a low entropy initial state of the universe (Popper (1956), Gold (1967)). In this manner, the second law becomes a selection principle for the boundary conditions of the universe.

Recently Hartle and Hawking have proposed a boundary condition for the wave function of the universe. In principle, the evolution of all fields in the universe should be derived from this single wave function. It should therefore explain the second law of thermodynamics.

Hawking has investigated the no-boundary proposal for the FRW model with a homogeneous massive scalar field $\varphi$ (Hawking 1983). He calculated the wave function using the saddle-point approximation to the path integral assuming that the the dominant contribution comes from solutions of the classical equations. The no-boundary proposal implies that the 3-geometries which close off the 4-geometries must have for regularity the following boundary conditions

$$a = 0; \quad \varphi = \varphi_0; \quad \dot{a} = 1; \quad \dot{\varphi} = 0 \tag{24.1}$$

† Present address: T-6, Theoretical Astrophysics, MS B288, Los Alamos National Laboratory, Los Alamos, NM 87545, USA. E-mail: laf@tdo-serv.lanl.gov.

$a$ is the radius of the 3-sphere. Hawking and Wu (1985) calculated the semi-classical wave function at small $a$ coming from parts of 4-sphere smaller than the equator. They used this to integrate the Wheeler-DeWitt equation and find the wave function on the whole of superspace. In the region with $a > 1/m|\varphi|$, the wave function oscillates and is thus of the WKB form. After a suitable coarse graining (see Habib & Laflamme (1990)), we can associate the phase of the wave function to the Hamilton-Jacobi function of classical general relativity. Thus it corresponds to a family of classical lorentzian trajectories which start with the intial conditions

$$a \approx 1/m|\varphi_0|; \quad \varphi \approx \varphi_0; \quad \dot{a} \approx 0; \quad \dot{\varphi} \approx 0 \qquad (24.2)$$

They have an inflationary period (for $\varphi_0 > 1$) until the scalar field oscillates. When it does so rapidly the universe behaves as a FRW with zero pressure perfect fluid. It expands to a maximum radius of order $a_m \approx \exp(9\varphi_0/2)$ and recollapses.

Hawking (1985) had initially suggested that the no-boundary proposal was picking out only trajectories which are symmetric with respect to the time of maximum expansion so that the classical solutions would return to the same boundary conditions, Eq. (24.2), at the end of the recollapsing phase. This would have implied that the arrow of time would have to reverse at the time of maximum expansion. The reversal of the arrow of time at the time of maximum expansion seems a rather academic question however it would suggest that something unusual should happen around black holes. When the Universe recontracts, it will, in its late stage, behave like the interior of a giant black hole. The no-boundary proposal might therefore have unusual predictions for fields around black holes. It is thus important to investigate the no-boundary proposal for this latter case.

In Section 2 we study the no-boundary proposal for 3-geometries with a metric which correspond to the interior of a black hole. We will show that in this case classical solutions corresponding to the no-boundary proposal are not symmetric with respect to the time of maximum expansion. Thus the background geometry provides a definite arrow of time (the cosmological arrow time). If we define an arrow of time through the Weyl tensor as proposed by Penrose, we can show that this arrow does not always point in the same direction as the cosmological one. In Section 3 we re-examine the homogeneous model of Hawking by looking at the behavior of small inhomogenous pertubations. We show that we can associate an arrow of time with them and that this arrow does not always point in the same direction as the cosmological one.

## 24.2 The $S^1 \times S^2$ model with a scalar field

Let's first investigate the case of a a black hole interior. The idea is to study a simple model for the Universe which has slightly more structure than the isotropic one discussed in the introduction. In particular I will focus on the model with homogenous 3-surfaces with topology $S^1 \times S^2$ and metric of the form

$$ds^2 = -N^2(t)dt^2 + c^2(t)dr^2 + b^2(t)d\Omega_2^2 \tag{24.3}$$

where the radial coordinate $r$ is assumed to be periodic and $d\Omega_2^2$ is the metric on the 2-sphere. The matter part will be taken to be a homogeneous massive scalar field minimally field $\varphi(t)$.

Using the Hilbert-Einstein action

$$I_g = \frac{1}{16\pi G} \int_{\mathcal{M}} d^4x \, (-g)^{1/2} R \; + \; \frac{1}{8\pi G} \int_{\partial\mathcal{M}} d^3x \, (h)^{1/2} K \tag{24.4}$$

and the massive scalar field action

$$I_\Phi = -1/2 \int_{\mathcal{M}} d^4x \, (-g)^{1/2} \, (g^{\mu\nu}\partial_\mu\phi\partial_\nu\phi + m^2\phi^2) \tag{24.5}$$

we can deduce the Wheeler-DeWitt equation

$$\left[ \frac{c}{2b^2}\frac{\partial^2}{\partial c^2} + \frac{1}{cb}\frac{\partial}{\partial b} - \frac{1}{b}\frac{\partial^2}{\partial c\partial b} + \frac{1}{cb^2}\frac{\partial^2}{\partial\varphi^2} + V(c,b,\varphi) \right] \Psi(c,b,\varphi) = 0. \tag{24.6}$$

for $V(c,b,\phi) = c/2 - cb^2m^2\varphi^2/2$. Note that the momenta constraints are trivially satisfied for this metric.

It is possible to calculate the wave function using the saddle point approximation for the no-boundary proposal. It implies that the 3-surface must be closed by a complex geometry with the scale factors $c$ or $b$ going to zero. The saddle points correspond to 3-geometries closed off by a 4-geometry with topology of a disc times a 2-sphere or a circle times a 3-sphere. In this essay we give the result for the latter case only (see Laflamme (1988) for the other case). At the point of zero volume the scale factors and scalar field must have the following behavior (for N=1)

$$c = c_0; \quad b = 0; \quad \varphi = \varphi_0; \quad \dot{c} = 0; \quad \dot{b} = 1; \quad \dot{\varphi} = 0. \tag{24.7}$$

The action for small volume (but large $\varphi_0$) can be written as

$$I_E \approx -cb(1 - \frac{m^2\varphi^2}{3}b^2)^{1/2}. \tag{24.8}$$

For large $\varphi_0$ the wave function will start oscillating at $b > 3/m|\varphi_0|$. As mentioned above we can after suitable coarse graining associate classical trajectories to an oscillating wave function if we identify its phase to the Hamilton-Jacobi function. In the no-boundary case with the above metric the Hamilton-Jacobi function corresponds to classical trajectories with initial conditions (for N=1)

$$c \approx 0; \quad b \approx \frac{\sqrt{3}}{m|\varphi|}; \quad \varphi \approx \varphi_0; \quad \dot{c} = \dot{c}_0 > 0; \quad \dot{b} \approx 0; \quad \dot{\varphi} \approx 0. \tag{24.9}$$

The classical lorentzian equation of motion are

$$b\ddot{b} + \frac{\dot{b}^2}{2} + \frac{1}{2} + \frac{b^2\dot{\varphi}}{2} - \frac{b^2m^2\varphi^2}{2} = 0 \tag{24.10a}$$

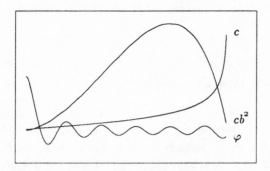

Fig. 24.1. Classical trajectories for the radius $c$, the volume $cb^2$ and the scalar field $\varphi$. The universe starts with an inflationary phase followed by matter dominated (dust) era and recollapse toward a singularity. The trajectories for $c$ is not symmetric with respect to the time of maximum expansion.

$$\ddot{c}b + b\ddot{c} + \dot{c}\dot{b} + cb\dot{\varphi}^2 - cbm^2\varphi^2 = 0 \qquad (24.10b)$$

$$\ddot{\varphi} + (\frac{\dot{c}}{c} + \frac{2\dot{b}}{b})\dot{\varphi} + m^2\varphi = 0. \qquad (24.10c)$$

and the constraint

$$b\dot{b}\dot{c} + \frac{cb^2}{2} + \frac{c}{2} - \frac{cb^2\dot{\varphi}^2}{2} - \frac{cb^2m^2\varphi^2}{2} = 0. \qquad (24.10d)$$

Typically the Universe begins in a de Sitter phase and expand exponentially until the scalar field starts to oscillate. When the oscillations become rapid the matter behaves like dust (the effective pressure is zero). The universe will then be similar to a dust-filled Kantowski-Sachs Universe, somewhat like a FRW solution, until it reaches a maximum 3-volume and starts to recollapse. As the 3-volume decreases to zero size, the universe behaves more and more like the 3-geometry of the interior of a black hole. The degree of inflation and the mass parameter of the black hole are a function of the intial value of the scalar field $\varphi_0$. A typical trajectory is shown in Figure 24.1.

It is important to notice that the trajectory for $c$ in Figure 24.1 is not symmetric with respect to the time of maximum volume. It increases both in the expanding phase and in the contracting one (the notion of expanding and contracting is not fundamental but merely a convenient way to distinguish the two different 3-geometries with same volume in Figure 24.1). In fact it is possible to show that no classical solutions with initial conditions (24.9) are symmetric with respect to the time of maximum expansion. This is more obvious using the coordinates $p, q$ and lapse $\tilde{N}$ defined as

$$c = e^p; \quad b = e^{-p+q}; \quad \tilde{N} = e^{-p+2q}. \qquad (24.11a,b,c)$$

The equations of motion for these variables in the Hamiltonian form reduces to

$$p' = \pi_p; \quad q' = -\pi_q; \quad \varphi' = -\pi_\varphi; \tag{24.12a,b,c}$$

$$\pi_p' = e^{-2p+4q}; \quad \pi_q' = e^{2q} - 2m^2\varphi^2 e^{-2p+4q}; \quad \pi_\varphi' = -m^2\varphi e^{-2p+4q} \tag{24.12d,e,f}$$

where prime denotes $d/\tilde{N}dt$. The initial conditions (24.9) requires that $p' > 0$, so (24.12a) and (24.12d) imply that $p$ must be monotonically increasing, because its first and second derivatives are initially positive. For large $\varphi_0$ we have $2p' \approx q'$. When the volume of the universe is large ($-p + 2q > 0$) and the $\varphi$ field oscillates rapidly, $q'$ will lag behind $2p'$ because of the first term on the right hand side of (24.12e). This term reverses the sign of $\pi_q'$ and the universe will recontract. Since $p$ is monotonically increasing, the first term on the right hand side of (24.12e) will dominate and the universe will then inexorably go to the 'cigar' singularity at $q = -\infty, (b = 0)$. An extensive numerical search for solutions which bounces at small volume did not produce any.

It is interesting to notice that no solution with the boundary conditions (24.9) do not deflate. This seems to question the hypothesis that inflation is a generic phenomena. Here with these very special boundary conditions we have that one part of the solution does have inflation and the other part not. So we would have for a given volume $cb^2$, 50% chance on being on the inflationary side, but if we use more general initial conditions then there are certainly solutions which do not have inflationary behavior on either side, thus reducing the above chance probability.

From the behavior of the different fields we can define an arrow of time which does not always point in the same direction as the cosmological arrow. For example the increase of the scale factor $c$ is monotonic along the trajectories anddoes define a definite arrow of time which does not always agree with the cosmological one (define by the behavior of the volume $cb^2$). We could also study the shear (which is a measure of the anisotropy) defined as

$$\sigma = \frac{1}{\sqrt{3}}\left(\frac{\dot{c}}{c} - \frac{\dot{b}}{b}\right) \tag{24.13}$$

and has a similar behavior to $c$. Another possibility is to study the Weyl tensor

$$W^{tr}{}_{tr} = \frac{1}{3}\left(\frac{\ddot{c}}{c} - \frac{\ddot{b}}{b} - \frac{\dot{c}\dot{b}}{cb} + \frac{\dot{b}^2 + 1}{b^2}\right), \tag{24.14}$$

with all other non-zero component being proportional to the above one. It was conjectured by Penrose (1979) that it might be related in a loose sense to gravitaitonal entropy. Again the Weyl tensor is asymmetric with respect to the maximum expansion. In fact the no-boundary proposal seems to incorporate Penrose's Weyl curvature hypothesis, one end of the classical trajectories has small value of Weyl tensor but not the other one.

Thus we can conclude that in this model it is impossible to have a global arrow of time which reverses at the time of maximum expansion, essentially because there do not exist any time-symmetric solutions (there are time-symmetric solutions but they do not correspond to the no-boundary proposal).

We have seen in that in the $S^1 \times S^2$ model the arrow of time coming from the radius $a$ does not reverse at the time of maximum expansion. It is of interest to see if this would also be applicable in more general models. It would be of interest to do an analysis similar to the one above using Bianchi IX following the work of Amsterdamski (1985) and study the behavior of the parameters $\beta_\pm$ in the dust phase having started with the no-boundary proposal. We conjecture that a behavior analogous to the one described above would happen.

## 24.3  The Local Arrow of Time

In the previous section we have seen in an homogeneous model that the no-boundary proposal picks up classical trajectories which are not symmetric with respect to the time of maximum expansion. Of more importance to the world around us is the local arrow of time determined by inhomogeneous perturbations. A fundamental puzzle about the arrow of time is that small subsystems in the universe seem to have all their arrow pointing in the same direction. In this section we turn to the implications of the no-boundary proposal for the local arrow of time.

It possible to gain information about this arrow by studying small perturbations around the homogenenous models described above. Halliwell and Hawking (1985) have described the wave function resulting from the no-boundary proposal for the FRW model with small inhomogeneous perturbations. They showed that the no-boundary proposal implies that inhomogeneous perturbations start in their ground state and lead to an almost scale free spectrum of density perturbations which are responsible for the structure of the universe.

It is possible to go further and study these pertubations in the matter dominated phase of the Universe assuming they remain linear. Halliwell and Hawking have expanded the Einstein-Hilbert action in terms of scalar, vector and tensor (with repect to rotation of the 3-sphere) harmonics around an homogenous FRW model with massive scalar field.

The tensor harmonics correspond to gravitons. They are decoupled from the scalar field. They essentially behave like an inhomogeneous massless scalar field. In the dust phase, they will oscillate with amplitude proportional to $1/a$ ($a$ being the scale factor of the universe. Their amplitude will be symmetric for most of the history of the Universe. However it is possible to show that at the end of the contracting phase they will not return to their ground state. This can be seen by showing that at the end of the inflationary phase, the gravitons outside the Hubble radius are not in the symmetric state of the dust phase. The behavior of gravitons is illlustrated in Figure 24.2.

The vector harmonics are pure gauge and thus not real degrees of freedom. It is

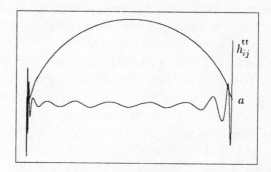

Fig. 24.2. Classical trajectories for the radius of the universe $a$ in the FRW model with scalar field. The amplitude of the graviton $h_{ij}^{TT}$ (the $n = 10$ mode) is time symmetric with respect of the maximum up to very small radius $a$.

possible to satisify the vectorial part of the momenta constraints by redefining the background variables (Shirai & Wada (1988)).

The remaining harmonics, the scalar ones, are the one of most interest. After solving the momenta constraints and the linearized Wheeler-DeWitt only one scalar degree of freedom is left at every point of the 3-surface. It is possible to show that the linear combination of Fourier modes $n$

$$z_n = \pi_\alpha f_n + \pi_\varphi(a_n + b_n) \tag{24.15}$$

are the real unconstrained degrees of freedom (it is a slight modification of Wada and Shirai's result so that it is well behaved at maximum expansion). $f_n, a_n$ and $b_n$ are the Fourier transform of the scalar field perturbations, the trace and traceless part of the scalar derived hamonics of the 3-metric perturbations. These are the degrees of freedom that should be considered however their equation of motion are rather complicated and it is easier and more intuitive to study the gauge invariant (with respect to linearised gauge transformation) variables (see the review of Mukhanov & al. (1990)). The gauge invariant scalar field perturbation and the gravitational potential are defined as

$$\delta\varphi_n^{gi} = (f_n + \varphi'\frac{(k_n - 3b_n)}{(n^2 - 1)})/\sqrt{6} \tag{24.16}$$

$$\Phi_n = (ag_n + [\frac{a(k_n - 3b_n')}{(n^2 - 1)}]')/a\sqrt{6} \tag{24.17}$$

and obey the coupled differential equation

$$\delta\ddot{\varphi}_n + 3H\delta\dot{\varphi}_n + [\frac{(n^2 - 1)}{a^2} + m^2 - 12\dot{\varphi}^2]\delta\varphi_n = -(4H\dot{\varphi} + 2m^2\varphi)\Phi_n \tag{24.18}$$

$$\ddot{\Phi}_n + 7H\dot{\Phi}_n + [\frac{(n^2 - 9)}{a^2} + 6m^2\varphi^2]\Phi_n = -6m^2\varphi\delta\varphi_n \tag{24.19}$$

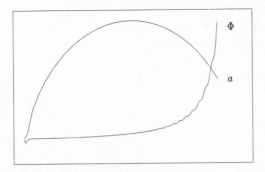

Fig. 24.3. Classical trajectories for the radius of the universe $a$ and the $n = 10$ mode of the gravitational potential $\Phi$. The latter defines an arrow of time which is not in the same direction as the cosmological arrow.

with the constraint

$$H\Phi_n + \dot{\Phi}_n = 3\dot{\varphi}\delta\varphi_n \qquad (24.20)$$

$H = \dot{a}/a$. It is possible to show that $z_n = \pi_\alpha\delta\varphi_n - \pi_\varphi\Phi_n$. The no boundary proposal implies that the perturbations are small at the onset of inflation. After a few e-fold of the radius of the Universe $a$, it is possible to use a $K = 0$ approximation (negligible spatial curvature) to find solutions to Eqs. (24.18)-(24.20). In the inflationary phase, modes outside the Hubble radius behave as $\delta\phi \approx A\dot{\varphi}$ and $\Phi \approx A$. During the inflationary phase the background scalar field $\varphi$ decreases and the universe turns into a dust FRW model when the scalar field starts oscillating. In the early dust phase, when the K=0 approximation is still reasonable, $\delta\phi \propto \dot{\varphi}a^{2/3}$. It is then possible to solve Eqs. (24.18), (24.19) by averaging the constraint (24.20) over a period of $2\pi/m$ to obtain

$$\langle\Phi_n\rangle \approx \frac{(2\sin^2\eta_d - 6\eta_d\sin\eta_d - 8\cos\eta_d + 8)}{(1 - \cos\eta_d)} \qquad (24.21)$$

where $\eta_d$ is the conformal time defined through the scale factor $a = a_m\sin^2(\eta_d)$. This function does not have any particular symmetry with respect to the time of maximum expansion ($a = a_m$). The behavior of the gravitional potential is illustrated in Figure 24.3. Details can be found in Hawking et al. (1992).

As in the $S^1 \times S^2$ there are fields (the inhomogeneous modes here) which give a well defined arrow of time which agrees with the cosmological arrow in the expanding phase but not in the contracting one.

### 24.4 Conclusion

The two models described in Sections 2 and 3 show that the no-boundary proposal incorporate a well defined arrow of time. It can be associated with either the Weyl

tensor or to the behavior of the small scale inhomogeneity of the universe. This
arrow of time agree with the cosmological arrow in the expanding phase of the
universe but not in the recollapsing one.

I have not however introduced entropy. The wave function is a pure state and
remains a pure state during evolution thus the entropy related to this state remains
zero. However in the no-boundary proposal the small perturbation are in their
ground state at the onset of inflation. Modes which come back inside the Hubble
radius in the matter dominated will be highly excited. Some of then will become
non-linear. The state of such modes if observations are restricted to a specific scale
would would become mixed and thus introduce entropy. The increase of entropy is
therefore a consequence of the specific initial condition of the universe.

## Acknowledgements

I would like to thank Stephen Hawking and Glenn Lyons for numerous conver-
sations and express my gratitude to Jonathan Halliwell, Juan-Perez Mercadez and
Wojtek Zurek, the organisers of the NATO workshop on "Physical Origins of Time
Asymmetry". I also acknowledge Peterhouse College (Cambridge) for support.

## Discussion

**Barbour**   I am not quite clear how you conceive time. From the way that Stephen Hawking
and you spoke, it seems to me that you have two basic elements in your scheme: superspace
and time. For you seem to speak of curves moving through time. Is it correct that time is
a fundamental independent element of your scheme?

**Laflamme**   What I mean by time is a relation between the different fields. Here for example,
the value of the volume of the universe could locally be thought as time and we look at
the behavior of the other fields or quantities as function of it. What you obtain from the
wave function of the universe is a set of 4-geometries. They are trajectories in superspace.
Time is a parameter along these trajectories. If you are asking why do we move along these
trajectories it is a question which I have not addressed and in fact has not been addressed
in this meeting so far, except possibly by Wheeler.

**Kuchař**   Is there any justification for taking the homogeneous mode of the universe to
be in a single quasiclassical state $\Psi \approx \exp(iS)$ rather than in the superposition $\Psi \approx \exp(iS) + \exp(-iS)$?

**Laflamme**   As soon as the wave funcion starts to oscillates rapidly, the interference term
between $\exp(iS)$ and $\exp(-iS)$ appears only on very short scale, related to their de Broglie
wavelength. Once you average over larger scales as suggested by Habib and myself then
you can neglect the interference. During the inflationary period this scale is very small.
Halliwell as also shown that this interference can be destroyed by summing over small
inhomogeneous perturbations.

**Kuchař**   Concerning Halliwell's argument that $\exp iS$ and $\exp -iS$ decohere: Cannot the same
argument be applied to any mode when one takes all the other modes as the environment?

If so, every mode would decohere. I thus find this justification of taking exp$iS$ instead of exp$iS$ + exp $- iS$ unconvincing.

**Laflamme**  In the example that Halliwell has worked out the environment were perturbations expanded to quadratic order and the different inhomogeneous modes do not interact with each other and thus would not decohere. If they become non-linear then they would probably decohere using his definition of decoherence.

**Starobinsky**  The problem with inflationary regimes in the Kantowski-Sachs metric with a massive field is no worse than in the closed Friedman world. In the former case, there still exists a set of trajectories of non-zero measure in phase space for which inflation is a transient attractor. These trajectories emerge from singularities.

**Laflamme**  What I have pointed out about the Kantowski-Sachs model is that with the very special boundary conditions corresponding to the no-boundary proposal there is one side of the trajectory which inflates and the other does not. I have the impression that this is generic for solutions which have inflation, when they will recollapse they will not deflate rather they will recollapse as dust models. Thus it seems that not more than 50% of spacetimes with a fixed 3-volume would inflate. To be more precise I would have to introduce a measure. However there doesn't seem to be agreement on which one is preferred. I am not denying that there exist trajectories which do inflate, I am only saying that they might not be generic.

# References

Amsterdamski P. (1985) Wave function of an anisotropic universe. *Physical Review*, **D31**, 3073.

Gold, T. (1967) *The Nature of Time*, Cornell University, Ithaca.

Habib, S. & Laflamme,R. (1990) Wigner Function and Decoherence in Quantum Cosmology. *Physical Review*, **D42**, 4056.

Halliwell J.J. & Hawking, S.W. (1985) Origin of structure in the universe. *Physical Review*, **D31**, 1777.

Hartle, J.B. & Hawking, S.W. (1983) Wavefunction of the universe. *Physical Review*, **D28**, 2960.

Hawking, S.W. (1984) The quantum state of the universe. *Nuclear Physics*, **B239**, 257.

Hawking, S.W. & Wu (1985). Numerical calculations of minisuperspace cosmological models. *Physics Letters*, **B107**, 15.

Hawking, S.W. (1985) Arrow of time in cosmology, *Physical Review*, **D32**, 2489.

Hawking, S.W., Laflamme, R. & Lyons, G. (1993) Origin of time asymmetry. *Physical Review* **D47**, 5342.

Laflamme, R. (1988) Time and quantum cosmology, Cambridge University Ph.D. thesis.

Laflamme, R. & Shellard, E.P.S. (1987) Quantum cosmology and recollapse. *Physical Review*, **D35**, 2315.

Lyons, G. (1992) Complex solutions for the scalar field model of the universe. *Physical Review* **D46**, 1546.

Mukhanov, V.F., Feldman, H.A. & Brandenberger, R.H. (1991) Theory of cosmological perturbations, Brown University report, Brown-HET-796.

Page, D. (1985) Will the entropy decrease if the Universe recollapse? *Physical Review*, **D32**, 2496.

*References*

Penrose, R. (1979) Singularities and time asymetry. In *General Relativity; An Einstein Centenary Survey*, Eds. S.W. Hawking & W.Israel, Cambridge University Press, Cambridge.

Popper, K.R. (1956) *Nature*, **177**, 538.

Shirai, I. & Wada, S. (1988) Cosmological perturbations and quantum fields in curved space. *Nuclear Physics*, **B303**, 728.

# 25

# Quantum Cosmology and Time Asymmetry

Jonathan J. Halliwell*

*Center for Theoretical Physics*
*Laboratory for Nuclear Science*
*Massachusetts Institute of Technology*
*Cambridge, MA 02139, USA*

## 25.1 Introduction

The subject of time asymmetry is one that has fascinated many minds going back to Boltzmann in the previous century†. From an early stage in the discussion, it was appreciated that the ultimate explanation of time asymmetry in the universe may only be found by appealing to the initial conditions with which the universe began. The explanation therefore seemed to lie beyond the domain of conventional classical cosmology, which is not properly equipped to deal with initial conditions. However, recent years have witnessed a considerable amount of development in the field of quantum cosmology, in which it is possible to at least address the issue of initial conditions in a meaningful way. The purpose of this contribution is ask how much the subject of time asymmetry has benefited from developments in quantum cosmology. The overall tone will be that of a review, rather than a presentation of new results, but on the way, I will try to put issues into context, comment on earlier work, and fill in some of the missing arguments.

Physics is generally held to be time-symmetric at a fundamental level. By fundamental, we mean at the level of field theories in fixed backgrounds. In particular, all field theories of interest are CPT invariant. At the coarser-grained macroscopic level of the real world, however, we see distinct asymmetry in time‡. There is a structural difference between past and future: the future is undetermined, we have no records of it, and we can influence it; the past is fully determined, we have records of it, and we cannot influence or change it. One says that the events that occur in the world have not just an ordering in time, but also a direction. This is to be contrasted with the spatial location of events, which have just an ordering. They may be characterized, for example, as being to the left or right of each other, but there is no structural difference between left and right (Reichenbach, 1971).

One can also look in the opposite direction, to a finer-grained, deeper level,

---

* Present address: Theory Group, Blackett Laboratory, Imperial College, London, SW7 2BZ, UK
† For general accounts of the topic of time asymmetry, see Davies (1976), Denbigh (1981), Gold (1967), Grünbaum (1973), Horwich (1987), Landsberg (1982), Reichenbach (1971) and Zeh (1989).
‡ See Amis (1991), for a striking account of this aspect of the world!

beyond that regarded by most as fundamental. Underlying field theory in fixed background spacetimes is the situation in which the spacetime metric is no longer fixed but becomes a quantum field. This is the level of quantum gravity. At this level, classical notions of spacetime break down, and there are strong indications that no notion of time exists. There have been attempts to find a fundamental time for quantum gravity, but all attempts have so far failed. The present evidence, at least in this author's opinion, suggests that there is no time in quantum gravity, and time as we know it only emerges under certain conditions. Here, we will also take the point of view (substantiated in what follows) that quantum gravity is time-symmetric, in the sense that when time emerges it does not appear with a preferred direction. The opposite point of view, that quantum gravity involves a fundamental time asymmetry has also been put forward (Penrose, 1979; Wald, 1980).

Returning to the issue of time asymmetry in the world about us, there are a variety of classes of phenomena in the world about us that exhibit a distinct asymmetry between the past and the future. Of these *arrows of time*, perhaps the two most important are:

(i) The thermodynamic arrow of time. This is usually described as the fact that the entropy of an isolated system generally increases, and can never decrease. However, in more practical terms it is the fact that in the world about us there exist a vast number of essentially isolated systems in states of low but increasing entropy.

(ii) The cosmological arrow of time. This is the fact that the universe is expanding, when the equations of motion suggest that a contracting universe is equally likely.

There are of course a number of other arrows of time, such as that associated with quantum measurement, the radiation arrow, and the psychological arrow. Here, we will assume that these arrows are subsumed under the thermodynamic arrow (although we acknowledge that accounting for them in terms of the thermodynamic arrow may be non-trivial). The fundamental problem of time asymmetry is to explain the origin of these arrows, and the possible connections between them, in terms of underlying time-symmetric dynamics.

In what follows, I will first of all go over the reasons why the thermodynamic arrow of time may be traced back to cosmological initial conditions. I will then discuss quantum cosmology, and the extent to which boundary conditions in quantum cosmology might explain the thermodynamic arrow of time, and its possible connections to the cosmological arrow. Throughout I will attempt to support the point of view that first of all, physics is, at its most fundamental level (*i.e.* that of quantum gravity), timeless. Time as an ordering parameter emerges only a certain coarse-grained level of description, at the level at which gravity appears classical, but all other fields may remain quantum. Secondly, that the direction of time may emerge at a coarser level of description from an underlying theory which is completely time-symmetric.

## 25.2 Branch Systems and the Thermodynamic Arrow

Our goal is to explain the thermodynamic arrow of time, and its possible correlation with the cosmological arrow. The thermodynamic arrow of time is essentially the fact that in the world about us there exist a vast number of isolated systems in states of low but increasing entropy. It is often stated that this is due to the universe starting out in a low entropy state, but the explicit line of reasoning leading to this claim is rarely given. The necessary line of reasoning is provided by the theory of *branch systems*, due to Reichenbach, which we now describe. We give here only the briefest account of this topic. For further details the reader is referred to Davies (1976, 1992), Denbigh (1981), Grünbaum (1973), Horwich (1987) and Reichenbach (1956).

A branch system is a system which was strongly coupled to the rest of the universe, or at least, to its immediate environment at some time in the past, and may also become so coupled again in the future, but for a substantial period of time is essentially isolated, *i.e.*, it is *quasi-isolated*. The universe is replete with such systems, and they are being created and destroyed all the time.

A classic example of a branch system is a piece of ice in a glass of water (Davies, 1976, 1992). The system begins its existence only when the ice was dropped in the water. After the system was created, the most significant thermodynamic processes are those going on between the ice and the water, not those between the system and its environment, and so the system is quasi-isolated. A characteristic feature of randomly created systems such as this one is that they are generally created out of equilibrium, in low entropy states, and their entropy will with high probability increase thereafter †.

The history of the universe is a ramifying saga of branch systems. Because clustering is entropically favourable for gravitational systems, the evolving cosmological fluid coalesces into the descending hierarchy of large scale structure – superclusters, clusters, galaxies, stars and planets. As each of these structures forms out of the fluid, they become essentially decoupled from it, except for coupling by gravity and the exchange of radiation, but these have essentially no effect on their local thermodynamics. If the "main branch" (the cosmological fluid at early times) from which the branch systems emerged began in a state of low entropy, then the branch systems will generally be created in low entropy states, and in their subsequent evolution their entropies will with high probability increase. It can be argued that the entropies in the vast majority of branches all increase in the same direction, and in parallel to the main branch. It is because of this process of extensive branching from

---

† By entropy, we mean a microscopically defined (Gibbs-Shannon) entropy with some suitable coarse-graining, as in the Ehrenfest urn model.

a low entropy initial state that we now see a very large number of quasi-isolated systems in states of low but increasing entropy. This explains the thermodynamic arrow of time, because we define it to be the direction of entropy increase in the vast majority of branch systems.

It is therefore certainly true that the assumed low entropy initial state of the universe is partly responsible for the thermodynamic arrow of time; but the theory of branch systems also plays an important role in that it shows how the general entropy increase of the entire universe is locally reiterated in the thermodynamic behaviour of quasi-isolated systems. It is, of course, yet to be explained *why* the universe began in a low entropy initial state. To explain this, we need to appeal to cosmology.

## 25.3  Cosmology

How can the low entropy initial state of the universe be explained? This could only be by the initial conditions with which the universe began. One might conjecture that there exist classical initial conditions in cosmology which would explain the low initial entropy of the universe. Such a proposal has been made by Penrose. His proposal was that the Weyl tensor should vanish at the initial singularity, but not at the final one (Penrose, 1975). There are, however, a number of arguments against such proposals. The main objection is that it seems most likely that the conditions at the beginning of the universe were so extreme that quantum gravitational effects were important. Imposing classical initial conditions in a regime not known to be accurately described by classical general relativity is therefore very dubious.

A further, perhaps more esoteric issue, is that in a sense one is putting in the time asymmetry by hand. Whilst there is no reason in principle why it should not be the case that the time asymmetry of the world arises from the asymmetry of the boundary conditions, it would be far more satisfying to see the asymmetry emerge without any explicitly asymmetric assumptions being built in. An example of how this may happen is contained in the time-symmetric cosmological models discussed below.

We will go on to discuss quantum cosmology and initial conditions later on, but it will first be useful to briefly describe classical cosmological solutions to the Einstein equations. For our purposes it is most convenient to work in the $(3 + 1)$ formalism of general relativity coupled to matter (York, 1983). In this formalism, one foliates the spacetime manifold into spacelike three-surfaces, which are taken to be closed. The dynamical variables one focuses on are the fields on a spacelike three-surface $t = constant$: the three-metric $h_{ij}(\mathbf{x})$ and the matter field which we take to be a scalar field $\Phi(\mathbf{x})$. In terms of these fields, the four-metric on spacetime is

$$ds^2 = -(N^2 - N^i N_i)dt^2 + 2N_i dx^i dt + h_{ij} dx^i dx^j \tag{25.1}$$

where $N$ and $N^i$, the lapse and shift, are essentially arbitrary, because they may be given arbitrary values by four-dimensional diffeomorphisms. Evolution is described by the Hamiltonian

$$H = \int d^3x \left[ N\mathcal{H} + N^i \mathcal{H}_i \right] \tag{25.2}$$

from which one obtains the Hamilton equations,

$$\dot{h}_{ij} = \frac{\delta H}{\delta \pi^{ij}}, \quad \dot{\Phi} = \frac{\delta H}{\delta \pi_\Phi}$$

$$\dot{\pi}^{ij} = -\frac{\delta H}{\delta h_{ij}}, \quad \dot{\pi}_\Phi = -\frac{\delta H}{\delta \Phi} \tag{25.3}$$

Here, $\pi^{ij}$ and $\pi_\Phi$ are the momenta conjugate to $h_{ij}$ and $\Phi$ respectively. Because the quantities $N$ and $N^i$ are arbitrary, there are also a set of constraints, which must be satisfied by the initial data,

$$\mathcal{H} = 0, \quad \mathcal{H}_i = 0 \tag{25.4}$$

In what follows, we will set $N = 1$ and $N^i = 0$, and so $t$ is proper time. Eqs. (25.3), (25.4) constitute the Einstein equations plus the evolution equations for matter.

The Einstein equations may be solved to yield a set of solutions, $(h_{ij}(\mathbf{x}, t), \Phi(\mathbf{x}, t))$. The solutions are conveniently represented as paths, parametrized by $t$, in the space of all three-metrics $h_{ij}(\mathbf{x})$ and matter fields $\Phi(\mathbf{x})$. This infinite dimensional space is often known as superspace. (The same nomenclature is often also applied to the space of three-metrics modulo three-dimensional diffeomorphisms, but the distinction will not be important here). An important feature of the paths in superspace is that the parameter $t$ is unphysical. It is not the same as the time parameter in, for example, non-relativistic theories. This is because general relativity is an example of a so-called parametrized theory – a theory in which time is already contained amongst the basic dynamical variables.

The Einstein equations are time-symmetric in the sense that if $(h_{ij}(\mathbf{x}, t), \Phi(\mathbf{x}, t))$ is a solution, then $(h_{ij}(\mathbf{x}, -t), \Phi(\mathbf{x}, -t))$ is also a solution. A solution is then said to be time-symmetric if there exists a parametrization of it and a value of the time parameter $t_0$ such that

$$h_{ij}(\mathbf{x}, t_0 + t) = h_{ij}(\mathbf{x}, t_0 - t), \quad \Phi(\mathbf{x}, t_0 + t) = \Phi(\mathbf{x}, t_0 - t) \tag{25.5}$$

This implies that $\dot{h}_{ij}(\mathbf{x}, t_0) = 0 = \dot{\Phi}(\mathbf{x}, t_0)$. (Recall that we are working in the gauge $N = 1, N^i = 0$. These conditions would be more complicated in more general gauges, but we will not go into that here). In superspace, time-symmetric paths come to a halt at the point parametrized by $t_0$ and then retrace their steps. The time-reversed trajectories are therefore physically indistinguishable from the original ones. The differ only in the sense that they are parametrized in the opposite direction.

A slightly more general notion of time-symmetric solutions is useful, and has been used. Many scalar field models of interest have a symmetry under $\Phi \rightarrow -\Phi$.

No quantities of interest recognize the sign of $\Phi$, so a more general definition of time-symmetric solutions is

$$h_{ij}(\mathbf{x}, t_0 + t) = h_{ij}(\mathbf{x}, t_0 - t), \quad \Phi(\mathbf{x}, t_0 + t) = -\Phi(\mathbf{x}, t_0 - t) \qquad (25.6)$$

These do not have $\dot{\Phi}(\mathbf{x}, t_0) = 0$, but have $\Phi(\mathbf{x}, t_0) = 0$. These solutions might reasonably be called "time-antisymmetric" solutions. The important point, of course, is that they have definite transformation properties under time reversal. These solutions do not retrace their steps in superspace, but are symmetric about $\Phi = 0$.

Given that the time parameter $t$ has no significance of its own, it is reasonable to ask for the significance of the operation of time-reversal. The point is that in practice a spacetime gauge condition is imposed, and this has the effect of tying the unphysical parameter $t$ to a physically significant quantity, *e.g.* the scale factor, or the trace of the extrinsic curvature. Through this condition, reversing the parameter time will reverse the physical time. In general, however, it is difficult to find a physically interesting quantity that changes monotonically with $t$ for the entire duration of the classical history, so such gauge conditions can be imposed only locally in time. This difficulty is of course intimately tied up with the problem of time in canonical geometrodynamics alluded to earlier.

Numerous authors have discussed time-symmetric cosmologies, in which the universe undergoes a completely symmetric expansion followed by contraction, with a complete microscopic reversal at the turnaround point (Cocke, 1967; Gell-Mann and Hartle, 1992; Gold, 1958, 1962; Schulman, 1992 and references therein). If the boundary conditions fix a suitably coarse-grained definition of entropy to be small initially and finally, this entropy generally increases during the expanding phase and decreases during the collapsing phase. It seems to be agreed that the psychological arrow of time is correlated with the thermodynamic arrow, so during the collapsing phase observers would perceive the universe to be expanding and entropy to be increasing. There would therefore be no observable distinction between expanding and collapsing phases. These models have a thermodynamic arrow of time of a *sectional* nature, *i.e.*, one that points in different directions for different sections of the history of the system. This is actually quite acceptable, and leads to no contradictions if the sections are long enough, a fact that was first recognized by Boltzmann.

The difficulty with the time-symmetric models is their implausibility. They require a very finely tuned set of boundary conditions, for which no explanation is offered. And amongst a number of other problematic features, it is difficult to imagine how a universe similar to the one in which we live, with large scale structures, stars and galaxies, might emerge, and then return to its initial state. However, they do have the very appealing feature that, as already noted, sections of the history of the universe may display a strong thermodynamic arrow of time, even though the underlying theory and its boundary conditions are completely time-symmetric. We shall see that this feature reappears in quantum cosmology, but in a somewhat different, far more plausible form.

## 25.4 Quantum Cosmology

In order to understand the reason for the low initial entropy of the universe, we need to address the issue of cosmological initial conditions. Because quantum gravitational effects are likely to have been important when the universe was very small, this issue is most properly addressed from a framework in which gravity is quantized. We are therefore led to study the quantization of general relativity coupled to matter for closed cosmologies, *i.e.* quantum cosmology†. It has to be stated at the outset that in adopting this course of action we must eventually run into the serious difficulties of quantum gravity. However, quantum gravitational considerations enter only in the justification of the boundary conditions, and the issue of time asymmetry is largely separate from the difficulties of quantum gravity.

The system we wish to quantize is described classically by Eqs. (25.2)-(25.4). One of its most salient features is that the Hamiltonian vanishes, as a consequence of the constraints. This feature is intimately related to the fact that the theory is a parametrized theory. It severely complicates the quantum theory, which generally relies rather heavily on being provided with an explicit time (Kuchař, 1989, 1992a, 1992b). Two approaches to this problem are normally contemplated:

(i) Solve the constraints (25.4) classically in a fixed gauge, thereby identifying the physical time and the (non-vanishing) physical Hamiltonian. Use these quantities to construct a time-dependent Schrödinger equation.

In this approach, one is assuming that a physical time exists, which is not guaranteed to be the case. To date, no-one has been successful in carrying through this program.

(ii) Employ the Dirac approach in which one quantizes first, and imposes the constraints afterwards.

This involves representing the quantum state of the system by a wave functional $\Psi[h_{ij}, \Phi]$, a functional on superspace. There is no explicit time-dependence, since as we have already indicated, time is already in there. There is no Schrödinger equation, only the constraints,

$$\hat{\mathcal{H}}\Psi[h_{ij}, \Phi] = 0, \quad \hat{\mathcal{H}}_i\Psi[h_{ij}, \Phi] = 0 \tag{25.7}$$

These are known respectively as the Wheeler-DeWitt equation and the momentum constraints. Because of the difficulties with the first approach, most approaches to quantum cosmology adopt the second approach. We shall do so here. This approach is consistent with the view that there is no time at the most fundamental level, and it

† For introductory/review material on quantum cosmology, see for example, Halliwell (1991), Hartle (1985, 1986, 1991) and Hawking (1984a).

is necessary to understand how time emerges from a quantum theory not possessing one.

The momentum constraints imply the invariance of the wave function under three-dimensional diffeomorphisms, and will play no role in what follows. The remaining constraint, the Wheeler-DeWitt equation is very difficult to solve in general, so it is necessary to make some approximations. A particularly convenient approximation is to restrict attention to the region of superspace in the immediate vicinity of homogeneity (and sometimes isotropy also) (Halliwell and Hawking, 1985; D'Eath and Halliwell, 1987; Fischler et al., 1985; Vachaspati and Vilenkin, 1988; Wada, 1986). This involves going from superspace coordinates $(h_{ij}(\mathbf{x}), \Phi(\mathbf{x}))$ to a set of coordinates $(q^\alpha, \delta\phi(\mathbf{x}))$. Here, the $q^\alpha$ denote a finite set of "large" homogeneous modes of the fields, known as "minisuperspace" modes. One could take them to be, for example, the scale factor and the large, homogeneous mode of a scalar field. The fields $\delta\phi(\mathbf{x})$ denote "small" inhomogeneous matter and gravitational perturbations about the homogeneous models, and are retained only up to quadratic order in the Hamiltonian. Although a somewhat crude approximation, probing but a small region of superspace, it is sufficient to discuss the initial growth of inhomogeneities and the associated thermodynamic arrow of time. For convenience, we will take the perturbations $\delta\phi(\mathbf{x})$ to be scalar field perturbations only.

With the above approximation, the Wheeler-DeWitt equation (after integration over the three-surfaces) takes the form

$$H\Psi(q^\alpha, \delta\phi) = \left(-\frac{1}{2}\nabla^2 + U(q) + H_m(q^\alpha, \delta\phi)\right)\Psi(q^\alpha, \delta\phi) = 0 \qquad (25.8)$$

Here $\nabla^2$ is the Laplacian in the minisuperspace modes, $q^\alpha$, and $H_m$ is the Hamiltonian of the matter perturbation modes $\delta\phi$ and is quadratic in them. This equation may be solved, for given boundary conditions, using a double expansion: a WKB expansion (effectively an expansion in the Planck mass, which is here set to unity), and an expansion about zero of the matter perturbation modes.

It is convenient to picture the behaviour of the solution in minisuperspace, the space of the $q^\alpha$'s. One typically finds the following. In certain regions of minisuperspace (sometimes, but not always, when the scale factor is very small) the wave function will be predominantly exponential in behaviour. This is like the under-the-barrier wave function in tunneling situations in quantum mechanics, and is therefore taken to imply that notions of classical spacetime (and in particular, notions of time) do not exist in such regions. On the other hand, there will be other regions in which the wave function is oscillatory in nature, and in these regions the $q^\alpha$'s are essentially classical, and classical spacetime is a valid notion.

In the oscillatory region we get solutions of the form

$$\Psi(q, \delta\phi) = C(q)\, e^{iS(q)}\, \psi(q, \delta\phi) \qquad (25.9)$$

or more generally, a sum of solutions of this form. Inserting this ansatz into the

Wheeler-DeWitt equation (25.8), one finds, from the lowest order terms, that $S$ must obey the Hamilton-Jacobi equation

$$\frac{1}{2}(\nabla S)^2 + U(q) = 0 \tag{25.10}$$

One can then argue (Halliwell, 1987) that the wave function corresponds to a set of classical solutions satisfying the first integral

$$\dot{q}^\alpha = p^\alpha = \nabla^\alpha S \tag{25.11}$$

This set of solutions, for given $S$, has far fewer free parameters than the general solution. One also obtains an equation indicating, loosely speaking, that $|C|^2$ provides a probability measure on this set of solutions. The higher order terms in the expansion yield the following equation for the matter mode wave functions,

$$i\frac{\partial \psi}{\partial t} = H_m \psi \tag{25.12}$$

where $\partial/\partial t \equiv \nabla S \cdot \nabla$. This is a Schrödinger equation along the classical trajectories of the minisuperspace modes, and one can expect the matter wave functions to be normalizable in the usual inner product. In the oscillatory region, therefore, we recover the functional Schrödinger formalism of quantum field theory for the perturbations on a classical minisuperspace background (Banks, 1985; Halliwell and Hawking, 1985; Laflamme, 1987).

Of course, there is much more to the emergence of classical behaviour than we have discussed here. Decoherence of the histories of the universe should really have been discussed prior to classical correlations, in order to be able to talk about histories of the universe in a meaningful way. It should also be shown that separate terms in a sum over WKB wave packets of the form (25.9) decohere (Halliwell, 1989). However, decoherence of the large modes is not particularly relevant to the issue of the arrow of time (although decoherence of the perturbation modes is – see below). It is expected that many of these things can be made a lot more precise using the decoherent histories approach (Gell-Mann and Hartle, 1990; Hartle, 1991).

Now suppose we have a set of boundary conditions which picks out a solution to the Wheeler-DeWitt equation. Two such boundary condition proposals will be discussed below. A particular solution to the Wheeler-DeWitt equation will pick out a particular solutions $S$ to the Hamilton-Jacobi equation in the oscillatory regime, and thus will define a particular set of classical trajectories for the minisuperspace modes.

More importantly, boundary conditions will pick out a particular solution $\psi$ to the Schrödinger equation for the matter modes. One can think of the boundary conditions on the entire wave function as implying boundary conditions on the perturbation wave functions at one or both ends of the minisuperspace trajectories, at the edge of the oscillatory regime. The question of the thermodynamic arrow of time concerns the issue of the entropy associated with the perturbations along these trajectories, and in particular, the question of whether or not the boundary conditions imply that it is low at one end of the trajectories. To proceed further, therefore, we need a more concrete notion of the entropy associated with the perturbations.

## 25.5 Entropy

Conspicuously absent from earlier discussions of time asymmetry in quantum cosmology has been precise notions of decoherence, entropy, entropy increase *etc.* What is needed is some kind of measure of the entropy of the inhomogeneous fluctuations $\delta\phi(\mathbf{x})$ as they evolve along the minisuperspace trajectories.

A standard definition of entropy is as follows. Introduce the density operator associated with the perturbation wave functions, $\rho_m = |\psi_m\rangle\langle\psi_m|$. Then define a coarse-grained density operator by $\tilde{\rho}_m = \mathrm{Sp}(\rho_m)$ where Sp denotes spur, a partial trace. The entropy of the reduced (or coarse-grained) density operator is then defined by

$$S(t) = -\mathrm{Tr}(\tilde{\rho}_m \ell n \tilde{\rho}_m) \qquad (25.13)$$

This entropy will change with time, $\dot{S}(t) \neq 0$ (unlike the entropy associated with $\rho_m$, which is zero for all time), but whether or not it generally increases depends on initial conditions.

The above construction is typically achieved by a "system-environment split" (Joos and Zeh, 1985; Zurek, 1982). For example, if $\delta\phi$ consisted of two interacting fields $\delta\phi = (\delta\phi_1, \delta\phi_2)$, one could regard $\delta\phi_1$ as the "system", and the reduced density operator would be obtained by tracing out over the "environment" $\delta\phi_2$). There are many examples of this sort of procedure in the literature (see, for example, Caldeira and Leggett, 1983).

From the operational point of view, this system-environment split is reasonable, since we can only ever observe a very small fraction of the potentially observable features of the universe. But this split is often the cause of some concern, in that it might seem that there is no reason in principle why one should not try to observe most, even all, of the potentially observable aspects of the universe. Put differently, the process of coarse-graining appears to involve a certain subjective element, and any consequent deduction concerning the behaviour of entropy may therefore depend on how one chooses to coarse grain, or even, whether or not one chooses to coarse grain at all (Denbigh and Denbigh, 1985).

The beginnings of a resolution to this problem may be found in the decoherent histories approach to quantum cosmology (Gell-Mann and Hartle, 1990). There, the object is to identify sets of histories of the universe that decohere, and therefore, to which probabilities may be assigned. These are asserted to be the only histories which have any meaning. In this formalism one can talk about completely fine-grained histories. These are histories in which one precisely specifies a complete set of commuting observables at every moment of time. However, it is readily shown that such histories do not decohere; probabilities cannot be assigned to them and they cannot be attached any meaning. This means that *some* amount of coarse-graining is necessary before one can talk about histories of the universe which have any meaning, and which one can begin to compare with the physical universe in which we live. This therefore to some degree alleviates the difficulty of the subjective nature of coarse graining, in that it shows that a completely fine-grained description of the universe is not allowed. But it does not remove all the difficulties in that we are not told how to carry out the coarse-graining, that is, what to ignore, and what to focus on.

This, then, is how one might proceed with a formal definition of entropy. However, one may proceed differently for the purposes of our present considerations. The issue of the thermodynamic arrow of time in the universe concerns gravitational clustering. The general increase in entropy in the universe is associated with the general tendancy of gravitational systems to evolve from smooth states to clustered states. We therefore need to explain why the universe started out in a smooth, unclustered state. This issue can in fact be appreciated by studying the perturbation wave functions directly, and in practice this is what was done by previous authors discussing the arrow of time in quantum cosmology.

The perturbation wave functions are described by the Schrödinger equation, (25.12). The Hamiltonian in (25.12) turns out to have the form of a sum over modes of time-dependent harmonic oscillator Hamiltonians,

$$H_m = \sum_{\mathbf{k}} H_m^{\mathbf{k}} \qquad (25.14)$$

At any moment of time, one can expand the wave functions in harmonic oscillator eigenstates. The fluctuations in density on the scale associated with the wave number $\mathbf{k}$ will be proportional to $\langle H_m^{\mathbf{k}} \rangle$, the expectation value of $H_m^{\mathbf{k}}$. This quantity is minimized when the fluctuation modes are in their ground state, and will be large when they are in an excited state. This means that if the fluctuations are in their ground state, one has a smooth, unclustered universe, whilst a clustered universe is represented by an excited state. Furthermore, because the Hamiltonian for the fluctuations is time dependent, a state which is initially the ground state will not remain so, but will evolve to a superposition of excited states. This corresponds to the classical notion that density fluctuations tend to grow with time. What this means for our present purposes is that a smooth initial state with a subsequent evolution to a clustered state is explained if the fluctuation wave functions start out in their ground states. It is this notion of being clustered or unclustered that we will use in what follows.

The association of these notions with the more formal definitions of entropy discussed above is rather loose, and would be an interesting issue for investigation†.

† For approaches in this direction, see Fukuyama and Morikawa (1989) and Morikawa (1989)

Underlying all of these considerations appears to be an as-yet undiscovered definition of gravitational entropy.

## 25.6 Time Reversal in the Wheeler-DeWitt Equation

In ordinary quantum mechanics, if $\psi(x, t)$ is a solution to the Schrödinger equation, then $\psi^*(x, -t)$ is also a solution. One therefore defines the time reversal operation to be

$$\psi^T(x, t) = \psi^*(x, -t) \tag{25.15}$$

The Wheeler-Dewitt equation has no obvious time parameter. But it is a real equation, so if $\Psi[h_{ij}, \phi]$ is a solution, $\Psi^*[h_{ij}, \phi]$ is also a solution. The most natural definition of time reversal is therefore

$$\Psi^T[h_{ij}, \phi] = \Psi^*[h_{ij}, \phi] \tag{25.16}$$

Because there is, strictly speaking, no notion of time in the Wheeler-DeWitt equation, the above definition is on the face of it suspect. However, as with many things in quantum cosmology, appeal to the semi-classical limit supports the use of this definition. For consider the WKB solution in the oscillatory (semi-classical) region, Eq. (25.9). The time reverse of this solution is

$$\Psi^T = C^*(q)\, e^{-iS(q)}\, \psi^*(q, \delta\phi) \tag{25.17}$$

It corresponds to the set of classical trajectories defined by

$$\dot{q}^\alpha = p^\alpha = -\nabla^\alpha S \tag{25.18}$$

which are nothing more than the time-reverse of the trajectories, (25.11). Furthermore, the matter modes evolve according to the Schrödinger equation,

$$-i\frac{\partial \psi^*}{\partial t} = H_m \psi^* \tag{25.19}$$

the complex conjugate of (25.12). Letting $t \to -t$ (but recalling that $H_m$ generally depends on $t$), we deduce that $\psi^*(-t)$ satisfies the Schrödinger equation along the time-reversed trajectories. In the semi-classical limit, therefore, we recover the expected properties of a time-reversed wave function.

Hawking (1985) proposed an alternative, not dissimilar definition of time reversal. His idea was to first split the three-metric $h_{ij}$ into a volume factor $h^{\frac{1}{2}}$ and the conformal three-metric, $\tilde{h}_{ij} = h^{-1/3} h_{ij}$, and then consider the wave function $\tilde{\Psi}[K, \tilde{h}_{ij}]$ obtained by a Laplace transform exchanging $h^{\frac{1}{2}}$ for $K$, the trace of the extrinsic curvature. He then defined the time reversal operation to be $\tilde{\Psi}^T[K, \tilde{h}_{ij}] = \Psi^*[-K, \tilde{h}_{ij}]$. Apart from the difficulty of actually defining the Laplace transform, this definition is tantamount to elevating $K$ to the status of an internal time parameter. Although this may be a good time parameter to use for many classical calculations, it is known that it cannot be a completely satisfactory definition of time at the quantum level (Kuchař, 1989, 1992a, 1992b).

Because the theory has a form of time-reversal invariance, time-symmetric solutions exist. The simplest such solution would, in the WKB region, have the form,

$$\Psi = C \, e^{iS} \, \psi + C^* \, e^{-iS} \, \psi^* \tag{25.20}$$

The first wave packet corresponds to a set of solutions defined by the first integral (25.11), with the perturbation wave functions evolving according to (25.12). Likewise, the second wave packet, the complex conjugate of the first, corresponds to (25.18) and (25.19), the time-reverses of (25.11) and (25.12). The entire wave function (25.20) corresponds to the union of the two sets of classical solutions, corresponding to (25.11) and (25.18), plus the perturbation wave functions on them. Most importantly, this set of solutions to which the entire wave function corresponds is a *time-symmetric set*, since every trajectory's time reverse is also contained in the set. Note that this does not mean that the individual trajectories are time-symmetric (in the sense of (25.5) and (25.6)), and in fact, in general they will not be, as was emphasized by Page (1985).

We therefore arrive at the central point. The underlying theory, and its boundary conditions may be time-symmetric, selecting a time-symmetric solution to the Wheeler-DeWitt equation of the form (25.20). The time asymmetry of the world emerges only when one goes from the time-symmetric ensemble of classical solutions defined by the wave function, to one particular solution corresponding to a possible history of the universe.

The situation is analogous to the phenomenon of spontaneous symmetry breaking in field theory. Suppose one has a system described by a Hamiltonian with a W-shaped double-well potential, symmetric about the origin. Both the Hamiltonian and the ground state of the system are symmetric under reflection about the origin. Yet the fact of the real world is that the system would be found to be either one or the other of the two wells, appearing to violate the reflection symmetry. Given this analogy, one might say that the time-symmetry of the laws of physics and its boundary conditions are observed to be spontaneously broken in nature.

Another parallel worth drawing is with time-symmetric statistical models satisfying identical initial and final conditions. There, single coarse-grained histories need not be time-symmetric, and generally are not, but averages of all quantities over the entire ensemble of trajectories are time-symmetric (Schulman, 1992 and references therein).

We have yet to see whether or not the classical trajectories corresponding to a given wave function exhibit a strong thermodynamic arrow of time. This depends on boundary conditions, which we discuss next.

## 25.7 Boundary Condition Proposals

Two particular boundary condition proposals have received the most attention over the last few years. We will not go into the motivation for these proposals, or their explicit statement. We will merely state the types of wave functions they pick out in the oscillatory region, for typical models. The first proposal is the so-called "no-boundary" proposal (Hawking, 1982, 1984b; Hartle and Hawking, 1983). It picks out a time-symmetric wave function, of the form (25.20). The second proposal is the "tunneling" proposal (Linde, 1984, Vilenkin 1982, 1984). In this proposal, the wave function in the oscillatory region is a single WKB wave packet of the form (25.17). The phase $S$ and the perturbation wave functions $\psi$ are the same in each of these proposals, but the prefactor $C$ is different (although this is irrelevant for present purposes). Most importantly, on calculating the perturbation wave function $\psi$, one finds that it is *in the ground state at one end* of the classical minisuperspace trajectories (and is generally in an excited state at the other end) (Halliwell and Hawking, 1985; Vachaspati and Vilenkin 1988). It follows from everything we have said so far, therefore, that both of these boundary conditions predict a strong thermodynamic arrow of time.

It is perhaps useful to elaborate on the wave function picked out by the no-boundary proposal. The boundary conditions give a low entropy state at one end of the set of trajectories. In one WKB component the trajectories appear to emerge from an *initial* state of low entropy; in the other, they appear to evolve into a *final* state of low entropy. With respect to the parameter $t$, the thermodynamic arrow of time appears to point in opposite directions in each WKB component. However, this is not a matter of concern. We as observers would always find ourselves in one trajectory of one WKB component, and there is no known way of detecting the existence of the other WKB component. In each component it is important only that there exists a strong thermodynamic arrow. There is nothing fundamental about the parameter $t$, and the words "initial" and "final" have no absolute significance. It only matters that entropy increases from one end of the trajectory to the other.

Let us now try to put the above result into context. If asked to pick end-point conditions at random for the fluctuations, the vast majority of such conditions would correspond to high entropy states, and only by picking very special end-point conditions could they be low entropy states. Generally, one could envisage the following three possibilities:

(i) It both ends were chosen to be typical, then they would be high entropy states. The fluctuations would therefore most probably have high entropy for much of their evolution, and would not exhibit a strong thermodynamic arrow of time. This seems to suggest that a generic boundary condition for the fluctuations in quantum cosmology will not imply a thermodynamic arrow of time. This means that quantum cosmology alone, without appeal to particular quantum boundary conditions, will not explain the thermodynamic arrow of time. This conclusion is supported by quantitative estimates of the entropy of the universe by Penrose (1979), who argued that even at the present epoch, the entropy of the universe is very much lower than it could be. This indicates that the universe is in a very special state now, and therefore evolved from an even more special state in the past.

(ii) If the boundary conditions were such that one end is a very special ordered state, and the other end generic, then entropy would go from low to high, going from one end to the other. Such a boundary condition would be sufficient to explain the thermodynamic arrow of time. This situation is realized by both the no-boundary and tunneling proposals.

(iii) Boundary conditions could be envisaged in which the fluctuations are in very special, ordered states at *both* ends of the classical trajectories. The classical histories would therefore have low entropy both initially and finally, but if they are long enough, the entropy would increase substantially moving away from either end-point, undergoing a turnaround somewhere in the middle. There would therefore be a thermodynamic arrow of time of a *sectional nature*, pointing in different directions at different sections of the trajectory. This is perfectly acceptable provided that each section is long enough. This situation could arise if the boundary conditions on the fluctuations are uniform everywhere along the boundary of the semi-classical regime. Boundary conditions in which the entropy is low at each end are most commonly associated with time-symmetric cosmologies, *i.e.*, time-symmetric classical solutions, but this is in no way assumed in the above. This possibility has been considered by Zeh (1983, 1986, 1988, 1989).

In the case of the no-boundary proposal, there was some initial confusion. It was at first thought that the proposal picks out solutions for which there was a correlation between low entropy and a small scale factor, therefore belonging to case (iii) (Hawking, 1985). This is no longer believed to be the case (Hawking, 1994; Page, 1985), and the no-boundary proposal is held to belong to case (ii).

Does any of this explain why the thermodynamic arrow of time is correlated with the expansion of the universe? In case (iii) the universe is in a very ordered state when small, so the direction of increasing entropy is always correlated with expansion. The thermodynamic and cosmological arrows are therefore correlated.

In the more likely case (ii), realized by the tunneling and no-boundary wave functions, we will find ourselves on a trajectory which has low entropy at one end, when the universe is small, and monotonically increasing entropy as the universe undergoes expansion and recollapse, ending in a high entropy state when the universe is small again. The thermodynamic and cosmological arrows are therefore not correlated.

A possible way to connect with the cosmological arrow in case (ii) is to invoke the anthropic principle†, together with the assumption that the universe underwent an early inflationary stage (Hawking, 1992). Inflation implies that the universe is extremely close to flatness, and thus, if closed, it will be a very long time indeed before it undergoes recollapse. The universe would undergo a heat death before recollapse, and thus lifeforms would not live to see the recollapse. Observers would therefore see the thermodynamic and cosmological arrows to coincide. This argument is not completely convincing, however, in that it is conceivable that life could survive through to the recollapsing phase.

Finally, we remark on the connection between the no-boundary and tunneling proposals, and the proposal of Penrose that the Weyl tensor vanishes initially (Penrose, 1975). The Weyl tensor is the source-free part of the Riemann curvature. For the present situation, in which we consider inhomogeneous perturbations about a homogeneous isotropic background, it depends only on the transverse traceless modes – the gravitational wave modes. The above analysis shows that these modes start out in their ground state, implying that at one end of the trajectories, the Weyl tensor starts out as small as it can be, consistent with the uncertainty principle. In this sense, the no-boundary and tunneling proposals imply the Penrose proposal, although not necessarily in the sense he intended.

## 25.8 Summary and Conclusions

The aim of this paper was to assess the contributions of quantum cosmology to the topic of time asymmetry. We took the point of view that a time asymmetric universe can emerge from a time-symmetric underlying theory, with time-symmetric boundary conditions.

We began by discussing branch systems, through which the thermodynamic arrow of time is traced back to an assumed low entropy initial state of the universe. Enquiring as to the explanation of this low entropy state, we turned to cosmology. Classical time-symmetric cosmological models can be found, in which a strong thermodynamic arrow of time may emerge, despite time-symmetric boundary conditions. These models are highly implausible, however, and no explanation is offered as to the origin of their boundary conditions.

Because of the extreme conditions encountered in the very early universe, quantum gravitational effects are likely to be important, and the issue of boundary conditions is best approached through quantum cosmology. We discussed the formalism of quantum cosmology, and the emergence of an approximately semi-classical regime, in which one can discuss the thermodynamic arrow of time along possible classical histories of the universe. We saw that the Wheeler-DeWitt equation admits time-symmetric solutions, corresponding to a time-symmetric set of classical solutions. Each individual solution, corresponding to a single possible history for the universe,

† See, for example, Barrow and Tipler (1986).

is generally not time-symmetric, and may exhibit a strong thermodynamic arrow of time. It is in going from the set to the particular member of the set that the time asymmetry of the universe may emerge.

The existence of a strong thermodynamic arrow of time along the trajectories is however contingent on boundary conditions. Quantum cosmology does not solve the issue of boundary conditions, but puts it into a form in which it can be reasonably addressed. A generic boundary condition in quantum cosmology will not imply a strong thermodynamic arrow of time, but boundary conditions in quantum cosmology which explain the thermodynamic arrow of time have been proposed. Not surprisingly, the necessary boundary conditions are special; but some of them may be argued to be natural in the context of quantum cosmology.

The no-boundary proposal of Hartle and Hawking is a time-symmetric boundary condition. A quantum theory of cosmology based on this boundary condition shares with the time-symmetric classical cosmologies the nice feature that a strong thermodynamic arrow of time may emerge, from completely time-symmetric boundary conditions, whilst avoiding their implausibility. In terms of the three cases given in Section 8 above, the classical time-symmetric cosmologies would lie in case (iii), whilst the no-boundary wave function lies in case (ii), one "plausibility class" higher†.

In conclusion, we have seen that that are two aspects of quantum cosmology that constitute significant contributions to the subject of time asymmetry: the construction of a framework in which cosmological boundary conditions can be discussed, and the emergence of particular boundary condition proposals which imply that the universe started out in the special state needed to explain the thermodynamic arrow.

## Acknowledgements

I am particularly grateful to Robert Brandenberger, Julian Barbour, Paul Davies, Fay Dowker, Robert Griffiths, Jim Hartle, Seth Lloyd, Jorma Louko, Don Page and Wojciech Zurek for useful conversations. I would also like to express my deep gratitude to my co-organizers, Juan Perez-Mercader and Wojciech Zurek. Organizing this meeting with them has been a great pleasure.

This work was supported in part by funds provided by the U.S. Department of Energy (DOE) under Contract No. DE-AC02-76ER03069 at MIT.

---

† This assumes that in the case of classical time-symmetric cosmologies, the boundary conditions are imposed microscopically on individual classical trajectories. One could envisage an ensemble, as in the time-symmetric statistical models mentioned earlier, in which boundary conditions are imposed on the ensemble. Such cosmologies would lie in case (ii).

## Discussion

**Hartle**   In order to answer the question of whether the expanding and contracting phases of the universe decohere, we are going to need the decoherence functional for sets of coarse-grained *histories* in quantum gravity (that both you, I and others have been investigating). The decoherence functional is a measure of decoherence in which the expanding phases and contracting phases can be distinguished. In its simplest form, it will involve the initial boundary condition that you have represented by a cosmological wave function. With this decoherence functional, we will be able to analyse such questions as to whether a boundary condition represented by a single wave function can result in time-symmetric cosmologies or whether time-symmetric conditions at both ends are needed as discussed in the contribution of Gell-Mann and myself. It is difficult to distinguish the two cases semiclassically because classical statistical boundary conditions imposed at two times can always be restated in terms of a boundary condition at one time since there is a unique classical evolution connecting conditions at one time with any other.

**Halliwell**   I generally agree with these remarks. Here, I have been discussing time asymmetry in quantum cosmology in terms of a single wave function. However, like most issues in quantum cosmology, the issue of time asymmetry would be best addressed using the decoherence functional. This is a topic for future research.

**Kuchař**   Time asymmetry is introduced by the $e^{iS}$ ansatz in the WKB approximation, or by the assumption that $e^{iS}$ and $e^{-iS}$ decohere. I do not see a convincing argument for this.

**Halliwell**   Arguments have been given in the past that $e^{iS}$ and $e^{-iS}$ decohere (*e.g.* Halliwell, 1989), although I did not give them here. These arguments are based on a wave function treatment, and are admittedly somewhat lacking. Again, the decoherent histories approach, when properly developed to apply to quantum cosmology, is likely to supply a more convincing argument.

**Teitelboim**   To what extent are you doing more than analyzing a linearized spin two field plus a scalar field on a classical background, or on a family of such backgrounds?

**Halliwell**   It is certainly true that the formalism of quantum cosmology reduces to that of quantum field theory on a family of backgrounds, and that most if not all of its predictions are through this semi-classical approximation. But there is more to it than this. First of all, the wave function for the full theory allows one to locate the classically forbidden regions in which the above semi-classical approximation would not be valid. There is nothing in the semi-classical approach itself that would indicate it to be invalid in these regions. Secondly, the theory provides a natural framework in which to address the boundary conditions issue.

**Griffiths**   Suppose the observational astronomers should eventually come to the (convincing) conclusion that the universe will not re-collapse. Would this require a completely different approach to quantum gravity? Would this mean that the direction of time asymmetry becomes trivial and obvious?

**Halliwell**   Quantum cosmology normally focuses on closed models of the universe, but open models can also be studied, and the question of time asymmetry still remains. One would still have to explain why the universe started out in the very special initial state needed to account for the thermodynamic arrow of time.

**Hu**   This is a comment directed to the discussions emergent from your talk, not necessarily implicating any inconsistency in your talk. I feel that the real issue of the emergence of the arrow of time is *not* on decohering the WKB branches, $e^{\pm iS}$, but how much of the information we decide to ignore or coarse-grain away. I am just thinking about the well-known cause of time-asymmetry in, say, classical statistical mechanics: the equation of motion is time-symmetric, but it becomes irreversible only at the point when you begin to work with a subsystem, coarse-graining the remaining variables. This is the case, for example, when one only looks at the dynamics of minisuperspace variables, and ignores the rest. More accurately, you will get a reversible dynamics for the system you are interested in if you *ignore* the rest, but you will get an irreversible dynamics for the same system if you take into account the *averaged effect* of the degrees of freedom you ignored. This is I think, the gist of the matter, and it applies to classical gravitational waves as well as quantum cosmology.

**Halliwell**  The process of coarse-graining is of course part of the story, but not all of it. In particular, coarse-graining alone does not break the symmetry under time-reversal; but coarse-graining is frequently accompanied by (possibly implicit) assumptions about boundary conditions, and it is these that lead to time-asymmetry.

**Barbour**  Has anyone, yourself included, ever thought of applying *criteria* for the selection of solutions to the Wheeler-DeWitt equation like those that Schrödinger applied to his time-independent equation? I know the parallel is imperfect, because of the indefinite metric in the Wheeler-DeWitt equation, but the two equations do look very alike. Let me remind you that Schrödinger's great insight, his belief that he had found the secret of quantum mechanics, rested crucially on the criteria of acceptability he imposed on his time-independent equation. Besides continuity, these were above all single-valuedness and boundedness.

**Halliwell**  Schrödinger's criterion of boundedness has physical content. It corresponds to the idea that the system (*e.g.* an atom) is localized in space, and the probability of finding it at infinity should be zero. There is no obvious motivation for the analogous situation in superspace, the configuration space of quantum cosmology. However, Kiefer has studied wave functions corresponding to single localized classical paths, the analogue of coherent states, and these have proved to be of some interest. Criteria of a very different sort were considered in a paper by Hartle and I, in which we imposed restrictions on the contour in the path integral representation of the wave function. (Halliwell, J.J. and Hartle, J.B. (1990) Integration contours for the no-boundary wave function of the universe. *Physical Review* **D41**, 1815-1834). These were largely motivated by the requirement that the wave function predict some very basic physical aspects about the universe we live in (*e.g.* that spacetime is classical).

# References

Amis, M. (1991) *Time's Arrow*, Harmony Books, New York.

Banks, T. (1985) TCP, quantum gravity, the cosmological constant and all that. *Nuclear Physics* **B249**, 332.

Barrow, J.D. and Tipler, F. (1986) *The Anthropic Cosmological Principle*, Oxford University Press, Oxford.

Caldeira, A. O. and Leggett, A. J. (1983) Path integral approach to quantum Brownian motion. *Physica* **121A**, 587.

Cocke, W. J. (1967) Statistical time symmetry and two-time boundary conditions in physics and cosmology. *Physical Review* **160**, 1165.

Davies, P. C. W. (1976) *The Physics of Time Asymmetry*, University of California Press, Berkeley, CA.

Davies, P. C. W. (1992), this volume.

D'Eath, P. D. and Halliwell, J. J. (1987) Fermions in quantum cosmology. *Physical Review* **D35** 1100.

Denbigh, K. G. (1981) *Three Concepts of Time*, Springer-Verlag, Berlin.

Denbigh, K. G. and Denbigh J. S. (1985) *Entropy in Relation to Incomplete Knowledge*, Cambridge University Press, Cambridge.

Fischler, W., Ratra, B. and Susskind, L. (1985) Quantum mechanics of inflation. *Nuclear Physics* **B259**, 730.

Fukuyama, T. and Morikawa, M. (1989) Two-dimensional quantum cosmology: directions of dynamical and thermodynamic arrows of time. *Physical Review* **D39**, 462.

Gell-Mann, M. and Hartle, J. B. (1990) Quantum mechanics in the light of quantum cosmology. In *Complexity, Entropy and the Physics of Information* (Santa Fe Institute Studies in the Sciences of Complexity, vol IX), Ed W.H.Zurek, Addison Wesley, Redwood City, CA.

Gell-Mann, M. and Hartle, J. B. (1992), this volume.

Gold, T. (1958) The arrow of time. In *11th Solvay Conference, Structure and Evolution of the Universe*, Stoops, Brussels.

Gold, T. (1962) The arrow of time. *American Journal of Physics* **30**, 403.

Gold, T. (ed.) (1967) *The Nature of Time*, Cornell University Press, Ithaca.

Grünbaum, A. (1973) *Philosophical Problems of Space and Time*, Reidel, Dordrecht.

Halliwell, J. J. (1987) Correlations in the wave function of the universe. *Physical Review* **D36**, 3626.

Halliwell, J. J. (1989) Decoherence in quantum cosmology. *Physical Review* **D39**, 2912.

Halliwell, J. J. (1991) Introductory lectures on quantum cosmology. In *Quantum Cosmology and Baby Universes, Proceedings of the 1989 Jerusalem Winter School*, Eds S. Coleman, J. Hartle, T. Piran and S. Weinberg, World Scientific, Singapore.

Halliwell, J. J. and Hawking, S. W. (1985) Origin of Structure in the Universe. *Physical Review* **D31**, 1777.

Hartle, J. B. (1985) Quantum cosmology. In *High Energy Physics 1985: Proceedings of the Yale Summer School*, Eds M.J.Bowick and F.Gursey, World Scientific, Singapore.

Hartle, J. B. (1986) Prediction and observation in quantum cosmology. In *Gravitation in Astrophysics*, Eds B. Carter and J. Hartle, Plenum, New York.

Hartle, J. B. (1991) The quantum mechanics of cosmology. In *Quantum Cosmology and Baby Universes, Proceedings of the 1989 Jerusalem Winter School*, Eds S. Coleman, J. Hartle, T. Piran and S. Weinberg, World Scienific, Singapore.

Hartle, J. B. and Hawking, S. W. (1983) Wave function of the universe. *Physical Review* **D28**, 2960.

Hawking, S. W. (1982) The boundary conditions of the universe. In *Astrophysical Cosmology*, Eds H.A.Brück, G.V.Coyne and M.S.Longair, Pontifica Academia Scientarium, Vatican City (*Pont. Acad. Sci. Varia.* **48**, 563).

Hawking, S. W. (1984a) Lectures on Quantum Cosmology. In *Relativity, Groups and Topology II, Les Houches Session XL*, Eds B. DeWitt and R. Stora, North Holland, Amsterdam.

Hawking, S. W. (1984b) The quantum state of the universe. *Nuclear Physics* **B239** 257.

Hawking, S. W. (1985) Arrow of time in cosmology. *Physical Review* **D32**, 2489.

Hawking, S. W. (1994), this volume.

Horwich, P. (1987) *Asymmetries in Time*, MIT Press, Cambridge, MA.

Joos, E. and Zeh, H. D. (1985) The emergence of classical properties through interaction with the environment. *Zeitung der Physik* **B59**, 223.

Kuchař, K. V. (1989) The problem of time in canonical quantization of relativistic systems. In *Proceedings of the Osgood Hill Meeting on Conceptual Problems in Quantum Gravity*, Eds A. Ashtekar and J. Stachel (Birkhauser, Boston).

Kuchar, K. V. (1992a) Time and Interpretations of Quantum Gravity. In Proceedings of the 4th Canadian Conference on General Relativity and Relativistic Astrophysics, eds. G Kunstatter, D. Vincent and J. Williams (World Scientific, Singapore).

Kuchař, K. V. (1992b), private communication.

Laflamme, R. (1987) Euclidean vacuum: justification from quantum cosmology. *Physics Letters* **B198**, 156.

Landsberg, P.T. (ed.) (1982) *The Enigma of Time*, Adam Hilger, Bristol.

Linde, A. D. (1984). Quantum creation of inflationary universe. *Il Nuovo Cimento* **39**, 401.

Morikawa, M. (1989). Evolution of the cosmic density matrix. *Physical Review* **D40**, 4023.

Page, D. N. (1985) Will entropy decrease if the Universe recollapses? *Physical Review* **D32**, 2496.

Penrose, R. (1979) Singularities and time asymmetry. In *General Relativity: An Einstein Centenary Survey*, Eds S. W. Hawking and W. Israel, Cambridge University Press, Cambridge.

Reichenbach, H. (1956) *The Direction of Time*, University of California Press.

Schulman, L. S. (1992), this volume.

Vachaspati, T. and Vilenkin, A. (1988) Uniqueness of the tunneling wave function of the universe. *Physical Review* **D37**, 898.

Vilenkin, A. (1982) Creation of universes from nothing. *Physics Letters* **B117**, 25.

Vilenkin, A. (1984) Quantum creation of universes. *Physical Review* **D30**, 509.

Wada, S. (1986). Quantum cosmological perturbations in pure gravity. *Nuclear Physics* **B276**, 729.

Wald, R. M. (1980) Quantum gravity and time reversibility. *Physical Review* **D21**, 2742.

York, J. (1983) The initial value problem and dynamics. In *Gravitational Radiation*, Eds N. Deruelle and T. Piran, North Holland, Amsterdam.

Zeh, H. D. (1983) Einstein nonlocality, spacetime structure and thermodynamics. In *Old and New Questions in Physics, Cosmology, Philosophy and Theoretical Biology*, Ed A. van der Merwe (Plenum).

Zeh, H. D. (1986) Emergence of classical time from a universal wave function. *Physics Letters* **A116**, 9.

Zeh, H. D. (1988) Time in quantum gravity. *Physics Letters* **A126**, 311.

Zeh, H. D. (1989) *The Physical Basis of the Direction of Time*, Springer-Verlag, Berlin.

Zurek, W. H. (1982) Environment-induced superselection rules. *Physical Review* **D26**, 1862.

# 26

# Time (A-)Symmetry in a Recollapsing Quantum Universe

H. D. Zeh

*Institut für Theoretische Physik*
*Universität Heidelberg*
*69 Heidelberg*
*Germany*

**Abstract**

It is argued that Hawking's 'greatest mistake' may not have been a mistake at all. According to the canonical quantum theory of gravity for Friedmann type universes, any time arrows of a general nature can only be correlated with that of the expansion. For recollapsing universes this seems to be facilitated in part by quantum effects close to their maximum size. Because of the resulting thermodynamical symmetry between expansion and (formal) collapse, black holes must formally become 'white' during the collapse phase (while physically only expansion of the universe and black holes can be observed). It is conjectured that the quantum universe remains completely singularity-free in this way (except for the homogeneous singularity) if an appropriate boundary condition for the wave function is able to exclude *past* singularities (as is often assumed).

## 26.1 Conditioned Entropy in Quantum Cosmology

Invariance under reparametrizations of time may be considered as a specific consequence of Mach's principle (which requires the absence of any preferred or 'absolute' time parameter). In quantum theory this leads to a time-independent Schrödinger equation (Hamiltonian constraint), since any parametrization of physical time (or 'clocks') would require the concept of a trajectory in configuration space. For example, in canonical quantum gravity the wave function of the universe is dynamically described by the 'stationary' Wheeler-DeWitt equation $H\Psi_{universe} = 0$ in superspace (the configuration space of geometry and matter). The conventional time dependence has then to be replaced by the resulting quantum correlations between all dynamical variables of the universe including those describing physical clocks, in particular the spatial metric (see Page and Wootters, 1983). However, this procedure leaves open the problem of how to formulate the asymmetry in time which is manifest in most observed phenomena.

For example, entropy as the thermodynamical measure of time asymmetry is defined in quantum theory as a functional of the density matrix $\rho$,

$$S = Trace\{\hat{P}\rho \ln(\hat{P}\rho)\} \quad . \tag{26.1}$$

This definition requires an appropriate 'relevance concept' or 'generalized coarse graining' which is represented by a 'Zwanzig projection' $\hat{P}$ (an idempotent operator on the space of density matrices – cf. Zeh, 1992). Well-known examples of relevance concepts are Boltzmann's neglect of particle correlations, or the neglect of all long-range correlations (quantum and classical) in the form of replacing the density matrix by a direct product $\hat{P}_{local}\rho := \Pi_i \rho_{\Delta V_i}$ of density matrices $\rho_{\Delta V_i}$ for separate volume elements $\Delta V_i$, each of them obtained from $\rho$ by tracing out the rest of the world (the 'environment'). The latter procedure gives rise to the usually presumed local concept of an entropy *density*. Under an appropriate Zwanzig projection, the density matrix in (26.1) may even represent a pure ('real') state, $\rho = |\psi ><\psi|$, which should, however, depend on some time variable in order to allow the entropy to grow.

Physical entropy, in contrast to the entropy of information, is objectively defined as a function of macroscopic variables (such as those characterizing density, volume, shape, position or temperature), regardless of whether they are known. Therefore, the wave function $\psi$ to be used in (26.1) cannot be identified with $\Psi_{universe}$ (which is a superposition of macroscopically different states), but must instead represent some 'relative state' (conditioned wave function) for the microscopic degrees of freedom with respect to 'given' macroscopic variables of the universe (including clocks). In the framework of a global quantum description, this state is understood as the 'present collapse component' (or as 'our Everett branch') that has resulted indeterministically from all measurements or measurement-like processes of the *past*. While the unitary part of von Neumann's dynamical description of measurements leads to a superposition of macroscopic 'pointer positions', its components can be considered as dynamically decoupled from one another once they have decohered. Measurements and decoherence represent the quantum mechanical aspect of time asymmetry (Joos and Zeh, 1985; Gell-Mann and Hartle, contributions to this conference) that also has to be derived from the structure of the Wheeler-DeWitt wave function.

A procedure for deriving the approximate concept of a time-dependent wave function $\psi(t)$ from the Wheeler-DeWitt equation has been proposed by means of the WKB approximation (geometric optics) valid for part of the dynamical variables of the universe. These variables may be those describing the spatial geometry (Banks, 1985), those forming the 'minisuperspace' of all monopole amplitudes on a Friedmann sphere (Halliwell and Hawking, 1985), or all macroscopic variables which define an appropriate 'midisuperspace'. For example, Halliwell and Hawking assumed that the wave function of the universe can approximately be written as a sum of the form

$$\Psi_{universe} \approx \sum_r e^{iS_r(\alpha,\Phi)} \psi_r(\alpha, \Phi; \{x_n\}), \tag{26.2}$$

where $\alpha = \ln a$ is the logarithm of the expansion parameter, $\Phi$ is the monopole amplitude of a massive scalar field which represents matter in this model, while the variables $x_n$ (with $n > 0$) represent all multipole amplitudes of order $n$. The exponents $S_r(\alpha, \Phi)$ are Hamilton-Jacobi functions with appropriate boundary conditions, while the relative states $\psi_r$ are assumed to depend only weakly on $\alpha$ and $\Phi$. If the corresponding orbits of geometric optics in minisuperspace are parametrized in the form $\alpha(t_r), \Phi(t_r)$, one may approximately derive from the Wheeler-DeWitt equation a Schrödinger type evolution

$$i\frac{\partial}{\partial t_r}\psi_r(t_r, \{x_n\}) = H_x\psi_r(t_r, \{x_n\}) \tag{26.3}$$

for the 'relative states' $\psi_r(t_r, \{x_n\}) := \psi_r(\alpha(t_r), \Phi(t_r), \{x_n\})$. It may apply within the limits of geometric optics along most parts of the trajectories on each WKB sheet $S_r(\alpha, \Phi)$, but one must keep in mind that this dynamical approximation does not *define* the states $\psi_r(t_r, \{x_n\})$ from which the entropy is to be calculated.

In order to be able to describe the dynamics of the observed quantum world, equation (26.3) must contain the description of the above-mentioned measurements and measurement-like interactions in von Neumann's unitary form

$$\psi_r \propto \left(\sum_k c_k \psi_k^S\right) \psi_0^A \underset{t_r}{\rightarrow} \sum_k c_k \psi_k^S \psi_k^A, \tag{26.4}$$

valid in the direction of 'increasing time'. For proper measurements the 'pointer positions' $\psi_k^A$ of the 'apparatus' $A$ must decohere through further 'measurements' by the environment, and thus lead to newly separated world branches, each one with its own corresponding 'conditioned (physical) entropy'. The formal entropy corresponding to the ensemble of different values of $k$ would instead have to be interpreted as describing 'lacking knowledge'.

This required asymmetry with respect to the direction of the orbit parameter $t_r$ means that (26.3) may be meaningfully integrated, starting from the wave function representing the present state of the observed world, only into the 'future' direction of $t_r$ (where it has to describe the entangled superposition of all outcomes of future measurements). In the 'backward' direction of time this calculation does *not* reproduce the correct quantum state, since the unitary predecessors of the non-observed components would be missing. This is particularly important if the trajectories are continued backwards into the inflationary era, or even into the Planck era where different trajectories in minisuperspace (and in the case of recollapsing universes even both of their 'ends') have to interfere with one another in order to form the complete boundary condition for the total Wheeler-DeWitt wave function (the 'intrinsic' initial condition).

Entropy is expected to grow in the same direction of time as that describing measurements. Any such asymmetry requires a *very special cosmic initial condition*; the existence of measurement-like processes in the quantum world requires essentially a non-entangled initial state (Zeh, 1992). Since the unitary dynamics (26.3) was derived as an approximation from the Wheeler-DeWitt equation, its initial condition for $\psi_r(t_r)$, too, must be derived from $\Psi_{universe}$. There are no free boundary conditions for trajectories or their relative states.

In order to obtain an appropriate asymmetry of the Wheeler-DeWitt wave function, it will be assumed in accordance with current models of the quantum universe that the Wheeler-DeWitt Hamiltonian for the gauge-free multipoles on the Friedmann sphere is of the form

$$2e^{3\alpha}H = +\frac{\partial^2}{\partial\alpha^2} - \frac{\partial^2}{\partial\Phi^2} - \sum_n \frac{\partial^2}{\partial x_n^2} + V(\alpha, \Phi, \{x_n\}), \qquad (26.5)$$

with a potential $V$ that becomes 'simple' (e.g. constant) in the limit $\alpha \to -\infty$. In his talk, Julian Barbour gave an example for how complicated the effective potential in configuration space becomes instead once the particle concept has emerged from the general quantum state of the fundamental fields. The hyperbolic nature of (26.5) defines an initial value problem with respect to $\alpha$ which then also allows one to choose a 'simple' (or symmetric) initial condition (SIC) for $\Psi_{universe}$ in the limit of small $a$. Its qualitative aspects may be illustrated by a WKB approximation with respect to $\alpha$ (Conradi and Zeh, 1991; Conradi, 1992)

$$\Psi_{universe}(\alpha, \Phi, \{x_k\}) \to \frac{1}{(-V)^{1/4}} \exp\left[\int_{-\infty}^{\alpha} \sqrt{-V(\alpha', \Phi, \{x_k\})}d\alpha'\right] \to \Psi(\alpha) \qquad (26.6)$$

for $\alpha \to -\infty$. The explicit form of the 'initial' wave function resulting from the no-boundary condition (Hartle and Hawking, 1983) is less obvious, but need not be different from (26.6).

If the initial simplicity of the relative states $\psi_r$ of (26.2) can be derived from this or some similar simple structure of the total wave function close to the singularity, this means that 'early times' (in the thermodynamical sense) must correspond to small values of $a$. However, *classical* trajectories in the minisuperspace spanned by $a$ and $\Phi$ return to small values of $a$ for closed universes with cosmological constant $\Lambda \leq 0$ (even though they are clearly not symmetric in the generic case – see Fig. 26.1).†
How, then, can one distinguish between the Big Bang and the Big Crunch? Or is

† The 'no-boundary' condition, defined as a boundary condition for the Wheeler-DeWitt *wave function*, is sometimes also used for deriving special 'initial' conditions for *trajectories* at *one* of their ends, which are then *classically* continued through all of their history (cf. Laflamme and Shellard, 1987). The required classical conditions at small values of $a$ are thereby often in violent conflict with the uncertainty relations. However, such a selection of *trajectories* is neither compatible with the usual probability interpretation of quantum mechanics, nor with the structure of the Wheeler-DeWitt wave function derived from its boundary condition. By no means should these trajectories be used to calculate 'corrections' to the wave function from which they were obtained as approximate and limited concepts. (Classically, the exceptional condition of a 'bounce' at small values of $a$, sometimes derived in this way, would describe the middle of a universe's history, not its beginning or end.)

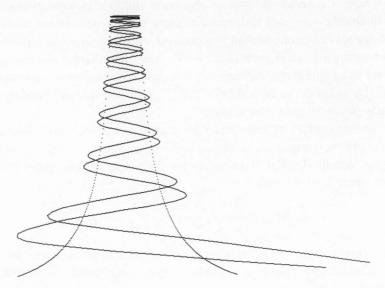

Fig. 26.1. Asymmetric classical trajectory in minisuperspace. (After Hawking and Wu, 1985 – see also Laflamme, this conference.) $a$ is plotted upwards, $\Phi$ from left to right. Dotted curve corresponds to $V = -a^4 + m^2 a^6 \Phi^2 = 0$. In more than two-dimensional minisuperspace, the trajectories need not intersect themselves. If the corresponding wave packets (Fig. 26.2) do not even overlap thereby, this would in reduced dimensions be described as their decoherence from one another.

that distinction really required for the definition of an arrow of time in quantum gravity?

The contributions of Murray Gell-Mann, Jim Hartle and Larry Schulman to this conference indicate that it is not, provided the considered universe is very young compared to its total lifetime. A symmetric (double-ended) low entropy condition for an assumed $\Psi_{universe}(t)$ would be allowed even if the latter obeyed a unitary time dependence (although it would then represent a very strong constraint). In quantum gravity, however, where there is no general time parameter $t$, one has to conclude that a 'simple' condition for $\psi_r(t_r)$ can either be derived from the boundary condition for $\Psi_{universe}$ at both ends of a turning quasitrajectory in minisuperspace, or at none. (Any asymmetric selection criteria for trajectories or their relative states – for example by means of a time-directed probability interpretation – would *introduce* an absolute direction of time 'by hand'.) In the second case, the asymmetry of the world would have to be explained as a 'great accident' occurring at one end. In the first case, all 'statistical' arrows of time must reverse their direction together with the expansion of the universe. Integrating (26.3) in the asymmetric sense of (26.4) beyond the cosmic turning point would precisely correspond to *presupposing* the quantum mechanical arrow to keep its direction.

If the concept of trajectories through minisuperspace were applicable at all for this purpose (cf. however Section 2), the derivation of *thermodynamically* asymmetric universes would require the existence of two extremely different regions at small values of $a$, together with a proof that *almost all* trajectories compatible with the structure of the correct Wheeler-DeWitt wave function have one of their ends in each of them. This would not only seem to be in conflict with the sensitivity of the trajectories to their initial conditions, but also with the statistical interpretation of entropy. In a correct quantum description, the broad 'initial wave packet' which represents the whole assumed low entropy region would, if exactly propagated through superspace according to the Wheeler-DeWitt equation and reflected from the repulsive curvature potential at large $a$, have to reproduce the complementary 'initial' wave packet that represents the high entropy region at the boundary of small $a$ without thereby interfering with the low entropy region. The condition of reflection (integrability for $a \rightarrow \infty$) restricts the otherwise complete freedom of choosing the *intrinsic* initial values (corresponding to the hyperbolic nature of the Wheeler-DeWitt equation) by a factor of $1/2$. Regardless of all open problems of dynamical consistency, no properties of the Wheeler-DeWitt Hamiltonian or in the no-boundary condition seem to indicate the existence of two that much contrasting regions for small values of $a$.

If the arrow of time is instead correlated with the expansion, the derived dynamics (26.3) for $\psi_r$ has always to be applied in the direction of growing values of $a$. In particular, considering the inflation of the early universe as 'causing' a low entropy state at one end of the trajectory only would be equivalent to *presuming* an arrow of causality in a certain direction of it (instead of deriving this asymmetry as claimed).

Notice that in quantum gravity there is no problem of consistency of the lifetime of the recollapsing universe with its supposedly much longer Poincaré cycles (or with the mean time intervals between two *statistical* fluctuations of cosmic size), as it would arise from the mentioned double-ended boundary conditions under deterministic (such as unitary) dynamics. The exact dynamics $H\Psi_{universe} = 0$, understood as an intrinsic initial value problem in the variable $\alpha$, constitutes a well-defined one-ended condition, while the reversal of the arrows of time described by the time-dependence $\psi_r(t_r)$ is facilitated by the required *corrections* to the derived unitary dynamics. These corrections have to describe recoherence and inverse branchings on the return leg.

I am thus trying to convince Stephen Hawking that he did *not* make a mistake[†] before he changed his mind about the arrow of time! Even in classical general relativity, the asymmetry of individual trajectories in minisuperspace (pointed out by Don Page, 1985) would not be sufficient for drawing conclusions on much stronger *thermodynamical* asymmetries.

---

[†] The title of Hawking's presentation at the conference was "My greatest mistake".

## 26.2 Reversal of the Expansion of the Universe in Quantum Gravity

Within the canonical quantum theory of gravity it appears therefore hardly *possible* for the arrow of time to maintain its direction when the universe starts recollapsing. However, the above picture of wave functions approximately evolving along separate WKB orbits in minisuperspace is not a sufficient representation of the dynamics described by the Wheeler-DeWitt equation – not even far outside the Planck region. As will be shown, the approximation of geometric optics does not justify the continuation of classical trajectories through the whole history of a universe. For example, a trajectory chosen to be compatible with the WKB approximation of the wave function at one end, and found to be incompatible with it at the other one, would *not* indicate an asymmetric arrow of time along this trajectory, but simply demonstrate that the concept of trajectories must have broken down in between.

Wave mechanically, trajectories have to be replaced by narrow wave packets which separately solve the wave equation. The exact dynamics for $\Psi_0(\alpha, \Phi)$ in minisuperspace (now replacing the approximation $e^{iS(\alpha,\Phi)}$) is described by

$$2e^{3\alpha}H\Psi_0(\alpha, \Phi) = \frac{\partial^2\Psi_0}{\partial\alpha^2} - \frac{\partial^2\Psi_0}{\partial\Phi^2} + [-e^{4\alpha} + m^2e^{6\alpha}\Phi^2]\Psi_0(\alpha, \Phi) = 0. \qquad (26.7)$$

The $\alpha$-dependent oscillator potential for $\Phi$ suggests the ansatz

$$\Psi_0(\alpha, \Phi) = \sum_n c_n(\alpha)\Theta_n\left(\sqrt{me^{3\alpha}}\Phi\right), \qquad (26.8)$$

where the functions $\Theta_n$ are the oscillator eigenfunctions. In the adiabatic approximation, the coefficients $c_n(\alpha)$ decouple dynamically,

$$\frac{d^2c_n(\alpha)}{d\alpha^2} + [-e^{4\alpha} + (2n+1)me^{3\alpha}]c_n(\alpha) = 0. \qquad (26.9)$$

In this case, coherent oscillator wave packets exhibit the least possible dispersion, and may therefore be expected to resemble the trajectories of geometric optics best.

As demonstrated by Kiefer (1988), the usual (here 'final' with respect to the intrinsic wave dynamics) condition of square integrability for $\alpha \to +\infty$ leads to the classically expected reflection of quasitrajectories from the repulsive curvature-induced potential $-e^{4\alpha}$. (Without such a condition, wave packets would not return at all.) For example, a further WKB approximation to (26.9), together with Langer's pasting to the exponentially decreasing WKB solutions at the classical turning point, leads to

$$c_n(\alpha) \propto \cos[\phi_n(\alpha) + n\Delta\phi] + \cos[\phi_n(\alpha) - n\Delta\phi + \delta_n] \qquad (26.10)$$

$$= \text{'expanding universe'} + \text{'collapsing universe'}, \qquad (26.11)$$

where the $\phi_n$'s are monotonic functions of $\alpha$ (approximately proportional to $n$) while $\delta_n = (\pi/4)m^2(2n+1)^2$ is the 'scattering' phase shift enforced by the 'final' (large $a$) condition. The two cosines correspond to the expanding and recollapsing parts of the histories of classical universes in minisuperspace. $\Delta\phi$ is the phase of

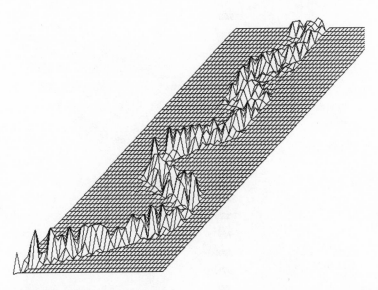

Fig. 26.2. Wave packet representing the trajectory of an expanding universe (first cosine of Eq. (26.10) only) for mass of scalar field $m = 0.2$ and mean excitation $\bar{n} = 600$, corresponding to $a_{max} = 240$. Plot range from left to right is $-0.19 < \Phi < 0.19$, while from bottom to top it is $50 < a < 150$. (The intrinsic structure of the wave packet is not resolved by the chosen grid size.)

the classical $\Phi$-oscillation at the point of maximum $\alpha$ (describing the asymmetry of the trajectory). If the constants of integration at our disposal from (26.9), which determine the size and phase of the coefficients $c_n$, are now chosen to form coherent states from the first cosine on the right-hand side of Eq. (26.10), these phase relations are then completely changed by the large phase shift differences $\delta_n - \delta_{n-1} \propto n$ resulting from the second cosine. While the term representing the expanding universe (Fig. 26.2) nicely resembles the corresponding part of a classical trajectory (Fig. 26.1), the reflected wave is smeared out over the whole allowed region (Fig. 26.3). This spreading must also be described by the corresponding Klein-Gordon current. From a sharp ($n$-independent) potential barrier in 'time' $\alpha$, the wave packets would instead be reflected without any dispersion.

This dispersion of the wave packet will become even more important for more macroscopic universes (higher mean oscillator quantum numbers $\bar{n}$), since the phase shift *differences* are proportional to $n$. The result depicted by Fig. 26.3 may therefore be expected to represent a generic property of Friedmann type quantum universes. Quasiclassical trajectories must then never be continued beyond the turning point in order to end in a well-defined region of high entropy. The wave mechanical continuation leads instead to a *superposition of many* recollapsing universes (each of which cannot be intrinsically distinguished from an expanding one). Cosmological quantum effects of gravity thus seem to be essential not only at the Planck scale!

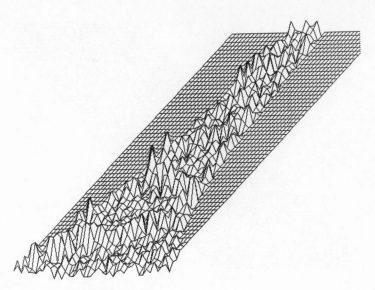

Fig. 26.3. Same wave packet as in Fig. 26.2 with recollapsing part (second cosine) added. The part of the wave packet representing the expanding universe of Fig. 26.2 is still recognizable.

The phase relations of the resulting superpositions of quasitrajectories on the return leg in minisuperspace are however destroyed by decoherence – now 'irreversibly' acting in the opposite direction of the trajectory (with increasing $a$ again) because of the (formally) final condition at the (formal) Big Crunch. (The phase shifts $\delta_n$ could as well have been put into the first cosine with a negative sign, since there is no absolute direction of probabilistic 'scattering' from one wave packet into the other. One has to be careful to avoid any notion of absolute time.) A related result has independently been obtained by Kiefer (1992b). This is further evidence that the unitary dynamics (26.3) cannot be continued along trajectories beyond the turning point at maximum $a$.

Although wave packets solving the Wheeler-DeWitt equation in minisuperspace can thus be *defined* to be intrinsically asymmetric, they are *physically* determined (as Everett branches) by their decoherence from one another. Wave packets in the *complete* configuration space (which never decohere, since they do not possess an environment) are not to describe the whole 'quantum world', but merely the (limited) causal connections which give rise to the latter's 'classical appearance'.

## 26.3 Black-and-White Holes

A formal reversal of the arrow of time (in particular if facilitated through quantum effects near the turning point of the universal expansion) must drastically affect the internal structure of black holes (Zeh, 1992). For comparison, consider black

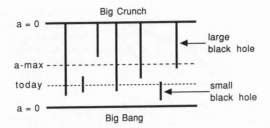

Fig. 26.4. Time-asymmetric classical universe with a homogeneous Big Bang only (Penrose, 1981).

holes which would form during the expansion of a time-asymmetric universe, and which are massive enough to survive the turning point (cf. Penrose's diagram in Fig. 26.4). If the arrow of time is now formally reversed along a (quasi)trajectory through mini- or midisuperspace in order to form a quasiclassical time-symmetric universe, black holes cannot continue 'losing hair' any further by radiating their higher multipoles *away* (by means of retarded radiation) when the universe starts recollapsing. They must instead grow hair by means of the now *coherently incoming* (advanced) radiation that has to drive the matter apart again.

The reversal of all arrows of time has of course to include the replacement of time-directed 'causality' by what would formally represent a 'conspiracy'. A mere reversal of the expansion would not by itself be able to 'cause' a reversal of the thermodynamical or radiation arrows without simultaneous reversal of the time-direction of this causation. The (fork-like) causal structure (see Zeh, 1992) must hence be contained in the dynamical structure of the universal wave function that results from the intrinsic initial condition by means of the Wheeler-DeWitt equation. Black holes must therefore formally disappear as 'white holes' during the recollapse phase of the universe.

This surprising fate of black holes thus seems to become important only in the very distant future (long after horizons and singularities may be expected to have formed in their interiors). However, our simultaneity with a black hole is not well defined because of the time translation invariance of the Schwarzschild metric. Fig. 26.5 shows a spherical black hole in Kruskal-type coordinates (a modified Oppenheimer-Snyder scenario) after translation of the Schwarzschild time coordinate t such that the turning point of the universal expansion is now at $t = 0$ (hence also at the corresponding Kruskal time coordinate $v = 0$). The resulting 'black-and-white hole' must then also exhibit a *thermodynamically symmetric* appearance, although it need not be symmetric in non-conserved microscopic or macroscopic properties ('hair'). If *past* horizons and singularities can in fact be excluded by an appropriate initial condition at the Big Bang (as it is claimed for the Weyl tensor hypothesis), the same conclusion must hold in quantum gravity also for *future* horizons and singularities.

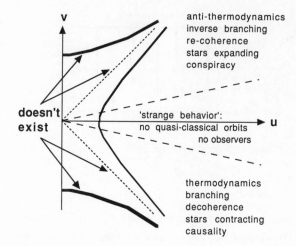

Fig. 26.5. 'Black-and-white hole' originating from a thermodynamically active (i.e., non-pathological) collapsing spherical matter distribution, with the Kruskal time coordinate $v = 0$ chosen to coincide with the time of maximum size of the universe. If the quantum effects studied in Section 2 are essential, this classical picture is not meaningful itself in the region of 'quantum behaviour' around $v = 0$. Only a probabilistic connection can then exist between its upper and lower parts.

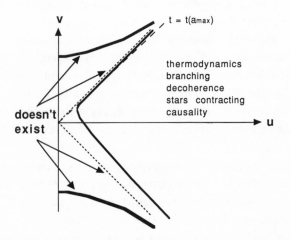

Fig. 26.6. Same black-and-white hole as in Fig. 26.5 considered from our perspective of a young universe.

One may therefore conjecture a completely singularity-free quantum world (i.e., a wave function vanishing at *all* singularities).

Fig. 26.6 shows the same situation as Fig. 26.5 from our perspective of a young universe (after a back-translation of the Schwarzschild time coordinate such that $t_{today} = 0$). From this perspective, the time coordinate $t = t_{turn}$ appears to be

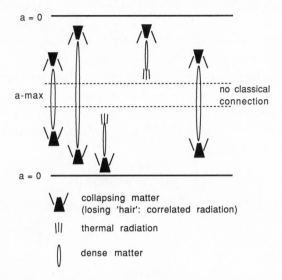

Fig. 26.7. Time-symmetric, singularity-free universe with black-and-white holes together with (small) black or white holes.

very 'close' to where one would expect the future horizon to form. The 'strange' thermodynamical and quantum effects now also appear to occur close to the horizon, thereby preventing it to form.

This reversal of the gravitational collapse cannot be *observed* from a safe distance, although it could be experienced by suicidal methods within relatively short proper times if a black hole were available in our neighborhood. If the black-and-white hole is massive enough, this kind of 'quantum suicide' must be quite different from the classically expected one by means of tidal forces. In a classical picture, travelling through a black-and-white hole may reduce the proper distance between the Big Bang and the Big Crunch considerably, but unfortunately we could not survive as information and memory *gaining* systems. This consideration should at least demonstrate that the classical (Kruskal-Szekeres) continuation of the Schwarzschild metric beyond the horizon is absolutely doubtful for thermodynamical and quantum mechanical reasons!

Before Stephen Hawking changed his mind about the time arrow in a recollapsing universe, he had conjectured (Hawking, 1985) that the arrow is reversed *inside* the horizon of a black hole, since "it would seem just like the whole universe was collapsing around one" (cf. also Zeh, 1983). This consequence would however not correctly describe the situation in a thermodynamically time-symmetric universe.

Penrose's black holes hanging like stalactites from the 'ceiling' (the Big Crunch) in Fig. 26.4 must now also become symmetric, as shown in Fig. 26.7. Black-and-white holes in equilibrium with thermal radiation (as studied by Hawking, 1976) would instead consist of thermal radiation at both ends. They would possess no 'hair' at all, neither to lose nor to grow. The classically disconnected upper and lower halves of Fig. 26.7 should rather be interpreted as two of the many Everett branches of the quantum universe, each of them representing an *expanding* quasiclassical world.

The absence of singularities from this quantum universe thus appears to be a combined thermodynamical and quantum effect. However, one may equivalently interpret the result as demonstrating that in quantum cosmology the thermodynamical arrow is a *consequence* of the absence of inhomogeneous singularities – a generalization (or symmetrization) of Penrose's Weyl tensor condition.

## Acknowledgment

I wish to thank H.D. Conradi and C. Kiefer for their critical reading of the manuscript.

## Discussion

**Hawking** Your symmetric initial condition for the wave function is wrong!

**Zeh** Do you mean that it does not agree with the no-boundary condition?

**Hawking** Yes.

**Zeh** It was not meant to agree with it, although we found it to be very similar to the explicit wave functions you gave in the literature for certain regions of minisuperspace. This is however not essential for my argument. It requires only that the multipole wave functions $\psi_r$ become appropriately 'simple' (low-entropic and factorizing) for small values of $a$ (as you too seem to assume, although only at that 'end' of the trajectory where you start your computation).

**Hawking** I think it is a mistake to to look at the no boundary condition in terms of solutions of the Wheeler Dewitt equation. The no boundary condition does not translate into a simple boundary condition on the wave function in superspace. Instead, one should look at semiclassical approximation. That is, one should look at complex solutions, that are saddle points in the path integral. These complex solutions will not be time symmetric. They will be non-singular at one end, but singular at the other end. The universe will be smooth and ordered at the non-singular end, but disordered at the singular end. So you have an arrow of time that does not reverse.

The only way one can use the no boundary condition, as a boundary condition, is by the semiclassical approximation.

I think you get a sum of a classical history, and its time reverse, only because you look at the wave function for a three geometry. But this is a bad thing to do, because a radius occurs twice in a universe that expands and collapses. You can overcome this problem in two ways. One is to express the wave function as a function of trace $K$, rather than the radius. Or you can look at the saddle point in the path integral. Either way separate a classical history from its time reverse. There is no question of coherence between them.

**Barbour** Did I understand you correctly to say that the criteria Kiefer used to obtain his solution was of the kind I call Schrödinger type, namely that there should be no blowing up of the wave function anywhere in the configuration space?

**Zeh**  Yes – if by blowing-up solutions you mean the exponentially increasing ones. Otherwise you would not be able to describe reflection (turning trajectories) by means of wave packets. I think this assumption corresponding to the usual normalizability is natural (or 'naive' according to Karel Kuchař) if the expansion parameter $a$ is considered as a dynamical quantum variable (as it should in canonical quantum gravity).

**Barbour**  Could it be that worries about the turning point are an artifact of the extreme simplicity of the model? Consider in contrast a two-dimensional oscillator in a wave packet corresponding to high angular momentum!

**Zeh**  The described quantum effects at the turning point are due to the specific Friedmann potential with an oscillator constant for $\Phi$ exponentially increasing with $\alpha$. They do not seem to disappear if added degrees of freedom possess similarly 'normal' potentials (e.g. polynomials multiplied by positive powers of $a$). This seems to be the case in Friedmann-type models.

**Kuchař**  Did you study decoherence between $\psi$'s corresponding to one $S$, or also the decoherence corresponding to different $S$'s?

**Zeh**  I expect decoherence to be effective between different trajectories in minisuperspace (cf. Kiefer, 1987), between macroscopically different branches of the multipole wave functions $\psi_r$ along every trajectory, and between different WKB sheets corresponding to different $S$'s except at very small and large $a$ (cf. Halliwell, 1989; Kiefer, 1992a). Otherwise equation (26.3) would not be valid as an independent approximation on different sheets.

**Griffiths**  In ordinary quantum mechanics of a closed system, I do not know how to make any sense out of it using the 'wave function of the closed system'. I need the unitary transformations that take me from one time to another. Is there any analogy of this in quantum gravity? For if not, it is hard to see how quantum gravity can be used to produce a sensible description of something like the world we live in.

**Zeh**  Your question seems to apply to quantum gravity in general. I think that it is sufficient for the wave function of the universe to contain correlations between all physical variables – including those describing clocks. In classical theory these correlations would be essentially unique, since they would be represented by the trajectories in the complete configuration space which remain after eliminating any (physically meaningless) time parameter. In quantum theory there are no trajectories that could be parametrized. These quantum correlations must of course obey 'intrinsic' dynamical laws as they are described by the Wheeler-DeWitt equation. From them one tries to recover the time-dependent Schrödinger equation (which has to describe the 'observed world') as an approximation when spacetime (the history of spatial geometry) is recovered as a quasiclassical concept.

**Lloyd**  Could you clarify how black holes would grow hair in the contraction phase? Is it through interference between incoming radiation and the Hawking radiation?

**Zeh**  Only the incoming (advanced) radiation is essential, since black holes can form by losing hair even if Hawking radiation is negligible. This is a pure symmetry consideration. A final condition which is thermodynamically and quantum mechanically (although not in its details) the mirror image in time of an initial condition that leads to black holes must consequently lead to their time-reversed phenomena. If Hawking radiation is essential (as for small mass), the black hole may disappear before $t(a_{max})$ is reached, but again before an horizon forms.

**Hawking** The no-boundary condition can only be interpreted by means of semi-classical concepts such as the saddle point method.

**Zeh** I would prefer to understand such a fundamental conclusion as the arrow of time in terms of an exact (even though incomplete) description. In particular, your opposite conclusion about the arrow of time seems to be introduced by the direction of computation (along the assumed trajectories) by using approximations, similar to how it is often erroneously argued in the theory of chaos by using 'growing errors' in the calculation for explaining the increase of 'real' physical entropy!

  – Did I understand you correctly during your talk that you – at the time when you made what you call your 'mistake' – also expected black holes to re-expand during the recollapse of the universe?

**Hawking** Yes. I did not understand black holes sufficiently until I changed my mind.

## References

Banks, T. (1985) TCP, Quantum Gravity, the Cosmological Constant, and all that ... *Nucl. Physics* **B249**, 332.

Conradi, H.D. and Zeh, H.D. (1991) Quantum cosmology as an initial value problem. *Phys. Lett.* **A154**, 321.

Conradi, H.D. (1992) Initial state in quantum cosmology. *Phys. Rev.* **D46**, 612.

Halliwell, J.J. (1989) Decoherence in quantum cosmology. *Phys. Rev.* **D39**, 2912.

Halliwell, J.J. and Hawking, S.W. (1985) Origin of structure in the Universe. *Phys. Rev.* **D31**, 1777.

Hartle, J.B. and Hawking, S.W. (1983) Wave Function of the Universe. *Phys. Rev.* **D28**, 2960.

Hawking, S.W. (1976) Black Holes and Thermodynamics. *Phys. Rev.* **D13**, 191.

Hawking, S.W. (1985) Arrow of Time in Cosmology. *Phys. Rev.* **D32**, 2489.

Hawking, S.W. and Wu, Z.C. (1985) Numerical Calculations of Minisuperspace Cosmological Models. *Phys. Lett.* **151B**, 15.

Joos, E. and Zeh, H.D. (1985) The Emergence of Classical Properties through Interaction with the Environment. *Z. Phys.* **B59**, 223.

Kiefer, C. (1987) Continuous measurement of mini-superspace variables by higher multipoles. *Class. Qu. Gravity* **4**, 1369.

Kiefer, C. (1988) Wave packets in mini-superspace. *Phys. Rev.* **D38**, 1761.

Kiefer, C. (1992a) Decoherence in Quantum Cosmology. In *Proceedings of the Tenth Seminar on Relativistic Astrophysics and Gravitation,* Eds. S. Gottlöber, J.P. Mücket and V. Müller, World Scientific.

Kiefer, C. (1992b) Decoherence in quantum electrodynamics and quantum gravity. *Phys. Rev.* **D46**, 1658.

Laflamme, R. and Shellard, E.P.S. (1987): Quantum cosmology and recollapse. *Phys. Rev.* **D35**, 2315.

Page, D.N. (1985) Will entropy increase if the Universe recollapses? *Phys. Rev.* **D32**, 2496.

Page, D.N. and Wootters, W.K. (1983) Evolution without Evolution: Dynamics Described by Stationary Observables. *Phys. Rev.* **D27**, 2885.

Penrose, R. (1981) Time Asymmetry and Quantum Gravity, In *Quantum Gravity 2*, Eds. Isham, C.J., Penrose, R. and Sciama, D.W., Clarendon Press.

H.D. (1983) Einstein Nonlocality, Space-Time Structure, and Thermodynamics, In *Old and New Questions in Physics, Cosmology, Philosophy, and Theoretical Biology*, Ed. van der Merwe, A., Plenum.

Zeh, H.D. (1992) *The Physical Basis of the Direction of Time*, Springer (second edition).

# 27

# The Emergence of Time and Its Arrow from Timelessness

J. B. Barbour

*College Farm, South Newington*
*Banbury, Oxon OX 15 4JG, England*

**Abstract**

An attempt is made to sketch a complete timeless theory of the universe and explain why time can nevertheless appear to flow in such a framework.

## 27.1 Is Time a Basic Concept?

During the Workshop, I conducted a very informal straw-poll, putting the following question to each of the 42 participants:

*Do you believe time is a truly basic concept that must appear in the foundations of any theory of the world, or is it an effective concept that can be derived from more primitive notions in the same way that a notion of temperature can be recovered in statistical mechanics?*

The results were as follows: 20 said there was no time at a fundamental level, 12 declared themselves to be undecided or wished to abstain, and 10 believed time did exist at the most basic level. However, among the 12 in the undecided/abstain column, 5 were sympathetic to or inclined to the belief that time should not appear at the most basic level of theory.

Although my straw-poll probably broke all rules of scientific opinion polling, and, as was pointed out to me, the question itself was likely to elicit the response I myself favoured (nonexistence of time at the fundamental level) - for what theoretical physicist will resist the challenge to reduce the world to the minimal number of basic concepts? - I think it was worth establishing that a clear majority was inclined to do away with time.

I also felt that, given the topic of the Workshop, this was a question on which we should be concentrating our minds more actively. I do believe that any theory of the world in which time is truly eliminated as a fundamental concept will have more startling consequences than many of even the most dedicated 'no-timers' appreciate. For example, a proposal made on the final day of our discussions for the cover design of the Workshop proceedings received widespread assent. Now the

405

implication of the proposed design was that the formalism of quantum gravity admits *pairs* of solutions corresponding to expanding and contracting universes. Such an assumption brings with it the problem of why we observe only an expanding world and not a contracting one too. However, as explained below, I suspect this problem is spurious, the result of a failure to exorcise the ghost of Newtonian absolute time from a context in which it is quite inappropriate.

My own contribution to the Workshop, a slightly edited version of which now follows, is an attempt, in part still very qualitative, to sketch a complete timeless theory of the world, including an account of how sentient beings within such an atemporal world could neverless perceive it as intensely temporal. I hope it will foster discussion and clarification of the fundamental issue of timelessness.

## 27.2  Outline of a Timeless Theory of the World

In the theory of the whole world, time is a redundant concept. One only needs the world's possible relative configurations. Think of them as pictures, a different one for each configuration. They form the *relative configuration space* of the universe, each point of which is a distinct structured whole. The essence of this approach is structure. There is nothing else *at all* but these structured wholes.

Our primitive idea of the passage of time derives from change. If I see one picture and then another, slightly different from the first, that is already enough to give the idea that time has passed. What I want to do in this talk is explore systematically a conceptual world in which time is totally banned as a basic concept and can be recovered as an effective concept solely from differences *between* configurations (at the level of classical physics) or from structure within a single configuration (in quantum physics).

I begin with classical physics.

A *history of the world* is a curve in its relative configuration space. We can define an action between points $A$ and $B$ of that history using data intrinsic to the points of the history and nothing else. Take any two configurations that differ just a little. Match, in some way, all the points of the one to all the points of the other. Choose some quantity that measures the difference between the corresponding pictures at a matched pair of points. Then add up all such quantities for all paired points. The result is a global, or integrated, difference for that trial matching. Do this for all possible trial matchings and choose the *best matching*, the one that makes the global difference extremal.

That defines an action between the compared neighbouring points. Moving along the history and adding the best matching differences as we go, we determine the action of that history between $A$ and $B$. One such history will have extremal action. I declare it to be the *classical history* of the world between $A$ and $B$.

Let us take all the pictures corresponding to the points of this classical history and throw them down in a confused heap. Since everything has been done using data intrinsic to the configurations, nothing will be lost. The pictures still tell the same story, and we can easily put them back in their 'right' order.

The advantage of this order is that the pictures tell their story in a manner easier to read. But we can do more. There's a dynamical reason, rooted in the extremalization of intrinsic differences, why the one picture follows the next. Let us bring that out more clearly. Let us start with the picture at one end, calling it the first. Let us then take the next one and move it around on top of the first until we find the position where they are in their best matching positions relative to each other. Then we take No. 3 and move it around on top of 2 until they too are locked into the best matching position. We go on like that all the way to the other end.

To keep them safely in these special positions, we could fit them onto a 'rod', which keeps them locked into the correct mutual positions.

There is one last thing we can do. We can space the pictures out along the rod in such a way that, as we move along it, the pictures seem to change in the *steadiest* possible way. When you do this properly with the mathematical equations (Barbour & Bertotti, 1982) there is a certain spacing which is uniquely convenient. It establishes distinguished spacings between the successive pictures.

Newton called these spacings the intervals of absolute time, but in the approach being developed here his time doesn't exist at all. It's a mere convenience to think of things that way.

The thing I most want to insist on is this. Newton supposed that the different configurations of the world are realized at different instants of time. That, I believe, is a pernicious misconception. The pictures do not occur *at* instants of time. They *are* the instants of time.

Now what about the conceptual 'rod' introduced to hold the pictures in the dynamically most revealing relative positions? Newton called it absolute space. Today we call it the rest inertial frame of reference. It's just as convenient and just as redundant.

Let me summarize: If from the very outset we consider the entire world, then all of dynamics, including general relativity (*ibid*), can be recovered in this timeless and frameless fashion. Pure configuration is enough. In particular, the passage of time is nothing but the *difference* between configurations, measured and parametrized in a more or less unique manner dictated by the same action that creates extremal histories.

What then are the consequences of the nonexistence of time? An important one is this: The classical theory we have constructed has no sense of direction coded into it. Curves in the configuration space do not carry the names *past* and *future* at their ends.

This simple fact has uncomfortable implications. Despite the widespread intuitive acceptance of the contrary, there is not a double set of possible motions of the complete universe with every motion going one way matched by another going the other.

In a timeless theory, that is simply wrong. Curves are curves. They don't have arrows on them. We do not have a BIG BANG nor a BIG CRUNCH nor even their superposition, the BIG BRUNCH, but simply curves in the configuration space of the world.

How then do we have the firm conviction we live in an expanding universe? Why indeed do I think I got up this morning?

It is no answer to say that in consciousness we are somehow directly aware of the passage of time and can therefore tell which way we travel through it. No: Not only dynamics but also the sense of the passage of time must come from the bare idea of configuration.

We need the concept of *time capsules*, which I first introduce qualitatively.

A *single* photograph of a geological section shows fossils at different levels. This is history compressed into now. For those trained to see it, the apparent evidence of temporal evolution from a less structured past over a vast period of time to a more structured present is literally set in rock. All this can be present in just one configuration. Any piece of rock on the earth is a time capsule. In fact, almost anything one can get ones hands on in the real universe is a time capsule, including all matter within stars (the chemical composition of which tells a history).

A formal definition will now be helpful. A *time capsule* is a single configuration (either of the entire universe or part of it) that seems to be the outcome of a dynamical process of evolution through time in accordance with definite laws. It appears to contain records of the past, and these records are mutually consistent. By means of these records contained within a single configuration, it is in principle possible to *date* the configuration (an example that illustrates the sense in which this is meant will be given later).

I believe that the ubiquity of time capsules has not been accorded the significance it warrants. My suggestion is that the belief in time and its pasage is *solely* a consequence of the fact that, at any instant, we find ourselves within a time capsule. If there were no time capsules, there would be no notion of time.

Indeed, it is a fact that what we experience psychologically is always a time capsule; for our memory is like a progress book, with snapshots taken every day and faithfully pasted in, one next to another, the brightest and clearest from what we think was yesterday, the ones in the supposed past getting fainter and fainter. All this is coded into the brain's now. It is at least plausible that what we take to be the direct perception of motion of images in consciousness is created in our mind by the juxtaposition of several different images, like successive movie stills, in a single structure.

I will not expand here further on this question, but I think it must be addressed and at least answered in outline to make plausible the idea that a completely timeless world could still be experienced as temporal.

Let me now start on this project.

I first introduce the *heap hypothesis*. There are two heaps: the heap of all possible configurations, the *heap of possibilities*, and the heap of realized configurations, the *heap of actualities*. I use the word *heap* because the individual objects in a heap are entities in their own right. They can be picked up and examined and have an intrinsic structure which exists independently of the fact that they belong to the heap.

We have two heaps. The next question is: what are our most basic theories, those of classical mechanics and quantum mechanics, actually telling us? The heap hypothesis is that these theories are simply rules to establish which configurations from the heap of possibilities go into the heap of actualities.

The fundamental law of classical mechanics, $\mathbf{F} = m\mathbf{a}$, tells us that the heap of actualities consists of a single curve lifted from the heap of possibilities. It says nothing about *which* particular curve that will be. It merely says there will be one and only one such curve of actualities.

How are we to save the appearances in such a framework? In particular, how are we to recover that most powerful appearance, the appearance that we move forward in time in one definite direction along such a curve?

I take up one picture at random from the curve. It may well be - and there is nothing in classical mechanics to deny the possiblity - that within this single picture there are self-sentient time capsules who deduce, from everything of which they are aware in their instant, that they occur at a certain time and place on that picture-carrying 'rod' which is such a useful device for representing in a transparent way the heap of actualities as it is conceived by classical physics.

What classical physics can never do is make these experiences, which do occur, seem at all likely. For histories containing time capsules of the type we need to explain our own experiences - to say nothing of the structure we see all around us - form the minutest fraction of the set of all possible histories. It is only a partial explanation to say that exceptional conditions at a boundary selected such a history of actualities. Classical physics can never provide a complete explanation and say why a particular solution is selected.

It is the presumed *linearity of time*, the assumption that instants are realized in a one-dimensional continuum, that makes classical physics forever impotent in this question. But if we once recognize that configurations do not occur at instants of time *but are the instants themselves* and reside, not in a puny one-dimensional line, but in a huge multidimensional space of configurations, the arena is transformed.

How are we to conceive quantum mechanics in this timeless arena? In ordinary quantum mechanics, the wave function is defined on the possible configurations, which are defined in a definite inertial frame of reference, at different times.

Nothing like this can happen in the wave mechanics of the world. There is neither time nor frame, just the heap of possibilities. The wave function $\Psi$ of the universe must be a function that takes values on the possible configurations in that heap, nothing else. This is, in fact, exactly the message of canonical quantum gravity (DeWitt, 1967). The Wheeler-DeWitt equation does not take the form of Schrödinger's time-dependent equation $i\hbar \partial \Psi / \partial t = \hat{H}\Psi$, but the form of his time- independent equation $\hat{H}\Psi = 0$. Moreover, the probabilities are for complete three-geometries and the values of the matter fields on them (i.e., for complete configurations of the universe), not for values of the metric on some underlying manifold.

What does the $\Psi$ obtained by solution of the Wheeler-DeWitt equation mean? I go for the so-called naive Schrödinger interpretation (Kuchař, 1992). A solution of the Wheeler-DeWitt equation puts a value of $\Psi$ and, with it, the Schrödinger density $\Psi^*\Psi$ on each configuration in the heap of possibilities. Let us then suppose that whoever or whatever creates the world puts a corresponding number of identical copies of that configuration into the heap of actualities.

The theory will have genuine predictive power if the Schrödinger density always has large values on configurations containing lots of time capsules with a structure such as we actually observe in the universe. For then, making the assumption that what we experience is probable, we shall have an explanation for our own actual experiences and the apparent emergence of time and its arrow from the timelessless of pure configuration, i.e., the theory predicts copious production of just the sort of configurations that we do actually experience.

Before outlining how this could work, let me draw the quantum parallel to my insistence that in classical mechanics there are just curves and neither BANG nor CRUNCH, which are simply not present in a timeless formalism.

In the quantum mechanics of subsystems of the world we are familiar with plane wave solutions of two forms: $e^{-i\omega t}e^{ikx}$ and $e^{-i\omega t}e^{-ikx}$. Out of them we can construct wave packets that move to the right and to the left. This has a good meaning in ordinary quantum mechanics because there is an external time and frame in which we can see, more or less literally, that the packets really do move in those directions.

But in the quantum mechanics of the universe it is quite common to set up a Wheeler-DeWitt equation for the scale factor $a$ of a Friedmann universe and obtain for it two fundamental solutions of the form $e^{-iS(a)}$ and $e^{iS(a)}$, where $S(a)$ is a solution of the corresponding Hamilton-Jacobi equation.

These are said to represent expanding and contracting universes, respectively. There are two severe problems with this.

First, there is no additional factor containing the time, which is what gives genuine motionic content to wave packets in ordinary quantum mechanics.

Second, the fact that the two solutions exist at all as simultaneously valid solutions is solely a consequence of the fact that the Wheeler-DeWitt equation has been taken to be a real equation but has been allowed to have complex solutions. However, I believe that a real equation should have real solutions, and then the above alleged distinction between expanding and contracting universes cannot be expressed. One may also ask whether a real equation gives a true representation of the full Wheeler-DeWitt equation for the universe when all known interactions are taken into account. In fact, I think the Wheeler-DeWitt equation must be essentially complex (Barbour,

1993). But a function and its complex conjugate cannot simultaneously be solutions of a complex linear equation. One way or another, the possibility of associating directions of motion with phase relationships in complex solutions seems to me most questionable if used in the context of the complete universe, though it is obviously valid for subsystems. On this particular point, my approach is radically different from the approach that seems to be taken thoughout the literature, including Halliwell's contribution to these proceedings.

To summarize, my conclusion is this: If we take the timeless approach seriously, then in neither the classical nor the quantum dynamics of the whole world does the theoretical formalism say anything about directions of motion of the complete world. Any conclusions of that type must be drawn from the *intrinsic structure of individual configurations*; they cannot be based on conjectured external structures. That is why the notion of time capsule is essential in a timeless theory. An arrow of time or motion can emerge only through the preferred selection of time capsules.

Now is that likely to happen? This brings me to what, so far as I know, is a new proposal (however, see Zeh, 1989). It is, I believe, the first attempt made at this workshop to find a genuine explanation of the arrow of time, as opposed to a description of it.

My conjecture is that the ultimate origin of the arrow of time is the *asymmetric structure* of the configuration space of the world. Let me note first that this configuration space is a vast and curious place. It has what in atomic physics Schrödinger (1926) called *natural boundaries*. In the cosmic context, these are, on the one side, the boundaries of zero size of the world and zero intensity of the fields in it. On the other side, they are the frontiers of infinite size and infinite field strength. Moreover, because the scale factor can have only positive values, the configuration space is decidedly lopsided! And that is just on the basis of the configurations that make up its points. We must also consider the effect of the variational principle defined on it. I think of the configuration space as a curiously shaped continent in the sea of nothing and the variational principle as putting a rich and even more asymmetric topography on that continent. Potentials are always represented as hills, valleys, wells, and walls. The configuration space of the world must be criss-crossed by the most extraordinary mountain ranges, ocean troughs, odd shaped obstacles, and so forth.

Above all, superimposed on everything is one prevailing direction, arrow if you like: from the small to the large and from low to high intensity. Pronounced structure and inhomogeneity (through still with a centre of symmetry that is absent in the cosmological case) is already clearly expressed in the picture of a far less exotic configuration space shown in Fig. 27.1. It is a projection of the configuration space of four particles on a line that interact through short-range forces. The fourth particle is at the origin. The pronounced structure is built into the configuration space by the existence of a natural origin, where all the particles sit on top of each other, and the barriers that spring up whenever two particles meet.

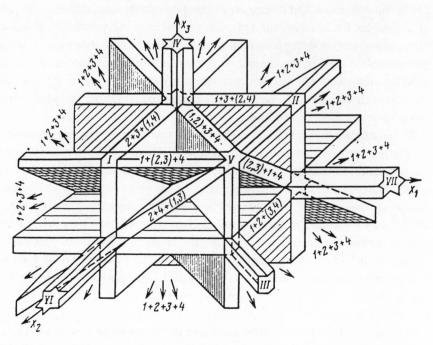

Fig. 27.1. Regions of (short-range) interaction of four particles in one-dimensional motion. Particle 4 is fixed at the origin. [Reproduced from: Zakhar'ev, B. N., Kostov, N. A., & E. B. Plekanov (1990) Exactly solvable single-channel and multichannel models (lessons in quantum intution). *Soviet Journal of Particles and Nuclei*, **21**, 384–405].

Now my conjecture: Any allowed solution of the Wheeler-DeWitt equation will be characterized by a very strong asymmetry which is imposed on it by the profound asymmetry of the configuration space on which it is defined. This asymmetry will be expressed in the preferential concentration of high values of the Schrödinger density on configurations that contain time capsules. If this is correct, the asymmetry of the world's configuration space is the ultimate origin of our belief that time exists, flows, and has an arrow at whose tip we now sit.

To make this more plausible, it will be necessary to show how the time-dependent Schrödinger equation can emerge together with classical worlds out of the timelessness of the Wheeler-DeWitt equation (discussed by Halliwell in these proceedings). It will also be necessary to explain how quantum mechanics can be interpreted on the basis of the two heaps - those of possibilities and those of actualities and to show how there is no need for collapse of the wave function in a cosmic context. That will also show why time is not an operator in ordinary quantum mechanics. If all this succeeds, the end result will be a modification of Everett's idea: not a many-worlds but a *many-instants* interpretation of quantum mechanics.

This task will be taken up elsewhere (Barbour unpublished).

*Note added in revision* (August 1992). Several months after the Workshop, I was lucky to have some extensive discussions with Dieter Zeh about the ideas expressed above. I should like to thank him for the suggestion, adopted here, to emphasize in the formal definition of a time capsule that it must contain *mutually consistent* records. He also drew my attention to many close parallels between my configuration-based many-instants interpretation of quantum mechanics (and my insistence that the notion of a past and a future arises exclusively from juxtaposition of mutually consistent records in the now) and J. S. Bell's gloss of Everett's many-worlds interpretation in his paper "Quantum mechanics for cosmologists" (Bell, 1981). In fact, Bell regarded the really novel element of Everett's theory as the "repudiation of the concept of the 'past' " and said that it could be considered "in the same liberating tradition as Einstein's repudiation of absolute simultaneity." Despite this, later in the same paper Bell declared Everett's replacement of the past by memories to be a "radical solipsism" and rejected it for that reason.

However, neither Everett's proposal nor mine is any more solipsistic than more conventional interpretations. They merely postulate a different external reality. It may also be noted that Bell did not discuss the Wheeler-DeWitt equation; had he done so, he might have been forced to take timelessness more seriously.

I would also like to take this opportunity to mention that, since the Workshop, I have discovered the notion of a time capsule (the significance of which I postulated purely speculatively at Huelva as the only possible way in which a notion of time can arise in a timeless context) is beautifully realized in Mott's account of the formation of straight tracks of alpha particles in cloud chambers (Mott, 1929). Mott's paper [the great importance of which for the interpretation of quantum mechanics is underlined by Bell (*ibid*)] shows clearly that solutions of the *time-independent* Schrödinger equation can be concentrated to an extraordinary degree on configurations that are time capsules in the sense of my definition. (Any photograph of a track in a cloud chamber is a time capsule. Moreover, it carries its *own date*, so to speak: from the configurations of the tracks in a photograph of a cloud chamber one can in principle determine the time that elapsed between the interaction event that produced the particles and the photographing of the tracks they produced.) It should be said that Mott obtained his solution by making important tacit assumptions (essentially, the assumptions under which time-independent scattering theory is equivalent to the time-dependent theory). I believe that the elucidation of the conditions under which Mott-type solutions can be obtained in the context of the Wheeler-DeWitt equation will cast much light on the origin of time and its arrow. This work is in hand.

Finally, I should like to thank Jonathan Halliwell for several helpful suggestions for revision of this paper.

## Discussion

**Gell-Mann**  Do the configurations come with an ordering?

**Barbour**  No, if by ordering you mean in a one-dimensional continuum, but yes if you mean ordering as the points of a multidimensional space. The example I give is the set of all possible relative configurations of $N$ particles in Euclidean space. Each distinct configuration of the $N$ particles is a point of the configuration space. Another example is DeWitt's superspace of all possible closed Riemannian three-geometries. Except at the frontiers of the exceptional configurations that exhibit symmetries, such configurations can be ordered.

**Unruh**  1) What is the measure you take over the configurations? The Born-Oppenheimer density means nothing unless you also have that measure. 2) What about alternatives like momenta, or do you believe that they are also figments of our imagination?

**Barbour**  1) As yet I have no definite proposal, but you are, of course, right that a choice must be made. I believe this is an important but secondary matter. 2) In the attempt to set up a timeless formulation of quantum mechanics, I do incline to think of configurations as ontologically primary and concepts like momentum as elements of the theory which explain why there is a high probability of realization of certain configurations. (For me, a theory such as quantum mechanics does not exist in any material sense, but I certainly would not regard it as a figment of my imagination.)

## References

Barbour, J.B. (1993) Time and complex numbers in canonical quantum gravity. *Phys. Rev.* **D47**, 5422-5430.

Barbour J.B. (unpublished) Time asymmetry in canonical quantum gravity. Submitted for publication. See also: Barbour, J.B., *Absolute or Relative Motion?* Vol. 2, Cambridge University Press, Cambridge (in preparation).

Barbour, J. B. and Bertotti, B. (1982) Mach's Principle and the structure of dynamical theories. *Proc. R. Soc. Lond.* **A382**, 295-306.

Bell, J. S. (1981) Quantum mechanics for cosmologists. In: *Quantum Gravity 2. A Second Oxford Symposium*, eds. Isham, C. J., Penrose, R, and Sciama, D. W., Clarendon Press, Oxford; reprinted in: Bell, J. S. (1987) *Speakable and Unspeakable in Quantum Mechanics*, Cambridge University Press, Cambridge.

Kuchař, K. V., (1992) Time and Interpretations of Quantum Gravity. In: *Proceedings of the 4th Canadian Conference on General Relativity and Relativistic Astrophysics*, eds. Kunstatter, G., Vincent, D., and Williams, J., World Scientific, Singapore.

Mott, N. F. (1929) The wave mechanics of α-ray particles. *Proc. R. Soc. Lond.* **A126**, 79-84; reprinted in: *Quantum Theory and Measurement*, eds. Wheeler, J. A., and Zurek, W. H., Princeton University Press, Princeton (1983).

DeWitt, B. S. (1967) Quantum theory of gravity. I. The canonical theory. *Phys. Rev.* **160**, 1113-1148.

Schrödinger, E. (1926) Quantisierung als Eigenwertproblem I-II. *Ann. Phys.* **79**, 361-376, 489-527.

Zeh, H. D. (1989) *The Physical Basis of the Direction of Time*, Springer, Berlin, p. 139.

# 28

# Wormholes and Time Asymmetry

Pedro F. González-Díaz[†]

*Instituto de Optica*
*Consejo Superior de Investigaciones Científicas*
*Serrano 121, 28006 Madrid, Spain*

## 28.1 Introduction

Wormholes are microscopic connections between two otherwise disconnected asymptotically flat regions of either the same universe, or two different universes (Hawking 1982, González-Díaz 1990). They can be described in Euclidean spacetime and represent a topology change by which an initial state which is flat space evolves into a final state which is made of flat space plus a given arbitrary number of baby universes (Hawking 1990).

I shall consider here the possible connection between the invariance properties of the quantum state of wormholes under time reversal and the origin of the observed time asymmetry in the whole set of ordinary matter fields at low energies in a nearly flat connected region of spacetime.

## 28.2 Correlations in Spacetime Fluctuations

In order to investigate the time-symmetry properties of wormholes, one first has to choose a suitable quantum state for them. It was first considered (Hawking 1988) that wormholes were in a pure state given by a wave function $\Psi[h_{ij}, \phi_0]$. However, this choice appears to be rather restrictive. An observer in the asymptotic region does not know the number of components on the inner boundary manifold which are required to divide the manifold. Thus, one had to sum over any number of such disjoint three-manifolds. The difficulty with using a wave function then comes from the fact that, even if we cut all components by a given disjoint cross section, the resulting four-manifold is not effectively divided. This follows from the feature that wormholes are actually microscopic fluctuations of the spacetime itself, and therefore baby universes which are branched off through a disjoint inner three-surface should be all correlated. In this case, some given information about the whole inner boundary becomes accessible to observers in the asymptotic regions

† Present address: Instituto de Mathemáticas y Física Fundamental, C.S.I.C., Serrano 121, 28006 Madrid, Spain

415

through gravitational flow lines from the whole baby universe sector. Such a connection renders the full wormhole four-manifold no longer effectively divided into disconnected parts. Thus, the most general quantum state of wormholes should be given by a density matrix representing a mixed state (González-Díaz 1991a). In any case, the quantum state of wormholes can be described by a path integral

$$Q.S. = \int_C d[g_{\mu\nu}]d[\phi]e^{-I[g_{\mu\nu},\phi]}. \tag{28.1}$$

If the inner manifold of the wormhole is simply connected, then (28.1) will represent a pure state wave function and $C$ will correspond to the class $C \equiv C_\Psi$ of paths associated with asymptotically flat four-geometries and asymptotically vanishing matter field configurations which match the prescribed data on a three-surface which divides the four-manifold in two parts. For the more general case where the inner wormhole manifold is nonsimply connected, (28.1) will represent a mixed state density matrix, and the integral is over the class $C \equiv C_\rho$ corresponding to the asymptotically flat Euclidean four-geometries and asymptotically vanishing matter field configurations which match the data on its three-surface and the orientation reverse of the data on its copy three-surface, these three-surfaces no longer dividing the four-manifold.

We present here the proposal that, if the quantum state of a single wormhole (i.e. a wormhole which inserts at just one point in the asymptotic regions) is given by a density matrix rather than a wave function, such a quantum state will be time asymmetric, and that this time asymmetry will be reflected in the evolution of all ordinary matter at low energies in the asymptotic regions. In what follows, we shall discuss some arguments in support of this proposal.

### 28.3  Wormhole-Induced Nonlocality

The effects of multiply connected wormholes on ordinary matter at low energies can be described in terms of a Hamiltonian density in Minkowski space given by (Coleman 1988a, González-Díaz 1992)

$$H = H_0 + \sum_i H_i^I A_i, \tag{28.2}$$

where $H_0$ is the usual local Hamiltonian for matter fields, the $H_i^I$'s form a linearly independent set of hermitian functions which represent the matter-field contribution to the effective wormhole-matter interaction, and the $A_i$'s are hermitian baby universe operators whose explicit dependence on Fock operators for single baby universes will depend on the inner topology of the microscopic connection (González-Díaz 1992).

The commutation relations for the field operators are derived from a local Lagrange density. Hence, prior to any interaction with wormholes, these operators should commute for spacelike separations. This property is obviously satisfied by

Hamiltonian $H_0$, but can only be preserved by the interaction terms $H_i^I$ for simply connected wormholes. In this case, one point on the asymptotic region is mapped into one point on any inner three-manifold during interaction. Clearly, this will not be the case for more complicated inner wormhole topologies, for which a single wormhole end at one point $x_0$ will be mapped into a given number of spacelike separated points on any inner wormhole cross section. It follows that, although the demand of locality on the asymptotic region will still imply commutativity of operators $H_0$ and $H_i^I$ for any nonzero spacelike separation in the case of simply connected wormholes, the commutativity for the interaction operators $H_i^I$ will not be preserved for all those spacelike separations which exactly match the whole set of proper distances between the disjoint three-manifold components on the inner boundary of a multiply connected wormhole. Hence, we have in general

$$\sum_i [H^I(x), H^I(y)] \neq 0, (x - y)^2 < 0. \tag{28.3}$$

Now, the demand of Lorentz invariance of the perturbation theory will imply that the total Hamiltonian $H$ must commute with itseft for spacelike separations, and hence leads to an overall demand of locality. Thus, since all $H_i^I$ must commute with all $A_i$, we have that in general

$$[A_i, A_j] \neq 0 \tag{28.4}$$

for multiply connected wormholes.

Following Coleman (1988a), we therefore conclude that, although simply connected wormholes preserve both causal locality and quantum coherence for matter fields in the asymptotic regions, that is no longer the case when the matter fields interact with multiply connected wormholes. Hence, for $CP$ and Lorentz invariant matter with positive energy, one must conclude that wormholes with nonsimply connected inner topologies induce a breakdown of $T$ invariance.

### 28.4 Wormholes as "Time Factories"

If one wants to deal with time invariance properties of wormholes properly, it appears most natural to redefine their quantum states in terms of a density matrix which is an explicit function of some suitable time notion. This can be achieved by replacing the dependence of the functional on the square root of the determinant of the three-metric on $S$ and $S'$ by a dependence on their respective conjugate momenta $K$ and $K'$, using the Laplace transform (González-Díaz 1991b)

$$\rho_L[\tilde{h}_{ij}, \phi_0, K; \tilde{h}'_{ij}, \phi'_0, K']$$

$$= N \int_0^\infty d[h^{\frac{1}{2}}]d[h'^{\frac{1}{2}}]e^{-\frac{1}{12\pi}(\int Kh^{\frac{1}{2}}d^3x + \int K'h'^{\frac{1}{2}}d^3x')}\rho[h_{ij}, \phi_0; h'_{ij}, \phi'_0], \tag{28.5}$$

where $N$ is a normalization constant, $\tilde{h}_{ij}$ and $\tilde{h}'_{ij}$ denote the metric on $S$ and $S'$, respectively, up to a conformal factor, and $\rho$ is given by (28.1). As an observer on the asymptotic region does not know anything about the number of components which are required to divide the manifold, the three-surfaces $S$ and $S'$ will be generally disjoint.

However, the Euclidean action $I[g_{\mu\nu}, \phi]$ is not bounded below (Gibbons, Hawking and Perry 1978) and one should deform the contour of integration in the path integral from Euclidean to complex metrics. For wormholes associated with a massless scalar field, $\phi$, conformally coupled to Hilbert-Einstein gravity and a conformal transformation defined by $\bar{g}_{\mu\nu} = \Omega^2 g_{\mu\nu}$, $\Omega = 1 + iy$, conjecturing validity of the positive action theorem also after introducing a wormhole quantum inner boundary will ensure convergence of the path integral provided we impose convenient boundary conditions on $y$: $y \to 0$ asymptotically, and $y \to y_0 \geq 1$ on the inner boundary (González-Díaz 1989). Actually, an uncertainty $\triangle a \geq$ Planck length becomes a necessary requirement to avoid zero-energy divergences in the density matrix, but this in turn implies that an originally closed universe can no longer remain closed which can only be ensured by resorting to complex metrics, and hence complex matter fields, also on the inner boundary.

Because the Laplace transform (28.5) is holomorphic for $\text{Re}(K) > 0$, one can analytically continue $\rho[\tilde{h}_{ij}, \phi_0, K]$ in $K$ and $K'$ to the Lorentzian values $K_L = iK$ and $K'_L = iK'$. Now, since under conformal transformation of the metric $\bar{g}_{\mu\nu} = \Omega^2 g_{\mu\nu}$, we have $\bar{K} = \Omega^{-1}K$ and $\bar{K}' = \Omega^{-1}K'$, it follows

$$\bar{K}_L = \frac{(1 \mp iy_0)K_L}{1 + y_0^2},$$                                   (28.6)

(and likewise for $\bar{K}'_L$) where the upper sign corresponds to complex metrics, and the lower sign, to complex conjugate metrics. The choice of time orientation on the inner three-manifold must be based on a causal analysis requiring the definition of suitable covering manifold (Hawking and Ellis 1973), i.e. it is required that not more than one of the considered several disjoint components of the three-boundary is joined at $S$ to its respective copy $S'$.

Using a conformal metric, after analytically continuing (28.5) in $K$ and $K'$ to their Lorentzian values, we obtain from (28.6)

$$\rho^{(c)}[\tilde{h}_{ij}, K_L, \phi_0; \tilde{h}'_{ij}, K'_L, \phi'_0] = N \int_{C^{(c)}} d[\bar{h}^{(c)\frac{1}{2}}]d[\bar{h}'^{(c)\frac{1}{2}}]$$

$$\times e^{\frac{1}{12\pi}[\frac{1}{1+y_0^2}\int d^3x(i+y_0)K_L\bar{h}^{(c)\frac{1}{2}} + \frac{1}{1+y_0'^2}\int d^3x'(i+y'_0)K'_L\bar{h}'^{(c)\frac{1}{2}}]} \rho^{(c)}[\bar{h}_{ij}, \bar{\phi}_0; \bar{h}'_{ij}, \bar{\phi}'_0]$$     (28.7)

$$\rho^{(cc)}[\tilde{h}_{ij}, K_L, \phi_0; \tilde{h}'_{ij}, K'_L, \phi'_0] = N \int_{C^{(cc)}} d[\bar{h}^{(cc)\frac{1}{2}}]d[\bar{h}'^{(cc)\frac{1}{2}}]$$

$$\times e^{\frac{1}{12\pi}[\frac{1}{1+y_0^2}\int d^3x(i-y_0)K_L\bar{h}^{(cc)\frac{1}{2}} + \frac{1}{1+y_0'^2}\int d^3x'(i-y'_0)K'_L\bar{h}'^{(cc)\frac{1}{2}}]} \rho^{(cc)}[\bar{h}_{ij}, \bar{\phi}_0; \bar{h}'_{ij}, \bar{\phi}'_0],$$     (28.8)

where the superscripts $(c)$ and $(cc)$ mean quantities obtained when, respectively, only complex metrics and only complex conjugate metrics are used. Integrations over contours $C^{(c)}$ and $C^{(cc)}$ respectively extend $\bar{h}^{(c)\frac{1}{2}}$ and the orientation reverse of $\bar{h}'^{(c)\frac{1}{2}}$ from 0 to $\infty$ in the complex plane for contour $C^{(c)}$, and $\bar{h}^{(cc)\frac{1}{2}}$ and the orientation reverse of $\bar{h}'^{(cc)\frac{1}{2}}$ also from 0 to $\infty$, but in the complex conjugate plane for contour $C^{(cc)}$. The lower integration limits have been chosen to be zero to allow for a total quantum freedom both for the three-geometry and slicing. Note that the values of the three-metrics, scalar fields and trace of the second fundamental form induced on the inner boundary are all still complex.

Applying complex conjugation $(*)$ and the operation of $K_L$-reversal to states (28.7) and (28.8), we obtain

$$\rho^{(c)*}[\tilde{h}_{ij}, -K_L, \phi_0; \tilde{h}'_{ij}, -K'_L, \phi'_0] = \rho^{(cc)}[\tilde{h}_{ij}, K_L, \phi_0; \tilde{h}'_{ij}, K'_L, \phi'_0] \qquad (28.9)$$

$$\rho^{(cc)*}[\tilde{h}_{ij}, -K_L, \phi_0; \tilde{h}'_{ij}, -K'_L, \phi'_0] = \rho^{(c)}[\tilde{h}_{ij}, K_L, \phi_0; \tilde{h}'_{ij}, K'_L, \phi'_0]. \qquad (28.10)$$

It follows that if the full density matrix contained equal contributions from metrics with a complex action and from metrics with a complex conjugate action, then that density matrix would be $T$ symmetric. However, fixing prescribed complex values for the boundary metric and scalar field makes the density matrix and its Laplace transform holomorphic functions, i.e. any contribution from metrics with complex conjugate action to the path integral in (28.7), or from metrics with a complex action to the path integral in (28.8) must be ruled out. It follows then that, fixing a complex value for the boundary metric and the boundary scalar field, contributions from complex conjugate four-metrics can never enter the path integral or its Laplace transform, and likewise, fixing the corresponding complex conjugate values for the boundary arguments will prevent any four complex-metric to contribute the path integral or its Laplace transform.

Since the Wheeler-DeWitt operator annihilates any pure-state wave function of wormholes, for such states we sum over both complex and complex conjugate four-metrics which induce the prescribed boundary real metric. Thus, for $CP$ invariant matter, a single wave function for wormholes, $\Psi$, should correspond to the two distinct density matrices $\rho^{(c)}$ and $\rho^{(cc)}$, and its Laplace transform to (28.7) and (28.8). Then,

$$\rho^{(c)}[\tilde{h}_{ij}, K_L, \phi_0; \tilde{h}'_{ij}, K'_L, \phi'_0] \neq \rho^{(c)*}[\tilde{h}_{ij}, -K_L, \phi_0; \tilde{h}'_{ij}, -K'_L, \phi'_0], \qquad (28.11)$$

$$\rho^{(cc)}[\tilde{h}_{ij}, K_L, \phi_0; \tilde{h}'_{ij}, K'_L, \phi'_0] \neq \rho^{(cc)*}[\tilde{h}_{ij}, -K_L, \phi_0; \tilde{h}'_{ij}, -K'_L, \phi'_0], \qquad (28.12)$$

and

$$\Psi[\tilde{h}_{ij}, K_L, \phi_0] = \Psi^*[\tilde{h}_{ij}, -K_L, \phi_0]. \qquad (28.13)$$

Eqns. (28.11) and (28.12) are a statement of $T$ noninvariance for the quantum state of wormholes, relative to independent observers in the regions which are connected by the wormhole. This would imply that, for an observer in one large region, the probability for creating baby universes is not the same as that for destroying them. The wormhole state representation $\Psi$ should be time-symmetric however; for (28.13) contains equal contributions from metrics with a complex action and from metrics with a complex conjugate action so that the wave function is real.

Having established that the quantum state of (multiply) simply connected wormholes is time (a)symmetric, we now turn to sketch the argument by which these properties are reflected in ordinary matter fields in the asymptotic region. We first note that the quantum states $\rho^{(c)}$ ($\rho^{(cc)}$): (i) are Lorentz invariant because the multiply connected baby universe spacetime is effectively connected to the asymptotic flat regions, and (ii) have positive (negative) definite energy as they are *off shell*, and associate with complex (complex conjugate) metrics relative to an observer in the given asymptotic region. Hence, for states $\rho^{(c)}$ and $\rho^{(cc)}$, the lack of $T$ invariance must be associated (Streater and Wightman 1964) to a breakdown of local causality in the baby universe sector. Now, by the discussion following (28.2), one should conclude again that nonsimply connected single wormholes induce a breakdown of time symmetry for all ordinary matter with positive energy which is $CP$ and Lorentz invariant. One should likewise expect that the interaction with simply connected wormholes does not lead to any violation of time symmetry in the low-energy matter field sector.

## 28.5 Convergence of the Density Matrix

The path integral for the quantum state of wormholes can be generally calculated by taking the real part of the propagator for the wave function from the data on a given three-surface $S$ at Euclidean time $\tau = \tau_1$. The propagating wave functional has been approximated in terms of harmonic oscillators for the gravitational and matter fields. In a perturbed Euclidean Friedmann metric where the perturbations are expanded in terms of spherical harmonics on the three-sphere, we obtain a propagator

$$K_0[a,f,0;a',f',\tau_1] = \sum_{m=0}^{\infty}\sum_{n=0}^{\infty} \Psi_{mn}[a,f]\Psi_{mn}[a',f']e^{-(E_n^{(f)}-E_m^{(a)})\tau_1}, \qquad (28.14)$$

where $a$ is the scale factor, $f$ is the scalar-field mode coefficient for the scalar harmonics, and $E_n^{(f)}$ and $E_m^{(a)}$ denote the energy levels of the matter field in the homogeneous lowest excited mode and the scale factor, respectively. The wave functions in (28.14) are given in terms of Hermite polynomials. Then, because $S$ and $S'$ may have any time separation $\tau_1$, a proper density matrix can in principle be obtained by integrating (28.14) over all values of $\tau_1$ (González-Díaz 1991b)

$$\rho_1[a,f;a',f'] = \text{Re}\int_0^{\infty} d\tau_1 K_0[a,f,0;a',f',\tau_1]. \qquad (28.15)$$

However, there is a crucial problem with (28.15). Since $\epsilon_{mn} = E_n^{(f)} - E_m^{(a)}$ may take on negative values, after integration over $\tau_1$, one would obtain a density matrix which is not positive definite. This actually means that the state is divergent for negative $\epsilon_{mn}$. This problem can really be related with the feature that one should introduce a density matrix just for observers in each of the two asymptotic regions, but not for observers in the two regions simultaneously. The integration limits in (28.15) are then valid only for $\epsilon_{mn} > 0$; i.e. for observers in just one of the two asymptotic regions, and time oriented forward. A so-defined density matrix would correspond to the functional $\rho^{(c)}$.

There will be another density matrix, obtained by reversing the $\tau_1$-time orientation in (28.15) when $\epsilon_{mn} < 0$,

$$\rho_2[a, f; a', f'] = Re \int_{-\infty}^{0} d\tau_1 K[a, f, 0; a', f', \tau_1], \qquad (28.16)$$

which is also positive definite and corresponds to the functional $\rho^{(cc)}$. Note that, at least for $CP$-invariant matter, (28.15) and (28.16) yield the same result, that is

$$\rho = \rho_1 = \rho_2 = Re \sum_{m=1}^{\infty} \sum_{n=0}^{m-1} \frac{\Psi_{mn}[a, f]\Psi_{mn}[a', f']}{\epsilon_{mn}}, \qquad (28.17)$$

where the lower limit in the sum over $m$ has been taken to be unity because of the discussion following (28.5).

This would be the quantum-gravity counterpart of the quantum-field feature that a particle with negative energy moving backward in time is physically equivalent to the particle with positive energy moving forward in time, because wormholes whose quantum state is given by a density matrix are off shell and, therefore, the Wheeler-DeWitt operator acting upon $\rho$ gives no longer zero.

### 28.6 Blackbody Probability in Quantum Gravity

There is an interesting consequence stemming from the statistical properties of wormhole states. The dynamics for a set of simply connected wormholes inserted into a single parent universe leads to a probability measure for the Coleman $\alpha$ parameters given by (Coleman 1988b)

$$\mu(\alpha) = P(\alpha)Z(\alpha), \qquad (28.18)$$

where the probability for the $\alpha$'s is

$$P(\alpha) \equiv P_\Psi(\alpha) = e^{-\frac{1}{2}D\alpha^2}, D = e^{S_w} \qquad (28.19)$$

($S_w$ being the Euclidean wormhole action), and the path integral $Z(\alpha)$ can be written as

$$Z(\alpha) = \int dg e^{-I(g,\lambda+\alpha)}, \qquad (28.20)$$

in which $\lambda$ collectively denotes all involved coupling constants. Had we started with a density matrix instead a wave function, the probability measure would still look like (28.18), but in this case (González-Díaz 1991c)

$$P(\alpha) \equiv P_\rho(\alpha) = \frac{1}{e^{\frac{1}{2}D\alpha^2} - 1}. \qquad (28.21)$$

This result raises up the possibility to interpret $\alpha^2$ and $D^{-1}$ as though if they respectively were the momenta of a *quantum field* in superspace and an *equilibrium temperature* for the whole system, and $Z(\alpha) = \frac{1}{2}\alpha^2$. Coleman law (28.19) would then correspond to the limit for large $D\alpha^2$ from (28.21), playing thus the role of a "Wien law" for quantum gravity, and the limit for small $D\alpha^2$ from (28.18) would give $D^{-1}$, i.e. the semiclassical probability $e^{-S_w}$ would then play the role of a "Rayleigh-Jeans law" for quantum gravity.

## Acknowledgements

The author wants to thank L.J. Garay, P.G. Tinyakov and S.W. Hawking for useful discussions. This work was supported by an Accion Especial of C.S.I.C., and a CAICYT Research Project N° 91-0052.

## Discussion

**Hartle**  I think I missed an elementary point. In the functional integral that defines the density matrix, if a complex metric contributes to the steepest descents approximation, why does not the complex conjugate metric also contribute?

**González-Díaz**  For equal contributions from complex and complex conjugate metrics, we have a pure state given by a wave function rather than a density matrix. The reason is that the real part of the analytically continued trace of the second fundamental form, which plays the role of causal time, is positive for complex metrics and negative for complex conjugate metrics. Equal contributions from configurations with complex and complex conjugate metrics destroys then all possible single-time correlations among disjoint three-manifolds, and results in a four-manifold divided in two disconnected parts. Two $T$-noninvariant density matrices can therefore be defined: $\rho^{(c)}$ for time orientation corresponding to observers in one of the two asymptotic regions, and $\rho^{(cc)}$ for the reversed time orientation corresponding to observers in the other asymptotic region.

## References

Coleman, S (1988a) Black Holes As Red Herrings. Topological Fluctuations and the Loss of Quantum Coherence. *Nuclear Physics*, **B307**, 867-882.

Coleman, S (1988b) Why There Is Nothing Rather Something. A Theory of the Cosmological Constant. *Nuclear Physics*, **B310**, 643-668.

Gibbons, G.W., Hawking, S.W. and Perry, M.J. (1978) Path Integrals and the Indefiniteness of the Gravitational Action. *Nuclear Physics*, **B138**, 141-150.

González-Díaz, P.F. (1989) On the Tolman-Hawking Wormhole. *Physical Review*, **D40**, 4184-4187.

González-Díaz, P.F. (1990) What Is Really Being Lost In Wormholes? *Physical Review*, *D42*, 3983-3996.

González-Díaz, P.F. (1991a) The Density Matrix of Wormholes. *Nuclear Physics*, **B351**, 767-777.

González-Díaz, P.F. (1991b) Topological Arrow of Time and Quantum-Mechanical Evolution. In *Proceedings of The II International Wigner Symposium*, ed. H.D. Doebner, World Scientific, Singapore.

González-Díaz, P.F. (1991c) Quantum Theory of Wormholes. *Modern Physics Letters* **A8**, 1089; Blackbody Distribution for Wormholes. *Classical and Quantum Gravity* (to appear).

González-Díaz, P.F. (1992) Regaining Quantum Incoherence for Matter Fields. *Physical Review*, **D45**, 499-506.

Hawking, S.W. (1982) The boundary Conditions of the Universe. In *Astrophysical Cosmology*, eds. H.A. Bruck, G.V. Coyne and M.S. Longair, Pontificia Academiae Scientiarum, Vatican City.

Hawking, S.W. (1988) Wormholes in Spacetime. *Physical Review*, **D37**, 904-910.

Hawking, S.W. (1990) Baby Universes. *Modern Physics Letters*, **A5**, 145-155; 453-464.

Hawking, S.W. and Ellis, G.F.R. (1973) *The Large Scale Structure of Space-Time*, Cambridge University Press, Cambridge.

Streater, R.F. and Wightman (1964) *PCT, Spin, Statistics and All That*, Benjamin, New York.

# 29

# Time, Mach's Principle and Quantum Cosmology[*]

T. Padmanabhan[†]

*Theoretical Astrophysics Group*
*T.I.F.R., Homi Bhabha Road*
*Bombay 400 005, India*

## Abstract

Newton's laws of motion appear simplest in an inertial frame when a specific time coordinate is used. Free particles remain unaccelerated in this frame. A hierarchy of limiting procedures ($G \to 0, c \to \infty, \hbar \to 0$) is required to obtain the Newtonian limit from the exact, quantum gravitational, description of the universe. I show that the inertial time coordinate and the conventional notion of particles will emerge only if restrictions are placed on the solution to Wheeler- DeWitt equation describing the universe.

The purpose of this note is to suggest that there could exist some interesting connections between the notion of time, wavefunction of the universe and a particular version of Mach's principle. Since each of these ideas is somewhat ill-defined (and interpreted differently by different people), it is probably best if I begin by clarifying how I intend to use these notions. Let me begin with the notion of Mach's principle.

Let S be a frame of reference in which distant "stars" [or galaxies, $\cdots$] are at rest. Newton's laws are usually stated in such a frame. A particle shielded from all external influences will follow an unaccelerated trajectory in this frame. This result can be restated as follows: Let $\mathbf{x}_1(t)$, $\mathbf{x}_2(t) \cdots \mathbf{x}_N(t)$ be the position vectors of N stars and let $\mathbf{x}(t)$ be the position of a test particle shielded from external influences. In a reference frame in which $\mathbf{x}_i(t) = \mathbf{x}_i(0)$ ['distant stars are fixed'], $\mathbf{x}(t)$ satisfies the equation $(d^2\mathbf{x}/dt^2) = 0$. Thus, by connecting up the local behaviour of test particles with the state of motion of distant matter, we have brought in (a particular version of) Mach's principle (Mach, 1912). In contrast, consider another frame $S'$ in which the distant stars are *not* at rest but moves according to the law $\mathbf{x}_i'(t) = (1/2)\mathbf{g}t^2$. In this frame - in which distant stars are not fixed - we cannot use the original version of Newton's law. However, it is easy to make a coordinate transformation which

---

† Present address: Inter-University Centre for Astronomy and Astrophysics, Pune University Campus, Ganeshkhind, Pune 411007, India.
* The author was not able to attend the workshop, but his paper was accepted by the editors for inclusion in these proceedings.

will bring these stars to rest and use Newton's law in such a frame. Transforming back we can find the equation of motion for the free test particle in our original frame $S'$. By this procedure we will find that $\mathbf{x}'$ satisfies the equation:

$$\frac{d^2\mathbf{x}'}{dt^2} = -\mathbf{g} \tag{29.1}$$

It is usual to call $S$ an 'inertial frame' and $S'$ a 'non-inertial frame'; and the acceleration $\mathbf{g}$ [in (29.1)], experienced by a test particle in $S'$, as due to a 'pseudo-force'. In its true form, Mach's principle does not distinguish between coherent motion of all the distant matter in the universe and a local transformation to a non-inertial frame.

Mach's principle, however, has another facet which is not often emphasized. To bring this up, consider again the frame in which $\mathbf{x}_i(t) = \mathbf{x}_i(0)$ - i.e. distant stars are fixed. Suppose *all* test particles, shielded from external influences, follow in this frame, a trajectory $\mathbf{x}(t)$, such that:

$$\frac{d^2\mathbf{x}}{dt^2} = -\alpha(t)\dot{\mathbf{x}} \tag{29.2}$$

Since the distant stars are now fixed, we will think of this frame as inertial. We are thus forced to conclude that the particles are experiencing a velocity-dependent drag force and are following an *accelerated* trajectory. This conclusion, however, is wrong. A transformation of the time coordinate from $t$ to $T$ such that

$$T = \int^t dt \left[\exp - \int \alpha(t)dt\right] \tag{29.3}$$

will change the trajectory to that of familiar free particle!

$$\frac{d^2\mathbf{x}}{dT^2} = 0 \tag{29.4}$$

We have here two coordinate systems for spacetime: $(\mathbf{x}, t)$ and $(\mathbf{x}, T)$ with $t$ and $T$ related by (29.3). Fixed stars stay fixed in both these frames but Newton's law picks up a pseudo-force in one of these frames.

In the usual discussions of Mach's principle, one never bothers about time transformations like the one in (29.3). This is because in Newtonian physics, there is an 'absolute time' which 'flows uniformly'. Motion is described using this particular time coordinate. Then - and, only then - the frame of fixed stars defines for us a useful inertial frame. If we have no information about the nature of the time coordinate then we cannot exclude pseudo-forces even in a frame of fixed stars.

This example raises some interesting questions. It is almost a miracle that we are gifted with such a time coordinate in which "motion appears to be simple". Where does such a time coordinate stem from? From cosmological observations, we know that matter on the very large scale does not have coherent motion. In other words, a frame with fixed stars does exist. *But why is it that in this particular frame of fixed*

*stars, test particles obey equation (29.4) rather than an equation like (29.1)?* I shall
argue that this requires an explanation.

If the nature was governed by strictly Newtonian laws then there would have been
no problem. As we said before, Newtonian physics permits arbitrary transformations
of the space coordinates **x** but forbids transformations of time (except for scaling
and translation, $t' = at + b$, under which Newton's law is invariant). By prescribing
an absolute time Newtonian physics has effectively bypassed this question.

But the world is not Newtonian; it is quantum mechanical and it is general
relativistic. In a universe obeying the laws of quantum mechanics and general
relativity there is no place for an absolute time coordinate. From such an exact
world, obeying the laws of relativistic quantum theory, we construct our approximate
'everyday world' by taking three limits. First we take the limit of weak gravitational
field ($G \to 0$); then we take the limit of non-relativistic quantum theory ($c \to \infty$);
finally we take the limit of classical physics ($\hbar \to 0$). In this limit, particles with
definite trajectories exist; so does a reference frame in which the space coordinates
of distant stars fixed. And what is more, there emerges a natural time coordinate in
which inertial motion with no pseudoforce is possible. This breaks the invariance of
the physical laws under the reparametrisation of time coordinate, which is present
at the higher levels of description of the theory. If the exact theory is fully invariant
under arbitrary time transformations, then a special time coordinate can emerge
only if it was introduced by hand. In what follows, we will examine how this
happens and what it means.

Let us begin by reversing the process and proceed from Newtonian mechanics
to more exact descriptions. A classical, free, particle has its quantum-mechanical
equivalent, described by the Schrödinger equation

$$i\hbar\frac{\partial\psi}{\partial t} = -\frac{\hbar^2}{2m}\nabla_x^2\psi \tag{29.5}$$

Arbitrary transformations of the time coordinate - of the form $t \to t' = F(t)$ - are
still forbidden. The Schrödinger equation retains the above form only when the
'sacred time coordinate' is used. The transition from (29.5) to Newtonian limit is via
the expectation value

$$< x > = \int \psi^* \hat{x} \psi d^3x \tag{29.6}$$

and Ehrenfest's theorem. In fact, the evolution equation for the expectation value

$$\left(\frac{\partial < x >}{\partial t}\right) = < \frac{i}{\hbar}[H, x] > \tag{29.7}$$

remains valid (for a given Hamiltonian) only if the time coordinate satisfying (29.5)
is used.

In the next stage - that of quantum field theory - the situation becomes trickier.
Let us suppose we are working in flat spacetime and that our particle is described

by a scalar field *operator* obeying the Klein-Gordon equation:

$$\left[ \frac{1}{c^2} \frac{\partial^2}{\partial t^2} - \nabla^2 + \frac{m^2 c^2}{\hbar^2} \right] \hat{\phi}(t, \mathbf{x}) = 0 \tag{29.8}$$

Note that (29.8) is an equation in Heisenberg picture for the *operator* $\hat{\phi}$ while (29.5) is an equation in Schrodinger picture for a c-number function $\psi$. To make the proper transition from the field theory to point quantum mechanics we have to first define the Fock basis corresponding to $\hat{\phi}$ in (29.8). Let $|0>$ and $|1_{\mathbf{k}}>$ be the vacuum and one-particle states of the quantum field theory described by $\hat{\phi}(t, \mathbf{x})$. One way of making the necessary identification between (29.8) and (29.5) will be to use the transition element $< 0|\hat{\phi}|1_{\mathbf{k}}>$:

$$< 0|\hat{\phi}(\mathbf{x}, t)|1_{\mathbf{k}}> = \frac{1}{\sqrt{2\omega_{\mathbf{k}}}} \exp(i\omega_{\mathbf{k}} t + i\mathbf{k}.\mathbf{x}); \quad \hbar\omega_{\mathbf{k}} = \sqrt{k^2 \hbar^2 c^2 + m^2 c^4} \tag{29.9}$$

In the non-relativistic limit of $c \to \infty$, we should identify the expression $< 0|\hat{\phi}|1_{\mathbf{k}}> e^{imc^2 t/\hbar}$ with the Schrodinger wave function $\psi(\mathbf{x}, t)$ (see e.g. Roman, 1968; p.86) . This is easily seen from noting that $< 0|\hat{\phi}|1_k > e^{imc^2 t/\hbar}$ has the limiting form

$$< 0|\hat{\phi}|1_{\mathbf{k}} > e^{imc^2 t/\hbar} \simeq \exp\left( i\mathbf{k}.\mathbf{x} - \frac{ik^2 \hbar}{2m} t \right) \tag{29.10}$$

which is just the free particle wave function with momentum $\hbar\mathbf{k}$. In other words, the correct limiting form for a quantum mechanical particle is obtained *only after we have defined the one-particle Fock state* $|1_{\mathbf{k}}>$. [It is also possible to reach the same conclusion by working in the Schrodinger picture of the field theory; but the analysis is simpler to understand in the Heisenberg picture]. As long as we work within the framework of special relativity and Lorentz transformations, the Fock basis is unique. Though time can be mixed with space in Lorentz transformations, inertial frames retain their identity. Hence we obtain the "inertial time" in the quantum mechanical limit.

This uniqueness is lost once we go beyond the realm of special relativity, and allow arbitrary coordinate transformations. Consider, for example, the Rindler frame in which the line element has the form

$$ds^2 = \left( 1 + \frac{gx}{c^2} \right)^2 c^2 d\tau^2 - dx^2 - dy^2 - dz^2 \tag{29.11}$$

As is well known the Fock basis defined using $\tau$ is not the same as the one defined using the inertial time $t$ (Fulling, 1973). The transition element constructed from the Rindler states has the following form:

$$_R < 0|\hat{\phi}(x, \tau)|1_k >_R = \frac{(\sinh \omega\pi)^{1/2}}{2\pi^2} e^{-i\omega\tau + ik_y y + ik_z z} K_{iv}(k(1 + gx)) \tag{29.12}$$

where $K_{iv}(z)$ is the modified Bessel function. In the non-relativistic limit, $(c \to \infty)$ the function $\psi =_R< 0|\phi(x, \tau)|1_k >_R e^{+imc^2 t/\hbar}$ satisfies the Schrodinger equation for a uniformly accelerated particle:

$$i\hbar \frac{\partial \psi}{\partial \tau} = -\frac{\hbar^2}{2m} \nabla^2 \psi + mgx\psi \tag{29.13}$$

Therefore, in the corresponding Newtonian limit, these particles obey equation (29.1) - with a pseudo-force! Suppose we had described our flat spacetime using $x$ and $\tau$. Then, in the appropriate non-relativistic limit all our test 'particles' [defined using the mode functions in (29.12)] will experience an acceleration. It is now clear that it is the choice of the ground state of the quantum field which leads to a preferred time coordinate in the classical limit. It is the definition of 'particle' which breaks the invariance under time transformations. *We choose the quantum vacuum state in such a way that "particles" in the classical limit experience no pseudo-force.*

In flat spacetime, of course, we could make a rule that we should always use the Minkowski time, rather than (29.11) say, to define the particles. In the exact theory, however, even flat spacetime has no special status. Flat spacetime happens to be one of the many possible solutions to vacuum Einstein's equations. In a quantum gravitational limit, we have to take into account the possibility that the universe cannot be described by any single metric but only by a wavefunction on the space of 3-geometries. We have to now see under what conditions we can recover the preferred Minkowski frame from the wavefunction of the universe.

Let us now see how this idea fits in with quantum cosmology. In the usual approaches, both the notion of the time *and* the functional Schrödinger equation for matter fields in a background metric emerge together as semi-classical constructs. This functional Schrödinger equation should contain, among other things, information about the ground (vacuum) state of the system and should allow one to define the notion of particles. Quite clearly, we may have to impose some constraints on the wavefunction of the universe to obtain the "correct" particles.

Let $\Psi[^3\mathscr{G}]$ be a solution of Wheeler-Dewitt (WD) equation satisfying some boundary conditions. Though people differ in their choice of the boundary conditions, they all would like to ensure the following: The wavefunction has a clear semi-classical limit in which it can be interpreted as describing a universe like ours. There has been considerable amount of discussion in the literature as to how the notion of time "emerges" from the wavefunction $\Psi$. In some special contexts, it is possible to obtain from the WD equation another - approximate - equation which may be interpreted as the functional Schrodinger equation for matter fields propagating in a specified curved background [see e.g. Lapchinsky and Rubakov, 1979; Banks, 1985; Hartle, 1986; Halliwell, 1987; Singh and Padmanabhan, 1989]. This equation has the general form:

$$iG_{abcd} \frac{\delta A}{\delta g_{cd}} \frac{\delta}{\delta g_{cd}} \psi[g;\phi] = \hat{H}[g;\phi]\psi. \tag{29.14}$$

Here $A[g]$ is the solution to the Einstein-Hamilton-Jacobi equation

$$G_{abcd} \frac{\delta A}{\delta g_{ab}} \frac{\delta A}{\delta g_{cd}} - \sqrt{g} \, {}^3R = 0 \qquad (29.15)$$

and $G_{abcd}$ is the metric in the superspace and ${}^3R$ is the curvature scalar of the three-space. The repeated indices imply not only summation over discrete indices but also integration over spatial coordinates. We have denoted the matter variables collectively by the symbol $\phi$ and $\hat{H}[g, \phi]$ is the Hamiltonian for the matter field in the background metric $g$. A time coordinate can now be identified by equating the left hand side of (29.14) with the "bubble-time" derivative defined on a spacelike hypersurface: We write

$$\frac{\delta}{\delta \tau} \equiv G_{abcd} \frac{\delta A}{\delta g_{ab}} \frac{\delta}{\delta g_{cd}}. \qquad (29.16)$$

This procedure allows one to produce a notion of time in the semi-classical limit. [It is even possible to generalize this notion and obtain a concept of time in the full quantum gravitational case, though we will not need this generalization in what follows; see Padmanabhan, 1990]. What is more, this procedure shows how the solution to the WD equation leads to a wavefunction(al) for quantum fields in curved spacetime. In particular, the 'ground state' of this Schrodinger equation will allow us to define the 'particles'.

The last feature might appear surprising and even incorrect. It is usually believed that the functional Schrodinger equation is generally covariant while the concept of a particle is not. Thus it may appear strange that the solution to the WD equation can lead to prescription for the ground state and a definition of 'particles'. The key to the resolution of this paradox lies in the following observation: In obtaining the semiclassical limit, we have chosen a particular form of the solution to the WD equation:

$$\Psi = \psi \, \exp\left(iA \left[{}^3\mathscr{G}\right]\right)$$

where $A[{}^3\mathscr{G}]$ is the solution to the Hamilton-Jacobi equation. In order to specify this solution uniquely, we need to specify $A$ by imposing suitable boundary conditions on the Hamilton-Jacobi equation. It is this choice which leads to a particular definition of time coordinate and - consequently - to the notion of particles.

It is possible to illustrate the above ideas in a fairly simple model for quantum cosmology. For this purpose, consider the minisuperspace with the line element

$$ds^2 = dt^2 - a^2(t)dx^2 - b^2(t)dy^2 - c^2(t)dz^2.$$

The solution to the WD equation in the absence of matter fields will be some wavefunction $\Psi[a, b, c]$. In the semiclassical limit, we would expect the wavefunction to be "peaked" at the vacuum solution to Einstein's equations. In this particular case, these solutions happen to be the Kasner universes which are described by the functions

$$a(t) \sim t^{p_1}, \quad b(t) \sim t^{p_2}, \quad c(t) \sim t^{p_3}$$

with $\Sigma p_i = \Sigma p_i^2 = 1$. Of these solutions, only three represent flat spacetime: This corresponds to chosing one of the three $p_i$'s to be unity and the others as zero. Notice, however, that a wavefunction which leads - in the semiclassical limit - to the choice $[a(t) = t, b = 1, c = 1]$ will pick out a *non-inertial* time coordinate in the flat spacetime. On the contrary, there does exist another solution to the vacuum Einstein equation, viz. $[a = b = c = 1]$, which does pick out the inertial time. If the wavefunction $\Psi[a, b, c]$ leads to this choice in the semiclassical limit, then we will obtain the inertial time.

This example illustrates clearly how the choice of $\Psi$ influences the choice of time coordinate and the definition of particles. [Notice that, we have not bothered to spell out in detail how any one configuration is favoured in the semiclassical limit; for our discussion, it is enough if some suitable procedure exists, say, e.g. one based on Wigner functions]. For the minisuperspace discussed above, the WD equation can be solved exactly to obtain the most general wavefunction of the form

$$\Psi[\Omega, \mathbf{q}] = \int \frac{d^2\mathbf{k}}{(2\pi)^2} A(\mathbf{k}) \exp(i\mathbf{k}.\mathbf{q} - i|\mathbf{k}|\Omega)$$

where

$$a = \exp 2(-\Omega + q_+ + \sqrt{3}q_-); \quad b = \exp 2(-\Omega + q_+ - \sqrt{3}q_-)$$

and $c = \exp 2(-\Omega - 2q_+)$. We have denoted by $\mathbf{q}$ the two dimensional vector $\mathbf{q} = (q_+, q_-)$. The function $A(\mathbf{k})$ has to be determined from the boundary conditions imposed on the wavefunction. One can easily verify that different choices of time coordinate emerges for different choices of $A(\mathbf{k})$. The detailed discussion of this correspondence will be presented elsewhere.

Thus the choice of boundary conditions for the wavefunction of the universe can lead to a choice for time coordinate and definition of particle in the low energy limit. Only a particular class of boundary conditions will ensure that the particles will follow the inertial trajectories in the Newtonian limit.

## References

T Banks, (1985), *Nucl. Phys.*, **B245**, 332.

S.A. Fulling, (1973) *Phys. Rev.*, **D7**, 2850.

J.J. Halliwell, (1987), *Phys. Rev.* **D36**, 3626.

J.B. Hartle, (1986) in *Gravitation in astrophysics*, eds. B. Carter and J.B. Hartle, (Plenum, New York).

V.G. Lapchinsky and V.A. Rubakov, (1979), *Acta. Phys. Polon.* **B10**, 1041.

E. Mach, (1912), *Die Mechanik in Ihrer Entwicklung Historisch Kritisch Dargestellet*, (Broachans, Leipzig).

T. Padmanabhan, (1990), *Pramana-Jour. of Phys.*, **35**, L199.

P. Roman, (1968) *Introduction to quantum field theory*, (Addison-Wesley, New York).

T.P. Singh and T. Padmanabhan, (1989), *Ann. Phys.*, **196**, 296.

# Part  Six

Time and Irreversibility in Gravitational Physics

# 30

## Times at Early Times

Jorma Louko[†]

*Department of Physics*
*Syracuse University*
*Syracuse, New York 13244–1130, USA*

### 30.1 Introduction

From the talks given at this meeting it is clear that issues of time, and in particular issues of time asymmetry, are considered by many workers to be deeply related to gravitational physics. It is also clear that we are far from a proper understanding of these issues. Therefore, much of the work on time in gravitational physics has been concentrated on arenas where these issues are present in a simplified, but hopefully not entirely unrealistic form. I would like to discuss one such arena: gravity in 2+1 spacetime dimensions. I shall review how the classical theory admits and in a sense consists entirely of Big Bang type cosmological models, and how the question of time asymmetry is treated in three different approaches to quantizing the theory. I shall end with some speculation on "fundamental" and "semiclassical" time variables in the different quantum theories.

### 30.2 Classical Solutions

It is well known that in three spacetime dimensions Ricci flatness implies the vanishing of the Riemann tensor. However, the condition that the metric be flat does not fix the global properties of the spacetime. This allows the theory to have nontrivial global degrees of freedom (Deser et al. 1984, Deser and Jackiw 1984, Moncrief 1989, Hosoya and Nakao 1990).

As an example, let us construct a simple family of non-Minkowski solutions to 2+1 gravity. Begin with the full 2+1 dimensional Minkowski space in the global coordinates $(T, X, Y)$ with the metric

$$ds^2 = -dT^2 + dX^2 + dY^2 \ . \tag{30.1}$$

Consider the region $T > |X|$, and perform in this region the coordinate transforma-

† Present address: Department of Physics, University of Wisconsin-Milwaukee, P.O. Box 413, Milwaukee WI 53201, USA

tion

$$
\begin{aligned}
T &= t \cosh \bar{x} \\
X &= t \sinh \bar{x} \\
Y &= \bar{y}
\end{aligned}
\tag{30.2}
$$

where the new coordinates $(t, \bar{x}, \bar{y})$ satisfy $0 < t, -\infty < \bar{x} < \infty, -\infty < \bar{y} < \infty$. The metric takes the form

$$
ds^2 = -dt^2 + t^2 d\bar{x}^2 + d\bar{y}^2 \ .
\tag{30.3}
$$

The constant $t$ surfaces are topologically $\mathbb{R}^2$ and the induced two-metric is flat, but the extrinsic curvature of these surfaces is nonvanishing. Let now $\lambda, \mu, a, b$ be real numbers satisfying $a\mu - b\lambda \neq 0$, and identify the points

$$
(t, \bar{x}, \bar{y}) \sim (t, \bar{x} + \lambda, \bar{y} + a) \sim (t, \bar{x} + \mu, \bar{y} + b) \ .
\tag{30.4}
$$

The constant $t$ surfaces become then tori, whose size and shape change in $t$, and the surface $t = 0$ turns from a coordinate singularity into a conical-type singularity, similar to that separating the Taub and NUT parts of Taub-NUT spacetime (Hawking and Ellis 1973).

Geometrically, the identifications (30.4) consist of quotienting Minkowski space by a discrete group of isometries, generated by two commuting Poincare transformations which are linear combinations of a translation in the $Y$ direction and a boost in the $(T, X)$ plane. The singularity on the surface $t = 0$ is a consequence of the fact that this group acts properly discontinuously in the domain $T > |X|$ but not in its closure $T \geq |X|$.

By the identification (30.4), we have obtained a four-parameter family of metrics. This becomes explicit if we introduce new spatial coordinates $(x, y)$ by

$$
\begin{aligned}
\bar{x} &= \lambda x + \mu y \\
\bar{y} &= a x + b y \ .
\end{aligned}
\tag{30.5}
$$

The identifications in the new coordinates are $(x, y) \sim (x + 1, y) \sim (x, y + 1)$, and the metric takes the form

$$
ds^2 = -dt^2 + t^2(\lambda dx + \mu dy)^2 + (a dx + b dy)^2 \ .
\tag{30.6}
$$

It can be shown that the metrics (30.6) include almost all solutions to 2+1 dimensional Einstein gravity with toroidal spatial topology (Moncrief 1989, Hosoya and Nakao 1990). The only solutions not included in (30.6) constitute a three-parameter family in which the metric on the torus is constant in time,

$$
ds^2 = -dt^2 + H_{ij} dx^i dx^j \ ,
\tag{30.7}
$$

where $H_{ij}$ is a symmetric positive definite matrix independent of $t$, and $\{x^i\} = (x, y)$ have the identifications given above. These solutions are clearly obtained from

Minkowski space by quotienting by a discrete group generated by two spatial translations.

For compact spatial topologies with genus $g$ two or higher, it can be shown that the solutions form a family with $12(g-1)$ parameters (Witten 1988, Moncrief 1989, Hosoya and Nakao 1990). It is however difficult to write these metrics in an explicit form. In what follows we shall only consider the case of toroidal spatial topology, $g = 1$.

The generic solutions (30.6) with toroidal spatial topology are not time-symmetric. At one end they have a singularity, and at the other end they expand to infinity. The exceptional solutions (30.7), on the other hand, are static and thus time-symmetric. We would now like to see how this classical time (a)symmetry is treated in three different approaches to quantization.

### 30.3 Two Metric Quantizations

In the classical Hamiltonian formulation in the metric variables, it is possible to explicitly eliminate almost all of the constraints of the theory by choosing a slicing of constant York time (Moncrief 1989, Hosoya and Nakao 1990). After this reduction in the toroidal case, one is left with "minisuperspace" metrics of the form

$$ds^2 = -N^2(t)dt^2 + h_{ij}(t)\,dx^i dx^j \ , \tag{30.8}$$

where the spatial coordinates $\{x^i\} = (x, y)$ have the same identifications as above. The action is given by

$$S = \int dt \ (p_\alpha \dot{q}^\alpha - N\mathscr{H}) \tag{30.9}$$

where $\{q^\alpha\}$ are three 'coordinates' parametrizing the symmetric positive definite matrix $h_{ij}$, $\{p_\alpha\}$ are the canonically conjugate momenta, and $\mathscr{H}$ is the superHamiltonian constraint. In the specific parametrization

$$h_{ij} = v \begin{pmatrix} 1/\eta & \xi/\eta \\ \xi/\eta & (\xi^2 + \eta^2)/\eta \end{pmatrix}_{ij} \tag{30.10}$$

where $v > 0$, $\eta > 0$ and $-\infty < \xi < \infty$, we have

$$\mathscr{H} = \left(\frac{8\pi G}{v}\right)\left[-v^2 p_v^2 + \eta^2 \left(p_\xi^2 + p_\eta^2\right)\right] \tag{30.11}$$

where $G$ is the 2+1 dimensional gravitational constant.

Let us now discuss two quantizations based on the action (30.9). For the restrictions implied by using such a classically reduced action, see for example Kuchař (1992).

Firstly, a straightforward application of Dirac's quantization method (Dirac 1964) leads to the Wheeler-DeWitt equation

$$\hat{\mathscr{H}}\Psi = 0 \ , \tag{30.12}$$

where the wave function $\Psi$ is a function of the coordinates $\{q^\alpha\}$, and $\hat{\mathscr{H}}$ is the quantum superHamiltonian operator obtained from the classical superHamiltonian through the operator substitution $p_\alpha \rightarrow -i(\partial/\partial q^\alpha)$ and a choice of the operator ordering. A variety of opinions exist as to how to relate this quantization to predictions of observable probabilities (Hawking and Page 1986, Ashtekar 1991).

The point of interest for us is that the Wheeler-DeWitt equation (30.12) is real, and at the level of this equation the quantization can thus be considered to be time symmetric (Barbour and Smolin 1988). There are *solutions* to the Wheeler-DeWitt equation that can be regarded as time asymmetric; however, given such a wave function, one can always find a real, time symmetric wave function by taking the real part. For example, a real $\cos(S)$ type wave function might be understood to correspond to a symmetric linear combination of ensembles of "expanding" and "contracting" universes (Halliwell 1994).

As a second approach, we reduce the action (30.9) classically further by choosing a time variable $\tau$ to be equal to the York time, the trace of the extrinsic curvature (York 1972). This amounts to taking $\tau = p_v$ and solving $\mathscr{H} = 0$ for $v$. The action then becomes

$$S = \int d\tau \left( \dot{\xi} p_\xi + \dot{\eta} p_\eta - H \right) \tag{30.13}$$

where the Hamiltonian $H$ is given by

$$H = \pm \frac{\eta \left( p_\xi^2 + p_\eta^2 \right)^{\frac{1}{2}}}{\tau} . \tag{30.14}$$

Quantization now leads to the Schrödinger equation

$$\hat{H}\psi = i \frac{\partial}{\partial \tau} \psi \tag{30.15}$$

where the wave function $\psi$ depends on $\xi$, $\eta$ and $\tau$, and $\hat{H}$ is obtained from $H$ by the operator substitutions $p_\xi \rightarrow -i\partial/\partial\xi$, $p_\eta \rightarrow -i\partial/\partial\eta$, and a choice of the factor ordering. One can then introduce a Schrödinger-type inner product compatible with the chosen factor ordering, and proceed to view this inner product as providing the theory with a probabilistic interpretation.

The difference between "expanding" and "contracting" universes now appears already at the classical level, in the choice of the sign in $H$ (30.14). Once this sign has been chosen, the resulting Schrödinger equation (30.15) is time asymmetric. One might thus be tempted to regard the Schrödinger equation (30.15) as a square root of the Wheeler-DeWitt equation (30.12). A problem here is, however, that the actions (30.9) and (30.13) are not equivalent at the classical level, not even if one allows both signs for $H$. The reason is that the York time is a good parameter for labelling the constant $t$ surfaces for the generic solutions (30.6), but it becomes degenerate for the exceptional solutions (30.7) in which the York time equals zero for every surface. Therefore, the exceptional classical solutions cannot be recovered from the reduced action (30.13). One consequence of this is the non-Hausdorff property of the space of the classical solutions (Hajicek 1986, 1989).

### 30.4 A Connection Quantization

Alternatives to metric quantizations have been developed by abandoning the role of the metric as the fundamental variable, and instead starting with an action which is expressed in terms of triads and connections. Two such closely related approaches are that of Witten (1988) and that of Ashtekar (Ashtekar et al. 1989, Ashtekar 1991). We shall here concentrate on Witten's approach.

In Witten's approach, the Einstein action is expressed in terms of the triad and the spin-connection. All the constraints turn out to be linear in the momenta, with an appropriate choice for the polarization, and the constraints can thus in principle be eliminated at the classical level. The resulting reduced action has an identically vanishing Hamiltonian, so that the coordinates on the reduced phase space are directly the classical constants of motion. Quantization is then in principle straightforward, although some subtleties arise from the fact that the reduced phase space consists of several connected components.

Consider now the specialization of this connection quantization to the case of toroidal spatial topology (Carlip 1990, 1992). After elimination of the constraints, the well-behaved component of the reduced phase space is four dimensional. One can introduce on this phase space four natural coordinates, which we suggestively denote by $(\lambda, \mu, a, b)$, such that the reduced action takes the form

$$S = 2 \int (a d\mu - b d\lambda) \ . \tag{30.16}$$

The nonvanishing Poisson brackets (denoted by curly braces) are thus

$$\{\mu, a\} = \{b, \lambda\} = \frac{1}{2} \ . \tag{30.17}$$

A natural quantization is obtained by replacing the Poisson brackets by commutators for the corresponding operators,

$$[\hat{\mu}, \hat{a}] = [\hat{b}, \hat{\lambda}] = \frac{i}{2} \ . \tag{30.18}$$

One can then choose a representation in which the "coordinates" $\hat{\lambda}$ and $\hat{\mu}$ are represented by multiplication and the "momenta" $\hat{a}$ and $\hat{b}$ by derivatives. In this representation the wave functions depend on $\lambda$ and $\mu$, and the Hilbert space is $L^2(\mathbb{R}^2)$.

We are interested in how spacetime metrics and their time asymmetry appear in this connection quantization. Consider first the classical connection theory defined by the action (30.16). As the notation suggests, the four phase space coordinates can be directly identified with the four constants of integration in the generic family

(30.6) of classical solutions to the metric theory (Carlip 1990). The connection theory thus contains all the generic solutions of the metric theory, but it contains also degenerate metrics of the form (30.6) where $a\mu - b\lambda = 0$. The exceptional family (30.7) of solutions to the metric theory is missing, however. This is related to the fact that the action (30.16) corresponds to only one of the connected components of the reduced phase space in the connection theory.

To discuss time asymmetry in the classical connection theory, a key observation is that the actions (30.13) and (30.16) are related to each other by the canonical transformation (Carlip 1990)

$$p_\xi d\xi + p_\eta d\eta - H d\tau = 2a\, d\mu - 2b\, d\lambda + dF \tag{30.19}$$

where the generating function $F$ of the canonical transformation is

$$F = \frac{(\mu - \xi\lambda)^2 + \eta^2\lambda^2}{\eta\tau} . \tag{30.20}$$

In particular, the expression of the Hamiltonian $H$ in terms of the connection variables is

$$H = \frac{\lambda b - a\mu}{\tau} . \tag{30.21}$$

Comparing with the metric expression for $H$ (30.14), we see that the connection theory contains copies of the reduced metric theory with both signs of the Hamiltonian, and in addition it contains a sector corresponding to the degenerate metrics. One can say that the classical connection theory contains both expanding and contracting classical universes, and it is able to distinguish between the two by the sign of (30.21).

What about the quantum connection theory? Given a wave function $\psi(\lambda, \mu)$, one can in principle compute expectation values of any operators constructed from the elementary operators $\hat{\lambda}$, $\hat{\mu}$, $\hat{a}$ and $\hat{b}$. The issue would be to find operators whose expectation values are of physical interest.

An interesting suggestion is to use the canonical transformation (30.19) to introduce in Witten's Hilbert space operators which would refer to quantities on a two-dimensional surface (Carlip 1990, 1992). To see the idea, consider a (possibly $\tau$-dependent) function $f(\tau)$ on the classical phase space of the reduced metric theory (30.13). By the canonical transformation (30.19), this function can be pulled back into a $\tau$-dependent function $f^*(\tau)$ on Witten's classical phase space. Some of the $\tau$-dependence of $f^*(\tau)$ comes from that in $f(\tau)$; some arises from the fact that the canonical transformation is explicitly $\tau$-dependent. One can now introduce in Witten's Hilbert space a $\tau$-dependent operator $\hat{f}^*(\tau)$ by replacing $\lambda$, $\mu$, $a$ and $b$ in the classical expression for $f^*(\tau)$ by the corresponding operators and adopting a factor ordering. The expectation value of $\hat{f}^*(\tau)$ could then be interpreted as the expectation value of the classical quantity $f(\tau)$. This is an expectation value referring to geometrical quantities measured on a single two-surface with a given value of $\tau$.

To make these ideas more precise, one would need to discuss in more detail the class of functions $f(\tau)$ for which this program could be satisfactorily completed, and in particular the issues of operator ordering when constructing $\hat{f}^*(\tau)$ from the elementary operators $\hat{\lambda}$, $\hat{\mu}$, $\hat{a}$ and $\hat{b}$. A further issue would be how to incorporate in the quantum theory the discrete classical symmetry known as modular invariance, originating from the freedom of performing discrete spatial coordinate transformations in (30.8). The requirement of quantum modular invariance occupies a central role in Carlip's recent discussions of relating the connection quantized theory to the metric variables (Carlip 1990, 1992).

## 30.5 Discussion: Classical and Semiclassical Spacetimes

In the Dirac quantized theory, where no explicit time variable occurs, it is often advocated that the connection to classical spacetimes and low energy quantum field theory should emerge through the semiclassical limit (Banks 1985, Halliwell and Hawking 1985). In spite of the appeal of this approach, it has proved difficult to give reasonably precise statements as to what are the assumptions under which such a semiclassical interpretation is supposed to be viable (Kuchař 1992). One source for ambiguity here is the question of the function space in which the total wave function is supposed to live. I would like to end by speculating how, in the model considered above, one might be able to introduce a semiclassical interpretation within the connection quantized theory, where some control over the assumptions could be maintained using the Hilbert space structure of this theory.

Let us first recall the first premise of the semiclassical interpretation. Suppose we have a wave function $\Psi$ which solves the Wheeler-DeWitt equation (30.12) and takes, at least in some region in the configuration space, the approximate semiclassical form $\exp(iS)$. It follows that $S$ is a particular solution to the Hamilton-Jacobi equation of the classical theory, and one can recover a particular two-parameter family of solutions to the classical equations of motion as the integral curves of the vector field $f^{\alpha\beta}\partial_\beta S$. Here $f^{\alpha\beta}$ is the inverse minisuperspace metric appearing in the superHamiltonian (30.11). The basic assertion of the semiclassical interpretation is that $\Psi$ is understood to correspond to this particular two-parameter family of classical spacetimes, in the region of the configuration space where the semiclassical approximation to $\Psi$ is good.

Now, could an analogous assertion be made in the connection quantized theory? From the correspondence between the reduced phase space and classical spacetimes (30.6), it is easy to see that the answer is positive: one can construct wave functions in the Hilbert space of the connection quantized theory that are arbitrarily sharply peaked on two-parameter families of classical solutions. As a simplest example, a wave function $\psi(\lambda, \mu)$ that is sharply peaked about $(\lambda_0, \mu_0)$ (say, a sharp Gaussian) would correspond to a family of spacetimes (30.6) in which $\lambda$ and $\mu$ are fixed to be respectively $\lambda_0$ and $\mu_0$, but $a$ and $b$ remain free.

There clearly exist families of classical solutions for which this procedure would not work. Examples would be families that do not define a local polarization in the phase space of the connection theory, i.e., families that would simultaneously fix both a coordinate and its conjugate momentum. This need not be a serious obstacle, however, since also in the Dirac quantized theory the semiclassical interpretation only gives families of classical solutions that project sufficiently smoothly into the configuration space.

Let us now turn to the second stage in the semiclassical interpretation, at which one is supposed to recover not only classical spacetimes but also the low energy quantum field theory for small perturbations. In the model considered above, one way to realize this in the Dirac quantized theory is to treat the nondiagonal component in the metric as a small perturbation. One seeks a wave function which would approximately factorize into a product of a semiclassical wave function in the diagonal theory and a perturbation wave function that would obey the Schrödinger equation on the diagonal background spacetimes.

Could this second stage be repeated in the connection quantized theory? One can certainly truncate the classical connection theory into diagonal metrics by setting $\mu = a = 0$, and in the resulting quantum theory one can find wave functions $\psi(\lambda)$ corresponding to one-parameter families of classical spacetimes. Such wave functions would now be the counterparts of the semiclassical wave functions in the Dirac quantized diagonal theory. To introduce the off-diagonal component as a small perturbation, one would then need to turn on the $\mu$-dependence in the wave function. It is however not obvious what one should assume at this stage about the $\mu$-dependence to mimic the approximate perturbation Schrödinger equation in the Dirac quantization. A possible avenue towards this question might be to find a sufficiently large set of operators $\left\{\hat{f}^*(\tau)\right\}$ of the kind discussed in Section 4, such that their classical counterparts $\{f(\tau)\}$ would be associated with the diagonal background spacetimes, and another set $\{g^*(\tau)\}$ whose classical counterparts would be associated with the nondiagonal degree of freedom.

## Acknowledgments

I would like to thank Abhay Ashtekar, Steve Carlip and Petr Hajicek for discussions. This work was supported in part by the Natural Sciences and Engineering Research Council of Canada, and by the National Science Foundation under Grant No. PHY 86-12424.

## Discussion

**Miller**  I am concerned with the identification of the cone singularity, in (2+1) dimensions, with the Big Bang singularity in (3+1) dimensions. This singularity is more like the coordinate singularity in the Taub universe and the subsequent Taub-NUT extension.

Perhaps you can give me another way of looking at this model so that I may make this identification?

**Louko** The singularity in (2+1) dimensions is indeed very similar to that in the Taub universe. Such singularities are weak but nevertheless real. The simplest example is probably the singularity occurring in Misner's two-dimensional model, discussed in Section 5.8 of Hawking and Ellis (1973).

**Page** Does the formalism give a clear answer to the question of the conditional probability of the shape of the 2-torus at a fixed volume?

**Louko** As outlined in Section 4, one can construct on the classical connection phase space functions $f^*(\tau)$ which characterize the shape and size of the torus. The volume, in particular, is given by the Hamiltonian (30.21). Assuming these functions can be consistently lifted into operators in the Hilbert space, their expectation values would give the expectation values of the shape and volume of the torus at a given value of the York time $\tau$. I am however not aware of a unambiguous way to obtain from these a $\tau$-independent conditional probability for the shape parameters, given just the volume.

An alternative approach to this question might be to reduce the constrained theory (30.9) by taking the volume $v$ as the time coordinate, and then to find the canonical transformation between the new reduced action and the connection theory. This might enable one to define in the connection Hilbert space operators in which the parameter is not the York time but the volume.

## References

Ashtekar, A. (1991) *Lectures on Non-perturbative Quantum Gravity*, World Scientific, Singapore.

Ashtekar, A., Husain, V., Rovelli, C., Samuel, J., and Smolin, L. (1989) *Class. Quantum Grav.* **6**, L185.

Banks, T. (1985) *Nucl. Phys.* **B249**, 332.

Barbour, J.B. and Smolin, L. (1988) Can quantum mechanics be sensibly applied to the Universe as a whole? Yale preprint (unpublished).

Carlip, S. (1990) *Phys. Rev.* **D42**, 2647.

Carlip, S. (1992), *Phys. Rev.* **D45**, 3584.

Deser, S., Jackiw, R., and 't Hooft, G. (1984) *Ann. Phys.* (N.Y) **152**, 220.

Deser, S. and Jackiw, R. (1984) *Ann. Phys.* (N.Y) **153**, 405.

Dirac, P.A.M. (1964) *Lectures in Quantum Mechanics*, Academic, New York.

Hajicek, P. (1986) *Phys. Rev.* **D34**, 1040.

Hajicek, P. (1989) *J. Math. Phys.* **30**, 2488.

Halliwell, J.J. (1994) Contribution to this volume.

Halliwell, J.J. and Hawking, S.W. (1985) *Phys. Rev.* **D31**, 1777.

Hawking, S.W. and Ellis, G.F.R. (1973) *The Large Scale Structure of Space-Time*, Cambridge University Press, Cambridge.

Hawking, S.W. and Page, D.N. (1986) *Nucl. Phys.* **B264**, 185.

Hosoya, A. and Nakao, K. (1990) *Class. Quantum Grav.* **7**, 163.

Kuchař, K.V. (1992) In *Proceedings of the 4th Canadian Conference on General Relativity and Relativistic Astrophysics*, edited by G. Kunstatter, D.E. Vincent and J.G. Williams, World Scientific, Singapore.

Moncrief, V. (1989) *J. Math. Phys.* **30**, 2907.

Witten, E. (1988) *Nucl. Phys.* **B311**, 46.

York, J.W. (1972) *Phys. Rev. Lett.* **28**, 1082.

# 31

# Abandoning Prejudices About Time Reparametrization Invariance

Claudio Teitelboim

*Centro de Estudios Científicos de Santiago*
*Casilla 16443, Santiago 9*

*School of Natural Sciences, Institute for Advanced Study*
*Olden Lane, Princeton, New Jersey 08540, USA*

## Abstract

The propagation amplitude of a system invariant under time reparametrizations may be expressed as a sum over histories in many different ways. One may choose to use histories in which "the time" moves only forward, or histories in which "the time" flows back and forth, or histories for which "the time" does not flow at all. Yet, the same answer is obtained. Thus one does not have to find a "good time variable" or "a good slicing" to define the path integral. In particular one may even use time-independent canonical gauges. However, in practice, it is more convenient to use "derivative gauges" such as the proper time gauge.

## 31.1 Introduction

In this report a rather narrow view of quantum mechanics will be taken. We will assume that the only object of interest is the propagation amplitude from one field configuration to another, or – a bit more generally – that amplitude with possible insertions of gauge invariant operators. All questions pertaining to the observer, or to the meaning of time in terms of measurements, will be ignored.

Once one focuses one's attention on the amplitude it is clear that any particular representation of it in terms of a sum over intermediate configurations is not to be given an especial physical meaning. As Schwinger put it many years ago [1] intermediate states are just a "mental construct". Therefore one should not be tied to prejudices about "the nature of time" if one is interested in evaluating the amplitude. Indeed, it turns out that one may – in principle – abandon all the prejudices and treat a generally covariant system just as an ordinary system endowed with an "internal" gauge symmetry.

Rather than elaborating on the general formalism, the essential point will be put through by means of simple examples in terms of a point particle. Then some comments about the case of more realistic generally covariant systems such as the

442

string and gravity will be made. A detailed study of the problem may be found in a paper written in collaboration with M.Henneaux and D.Vergara [2]. The present report is based on that joint work.

## 31.2 Abandoning Prejudices

*Prejudice 1. A Zero Hamiltonian is the Distinguishing Mark of a Generally Covariant System.*

In cases found in practice the action, in Lagrangian form say, is invariant under reparametrizations of the time variable when the $q$'s transform as scalars. That is, under

$$
\begin{aligned}
\tau \to \bar{\tau} &= \tau - \epsilon(\tau) \\
q(\tau) \to \bar{q}(\tau) &= q(\tau) + \delta q(\tau)
\end{aligned}
$$
(31.1)

with

$$
\delta q(\tau) = \dot{q}(\tau)\, \epsilon(\tau)
$$
(31.2)

the Lagrangian transforms as a scalar density

$$
L(\bar{q},\, \dot{\bar{q}}) = L(q,\, \dot{q}) + \delta L
$$
(31.3)

where

$$
\delta L = \frac{d}{dt}(\epsilon L)
$$
(31.4)

However, if one expands (31.3) using (31.1) and (31.2), one finds

$$
\begin{aligned}
\delta L &= \frac{\partial L}{\partial q}\dot{q}\epsilon + \frac{\partial L}{\partial \dot{q}}\frac{d}{d\tau}(\dot{q}\,\epsilon) \\
&= \left(\frac{\partial L}{\partial q}\dot{q} + \frac{\partial L}{\partial \dot{q}}\ddot{q}\right)\epsilon + \frac{\partial L}{\partial \dot{q}}\,\dot{q}\,\dot{\epsilon} \\
&= \frac{dL}{d\tau}\epsilon + \frac{\partial L}{\partial \dot{q}}\dot{q}\,\dot{\epsilon} \\
&= \frac{d}{d\tau}(\epsilon L) - L\dot{\epsilon} + \frac{\partial L}{\partial \dot{q}}\dot{q}\,\dot{\epsilon} \\
&= \frac{d}{d\tau}(\epsilon L) + H\dot{\epsilon}
\end{aligned}
$$
(31.5)

But this must be equal to (31.4). Hence

$$
H = 0
$$
(31.6)

It should be apparent from the above derivation that the result $H = 0$ does not follow just from the invariance of the action (Eq. (31.4) for the Lagrangian. It also needs that the coordinates transform as scalars (Eq. (31.2)).

Now, we want to keep (31.4), but there is nothing essential about (31.2). In fact, it suffices to define a new variable, say

$$q'(\tau) = q(\tau) - \tau \tag{31.7}$$

which transforms inhomogeneously ("as a connection")

$$\delta q'(\tau) = \dot{q}'(\tau) \, \epsilon(\tau) + \epsilon(\tau) \tag{31.8}$$

to achieve a non-zero Hamiltonian

$$H' = \frac{\partial L}{\partial \dot{q}'} = -p' \neq 0 \tag{31.9}$$

Of course, this is just the statement that the Hamiltonian is not invariant under time-dependent canonical transformations. We therefore conclude that there is nothing intrinsic connecting time reparametrization invariance with a vanishing canonical Hamiltonian. So much for Prejudice 1.

*Prejudice 2. Time Reparametrizations are different from other Gauge Symmetries in that they must vanish on the boundaries to make the action invariant.*

One normally writes the action as the time integral of the Lagrangian

$$S' = \int_{\tau_1}^{\tau_2} L \, d\tau \tag{31.10}$$

Then it follows from (31.4) that

$$\delta S' = \epsilon(\tau_2) L(\tau_2) - \epsilon(\tau_1) L(\tau_1) \tag{31.11}$$

and therefore the action is not invariant unless

$$\epsilon(\tau_2) = \epsilon(\tau_1) = 0 \tag{31.12}$$

On the other hand, for other symmetries, such as the gauge symmetry of the Maxwell theory, one finds that the Lagrangian is strictly invariant (rather than being invariant up to a divergence) and therefore the action is invariant even when (31.12) does not hold.

Just as with Prejudice 1, there is again here an unnecessary assumption. The action does not have to be the time integral of the Lagrangian. It is possible, and often necessary, to add boundary terms. This is the case here too. By adding a suitable boundary term, and also by appropriately identifying what is to be fixed at the endpoints, one may make the action for a generally covariant system gauge invariant even at the endpoints.

There is no general formula for the bounday term. It needs to be constructed system by system. Moreover, its construction basically needs to know the solution of the equations of motion. Therefore, in practice, the fully invariant action can only be written in closed form for generally covariant systems that are exactly soluble. For this reason, for realistic theories, one is better off separating the gauge invariance "inside" from that of the boundaries. One may then deal with the former in terms of the usual action (the time integral of the Lagrangian) and write an expression for the amplitude in terms of a non-canonical gauge, such as the proper time gauge [3], which does not affect the endpoints.

Since our point here is one of principle it will be demonstrated in the very simple example of the non-relativistic free particle. One may describe the histories of the particle in parametric form. The position $q$ and the Newtonian time $t$ are given in terms of the arbitrary parameter $\tau$. The Lagrangian is

$$L = \frac{1}{2} \left(\frac{dq}{dt}\right)^2 \left(\frac{dt}{d\tau}\right)^{-1} \tag{31.13}$$

It transforms as a density (Eq. (31.4)) if $q$ and $t$ transform as scalars. The canonical Hamiltonian is therefore zero.

To proceed further it is convenient to pass to the Hamiltonian form of the action

$$S' = \int_{\tau_1}^{\tau_2} \left(p_t \dot{t} + p\dot{q} - N\mathscr{H}\right) d\tau \tag{31.14}$$

All the information about the system is then contained in the constraint generator

$$\mathscr{H} = p_t + \frac{1}{2}p^2 \tag{31.15}$$

Here $p_t$ and $p$ are the momenta canonically conjugate to $t$ and $q$ respectively and $N$ is a Lagrange multiplier.

The action (31.14) is the Hamiltonian version of (31.10) with the Lagrangian (31.13) and, therefore, is reparametrization invariant only when (31.12) holds. One may, however, write an improved action which is also invariant at the endpoints. It is

$$S = S' + B \tag{31.16}$$

where $B$ is a boundary term given by

$$B = -\frac{1}{2}\left[t(\tau_2) - t_2\right]p^2(\tau_2) + \frac{1}{2}\left[t(\tau_1) - t_1\right]p^2(\tau_1) \tag{31.17}$$

The procedure for constructing $B$ is explained in the Appendix.

The improved action (31.16) has an extremum when the equations of motion hold, provided that the appropriate functions are fixed at the endpoints. For the original action (31.14) one simply fixed

$$t(\tau_2) = t_2, \quad q(\tau_2) = q_2 \tag{31.18a}$$

$$t(\tau_1) = t_1, \quad q(\tau_1) = q_1 \tag{31.18b}$$

but, for (31.16) one must fix instead the gauge invariant combinations

$$\begin{aligned}(q - pt + pt_2)(\tau_2) &= q_2 \\ (q - pt + pt_1)(\tau_1) &= q_1\end{aligned} \tag{31.19}$$

Note that only one function is fixed at each end point for (31.16), whereas two functions were fixed at each endpoint for (31.14). However the one function fixed in the case of (31.16) depends on two parameters $(t_1, q_1)$ at the lower endpoint, and on two parameters $(t_2, q_2)$ at the upper endpoint. This will make the path integrals over (31.12) and (31.16) to both depend on the same arguments, and – actually – be identical. So much for Prejudice 2.

*Prejudice 3. In the Path Integral for a System invariant under time reparametrizations the gauge must be fixed so that a particular coordinate, "the time", increases monotonically along the allowed histories, as one proceeds from the initial configuration to the final one.*

For the non-relativistic particle, "the time" par excellence, is the coordinate $t$. A good way to see that it is not necessary for $t$ to increase is to show how the correct amplitude is obtained in a gauge in which "the time", most blatantly, does not flow monotonically, namely

$$t(\tau) = 0 \,, \text{ for all } \tau \tag{31.20}$$

If one sets $t(\tau_1) = 0$ and $t(\tau_2) = 0$ in (31.16), one obtains

$$S\big[q(\tau),\, p(\tau)\big] = \int_{r_1}^{r_2} d\tau (p\dot{q}) + \frac{1}{2}t_2\, p^2(\tau_2) - \frac{1}{2}t_1\, p^2(\tau_1) \tag{31.21}$$

on the surface $t = 0$, $p_t + \frac{p^2}{2} = 0$. The path integral is thus

$$K = \int Dq\, Dp\, Dt \prod_{\tau} \delta(t)\delta(p_t)\exp iS \tag{31.22}$$

(the Faddeev-Popov determinant is unity), with $S$ given by (31.21). The paths in (31.22) are subject to the boundary conditions (31.19), which read here ($t = 0$)

$$(q + pt_1)(\tau_1) = q_1 \tag{31.23}$$

$$(q + pt_2)(\tau_2) = q_2 \tag{31.24}$$

The integration over $t$ and $p_t$ is direct and yields unity. Hence (31.22) becomes

$$K = \int Dq\, Dp\, \exp i\left\{ \int_{\tau_1}^{\tau_2} d\tau (p\dot{q}) + \frac{1}{2}\big[t_2 p^2(\tau_2) - t_1 p^2(\tau_1)\big] \right\} \tag{31.25}$$

By making the canonical change of integration variable

$$Q = q + pt_1 + p\frac{t_2 - t_1}{\tau_2 - \tau_1}(\tau - \tau_1) \tag{31.26}$$

$$P = p \tag{31.27}$$

and the time rescaling $\bar{\tau} = \frac{t_2 - t_1}{\tau_2 - \tau_1}(\tau - \tau_1) + t_1$, one can recast (31.25) in the form

$$K = \int DQDP\, \exp i\left[ \int_{t_1}^{t_2} d\bar{\tau}\left( P\frac{dQ}{d\tau} - \frac{P^2}{2} \right) \right] \tag{31.28}$$

with

$$Q(\bar{\tau} = \tau_1) \; = \; q_1 \qquad Q(\bar{\tau} = \tau_2) \; = \; q_2 \tag{31.29}$$

This is just the standard path integral for the non-parametrized particle, known to be equal to

$$K \; = \; [2\pi i(t_2 - t_1)]^{-1/2} \exp \frac{i(q_2 - q_1)^2}{2(t_2 - t_1)} \tag{31.30}$$

as it should.

The situation in the gauge $t = 0$ illustrates vividly the point that intermediate states are a "mental construct". Indeed if one looks at expression (31.25) for the action one sees that there is no Hamiltonian for $q$. Thus all the phase of the wave function is acquired "suddenly" at the endpoints (through the boundary term), rather than being gained monotonically as one proceeds from the initial state to the final state. Other gauges correspond to different ways of spreading the phase. In some cases the total phase will be a sum of a "spread" and a "sudden" portion. This happens for example for the gauge $t = p\tau$, for which time moves forward when $p > 0$ and backwards for $p < 0$. None of this matters if one is just interested in the resulting amplitude, which is the same for all cases. So much for Prejudice 3.

### 31.3 Causality. Proper Time

The amplitude (31.30) solves the constraint equation

$$\hat{\mathcal{H}}K \; = \; \left(\frac{1}{i}\frac{\partial}{\partial t} - \frac{1}{2}\frac{\partial^2}{\partial t^2}\right)K \; = \; 0 \tag{31.31}$$

because one path integrated a gauge invariant action in a gauge invariant manner. However, $K$ does not vanish for $t_2 < t_1$. This may may be described by saying that the amplitude is not causal. To obtain a causal amplitude one multiplies (31.30) by a step function of $(t_2 - t_1)$

$$K_+ \; = \; \theta(t_2 - t_1)K \tag{31.32}$$

which makes the amplitude vanish for $t_2 < t_1$ and also has the effect of making it not obey the constraint anymore

$$\hat{\mathcal{H}}K \; = \; \frac{1}{i}\delta(t_2 - t_1) \tag{31.33}$$

The difference between the gauge invariant amplitude and the causal amplitude has nothing to do with gauge fixing. Indeed one can also express $K_+$ as a path integral, which can be computed in any gauge, just as $K$. However, the path integral giving $K_+$ includes the insertion of the "gauge invariant operator" $\theta(T)$. That is, instead of $\exp(iS)$ one integrates $\theta(T)\exp(iS)$.

The functional $T$, called the "proper time" elapsed between the initial and the final states [3]- a terminology borrowed from the relativistic case- is given by

$$T \; = \; \int_{\tau_1}^{\tau_2} N d\tau \; + \; t_2 - t(\tau_2) - t_1 + t(\tau_1) \tag{31.34}$$

where N is the Lagrange multiplier appearing in the action (14). Expressed in the form (31.33) the proper time is invariant even under reparametrizations that do not vanish at the endpoints.

Whether one should "choose" the gauge invariant amplitude $K$ or the causal amplitude $K_+$, depends on criteria that go beyond the present discussion. One must know what question one is going to ask, or, in other words, one must go into the interpretation of the theory. See [3] for a related discussion.

## 31.4 Comments

One may, in principle, repeat the analysis described above for any generally covariant system. In other words one may treat general covariance as an "ordinary gauge symmetry". This means that one may use the gauge $tr\pi = 0$ for gravity in a compact space, or $x^+ = 0$ for the string. These would be gauges for which "the time does not flow". The phase change would then be "sudden" and accounted for by the boundary term. However, as stated above and illustrated in the Appendix, the construction of the boundary term, and of what is fixed at the boundaries, requires the integration of the infinitesimal gauge transformations. For general relativity – just as for the point particle – this amounts to knowing the general solution of the equations of motion, which is not tractable. It is important however to realize that the difficulty is a practical one rather than one of principle.

## Acknowledgements

The author is much indebted to Prof. Marc Henneaux for many illuminating discussions. He is especially grateful to him and to Dr. J. David Vergara for their kind permission to report at the "Physical Origins of Time Asymmetry" meeting on joint unpublished work. Appreciation is expressed to the meeting organizers and sponsors and, especially, to Prof. Juan Pérez-Mercader for the warm hospitality at Mazagon. The work on which this report is based was supported in part by grants 0862/91 and 0867/91 of FONDECYT(Chile), by a European Communities research contract, and by institutional support provided by SAREC (Sweeden) and Empresas Copec (Chile) to the Centro de Estudios Cientificos de Santiago.

## Appendix: Construction of the Boundary Term

The key idea is the important concept of "gauge invariant extension". Any time that one has a set (group) of transformations one may define an invariant under it as follows. Consider an object on which the transformations act. Go to a particular frame, selected by a gauge condition that sets equal to zero some of the components of the object. There will be as many components set equal to zero as there are independent transformation parameters. There is a transformation that takes the particular frame to a generic frame. It will always be possible to express its parameters as functions of the components in the generic frame. Once this is done one can, in turn, express those components that were not set equal to zero in the particular frame in terms of the components in the generic frame. The functions thus obtained are invariant by construction and are called invariant extensions (from the particular frame).

This procedure is often used, without even being noticed, as shown by the following examples:

(a) *The rest mass is the (Lorentz) invariant extension of the energy from the rest frame.*

Consider, for simplicity, two spacetime dimensions. The rest frame is selected by $(E', p') = (m, 0)$. The Lorentz transformation that takes the rest frame to a generic frame reads

$$E = E' \cosh \alpha + p' \sinh \alpha = m \cosh \alpha \tag{A.1a}$$

$$p = E' \sinh \alpha + p' \cosh \alpha = m \sinh \alpha \tag{A.1b}$$

Thus one may express $\alpha$ in terms of $E$, $p$ as

$$\alpha = \operatorname{arctanh} \frac{p}{E} \tag{A.2}$$

which in turn yields for $m$

$$m = \sqrt{E^2 - p^2} \tag{A.3}$$

the standard expression for the rest mass in terms of the energy and momentum in any frame.

(b) *The transverse components of the electromagnetic vector potential are the gauge invariant extension of the vector potential in the Coulomb Gauge.*

The Coulomb Gauge is given by $\nabla \cdot \vec{A}' = 0$. The transformation that takes the "Coulomb Gauge Frame" to a generic gauge frame reads

$$\vec{A}' = \vec{A} + \nabla \Lambda \tag{A.4}$$

with

$$\Lambda = -\frac{1}{\nabla^2} \nabla \cdot \vec{A} \tag{A.5}$$

If we now insert this value of $\Lambda$ back in (A.4) we find

$$\vec{A}' = \vec{A} - \frac{1}{\nabla^2} \nabla (\nabla \cdot \vec{A}) \tag{A.6}$$

These are the transverse components of $\vec{A}$ in a generic frame. They are gauge invariant and there are only two of them because they obey the *identity* $\nabla \cdot \vec{A}' = 0$

Once the concept of gauge invariant extension is at hand, the procedure to follow suggests itself. One writes the action in its original form (31.14) which has no boundary term, and in which two variables are fixed at each endpoint [Eq. (31.18)].

Since that action, and those boundary conditions, are not gauge invariant at the endpoints, one is in the "particular frame" (at the endpoints) – whose variables were denoted with a prime in the above examples.

One then makes a gauge transformation to go to the generic frame. The transformation will effectively be used only at the endpoints (the action is already gauge invariant in the interior). Thus one can continue it arbitrarily inside. One can then express the transformation in terms of what was fixed in the original action. Once the transformation is found it can be used to express the histories in the particular frame in terms of the histories in the generic frame. Since the gauge transformation is a canonical transformation, the action in the generic frame will differ in form from the action in the old frame by a contribution at the endpoints. This gives the boundary term.

It remains to decide what will be fixed at the endpoints in the generic frame. Part of what was fixed in the particular frame was just a way of specifying the frame. The other variables, when invariantly extended, are what is fixed in the generic frame. Thus one will fix "the gauge invariant content of what was fixed in the particular frame". The conditions at the endpoints are thus less in number than the total number of $q$'s, but these conditions depend on as many parameters as there are $q$'s. These parameters show up quantum mechanically as the arguments of the amplitude resulting from path integral.

Let us now do this for our example of the point particle. The infinitesimal gauge transformation in Hamiltonian form reads

$$\delta q = \epsilon[q, \chi] \; = \; \epsilon p, \quad \delta t = \epsilon[t, \; \chi] = \epsilon \tag{A.7a}$$

$$\delta p \; = \epsilon[p, \chi] \; = \; 0, \quad \delta p_t \; = \; \epsilon[p_t, \chi] \; = \; 0 \tag{A.7b}$$

$$\delta N \; = \; \dot{\epsilon} \tag{A.7c}$$

This transformation is Abelian and can be easily integrated. The integrated gauge transformation has the same form. Of the boundary conditions (31.18) one may consider $t(\tau_1) = t_1$ and $t(\tau_2) = t_2$ as fixing the gauge at the endpoints. The gauge parameter can then be solved for as

$$\epsilon(\tau) \; = \; t(\tau) - t'(\tau) \tag{A.8}$$

which yields, at the endpoints

$$\epsilon(\tau_1) = t(\tau_1) - t_1, \; \epsilon(\tau_2) \; = \; t(\tau_1) - t_2 \tag{A.9}$$

Only the value of $\epsilon$ at the endpoints will be relevant. One may continue it inside, for example, by a linear interpolation between $\tau_1$ and $\tau_2$.

$$\epsilon(\tau) \; = \; t(\tau_1) - t_1 + \frac{\tau - \tau_1}{\tau_2 - \tau_1} \left[ t(\tau_2) - t_2 - t(\tau_1) + t_1 \right] \tag{A.10}$$

When (A.10) is inserted back into (A.7) one finds

$$q'(\tau) = q(\tau) - \epsilon(\tau)p(\tau) \tag{A.11}$$

and corresponding expressions for the other variables. If the action written in terms of the primed variables is taken to be (31.14) one finds that when expressed in terms of the unprimed variables it acquires the boundary term (31.17). Finally setting $\tau = \tau_1$ and $\tau = \tau_2$ in (A.11), one obtains the boundary conditions (31.19).

## Discussion

**York**  When you use "strange" gauge or slicing conditions like $t = 0$ for a particle or trace $(K) = 0$ for a compact three-geometry, and yet obtain correct results, is it true that you have succeeded by encoding special non-local information into the boundary terms?

**Teitelboim**  Yes, you may put it that way. Although the boundary term is local (i.e. depends on the endpoints only) its construction needs the solution of the equations of motion.

**Mottola**  As I understood your earlier papers on path integral quantization of gravity you were always emphasizing that only those gauges which involve $\dot{N}$ are allowed, and these lead to the *causal* propagator $K(2,1)$ which does *not* satisfy the constraint (Wheeler-DeWitt equation) because of a $\delta$-function source. Have you now changed your opinion on this question of allowed gauge conditions for $N$? What principle determines what are the allowed or "good" gauge condition in a theory involving a reparameterization of time (gauge) symmetry? Certainly $\mathcal{H}\Psi = 0$ or $\mathcal{H}K = 0$ and $\mathcal{H}_2 K(2,1) = \delta(2,1)$ are different equations. Which is correct?

**Teitelboim**  I had stated earlier that the gauge conditions must involve $\dot{N}$. This is so if one uses the "standard action" (without boundary term). In practice, for a realistic system this is the only open option. However, in principle, by appropriately modifying the action through the addition of a boundary term one could even use canonical gauges. So, yes, I have mellowed in the sense that in what concerns gauge fixing the difference between general covariance and "internal gauge symmetries" appears to me now to be more of a practical nature than of a fundamental kind.

The second part of your question refers to the difference between $K$ and $K_+$. As explained in the text this issue is independent from considerations about gauge fixing. The possibility of defining the propagator $K_+$ depends crucially on the form of the constraint algebra (in gravity and the string), where the Hamiltonian constraint $\mathcal{H}$ appears in a very different way from the spatial generators $\mathcal{H}_i$. This makes it possible (and natural) to select a *unique* propagator by restricting $N$ to be positive in the proper time gauge. It does not seem sensible, or useful, to attempt something similar in, say, Yang-Mills, by restricting the range of integration of one component of the multiplier $A_0^a$. Just as one does not think, in gravity itself, of restricting the range of a component of the shift vector. Thus, in this context, generally covariant systems do appear to be different from those with internal gauge symmetries.

## References

[1] J. Schwinger *Quantum Kinematics and Dynamics*, Benjamin, N.Y. (1970), p. 30.
[2] M. Henneaux, C. Teitelboim and J.D. Vergara, *Nucl. Phys.* **B387**, 391 (1992).
[3] C. Teitelboim, *Phys. Rev.*, **D25**, 3159 (1982), *Phys. Rev. Lett.* **50**, 705 (1983).

# 32

## Physics in Knots

Daniel Holz

*Dept. of Physics, Princeton University*
*Princeton, New Jersey 08544, USA*


Arkady Kheyfets

*Dept. of Mathematics, North Carolina State University*
*Raleigh, North Carolina 27695-8205, USA*


Warner A. Miller

*Theoretical Division, Los Alamos National Laboratory*
*Los Alamos, NM 87545, USA;*
*and Phillips Laboratory, Kirtland AFB*
*NM 87117-6008, USA*


John A. Wheeler

*Dept. of Physics, Princeton University*
*Princeton, New Jersey 08544, USA*

### 32.1 A-B Phase as Definer of Field

Physicists today are striving for the rapprochement of the concepts of spacetime and the quantum (Kuchař 1992). Perhaps the most surprising and recent consequence of this endeavor is the emergence of a link between information-theory and spacetime geometry (Zurek 1991; Wheeler 1982, 1984). No examples better illustrate the tie than those cited in Wheeler's *"It from Bit"* thesis (Wheeler 1990), among them Bekenstein's discovery of the proportionality between horizon area and black-hole entropy, and Bekenstein's conjecture (and subsequent development by Hawking, Christodoulou, Thorne and Zurek, Unruh etc.) regarding the entropy/information content of a black hole (Bekenstein 1972, 1973, 1980; Hawking 1974, 1975, 1976).

Three clues offer themselves as to the information-theoretic foundation of quantum theory. First, Stueckelberg (Stueckelberg 1960) proved that the principle of complementarity inevitably leads to complex numbers $[x, p] = \sqrt{-1}\,\hbar$. Wootters showed – following R. A. Fisher – that the distinguishability of two outcomes expresses itself not in probability but rather in the complex probability amplitude (Wootters 1980, 1981). Finally, Aharonov-Bohm and Anandan showed how to use phase as definer of field. The phase change of an electron's wave function around a closed loop encircling a flux of magnetic field is given by,

$$\delta(\phi_{\text{phase}}) = \left( \begin{array}{c} \text{phase} \\ \text{change} \end{array} \right) = \frac{e}{\hbar} \oint_{\alpha} A \cdot dl$$

$$= \frac{e}{\hbar} \int \int B \cdot dS, \tag{32.1}$$

where $A$ is the electromagnetic potential (or connection form), and the line integral is around the perimeter of the loop $\alpha$. This can be realized experimentally via a bit-by-bit recording on a photographic emulsion (Fig. 32.1). One can thus define the field via the bit-by-bit measurement of the interference pattern before and after the phase change. Wheeler asks, "Can all of nature be expressed fundamentally by such a bit-by-bit quantum measurement procedure?" If spacetime curvature is the measure of gravity, as Einstein's theory tells us, and if the neutron interference experiments of Werner (Colella, Overhauser, and Werner 1975) measure gravity, then there must exist an it-from-bit description of spacetime curvature. In the present report we work out the relevant it-from-bit means to measure spacetime curvature. It expresses itself as

$$\delta(\phi_{\text{phase}}) = \left( \begin{array}{c} \text{phase} \\ \text{change} \end{array} \right) = \frac{1}{\hbar} \oint_{\alpha} \mathbf{A} = \frac{1}{\hbar} \oint_{\alpha} A_a^i \, dx^a$$

$$= \frac{1}{\hbar} \int \int d\mathbf{A} = \frac{1}{\hbar} \int \int A_{a,b}^i \, dx^a \wedge dx^b \tag{32.2}$$

where $\mathbf{A} = A_a^i \, dx^a$ is the 1-index Ashtekar connection related to the Ashtekar self-dual 4-dimensional connection ${}^4 A_a^{JK}$ via $A_a^i = \epsilon^{0i}{}_{JK} {}^4 A_a^{JK}$.

Here the integration goes around the boundary of the region encompassed between the two neutron beams. In Line 1 of Eq. (32.2) the integrand is the complex connection which Ashtekar introduced in 1985 to transform the independent variable in the standard Schrödinger-like WDW geometrodynamic wave equation from 3-geometry to knot class. The real part of this connection is just the 3-dimensional dual of the spin connection of the spacelike triad field $A_a^i = \epsilon^i{}_{jk} \omega_a^{jk}$, while its imaginary part is related to the extrinsic curvature $K^b{}_a$ of a 3-slice via $A_a^i = e_b^i K^b{}_a$. In line 2 the integral runs, not around the periphery of the area embraced by the neutron beams but over the area itself, and the integrand $d\mathbf{A} = A_{a,b}^i \, dx^a \wedge dx^b$ is an analog of the magnetic field of electrodynamics. The precise physical nature of these equations is not yet known.

Ashtekar's formulation of general relativity provides a new Aharonov-Bohm-like description of the spacetime geomtery. The phase difference around a loop (analogue of the vector-potential line integral in electrodynamics) serves as fundamental operator in this quantum theory. Here we attempt to translate this new language into the more familiar description of Einstein's theory. We show that the deformation of the 1-index loop variable is related to the pure spatial component of the Einstein tensor. This provides the first geometric description of the Ashtekar $T$ variable, and is an essential step toward our goal of transferring between the quantum probability

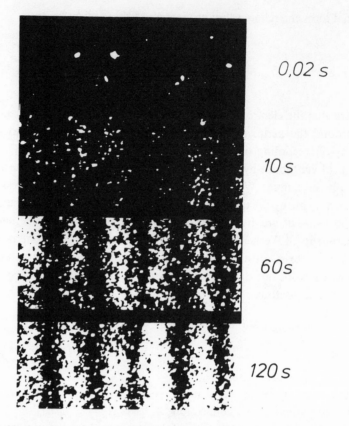

Fig. 32.1. An example of *It-from-Bit* as it applies to an electron biprism experiment. The electron biprism interference patterns taken at four different exposure times show the statistical aspect of electron accumulation. This experiment was performed by Hannes Lichte at the Institut für Angewandte Physik, Universität Tübingen, Germany.

amplitude over 3-geometries in superspace $\Psi(\mathscr{G}^{(3)})$ and the probability amplitudes over the Ashtekar knot classes $\Psi(K)$.

It is not our intention here to provide a complete review of Ashtekar's formulation (this has recently been done by Rovelli 1991). We intend in Section 2 to highlight the Ashtekar approach, and in Section 3 to give a precise geometric interpretation of the 1-index loop variable. We include in the Appendix a derivation of Section 3's main result using the notation commonly found in the literature.

## 32.2 Essential New Features of the Knot Description of Gravity

Ordinarily, general relativity is described in Lagrangian formulation by metric and its Levi-Civita connection. The variables of its Hamiltonian description are the six components of three-metric ($\gamma$) for each three-dimensional space-like section and its conjugate ($\pi$) moments closely related to the extrinsic curvature ($K$) of these sections. In Ashtekar's formulation of general relativity, the Lagrangian description uses as variables the component of a soldering form determined by a tetrad field $e_\mu^I$, and the self-dual part of the spin connection of this field,

$$^4A_\mu^{MN} = -\frac{i}{2}\epsilon_{IJ}^{MN}\, {}^4A_\mu^{IJ}, \tag{32.3}$$

where $\epsilon_{IJ}^{MN}$ is the completely antisymmetric tensor. The natural appearance of $i = \sqrt{-1}$ in such a description is solely a consequence of the properties of duality operators on four-dimensional spaces with a Lorentz signature metric. The self-dual connection given in Eq. (32.3) can be written in terms of the familiar spin connection $\omega$ as follows

$$^4A_\mu^{IJ}[\omega] = \omega_\mu^{IJ} - \frac{i}{2}\epsilon_{MN}^{IJ}\, \omega_\mu^{MN}. \tag{32.4}$$

The curvature (and Yang-Mills field strength) of this self-dual spin connection is given by

$$F_{\mu\nu}^{IJ}[^4A[\omega]] = R_{\mu\nu}^{IJ}[\omega] - \frac{i}{2}\epsilon_{MN}^{IJ}R_{\mu\nu}^{MN}[\omega]. \tag{32.5}$$

The Hamiltonian description of Ashtekar's formulation is obtained from the Lagrangian description via a Legendre transformation, leading to the introduction of a new set of variables, called the connection variables. The connection variables are the densitized triads (densitized three-dimensional parts of tetrads) (see Appendix I, Eqs. (A.6), (A.7)) and conjugate momentum equal to the Ashtekar connection multiplied by the imaginary unit,

$$p_a^i = \frac{\partial L}{\partial \tilde{E}_i^a} = iA_a^i[e]. \tag{32.6}$$

The Ashtekar connection itself is defined as a three-dimensional projection of the self-dual connection used in the Lagrangian formalism (see Eq. (A.8)). The real part of this momentum is closely related to the extrinsic curvature (i.e. standard ADM momentum) Re $p_{(a}^i e_{b)}^i = k_{ab}$, while the imaginary part of the momentum is completely determined by the three-dimensional connection of the densitized field triad

$$\text{Im } p_a^i = \epsilon_{jk}^i w_a^{jk}[e]. \tag{32.7}$$

The transition to these Ashtekar variables leads to a constraint Hamiltonian formalism similar to the ordinary ADM formalism. The Hamiltonian theory includes the usual gauge constraint, vector constraint, and Hamiltonian constraint. The gauge and vector constraint can be united in the diffeomorphism constraint. In addition, a new constraint, the so-called reality condition, is imposed. All the constraints, including the reality condition, are of the first class. The constraints in new variables can all be written in polynomial form, and look much simpler than in the standard ADM formulation. The gauge, vector, and Hamiltonian constraint are maximally quadratic.

The resulting Hamilton-Jacobi equation on the Ashtekar superspace of connections does not have a potential term, which should lead to considerable simplification in its solution. Although the complete solution of this equation is as yet unknown, several exact solutions have been obtained, leading to a natural description of general relativity in terms of the Wilson loops of the Ashtekar connection and the Wilson loop generalizations known as T-variables.

To facilitate a description of the Wilson loop of the Ashtekar connection and T-variables, we introduce a basis $\{\tau_i\}_{i=1}^3$ of the Lie algebra of the internal symmetry group (the group of three-dimensional rotation). In most of the literature on loop variables, the basis $\tau_i$ consists of the Pauli matrices $\tau_i{}^A{}_B$ (cf. Appendix) and

$$A_a = A_a^i \tau_i,$$
$$\tilde{E}^a = \tilde{E}^{ai}.\tau_i \tag{32.8}$$

The operator of parallel transport around a closed curve $\alpha : [0, 2\pi] \to \Sigma$ in the 3-dimensional physical space $\Sigma$ can then be expressed as

$$U_\alpha(0, 2\pi) = \left(\mathscr{P}e^{\int_\alpha A}\right) = \left(\mathscr{P}e^{\int_0^{2\pi} ds \dot{\alpha}^a(s) A_a^i(\alpha(s)) \tau_i}\right), \tag{32.9}$$

where $s$ is the parameter of the curve $\alpha$, $\alpha^a(s)$ are coordinates of the point on the curve $\alpha$ determined by the value of $s$, $\dot{\alpha}^a(s) = \frac{d\alpha^a}{ds}$, and $\mathscr{P}$ means path ordering.

The Wilson loop can be expressed as

$$T[\alpha] = \mathrm{Tr} U_\alpha(s), \tag{32.10}$$

where $U_\alpha$ is the operator of parallel transport around the loop $\alpha$ starting from the point $s$. The Wilson loop does not depend upon the choice of $s$. The generalizations of the Wilson loop with one or more indices can be defined using the operator of parallel transport and the triad $\tilde{E}^a$. For example, the one-index T-variable is given by

$$T^a[\alpha](s) = \mathrm{Tr}[U_\alpha(s)\tilde{E}^a(\alpha(s))]. \tag{32.11}$$

The known solutions of the Hamilton-Jacobi equations can all be expressed in terms of T-variables. In order for these solutions to satisfy the diffeomorphism constraint equations, they should depend not on the loops themselves but on the knot classes of the loops.

In quantum theory the Hamilton-Jacobi equation is replaced by the Schrödinger-Ashtekar equation and the state functionals are defined on the knot classes rather than on loops.

## 32.3 The One Index Loop Variable and the Einstein Tensor

Gravity manifests itself through geometry, the curvature of spacetime. This curvature displays itself as the rotation of vectors parallel transported around loops. Therefore,

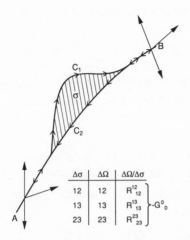

| Δσ | ΔΩ | ΔΩ/Δσ |
|----|----|-------|
| 12 | 12 | $R^{12}_{12}$ |
| 13 | 13 | $R^{13}_{13}$ |
| 23 | 23 | $R^{23}_{23}$ |

$\left. \vphantom{\begin{matrix}1\\2\\3\end{matrix}} \right\} -G^0_0$

Fig. 32.2. The one-index loop variable is expressed, for an infinitesimal loop $\sigma$ (shaded region), as a rotation operator, and can be examined by the parallel transport of a triad (initially at point $A$) around a closed curve $C = \partial\sigma$. The triad is transported from $A$ to $B$ along spacelike curve $C_1$, and then back to $A$ along spacelike curve $C_2$. The triad will ordinarily return to $A$ with timelike components. Nevertheless, the spatial components of the "initial" and "parallel-transported" triad at $A$ are related by a rotation operator ($\Delta\Omega$). We show that the trace of the differential of the one-index loop variable with respect to infinitesimal changes in the area $\sigma$ is precisely the $G^0_0$ component of the Einstein tensor.

to understand the mechanics of the one-index loop variable it is useful to examine the rotation experienced by a triad of space-like vectors when they are carried in parallel transport around an infinitesimal loop, $\sigma$. The Taylor expansion of the 1-index loop variable yields such an operator.

If we parallel transport the triad along a space-like curve $C$, shown in Fig. 32.2, from the point $A$ to the point $B$, and then transport it back along the other curve ($C_2$) to $A$, the triad will ordinarily return rotated. This rotation is given by the standard rotation opertor,

$$\left( \begin{matrix} \text{Infinitesimal} \\ \text{Rotation Operator} \end{matrix} \right) = \Delta\Omega_{AB} = -R^{AB}{}_{CD} \, \Delta\sigma^{CD}, \tag{32.12}$$

where $R$ is the four-dimensional Riemann curvature tensor and $\Delta\sigma^{CD}$ is the bivector representing the area spanned by curves $C_1$ and $C_2$. Ordinarily, the transported triad of vectors will be rotated out of the space-like hypersurface it originated in; however, as in any Hamiltonian formulation, it is customary to consider the spatial components of the bivector of rotation. One does so with the one-index loop variable in Ashtekar's formulation, and we will do so here. The rotation bivector can be described as a rotation vector in space. In particular, the rotation vector is oriented with components given by the 1-form

$$\omega_E = \epsilon_{EAB} \Delta\Omega^{AB}. \tag{32.13}$$

and similarly the components of the orientation of the shaded loop in Fig. (32.2) are given by the 1-form

$$\omega_F = \epsilon_{FCD}\,\Delta\sigma^{CD}. \tag{32.14}$$

There are only three independent loops at each point. Without loss of generality these can be chosen to correspond to the triad directions, giving

$$\Delta\sigma^{12}, \Delta\sigma^{13} \text{ and } \Delta\sigma^{23}. \tag{32.15}$$

It is clear how to construct the pure spatial components of the Einstein tensor from the above elements. If we take the differential of the rotation operator with respect to the infinitesimal area, we obtain components of the Riemann tensor, some of which form the $G^{00}$ tensor

$$\frac{\Delta\Omega^{AB}}{\Delta\sigma^{CD}} = -R^{AB}{}_{ED}. \tag{32.16}$$

If we contract the vector components of the rotation with the vector components of the loop orientation, we are left with precisely the three components of the Riemann tensor necessary to construct $G^{00}$,

$$R^{12}{}_{12}, \ R^{13}{}_{13}, \text{ and } R^{23}{}_{23}. \tag{32.17}$$

The trace that appears in the 1-index loop variable completes the derivation; namely,

$$G^{00} = R^{12}{}_{12} + R^{13}{}_{13} + R^{23}{}_{23}. \tag{32.18}$$

We can compare this procedure to the deformation of the 1-index loop variable. For infinitesimal (small) loops the 1-index $T$-variable can be related to the more familiar constructions used in general relativity. Let us suppose that $\alpha_{x,c,\sigma}$ is a loop around a point $x$, orthogonal to the direction $c$ and having the area $\sigma$ (small). It is easy to see that in this case (cf. Appendix)

$$\Delta\Omega_\alpha = \Delta\sigma R_{ab}\epsilon^{abc} + O(\sigma). \tag{32.19}$$

So that the operator

$$\Delta\Omega_\alpha \approx \Delta\sigma R_{AB}\epsilon^{abc} \tag{32.20}$$

expresses the differential effect of parallel transport around the loop $\alpha$ of area $\Delta\sigma$. It is also clear from the definition of $\Delta\Omega_\alpha$ that it acts on dual vectors rather than on vectors themselves. If one considers a 1-index $T$-variable on this loop, the differential of $T^a$ can be identified with the (dualized) rotation produced by parallel transport around the surface of unit area. The derivative $dT^{a(c)}/d\sigma$ of the 1-index $T$-variable contains implicitly the second index describing the direction $c$ orthogonal to the loop $\alpha$. Here,

$$\frac{dT^{a(c)}}{d\sigma} = \frac{dT^a}{d\sigma^c} = \frac{\Delta\Omega^a}{\Delta\sigma^c}. \tag{32.21}$$

The contraction of this derivative with respect to the spatial metric is precisely the $G^{00}$ Einstein tensor

$$G^{00} = g_{ac} \frac{dT^{a(c)}}{d\sigma}.$$

(32.22)

## 32.4 Program, Problems, and Promise

If knots or loops do provide a complete solution to the Schrödinger-Ashtekar equation, then we are challenged to translate them into information-theory content á la Anandan and Bohm-Aharonov. However, we are just at the beginning of this endeavor, and there are still serious obstacles to overcome. First, we must decide if knots in the Ashtekar formalism gives a complete solution. Second, we note that the countable infinity of solutions obtained thus far have a singular character. It is not clear how to overcome this difficulty. Nevertheless, we have not seen a more promising theory for the unification of the quantum with gravitation. There are research avenues open (Wheeler et al. 1991):

(i) Redo standard quantum gravity in Claudio Teitelboim's version of Mandelstam's connection-free description of relativity, and inject that formalism into the Knot representation.

(ii) Decide, yes or no, whether knots either in Claudio Teitelboim's or Ashtekar's formalism, or both, give a complete solution.

(iii) Provide a Hamiltonian formulation of general relativity from $\partial \circ \partial \equiv 0$ via the Mandelstam-Teitelboim approach.

(iv) Provide correspondence to continuum general relativity as we know it (as one builds a "Bohr-orbit" wavepacket out of Schrödinger solutions).

(v) Determine where and how the observer interfaces with the quantum gravitational system.

(vi) Determine implications for cosmology ("Time" ends? System necessarily closed? Equivalent of foam-like structure?).

We have clearly just begun to examine the ramifications of this reformulation of general relativity. We have taken one step toward the lofty goals outlined above by identifying the deformation of the one-index loop variable with the pure spatial component of the Einstein field tensor. We do not know if this new approach will provide a way station toward the unification of the quantum and gravitation, but the limited results obtained thus far seem to warrant further investigation. In particular, (1) the wave equation translates into an appreciably simpler looking equation for $\phi(\alpha)$, (2) there exists a countable infinity of exact solutions,

$$| \Psi > = \sum_{\substack{\text{regular K} \\ \text{no intersections} \\ \text{no corners}}} C_k | K >,$$

(32.23)

(3) knots are only well defined in three dimensions, and (4) we now have a relationship between the Einstein tensor and the differential of the one index loop variable, Eq. (32.22).

## Acknowledgements

Much of this material was obtained from discussions at Princeton University on Thursday and Friday 14-15 Nov. 1991 among J. David Brown, Claudio Teitelboim, and the authors, assisted by a visit in September to Abhay Ashtekar and Lee Smolin and by discussions in October at this meeting with Karel V. Kuchař and Claudio Teitelboim, and by phone calls on Thursday 14 Nov. 1991 to Gary Horowitz, Theodore S. Jacobsen and Stanley Mandelstam. We wish to express our gratitude to the organizers of the conference and to the hospitality of all involved. We also wish to acknowledge the hospitality of the Aspen Center for Physics, where this manuscript was completed. This work was supported in part by AFOSR grant No. 91NP025.

## Appendix I

We show here that the one-index loop variable $T^a(\alpha)$ is related to the E. Cartan moment of rotation via

$$g_{ac} \frac{dT^{a(c)}}{d\sigma} = G^{00}$$

where $\sigma$ is the area spanned by an infinitesimal loop $\alpha$ orthogonal to the direction $c$.

In what follows capital roman indices from the beginning of the alphabet $(A, B, \ldots)$ are spinor indices taking values $1, 2$. Lower case roman indices $(a, b, c, \ldots)$ are space indices (spatial part of spacetime) taking values $1, 2, 3$. Capital roman indices from the middle of the alphabet $(I, J, K, \ldots)$ are 4-dimensional internal indices with $(i, j, k, \ldots)$ being their 3-dimensional part.

For completeness, we include here the definition of the loop variable. Namely, for the loop $\alpha$

$$T[\alpha] = \mathrm{Tr}\,\mathscr{P}e^{\oint_\alpha A} = \mathrm{Tr}\, U_\alpha(0, 2\pi) \tag{A.1}$$

$$T^a[\alpha](s) = \mathrm{Tr}\left[ U_\alpha(s)\, \tilde{E}^a(\alpha(s)) \right] \tag{A.2}$$

where

$$
\begin{aligned}
U_\alpha(0, 2\pi) &= \left( \mathscr{P}e^{\oint_\alpha A} \right)^A{}_B \\
&= \left( \mathscr{P}e^{\int_0^{2\pi} ds\, \dot{\alpha}^a A_a^i(\alpha(s))\tau_i} \right)^A{}_B \\
&= 1^A{}_B + \int_0^{2\pi} ds\, \dot{\alpha}^a A_a^i(\alpha(s))\, \tau^A{}_B + \cdots
\end{aligned}
\tag{A.3}
$$

and $\tau_i$ $(i = 1, 2, 3)$ are the Pauli matrices divided by $\sqrt{2}$

$$
\begin{aligned}
\tau_1 &= \frac{1}{\sqrt{2}}\sigma_x = \frac{1}{\sqrt{2}}\begin{pmatrix} 0 & 1 \\ 1 & 0 \end{pmatrix} \\
\tau_2 &= \frac{1}{\sqrt{2}}\sigma_y = \frac{1}{\sqrt{2}}\begin{pmatrix} 0 & -i \\ i & 0 \end{pmatrix} \\
\tau_3 &= \frac{1}{\sqrt{2}}\sigma_z = \frac{1}{\sqrt{2}}\begin{pmatrix} 1 & 0 \\ 0 & -1 \end{pmatrix}
\end{aligned} \tag{A.4}
$$

so that

$$
\mathrm{Tr}(\tau_i{}^A{}_B\ \tau_j{}^B{}_C) = \delta_{ij}. \tag{A.5}
$$

$U_\alpha(s)$ of equation (A.2) is defined by (A.3) with the agreement that the zero of the angle variable is aligned with the value $s$ of the path parameter.

Other notation used in equations (A.1) and (A.2) are

$$
A_a{}^A{}_B = A_a^i \tau_i{}^A{}_B; \qquad \tilde{E}^{aA}{}_B = \tilde{E}^{ai}\tau_i{}^A{}_B \tag{A.6}
$$

with

$$
\tilde{E}_I^a = \sqrt{q}E_I^a; \qquad E_I^a = e_I^a - N^a e_I^0; \qquad N^a = e_I^a e_0^I; \qquad N = \frac{e}{\sqrt{q}} \tag{A.7}
$$

where $e_I^a$ is coming from the tetrad field and $q$ is the determinant of the 3-metric $g_{ab}[e]$.

The one-internal-index connection $A_I^a$ and the curvature $F_{ab}^I$ of it are related to ${}^4A_a^{JK}$ and $F_{ab}^{JK}$ via

$$
\begin{aligned}
A_a^I &= \epsilon^I{}_{JK}\ {}^4A_a^{JK} \\
F_{ab}^I &= \epsilon^I{}_{JK}\ {}^4F_{ab}^{JK}
\end{aligned} \tag{A.8}
$$

where the symbol $\epsilon^I{}_{JK}$ is defined by

$$
\epsilon^{IJK} = e_L^0 \epsilon^{LIJK}. \tag{A.9}
$$

Very often the gauge is fixed so that $e_i^0 = 0$. Thus

$$
\epsilon^{IJK} = \epsilon_{0IJK}. \tag{A.10}
$$

Also, one can perform calculations in a Riemannian normal coordinate system with the origin at the point where calculations are performed. In this case notation analogous to (A.10) can be used for spacetime indices as well

$$
\epsilon^{abc} = \epsilon^{0abc} \tag{A.11}
$$

with $\epsilon^{abc} = \pm 1$. Also, in this case, $\sqrt{q} = 1$. We use such coordinates unless otherwise stated explicitly.

To calculate $\frac{dT^a}{d\sigma}$ we use the expansion

$$
U_{\alpha_{x,c,\sigma}} = 1 + \sigma F_{ab}\epsilon^{abc} + O(\sigma) \tag{A.12}
$$

of area $\sigma$ orthogonal to the direction $c$. Internal and spinor indices are suppressed in (A.12). Restoring them, we come up with

$$
\begin{aligned}
U^A{}_B &= 1^A{}_B + \sigma F^i_{ab} \epsilon^{abc} \tau_i{}^A{}_B + O(\sigma) \\
&= 1^A{}_B + \sigma \epsilon^{0i}{}_{jk} F^{jk}_{ab} \epsilon^{0abc} \tau_i{}^A{}_B + O(\sigma).
\end{aligned}
\tag{A.13}
$$

Substituting (A.13) and (A.6) – (A.7) (with gauge $e^0_i = 0$, and $\sqrt{q} = 1$) in the definition (A.2) we come up with

$$
\begin{aligned}
\frac{dT^{a(c)}}{d\sigma} &= \epsilon^{0i}{}_{jk} F^{jk}{}_{db} \epsilon^{0dbc} e^a_m \operatorname{Tr}\left(\tau_i{}^A{}_B \, \tau^{mB}{}_C\right) \\
&= \epsilon^{0i}{}_{jk} F^{jk}{}_{db} \epsilon^{0dbc} e^a_m \delta^m_i \\
&= \epsilon^{0i}{}_{jk} F^{jk}{}_{db} \epsilon^{0dbc} e^a_i.
\end{aligned}
\tag{A.14}
$$

Contracting (A.14) in indices $a$ and $c$ finally yields

$$
\begin{aligned}
g_{ac} \frac{dT^{a(c)}}{d\sigma} &= g_{ac} \epsilon^{0i}{}_{jk} F^{jk}{}_{db} \epsilon^{0dbc} e^a_i \\
&= g_{ac} \epsilon^{0a}{}_{fg} F^{fg}{}_{db} \epsilon^{0dbc} = G^{00}.
\end{aligned}
\tag{A.15}
$$

## Discussion

**Kuchař**  The relation between the affine connection representation and the loop representation was recently clarified by Ashtekar and Isham. The relation between the metric and the affine connection representation is similar to that between the position and the momentum representation.

**Hartle**  How does Planck's constant, that sets the scale of the quantum fluctuations of spacetime, come into this?

**HKMW**  Rather than adopting the representation of quantum gravity using the canonical variables $\tilde{E}$ and the self-dual connection $A$, Ashtekar et. al. adopt the non-canonical representation using loop variables ($T$ variables). The commutator is then of the form $[T[\alpha], T^{(a)}[\alpha](s)] = \hbar F[\alpha, s]$, where $F$ degenerates to a delta-function for a canonical representation. Chris Isham has argued that non-canonical algebras are better suited for non-perturbative quantization; however, it is not at all clear to us what the implications of this are to the theory of gravity.

**Mottola**  I just wanted to comment that there was a similar proposal in Yang-Mills theory to discuss the configuration space of QCD in a gauge-invariant loop representation (by Migdal, for example). Though elegant and providing a denumerable infinity of loops in a discrete lattice, it was found that in the continuum the irregular and tiny loops are the important ones and the mathematics was unable to deal with these nonsmooth self-intersecting loops. Can one expect the situation in gravity to be any better behaved or tractable?

**HKMW**  This is an excellent question and should be looked into. It is not clear to us that the gravitational field will be more tractable than QCD.

# References

Aharonov, Y. and Bohm, D. (1959) Significance of electromagnetic potentials in quantum theory. *Phys. Rev.* **115**, 485.

Anandan J. (1988) Comment on geometric phase for classical field theory. *Phys. Rev. Lett.* **60**, 2555.

Anandan, J. and Aharonov Y. (1988) Geometric quantum phase angles. *Phys. Rev.* **D38**, 1863.

Atiyah M. (1990) *The geometry and physics of knots*, Cambridge University Press.

Ashtekar A. (1990) Old problems in the light of new variables. In *Conceptual Problems in Quantum Gravity*, A. Ashtekar and J. Stachel (eds.) Birkhäuser, Boston.

Ashtekar A. (1991) *Lectures on Non-perturbative Canonical Gravity*, Singapore: World Scientific.

Bekenstein, J. D. (1972) Black holes and the second law. *Nuovo Cimento Lett.* **4**, 737.

Bekenstein, J. D. (1973) Generalized second law of thermodynamics in black-hole physics. *Phys. Rev.* **D8**, 3292.

Bekenstein, J. D. (1980) Black-hole thermodynamics. *Physics Today* **33**, 24.

Colella R., Overhauser A. W., and Werner S. A. (1975) Observation of gravitationally induced quantum interference. *Phys. Rev. Lett.* **34**, 1472

Hawking, S. W. (1974) *Nature* **248**, 30.

Hawking, S. W. (1975) *Commun. Math. Phys.* **43**, 199.

Hawking, S. W. (1976) *Phys. Rev.* **D13**, 191.

Kauffmann L. H. (1991) *Knots and Physics*, Singapore: World Scientific.

Kheyfets A. and Miller W. A. (1991) The boundary of a boundary principle in field theories and the issue of austerity of the laws of physics, *J. Math. Phys.* **32**, 3168.

Kheyfets A. and Miller W. A. E. Cartan moment of rotation in Ashtekar's self-dual representation of the gravitation theory, *J. Math. Phys.*, to appear.

Kuchař, K. V. (1992) Time and interpretations of quantum gravity. In *Proceedings of the 4th Canadian Conference on General Relativity and Relativistic Astrophysics*, eds. G. Kunsatter, D. Vincent and J. Williams, Singapore: World Scientific.

Mandelstam S., (1962) Quantization of the gravitational field, *Annals of Phys.* **19**, 25.

Rovelli C. (1991) Ashtekar's formulation of general relativity and loop space non-perturbative quantum gravity: a report, *Class. Quant. Grav.* **8**, 1613.

Stueckelberg, E. C. G. (1960) Quantum theory in real Hilbert space. *Helv. Phys. Acta.* **33**, 727.

Werner, Samuel A. (1986) Neutron Interferometry at Missouri. *Annals of the New York Academy of Sciences*, 147.

Wheeler J. A. (1982) The computer and the universe. *Int. J. Theor. Phys.* **21**, 557.

Wheeler J. A. (1984) Bits, Quanta, Meaning. In *Problems in Theoretical Physics*, eds. Giovannini, A., Mancini, Marinaro, Salerno: Univ. of Salerno Press.

Wheeler J. A. (1990) Information, Physics, Quantum: The Search for Links. In *Complexity, Entropy, and the Physics of Information*, ed. W.H. Zurek, Redwood City, CA: Addison-Wesley Publishing Company.

Wheeler J. A. (1988) World as system self-synthesized by quantum networking, *IBM J. of Research and Development*, **32**, 4.

Wheeler J. A., Brown J. D., Holz D., Kheyfets A., Miller W. A., Teitelboim C. (1991) Unpublished notes from a meeting at Princeton University on 14-15 November, 1992.

Wootters, W. K. (1980) The acquisition of information from quantum measurements. Ph.D. dissertation, Univ. of Texas at Austin.

Wootters, W. K. (1981) Statistical distribution and Hilbert space. *Phys. Rev.* **23**, 357.
Zurek, W. H., ed. (1990) *Complexity, Entropy, and the Physics of Information.* Redwood City,
CA: Addison-Wesley Publishing Company.

# 33

# Temperature and Time in the Geometry of Rotating Black Holes

## J. David Brown and James W. York, Jr.

*Institute of Field Physics and*
*Theoretical Astrophysics and Relativity Group*
*Department of Physics and Astronomy*
*The University of North Carolina*
*Chapel Hill, NC 27599-3255, USA*

## 33.1 Introduction

Let us consider a problem with, ostensibly, no time, a problem of thermodynamic equilibrium involving black holes. Whatever the results of the ambitious programs described in this volume, it seems likely that black–hole entropy (Bekenstein 1973) and its place in the larger structure of physics will play an important role in understanding the physical origins of time–asymmetry. Black–hole radiance (Hawking 1975) involves in a relatively concrete context many of the themes that have appeared repeatedly in this Workshop: quantum theory, gravity, statistical and thermal physics, and, perhaps, information theory in some not–yet–clear form. The nature of thermodynamic equilibrium involving black holes, and, more generally, in self–gravitating systems, must be formulated as precisely as possible, we believe, before non–equilibrium processes involving gravity, with corresponding increase of entropy and appearance of time–asymmetry, can be described with confidence. For these reasons it is worthwhile to pursue a well–posed mathematical treatment of the equilibrium case.

The present work began with a clarification (York 1986; Whiting and York 1988) of two very different interpretations of the Euclidean Schwarzschild solution, either as a stationary configuration for the canonical partition function describing a physical black hole (Gibbons and Hawking 1977), or as an instanton characterizing black hole nucleation (Gross *et al.* 1982). The question arose because of the belief that the relevant heat capacity derived from the partition function was always negative. However, it can have either sign, and this fact enabled a reconciliation of the two points of view. That work has progressed in a continuing program of studies, part of which we shall now describe briefly. See Brown *et al.* (1991a, 1991b, 1993b) and references therein for details.

## 33.2 Physical Picture

Gravitational thermodynamics has a number of characteristic features that arise from the long–range, unscreened nature of the gravitational interaction. Gravitating systems are never truly isolated in the traditional sense of thermodynamics. Although ordinary systems are never actually perfectly isolated either, in practice this leads to small and usually unimportant errors. In contrast, in gravitating systems the failure of perfect isolation has crucial implications, leading to the more prominent role of boundary conditions and the use of finite systems. It should be noted that the use of finite systems, irrespective of gravity, already implies the inequivalence of the various statistical ensembles as well as possible modifications of ordinary thermodynamic relations (Hill 1963, 1964; Horwitz 1966).

Problems of gravitational thermodynamics have to be formulated globally because of the infinite range of the gravitational field, even though at a given spacetime event the principle of equivalence allows us formally to ignore gravity in writing the local thermodynamic laws of gravitating matter (apart from possible quantum corrections depending on the curvature, which we shall not consider explicitly here). Thermodynamic properties of the gravitational field reveal themselves globally, or, alternatively, quasi–locally, subject to definite boundary conditions. (We shall use the term "global" to include "quasi–local with definite boundary conditions".) Black–hole thermodynamics is an extreme case of gravitational thermodynamics and this is where we shall focus attention in this discussion. Here the requirement of a global viewpoint is even more evident. One certainly cannot consider subdividing a black hole! Many of the unusual features of black–hole thermodynamics can be expected to carry over to the case of massive gravitating systems that contain no black holes.

Physical systems are, of course, quantum mechanical by nature, so to speak of a system as containing a black hole is to assume the system can be approximated reasonably well by a single classical configuration. In a "semiclassical" approximation, the system can be viewed as containing a black hole surrounded by radiation described by a non–zero renormalized stress–energy–momentum tensor $< T_{\mu\nu} >$. We shall refer to this radiation as the "environment" of the hole, and assume that it extends to some (possibly large) radius $r_0$. The black hole plus environment make up the thermodynamic system to be studied. In this paper we focus on the case of thermodynamic equilibrium, which implies, in particular, maintaining appropriate boundary conditions at $r_0$. We shall *not* refer to the radiation environment of the hole as a "heat bath", for this terminology brings to mind, misleadingly, a "heat reservoir" as in traditional statistical thermodynamics, whose function is to maintain a definite temperature in the system, as in a canonical ensemble. Once one sees that the radiation–filled environment of a black hole is not to be regarded as a heat reservoir, the argument that only the microcanonical ensemble for black holes can be stable (Hawking 1976; Gibbons and Perry 1978) loses its force (York 1986).

The role of a heat reservoir in contact with the system at radius $r_0$, if one is needed, is to maintain certain boundary conditions to the accuracy permitted by physical principles. The reservoir is not thought of as part of the system in this view. It can have a stress–energy tensor entirely different from that of the environment. As one example, meant to be illustrative of principle rather than representative of what is expected in the natural world, consider a heat reservoir for the black hole problem as a large hollow "steel ball" with inner radius $r_0$, suitably instrumented and equipped to maintain specified boundary conditions at $r_0$. If necessary, the thick shell could include "rocket ships" to support it against being ruptured and pulled into the black hole. In such a case the radial members of the reservoir would be in a state of tensile, rather than compressive, stress. The traditional view of the foundations of statistical thermodynamics would hold that the meta–system, black hole plus environment plus reservoir, if regarded as an insulated (microcanonical) structure with fixed energy, angular momentum, *etc.*, should be in a state of extremal entropy. If the system is found to be stable, then so should the meta–system. All our results are consistent with this view.

The distinctions we have described above are required precisely because black–hole thermodynamics is indeed global. Thus, even when equilibrium is assumed under the *approximation* $< T_{\mu\nu} >= 0$ between the horizon and the boundary radius $r_0$, the space around the hole, which is classically a vacuum, is nevertheless part of the system because it has non–trivial properties. This point is especially clear when one considers a rotating black hole and the dragging of local frames in the space around it. The role of this effect in black–hole thermodynamics (Brown *et al.* 1991a, 1991b) will be described below.

It is clear that an equilibrium system with $< T_{\mu\nu} > \neq 0$ cannot have a gravitational field that is asymptotically flat in spacelike directions, with bounded total energy, unless the system itself is finite, that is, spatially bounded, and is enclosed by a perfectly insulating (microcanonical) boundary $B$. The ordinary thermodynamic limit does not exist because as the boundary radius $r_0$ increases, the environment will eventually collapse onto the existing hole, thus forming a larger one. In considering a modified thermodynamic limit, one finds from rough estimates that $E^5 \sim r_0^3$, where $E$ is the total energy of the system (Hawking 1976; Gibbons and Perry 1978). The equilibrium is stable only when most of the energy of the system is the black hole's mass, so the hole's environment is not suitable as a heat reservoir and, as we have stated, it is not considered as such. More precise calculations, including backreaction effects, confirm these results qualitatively whenever $r_0$ is large compared to the gravitational radius of the black hole, which in turn is large compared to the Planck length (York 1985).

The various *canonical* ensembles that can be defined are of great interest because they link statistical mechanics and quantum gravity through Feynman's prescription for the partition function as a path integral in "periodic imaginary time" (Feynman and Hibbs 1965). In these cases there is no "asymptotically flat spatial infinity" pertaining to the thermodynamic system because of the required intervention of a heat reservoir to maintain canonical boundary conditions. The heat reservoir, and the boundary conditions it maintains, represent the connection between the self–gravitating thermodynamic system and the rest of the universe. Thus, the pioneering effort of Gibbons and Hawking (1977) to implement the functional integral approach, using a canonical ensemble, was flawed by the use of asymptotically flat boundary conditions ($r_0 = \infty$). As a result, they necessarily found a negative heat capacity that spoiled their approach to the canonical ensemble. However, by using $r_0 < \infty$ one finds that the heat capacity can also be positive (York 1986). To go a step further, one knows that in a canonical ensemble one's task is to *find* from the partition function the energy as a function of a given temperature, not to assume it from the start. This was carried out for any $r_0$ by Whiting and York (1988), who used a stationary phase approximation.

### 33.3 Thermodynamic Equilibrium and Gravity

The spacetime associated with a self–gravitating system in thermodynamic equilibrium is time–independent in the sense that it possesses a timelike Killing vector field $\partial/\partial t$ that can be either orthogonal to the "preferred" spacelike slices $t = constant$ (static) or not (stationary), the latter case being associated with non–vanishing angular momentum. In either case, such systems are spatially inhomogeneous. The temperature, "chemical potentials", and other "intensive" variables vary in space but not in time (Tolman 1930). This has very interesting implications for a system containing a black hole *and for other self–gravitating systems as well*. Gravity breaks the translational symmetry of space that is typically (implicitly) assumed in standard thermodynamics. As a consequence of this symmetry–breaking, there does not exist an Euler relation (of degree one) relating the intensive and extensive variables. It is then not difficult to show that there is no Gibbs–Duhem formula relating variations of the intensive variables. This alters the usual "Gibbs phase rules". This point was apparently first appreciated, in the particular case when a Schwarzschild black hole and flat–space radiation constitute the system, by Wright (1980). However, the phenomenon is general for self–gravitating systems.

The absence of Gibbs–Duhem relations means that variations of all the intensive variables, as well as variations of the corresponding extensive variables, are linearly independent even when variations are restricted to comparison of equilibrium states. As a result, there is a global non–degenerate symplectic or canonical Hamiltonian structure associated with thermodynamically conjugate pairs in general relativity, in contrast with the usual flat space, flat spacetime, or local formulations of thermodynamics. Furthermore, this thermodynamical conjugacy follows directly from the action integral of general relativity (Brown *et al.* 1990; Brown and York 1993b). The conjugacy one obtains is that associated with the dimensionless Massieu potentials (Callen 1985), except that it is not degenerate. The Massieu potentials are the usual assortment of "free energies" divided by $kT$ and the entropy divided by Boltzmann's constant $k$. Massieu potentials correspond to the gravitational action, with suitable boundary terms, divided by Planck's constant $\hbar$.

As a consequence of the features described above, one has canonical conjugacy between the energy $E$ and $\beta = (kT)^{-1}$, between the angular momentum $J$ and $\beta\omega$, where $\omega$ is an angular velocity, between the electric charge $Q$ and $\beta\phi$, where $\phi$ is the electrostatic potential, *etc.* In general relativity, all the traditional constants of motion, or "extensive" variables $E$, $J$, $Q$, ... are given by integrals over the closed spatial two–surface boundary of the system. Likewise, the conjugate "intensive" variables $\beta$, $\beta\omega$, $\beta\phi$, ... are also defined by boundary data. The properties (1) non–degenerate symplectic structure, and (2) all thermodynamic variables defined as boundary data, enable a widening and strengthening of Feynman's idea of functional integration as both a technical means of obtaining partition functions and a conceptual unifier of quantum and statistical physics.

In gravity, for a completely isolated system, one can express the density of states $v(E, J, Q,...)$ as a Lorentzian functional integral and, for a completely open system, one can express the partition function $Z(\beta, \beta\omega, \beta\phi,...)$ as a "Euclidean" functional integral (Brown and York 1993b). All other partition functions can be obtained systematically from either of these two. (In statistical thermodynamics without gravity, the partition function $Z(\beta, \beta\omega, \beta\phi,...)$ for a completely open system has been stated not to exist (Prigogine 1950).) We believe that the Lorentzian "microcanonical functional integral" may well be a fundamental one through which stronger links joining gravity, statistical physics, "maximum entropy", and the quantum can be forged.

### 33.4 Rotating Black Hole

Consider a rotating, stationary, axisymmetric black hole with metric $g_{\mu\nu}$ given by

$$ds^2 = -N^2 dt^2 + h_{rr} dr^2 + h_{\theta\theta} d\theta^2 + h_{\varphi\varphi}(d\varphi + \omega\, dt)^2 \,, \qquad (33.1)$$

where the lapse function $N$, the shift vector $V^i = (0, 0, \omega)$ with $i = r$, $\theta$, $\varphi$, and the spatial metric $h_{ij}$ of the slices of constant time depend only on $r$ and $\theta$. For later convenience, we work in a spatial coordinate system rotating with the horizon's constant angular velocity $\omega_H^{\star}$ with respect to the distant stars ("$\star$"); thus $\omega = \omega_H^{\star} - \omega_Z^{\star}(r, \theta)$, where $\omega_Z^{\star}$ is the non-uniform angular velocity of the observers at rest in the "preferred" slices of constant time $t$. Because the angular Killing vector $\partial/\partial\varphi$ lies in these slices, these "Eulerian" observers (hydrodynamical terminology) have zero angular momentum per unit mass; hence, they are frequently called ZAMOs ("$Z$"). This property follows from observing that their unit timelike four–velocity field is $u^{\mu} = N^{-1}(1, -V^i)$ and that $g(u, \partial/\partial\varphi) \equiv u_{\varphi} = 0$.

The physical significance of the rotating spatial coordinates is that they are co–moving with the elements of the "radiation fluid" environment that is required to equilibrate the black hole. That this environment must rotate rigidly *with respect to the distant stars* at the rate $\omega_H^{\star}$ can be shown using local thermodynamic arguments based on extremal entropy in the manner of Israel and Stewart (1980), or similarly to equations (33.4)–(33.6) (only!) in Zannias and Israel (1981). A quantum field–theoretic argument establishing this same rigid rotation was given by Frolov and

Thorne (1989). Because the environment cannot rotate with finite angular velocity for a too–large radius, the system's boundary $B$ must lie inside the speed–of–light surface. Again, we see the necessity of using a finite system.

The world lines of the elements of the environment are parallel to the timelike "co–rotating" Killing vector $\partial/\partial t$ (Carter 1979). The unit four–velocity $\bar{u}^\mu$ of the radiation elements is given by $(\partial/\partial t)^\mu = \bar{N}\bar{u}^\mu$ where $\bar{N}^2 = -g_{tt}$. Local observers with four–velocity field $\bar{u}^\mu$ are called "Lagrangian" observers because they are comoving with the fluid environment. They see a locally isotropic thermal distribution of quanta (Frolov and Thorne 1989) and *seem* to be the natural observers to describe the equilibrium of a system containing a black hole, in imitation of the usual local formulation of thermodynamic laws in general relativity. But because $\bar{u}_{[\mu}\nabla_\nu \bar{u}_{\sigma]} \neq 0$, there is no way for them globally to synchronize their clocks ("barber pole effect") and therefore there is no global time–slicing with respect to which they are at rest. Thus, while Lagrangian observers may seem "ideal" for local physics and the "recognition" of equilibrium thermal properties, they are not actually suitable for defining global properties of the system, such as black–hole entropy and its changes. Consequently, the local temperature $\bar{\beta}^{-1} = \bar{T}$ they measure (Frolov and Thorne 1989, Eq. (4.3)) is not a suitable thermodynamic temperature for the system. (This implies that Zaslavskii's (1991) Lagrangian treatment of the rotating black hole is, in our view, unsatisfactory.)

Using the thermodynamic arguments that establish "rigid rotation", one also finds that $\bar{\beta}\bar{N}^{-1}$ must be a constant; this is the Tolman (1930) result for the red–shifting of Lagrangian temperature. For a system with a black hole, the constant is $2\pi\kappa^{-1}$, where $\kappa$ is the surface gravity of the black hole in absolute units where $G = \hbar = c = k = 1$. What is the globally correct Eulerian or ZAMO temperature? Following Israel and Stewart (1980), define the inverse temperature four–vector $\bar{\beta}^\mu = \bar{\beta}\bar{u}^\mu$, which is easily shown to satisfy Killing's equation. Define the spacelike velocity of the environment's elements with respect to the ZAMOs in ZAMO proper time: $\hat{V}^\mu = (0, \hat{V}^i)$ with $\hat{V}^i = N^{-1}V^i$ given by the quotient of shift vector by lapse function. Note that $\hat{V}^\mu u_\mu = 0$ and $u^\mu + \hat{V}^\mu = N^{-1}(1,0)$. Then (see, for example, Smarr and York (1978)) we have

$$\bar{u}^\mu = \gamma(u^\mu + \hat{V}^\mu)\,, \tag{33.2}$$

$$\bar{\beta}^\mu = \bar{\beta}\bar{u}^\mu = (\bar{\beta}\gamma)u^\mu + (\bar{\beta}\gamma)\hat{V}^\mu \equiv \beta u^\mu + \beta\hat{V}^\mu\,, \tag{33.3}$$

where $\gamma$ is the local kinematical special–relativistic factor $\left(1 - h_{ij}\hat{V}^i\hat{V}^j\right)^{-1/2} = -g(u,\bar{u})$. From (33.3) one can identify the ZAMO–measured local proper inverse temperature as $\beta = \bar{\beta}\gamma$. This result is obtained by considering the Doppler effect for zero angular momentum (transverse) photons. The second term in (33.3) corresponds to the presence of a purely kinematically induced dipole "blue/red" longitudinal Doppler shift that a ZAMO would "see" as the radiation environment flows past (kinematically analogous to a determination from Earth of the 2.7K "isotropic" cosmic microwave background). From the definitions of $\bar{N}$, $N$, and $V^i$, it is easily shown that $N = \gamma\bar{N}$, from which we establish that the red–shift law for ZAMO–measured temperature is

$$\beta N^{-1} = \left(\gamma\bar{\beta}\right)\left(\gamma\bar{N}\right)^{-1} = \bar{\beta}\bar{N}^{-1} = 2\pi\kappa^{-1} \,. \tag{33.4}$$

The result (33.4) has a geometrical interpretation, as well as the direct physical one that at fixed $r$ and $\theta$, a ZAMO detects a lower temperature than a Lagrangian ("moving") observer ($\gamma > 1$), except at the poles ($\theta = 0, \pi$).

The geometrical interpretation is described in real Lorentzian– signature spacetime as an analog of the aging–of–identical–twins effect that is well known in special relativity. Fix $r$ and $\theta$ (not at the north or south poles) and consider two identical twins: call them "Euler" and "Lagrange". Euler stays home while Lagrange follows a circular orbit at constant angular velocity $\omega_H^*$ and returns home after the passage of $P$ (physical period of trip) units of "universal" time $t$. The proper time duration of Euler's world line is, restoring ordinary units,

$$\beta' = \int_0^P N\,dt = NP = \int_0^P \gamma\bar{N}\,dt = \gamma\bar{N}P = \gamma\bar{\beta}' \,, \tag{33.5}$$

where $\bar{\beta}'$ is the total proper time for Lagrange's trip. Note that $\beta' > \bar{\beta}'$: the stay–at–home ZAMO ("inertial") twin Euler has aged more than Lagrange. The time $\beta'$ (in seconds) is converted formally to an inverse temperature $\beta$ (in $K^{-1}$) according to $\beta \leftrightarrow \beta'k\hbar^{-1}$. Thus from (33.5) we recognize the anology between the ZAMO and Lagrangian temperatures and the Lorentzian "twin effect".

The geometrical interpretation also can be associated directly with the complexification $N \to -iN$, $V^i \to -iV^i$, of the metric (1) whose corresponding action is used to obtain the stationary phase approximation to the partition function (Brown *et al.* 1991). (In another context, Mellor and Moss (1989) have also considered particular complex stationary black hole metrics.) In the complex geometry, in order to obtain an everywhere regular geometry, the slices defined by the scalar $t$ are periodically identified with a period $P$ given by $2\pi c\kappa^{-1}$ ($\kappa$ contains the gravity constant $G$) yielding a topology $S^2 \times R^2$ for the complexified spacetime (Gibbons and Perry 1978; Gibbons and Hawking 1977). Multiplying (33.5) by $k\hbar^{-1}$, we obtain the temperature relation in terms of the proper lengths (in $K^{-1}$) of the Lagrangian and Eulerian curves in the complex geometry. (The topology of the three–boundary $^3B$ of the complex spacetime is $S^2 \times S^1$, where $S^2$ is the topology of the axisymmetric two–boundary $B$. A Lagrangian curve in $^3B$ can be closed with the identification of points $(t, r, \theta, \varphi)$ and $(t + 2\pi\kappa^{-1}, r, \theta, \varphi)$, and an Eulerian curve can be closed with the identification of $(t, r, \theta, \varphi)$ and $(t + 2\pi\kappa^{-1}, r, \theta, \varphi - 2\pi\kappa^{-1}\omega)$. The difference is physically irrelevant.

Finally, let us point out a further aspect of characterizing the system by the ZAMO–measured temperature $T = \beta^{-1}$ at the boundary $B$. In the statement of the thermodynamic law of equilibrium change

$$dS = \beta \, dE - \beta \omega \, dJ + \cdots \qquad (33.6)$$

for the system, the temperature will not be correctly identified without a corresponding correct identification of $\omega$. Therefore, let us consider a process whereby energy is injected into the system from outside and through $B$ such that the angular momentum $J$ is unaffected. For this to occur, the $\varphi$–component of the four–momentum per unit mass injected, $w_\mu$, must vanish. Clearly, this means that

$$\left. (d\varphi/dt)\right|_B = \left. (w^\varphi/w^t)\right|_B = \left. -(g_{t\varphi}/g_{\varphi\varphi})\right|_B = \left. -\omega\right|_B . \qquad (33.7)$$

Therefore, the injection must be described by a four–velocity "aimed at the rotation axis" by the ZAMOs. Thus, the ZAMO–measured variables are singled out.

That the angular velocity $\omega_H^*$ of the system is spatially uniform while the "chemical potential" $\omega$ depends on the location of $B$ deserves comment. The former is necessary for maximum entropy of the system subject to conservation of total energy and angular momentum, while the latter refers to injection of matter from *outside* the system that conserves its total angular momentum but not its total energy $E$.

In Brown *et al.* (1991a, 1991b), it was shown that the complexified action of general relativity singles out the proper ZAMO–measured boundary values $\beta$ and $\hat{\omega} = N^{-1}\omega$ at $B$, as well as their respective conjugates which are also boundary data at $B$. One also obtains directly the Bekenstein–Hawking value $(A_H/4)(kG^{-1}\hbar^{-1})$ for the black–hole entropy, where $A_H$ is the area of the event horizon. The harmonious relationships among the thermodynamics implied by extrema of the "classical" action, the quantum principle, and applications of the maximum entropy principle are striking. Coupled with the fact that general relativity enables one to define all the microcanonical and canonical ensembles in terms of "black box" problems specified completely by boundary data and appropriate "actions", our results suggest "general relativity as thermodynamics" and may imply that general relativity can be derived from some presumably deeper underlying framework such as string theory, information theory, or both.

## Discussion

**Wheeler**   Your suggestion that general relativity may be derived someday from entropy or information–theoretic considerations is truly inspiring. What do you foresee as the next step along the road?

**B&Y**   Unlike what our questioner appears to be able to do, *we* do not feel able to "leap tall buildings in a single bound". We hope next to turn our attention to non–equilibrium processes (dynamics) to see if the idea of "spacetime geometry as statistical thermodynamics" seems still to be valid.

**Teitelboim**   There are "lower dimensional black holes" that come out of string theory. Have you looked at those from your point of view?

**B&Y** We have considered the possibility, but those two-dimensional black holes do not arise as extrema of some classical action in the usual way. It is therefore difficult to see how our methods, in which the classical action functional plays a central role, might be applied.

**Hartle** Do your results extend to the Weyl, static, axisymmetric, distorted black holes?

**B&Y** Yes. [See Geroch, R. and J.B. Hartle (1982) Distorted black holes. *Journal of Mathematical Physics*, **23**, 680–692.] We also expect to be able to treat stationary, axisymmetric, distorted black holes. We have not used any explicit properties of the Kerr metric in this discussion. Similarly, we have not used the Kerr–Newman fields explicitly in our referenced recent work, which includes the Maxwell field.

**Griffiths** Suppose there is a big star which blows up to form a black hole. Before the explosion I know how to find the total entropy in the part of the universe near this star—adding it up cubic meter by cubic meter. After the explosion I can calculate the entropy outside the black hole (or far away from it) as before. If I then add in the entropy you calculate, has the entropy of this part of the universe gone up?

**B&Y** Yes. Bekenstein's "generalized second law", that the entropy of all the matter outside black holes, plus the entropy of black holes, is non-decreasing, appears to be valid when Hawking's results on black-hole radiance are taken into account.

## References

Bekenstein, J.D. (1973) Black holes and entropy. *Physical Review* **D7**, 2333–2346.

Brown, J.D., G.L. Comer, E.A. Martinez, J. Melmed, B.F. Whiting, and J.W. York (1990) Thermodynamic ensembles and gravitation. *Classical and Quantum Gravity* **7**, 1433–1444.

Brown, J.D., E.A. Martinez, and J.W. York (1991a) Complex Kerr–Newman geometry and black–hole thermodynamics. *Physical Review Letters* **66**, 2281–2284.

Brown, J.D., E.A. Martinez, and J.W. York (1991b) Rotating black holes, complex geometry, and thermodynamics. In *Nonlinear Problems in Relativity and Cosmology*, Eds J.R. Buchler, S.L. Detweiler, and J.R. Ipser, vol. 631 of the Annals of the New York Academy of Sciences, New York.

Brown, J.D. and J.W. York (1993a) Quasilocal energy and conserved charges derived from the gravitational action. *Physical Review* **D47**, 1407–1419.

Brown, J.D. and J.W. York (1993b) Microcanonical functional integral for the gravitational field, *Physical Review* **D47**, 1420–1431.

Callen, H.B. (1985) *Thermodynamics and an Introduction to Thermostatistics*, second edition, John Wiley & Sons, New York.

Carter, B. (1979) The general theory of the mechanical, electromagnetic and thermodynamic properties of black holes. In *General Relativity*, Eds S.W. Hawking and W. Israel, Cambridge University Press, Cambridge.

Feynman, R.P. and A.R. Hibbs (1965) *Quantum Mechanics and Path Integrals*, McGraw–Hill, New York.

Frolov, V. and K.S. Thorne (1989) Renormalized stress–energy tensor near the horizon of a slowly evolving, rotating black hole. *Physical Review* **D39**, 2125–2154.

Gibbons, G.W. and S.W. Hawking (1977) Action integrals and partition functions in quantum gravity. *Physical Review* **D15**, 2752–2756.

Gibbons, G.W. and M.J. Perry (1978) Black holes and thermal Green functions. *Proceedings of the Royal Society* (London), **A358**, 467–494.

Gross, D.J., M.J. Perry, and L.G. Yaffe (1982) Instability of flat space at finite temperature. *Physical Review* **D25**, 330–355.

Hawking, S.W. (1975) Particle creation by black holes. *Communications in Mathematical Physics* **43**, 199–220.

Hawking, S.W. (1976) Black holes and thermodynamics. *Physical Review* **D13**, 191–197.

Hill, T.L. (1963) *Thermodynamics of Small Systems*, Part 1, W.A. Benjamin, New York.

Hill, T.L. (1964) *Thermodynamics of Small Systems*, Part 2, W.A. Benjamin, New York.

Horwitz, G. (1966) Statistical mechanics of finite systems: Asymptotic expansions. I. *Journal of Mathematical Physics* **7**, 2261–2270.

Israel, W. and J.M. Stewart (1980) Progress in relativistic thermodynamics and electrodynamics of continuous media. In *General Relativity and Gravitation*. II. Plenum Press, New York.

Mellor, F. and I. Moss (1989) Black holes and gravitational instantons. *Classical and Quantum Gravity* **6**, 1379–1385.

Prigogine, I. (1950) Remarque sur les ensembles statistiques dans les variables pression, temperature, potentiels chimique. *Physica* **16**, 133–134.

Smarr, L. and J.W. York (1978) Kinematical conditions in the structure of spacetime. *Physical Review* **D17**, 2529–2551.

Tolman, R.C. (1930) On the weight of heat and thermal equilibrium in general relativity. *Physical Review* **35**, 904–924.

Whiting, B.F. and J.W. York (1988) Action principle and partition function for the gravitational field in black hole topologies. *Physical Review Letters* **61**, 1336–1339.

Wright, D.C. (1980) Black holes and the Gibbs–Duhem relation. *Physical Review* **D21**, 884–890.

York, J.W. (1985) Black hole in thermal equilibrium with a scalar field: The back–reaction. *Physical Review* **D31**, 775–784.

York, J.W. (1986) Black hole thermodynamics and the Euclidean Einstein action. *Physical Review* **D33**, 2092–2099.

Zannias, T. and W. Israel (1981) Local thermodynamics and stress tensor of the Hawking radiation. *Physics Letters* **86A**, 82–84.

Zaslavskii, O.B. (1991) Canonical ensemble for rotating relativistic systems. *Classical and Quantum Gravity* **8**, L103–L107.

# 34

# Fluctuation, Dissipation and Irreversibility in Cosmology

B. L. Hu

*Department of Physics, University of Maryland*
*College Park, Maryland 20742, USA*

## Abstract

We discuss the appearance of time-asymmetric behavior in physical processes in cosmology and in the dynamics of the Universe itself. We begin with an analysis of the nature and origin of irreversibility in well-known physical processes such as dispersion, diffusion, dissipation and mixing, and make the distinction between processes whose irreversibility arises from the stipulation of special initial conditions, and those arising from the system's interaction with a coarse-grained environment. We then study the irreversibility associated with quantum fluctuations in cosmological processes like particle creation and the 'birth of the Universe'. We suggest that the backreaction effect of such quantum processes can be understood as the manifestation of a fluctuation-dissipation relation relating fluctuations of quantum fields to dissipations in the dynamics of spacetime. For the same reason it is shown that dissipation is bound to appear in the dynamics of minisuperspace cosmologies. This provides a natural course for the emergence of a cosmological and thermodynamic arrow of time and suggests a meaningful definition of gravitational entropy. We conclude with a discussion on the criteria for the choice of coarse-grainings and the stability of persistent physical structures.

## 34.1 Introduction

In this talk I would like to discuss the nature and origin of irreversibility in time, or, the 'arrow of time' in cosmology. This includes physical processes in the Universe, as well as the dynamics of the Universe itself. I will use examples from modern cosmological theories since the sixties: i.e. the 'standard' cosmology (Peebles, 1971; Weinberg, 1972); the chaotic (Bianchi) cosmology (Misner, 1969; Ryan and Shepley, 1975), the inflationary cosmology (Guth, 1981; Albrecht and Steinhardt 1982; Linde 1982), the semiclassical cosmologies (Hu, 1982; Parker, 1982; Hartle, 1983) and to a lesser extent, quantum cosmology (Wheeler, 1967; DeWitt, 1968; Misner,

1972; Hartle and Hawking, 1983; Vilenkin, 1986; Halliwell 1993). (For a layman's introduction to these theories, see, e.g., Hu, 1987.)

There are many ways irreversibility shows up in ordinary physical processes. I shall in the first part of my talk present some well-known examples (such as dispersion, diffusion, dissipation and phase mixing) and discuss the nature and origin of irreversibility in them. Distinction between dissipative processes (which are always irreversible) and irreversible – or 'apparently' irreversible processes (which are not necessarily dissipative) is highlighted. I'll then use the insights gained here to discuss certain aspects of chaotic and inflationary cosmology. In the second section I'll discuss some not-so-well-known but important examples involving quantum field processes such as vacuum fluctuation and particle creation and discuss the origin of time-asymmetry in them. This touches on basic questions like the statistical nature of the vacuum, which underlies novel processes like the Hawking and Unruh effects discovered in the seventies (Bekenstein, 1973, 1974; Hawking, 1975; Davies, 1975; Unruh, 1976). In the third and fourth sections I shall discuss how these quantum processes influence the structure and dynamics of the early Universe. We show that a statistical mechanical interpretation of these so-called cosmological 'backreaction' processes is possible: they are manifestations of a fluctuation-dissipation relation involving quantum fields. In this semiclassical theory it is the fluctuation of the quantum field which brings about dissipation in the spacetime dynamics. With this understanding I shall suggest some ways to examine the notion of gravitational entropy (Penrose, 1979)– from the entropy of gravitational fields to that of spacetimes. As for quantum cosmology, where spacetime and matter are both quantized, I only indicate how the basic ideas and methods in statistical mechanics adopted above to discuss irreversibility in cosmological processes can also be fruitfully applied to address issues in quantum cosmology (Hu, 1991a), but I'll shy away from extrapolations, because many concepts remain ill-defined or ambiguous (See, e.g. Ashtekar and Stachel, 1991; Isham, 1991). On the issue of the origin of time in quantum gravity, see, e.g., Kuchar (1992). For a discussion of time asymmetry in quantum cosmology, see the contributions of Halliwell, Hartle, and Hawking in this volume.

In the conclusion I summarize the key observations. The emphasis of this talk is to put many cosmological phenomena on the same footing as ordinary statistical processes and to try to understand their meaning in terms of basic concepts in theoretical physics. We reach the conclusion that time asymmetry in cosmology is attributable to the same origins as those observed in ordinary physical processes; i.e., they are determined by the way one stipulates the boundary conditions and initial states, the time scale of observation in comparison with the dynamical time scale, how one decides what the relevant variables are and how they are separated from the irrelevant ones, how the irrelevant variables are coarse-grained, and what assumptions one makes and what limits one takes in shaping the macroscopic picture from one's imperfect knowledge of the underlying microscopic structure and dynamics. Note that here I try only to explain HOW time-asymmetry arises from the imposition of certain conditions or taking certain approximations, but do not pretend to explain WHY the Universe had to start in some particular condition, e.g., smooth, or low gravitational entropy state according to Penrose (1979) or a state defined by the no-boundary condition of Hartle and Hawking (1983), which can by design hopefully 'explain' time-asymmetry. When it comes to comparing philosophical inclinations my personal preference is that there should be no special initial state (Misner, 1969). The challenge would be to explain the present state of our Universe as a plausible and robust consequence of evolution from a wide variety of arbitrary initial states.

The material in the first part of my talk is old, as old as non-equilibrium statistical mechanics itself. The second part's results are known but more recent–from the work of quantum field theory in curved spacetimes applied to semiclassical cosmology. So I shall spend less time on them. The third and fourth parts contain new results, specifically, i) the existence of a fluctuation-dissipation relation for dynamical quantum fields at zero-temperature (thus under non-equilibrium conditions and detached from thermal considerations, where most previous discussions of this relation are premised upon) (Hu, Paz and Zhang, 1992, 1993a). ii) the appearance of dissipative dynamics in an effective Wheeler-DeWitt equation for the minisuperspace variables in quantum cosmology (Sinha and Hu, 1991; Hu, Paz and Sinha, 1993). Dissipation in quantum fields and semiclassical gravity has been discussed before (Hu, 1989, where references to earlier work on these issues can be found). The main emphasis in this talk is dissipation and irreversibility, the properties of noise and fluctuation which underlie many important quantum statistical field processes are only briefly touched on. Some ideas mentioned in my talk are not discussed here. These are: decoherence and dissipation in quantum cosmology (for a general discussion of the interrelation of these processes, see Hu 1991a; for specific models, see Calzetta 1991, Calzetta and Mazzitelli, 1991, Paz and Sinha, 1991, 1992), noise and fluctuations in semiclassical cosmology (Hu, Paz and Zhang, 1993c; Calzetta and Hu, 1993b), coarse-graining in spacetime and gravitational entropy (Hu, 1983, 1984; Hu, 1993; Calzetta and Hu, 1993a; Hu and Sinha, 1993).

## 34.2 Irreversibility and Dissipation: Examples from Well-known Processes

Let me begin by examining a few text-book type examples of irreversible processes to illustrate their different natures and origins.

### 34.2.1 Dispersion (Case A)

Consider the trajectory of a particle colliding with fixed hard spheres (Ma, 1985, Section 26.5). Assume that the spheres are disks with radius $a$. The particle moves

with constant velocity $v$ and has mean free distance $\lambda >> a$ (dilute gas approximation). The trajectories of this particle are of course reversible in time. However, if
the incident angle of the particle on the first scattering is changed by $\delta\theta(0)$ initially,
then after many collisions

$$|\delta\theta(t)| \geq e^{t/\tau}|\delta\theta(0)|, \quad \tau = (\lambda/v)/\ln(2\lambda/a) \tag{34.1}$$

At sufficiently long time, $|\delta\theta(t)| \approx 1$, the exit direction is randomized by the accu-
mulated error. The asymmetry in the initial and final conditions of the congruence
comes from the accumulation and magnification of the uncertainty in the initial
conditions due to the collisions, even though the dynamical laws governing each
trajectory is time-symmetric. To trace a particular trajectory backwards in time after
a large number of collisions requires an exponentially increasing degree of precision
in the specification of the initial condition.

This situation occurs in the inflationary cosmology, in which the scale factor of
the Universe grows rapidly $a(t) \sim e^{Ht}$ for a certain period of time in the early
history. Any initial small disturbance with some functional dependence on $a(t)$ will
differ exponentially in time. Indeed this is what gives the desirable properties of
inflationary cosmology in, say, addressing the flatness and horizon problems. The
apparent irreversibility of inflation is also of this nature: not in the dynamics, but
in the inbalance of the initial and final conditions. (See, e.g., Page, 1984.)

This simple phenomenon is amply illustrated by the many sophisticated results of
modern chaotic dynamics. There, the divergence of neighboring trajectories in phase
space or parameter space is an intrinsic property of the nonlinear Hamiltonian of
the system, not a result of coarse-graining (which is implicit in, say, the postulate
of molecular chaos in Boltzmann's treatment of gas kinetics.) The evolution of an
ensemble of such systems at some finite time from the initial moment often appears
to be unrelated to ('forgetful' of) their initial conditions, not because the individual
systems are insensitive to the initial conditions (as in dissipation) but because they
are overly sensitive to them to make an accurate prediction of each system almost
impossible. It is in this sense that these systems manifest irreversibility.

Chaotic dynamics also appears in cosmology, one example is the dynamics of
the mixmaster (diagonal Bianchi Type IX) Universe (Misner, 1969). The chaotic
behavior is associated with the divergence of trajectories which describe different
world histories in the minisuperspace (Misner, 1972) parametrized by the shape
parameters $(\beta_+, \beta_-)$, while the deformation parameter $\alpha$ plays the role of time in
quantum cosmology. This was pointed out by Lifshitz and Khalatnikov (1971),
Barrow (1982), Bogoiavlenskii (1985), and many others (for a recent work, see,
e.g., Berger, 1992). The collision of the 'world particle' is now with the moving
'walls' arising from the anisotropic 3-curvature of the homogeneous space. One can
define quantities like 'topological entropy' to measure the trajectory instability of this
nonlinear system. It is of interest to see if the trajectories in the minisuperspace will
exhibit mixing properties, in which case all configurations of the Universe at a later

time can be equally accessible from arbitrary initial conditions. If the trajectories distribute unevenly in certain regions it will also be interesting to distinguish the set of initial conditions which give rise to such distinct behaviors. Notice that, by contrast, in the presence of dissipative mechanisms, as we will discuss in Example C, the trajectories in the minisuperspace will indeed evolve to a particular region around the origin, which corresponds to the Friedmann Universe. This signifies the dissipation of anisotropy, a necessary condition for the implementation of the chaotic cosmology program.

### 34.2.2 Diffusion (Case B)

Let us look at some simple examples in kinetic theory: gas expansion, ice melting and ink drop in water. These are irreversible processes simply because the initial states of $10^{23}$ molecules on one side of the chamber and a piece of ice or ink drop immersed in a bath of water are highly unlikely configurations out of all possible arrangements. These initial conditions are states of very low entropy. The only reason why they are special is because we arrange them to be so. For these problems, we also know that the system-environment separation and interaction make a difference in the outcome. In the case of the expanding gas, e.g., for free expansion the change of entropy: $\delta S_{system} > 0$ whereas for isothermal expansion: $\delta S_{system} = -\delta S_{environ} > 0, \delta S_{tot} = 0$.

Another important factor in determining whether a process is irreversible is the time scale of observation compared to the dynamical time scale of the process. We are all familiar with the irreversible process of an ink drop dispersing in water which happens in a matter of seconds, but the same dye suspension put in glycerin takes days to diffuse, and for a short duration after the initial mixing (say, by cranking the column of glycerin with a verticle stripe of dye one way) one can easily 'unmix' them (by reversing the direction of cranking, see, e.g., Heller, 1960). Diffusion is nevertheless an intrinsically irreversible process.

In evolutionary cosmology, the significance of any physical processes is evaluated in comparison with the Hubble expansion ($H = \dot{a}/a$, where $a$ is the scale factor). Those with characteristic time scales shorter than the Hubble time ($H^{-1}$) could have enough time to come to equilibrium with the environment, whence one can assign some temperature to the mixture and use thermodynamical descriptions. Thus in the radiation-dominated era ($a \sim t^{1/2}$) one usually refers to the temperature of the ambient photon gas as the temperature of the Universe. However, for weakly interacting particles like neutrinos and gravitons which are rarely collison-dominated, kinetic equations are needed to describe their transport processes. For quantum processes such as particle creation from the vacuum occurring at the Planck time $t_{pl} = 10^{-43}$ sec, they are intrinsically nonequilibrium quantum processes which require a statistical field-theoretical description. By the same token, when the background spacetime expands very rapidly, as during the vacuum-energy-dominated inflation epoch ($a \sim e^{Ht}$), the ordinary pratice of describing the phase transition with finite temperature theories may prove to be rather inadequate. Such are the ways how time-scales and the time dependence of the scale factor enter in cosmological processes. Now what about the time-reversible behavior of $a(t)$ itself?

It is often assumed that the dynamics of the Universe in the contraction phase (say, in a closed Friedmann model) is identical with the expansion phase, because the Einstein equation is time-reversal invariant. (Of course more coalescing and greater inhomogeneity will appear in the contraction phase due to the phase-space difference). One can ask: How about deflation– Is deflation during the contracting phase just as likely to happen as inflation in the expanding phase? The answer to this question depends not on the dynamics, as all cosmological models based on Einstein's theory are time-reversal invariant, but on the initial conditions. Specifically, can the conditions conducive to these different behaviors exist with equal likelihood in the expansion and contraction phases for these universes? The radiation-dominated condition responsible for the Friedmann-class of behavior can be assumed to hold approximately at the beginning of the contracting phase just as in the expanding phase. However, the vacuum-dominated condition may not be so. This is because inflation is associated with phase transition–be it via nucleation ('old') or spinodal decomposition ('new')– which is not necessarily time-symmetric. To answer this question one should analyze the probability for vacuum energy dominance to occur as the temperature of the Universe increases during contraction, as the broken symmetries are restored, and as the curvature and inhomogeneities of spacetime grow in the approach towards the big crunch. Recent results suggest that deflation is less likely (Goldwirth 1991).

### 34.2.3 Dissipation (Case C)

There are two basic models of dissipation in non-equilibrium statistical mechanics: the Boltzmann kinetic theory of dilute gas, and the (Einstein-Smoluchowsky) Langevin theory of Brownian motion. Each invokes a different set of concepts, and even their relation is illustrative. In kinetic theory, the equations governing the $n$-particle distribution functions (the BBGKY hierarchy) preserve the full information of an $n$ particle system. It is in ignoring (more often restricted by the precision of one's observation than by choice) the information contained in the higher-order correlations which brings about dissipation and irreversibility in the dynamics of the lower-order correlations. (Zwanzig, 1961; Prigogine, 1962; Balescu, 1975; de Groot, van Leeuven and van Weert, 1980; Calzetta and Hu, 1988). For the Brownian motion problem modeled, say, by a set of coupled oscillators with one oscillator (mass $M$) picked out as the Brownian particle and the rest (with mass $m$) serving as the bath (Rubin, 1960; Ford, Kac and Mazur, 1963; Feynman and Vernon, 1963; Caldeira and Leggett, 1983; Hu, Paz and Zhang, 1992, 1993; Hu and Matacz, 1993). Dissipation in the dynamics of the system arises from ignoring details of the bath variables but only keeping their averaged effect on the system (this also brings about a renormalization of the mass and the natural frequency of the Brownian particle). Usually one assumes $M >> m$ and weak coupling of the system and the bath to simplify calculations. The effect of the bath can be summarized by its spectral density function, which is not unique to any particular bath. In both of these models, as well as in more general cases, the following conditions are essential for the appearance of dissipation (Hu, 1989, 1990; Calzetta, 1989, 1991):

a) *system-environment separation.* This split depends on what one is interested in: it could be the slow variables, the low modes, the low order correlations, the mean fields; or what one is restricted to: the local domain, the late history, the low energy, the asympototic region, outside the event horizon, inside the particle horizon, etc. We shall bring up this issue again at the end of this talk.

b) *coupling.* The environment must have many degrees of freedom to share and spread the information from the system; its coupling with the system must be effective in the transfer of information (e.g., non-adiabatic) and the response of the coarse-grained environment must be sufficiently non-systematic that it will only react to the system in an incoherent and retarded way.

c) *coarse-graining.* One must ignore or down-grade the full information in the environmental variables to see dissipation appearing in the dynamics of the open system. (The time of observation enters also, in that it has to be greater than the interaction time of the consitituents but shorter than the recurrence time in the environment). Coarse-graining can be the truncation of a correlation hierarchy, the averaging of the higher modes, the 'integrating out' of the fluctuation fields, or the tracing of a density matrix (discarding phase informations). See the last section for more discussions on this point.

d) *initial conditions.* Whereas a dissipative system is generally insensitive to the initial conditions in that for a wide range of initial states dissipation can drive the system to the same final (equilibrium) state, the process is nevertheless possible only if the initial state is off-equilibrium. The process manifests irreversibility also because the initial time is singled out as a special reference point when the system is prepared in that particular initial state. Thus in this weaker sense, dissipation is also a consequence of specially prescribed initial conditions.†

While the original combined system and environment still preserve the unitarity of motion, and its entropy remains constant in time, under these approximations, the subsystem becomes an open system, the entropy of the open system (constructed

---

† Note the distinction between these cases: If one defines $t_0$ as the time when a dissipative dynamics begins and $t_1$ as when it ends, then the dynamics from $t_0$ to $-t$ is exactly the same as from $t_0$ to $t$, i.e., the system variable at $-t_1$ is the same as at $t_1$. This is expected because of the special role assigned to $t_0$ in the dynamics with respect to which there is time-reversal invariance, but it is not what is usually meant by irreversibility in a dissipative dynamics. The arrow of time there is defined as the direction of increase of entropy and irreversibility refers to the inequivalence of the results obtained by reversing $t_0$ and $t_1$ (or, for that matter reversing $t_0$ and $-t_1$), but not between $t_1$ and $-t_1$. The time-reversal invariance of the H-theorem has the same meaning.

from the reduced density matrix by tracing out the environmental variables) increases in time, and irreversibility appears in its dynamics.

Both irreversible (but non-dissipative) processes and dissipative (and irreversible) processes depend on the stipulation of special initial conditions. The difference is that the former depends sensitively so, the latter insensitively. Dissipative processes involve coarse-graining while non-dissipative processes do not. However, both type of irreversible processes (Case B and C) can entail entropy generation (even in Case A one can associate some mathematical entropy to describe the divergence of the trajectories). Irreversible processes described by the second law is what usually defines the thermodynamic arrow of time.

In the context of dissipative processes, it is important to distinguish dissipation from phase mixing, which, though sometimes called damping (e.g. Landau damping) and has the appearance of an irreversible process, is actually reversible.

### 34.2.4  Phase Mixing (Case D)

Two well-known effects fall under this category: Landau damping and spin echo (e.g., Balescu, 1975, Section 12.2; Ma, 1985, Section 24.3). Let us examine the first example. In the lowest order truncation of the BBGKY hierarchy valid for the description of dilute gases, the Liouvillian operator $L$ acting on the one-particle distribution function $f_1(r_1, p_1, t)$ is driven by a collision integral involving a two-particle distribution function $f_2(r_1, p_1, r_2, p_2, t)$:

$$[\frac{\partial}{\partial t} + \frac{\mathbf{p}_1}{m} \cdot \nabla_{r_1} + \mathbf{F}(r_1) \cdot \nabla_{p_1}] f_1(\mathbf{r}_1, \mathbf{p}_1, t) = \left(\frac{\partial f_1}{\partial t}\right)_{coll} \qquad (34.2)$$

$$(\frac{\partial f_1}{\partial t})_{coll} = (\frac{N}{V}) \int [\nabla_{r_1} V(\mathbf{r}_1, \mathbf{r}_2)] \cdot \nabla_{p_1} f_2(\mathbf{r}_1, \mathbf{p}_1, \mathbf{r}_2, \mathbf{p}_2, t)] d^3 r_2 d^3 p_2$$

The molecular chaos ansatz assumes an initial uncorrelated state between two particles (a factorizable condition): $f_2(1, 2) = f_1(1) f_1(2)$, i.e., that the probability of finding particle 1 at $(r_1, p_1, t)$ and particle 2 at $(r_2, p_2, t)$ at the same time t is equal to the product of the single particle probabilities. When this condition is assumed to hold initially and finally in a collision processes, (but the two collision partners are assumed to be correlated within the short range of the interaction force), one gets the Boltzmann equation. However, for long-ranged forces such as the Coulomb force in a dilute plasma gas where close encounters and collisions are rare, the factorizable condition can be assumed to hold throughout. In such cases the kinetic equation becomes a Vlasov (or collisionless-Boltzmann) Equation: (e.g., Balescu, 1975; Kreuzer, 1981)

$$\{\frac{\partial}{\partial t} + \frac{\mathbf{p}_1}{m} \cdot \nabla_{r_1} + [\mathbf{F}(\mathbf{r}_1) - \nabla_{r_1} \bar{\Phi}(\mathbf{r}_1, t)] \cdot \nabla_{p_1}\} f_1(\mathbf{r}_1, \mathbf{p}_1, t) = 0 \qquad (34.3)$$

Here

$$\bar{\Phi}(\mathbf{r}_1, t) = (\frac{N}{V}) \int V(\mathbf{r}_1, \mathbf{r}_2) f_1(\mathbf{r}_2, \mathbf{p}_2, t) d^3 r_2 d^3 p_2 \qquad (34.4)$$

is the mean field potential experienced by any one particle produced by all other particles. It is determined by the density excess over the equilibrium value. The effect of the mean field potential is similar to the Debye-Huckel screening in dilute electrolyte systems. The dependence on $f_1$ makes the Vlasov equation nonlinear: Equations (34.3) and (34.4) have to be solved in a self-consistent way. This is analogous to the Hartree approximation in many-body theory. Note that the Vlasov equation which has a form depicting free streaming is time-reversal invariant: The Vlasov term accounting for the effect of the averaged field does not bring about entropy generation. This mean-field approximation in kinetic theory, which yields a unitary evolution of reversible dynamics, is, however, only valid for times short compared to the relaxation time of the system in its approach to equilbrium. This relaxation time is associated with the collision-induced dissipation process.

Landau damping in the collective local charge oscillations described by the Vlasov equation is only an apparently irreversible processes. The appearance of 'damping' depends critically on some stipulated special initial conditions. This damping is different from the dissipation process discussed in Case C, in that the latter has an intrinsic time scale but not the former, and that while dissipation depends only weakly on the initial conditions, mixing is very sensitive to the initial conditions. A more appropriate name for these processes is 'phase mixing' (Balescu 1975). Spin echo is a somewhat different example of phase mixing.

From all of the above examples we see that irreversibility and dissipation involve very different causes. The effect of interaction, the role of coarse-graining, the choice of time-scales, and the specification of initial conditions in any process can give rise to very different results. In the next section we shall use these examples to illustrate the statistical properties of quantum field processes in the early Universe.

## 34.3 Fluctuations and Irreversibility: Examples from Cosmological Particle Creation

We see in the above the many origins of irreversibility and the distinction between dissipative and irreversible processes. Let us continue exploring these conceptual issues now by adding an additional dimension, fluctuations – both quantum and thermal fluctuations. These refer to statistical variations from the mean – the vacuum or the background field in the case of quantum fluctuations, the equilibrium state or the mean field in the case of thermal fluctuations. (Their relation is an interesting issue in itself, involving the relation of thermal and classical, the crossover from quantum to thermal, the kinematric and geometric effects etc. See e.g. Hu and Zhang, 1992, 1993; Anderson and Halliwell, 1993; Calzetta and Hu, 1993c; Sciama, 1979.) Processes involving fluctuations play important roles in cosmology. Examples are: Fluctuations in background spacetimes induce density contrasts as seeds for galaxy formation (Bardeen, 1992; Mukhanov et al., 1993; Hu, Paz and Zhang, 1993b); parametric amplification of vacuum fluctuations leads to particle creation in the early Universe (Parker, 1969; Zel'dovich, 1970); fluctuations of quantum fields bring about phase transitions in the inflationary cosmology (Guth, 1981; Sato, 1981; Linde, 1982; Albrecht and Steinhardt, 1982); thermal fluctuation (noise)-induced phase transitions (the Kramer process). Even the creation of the Universe (and its babies!) has been attributed to fluctuations of spacetime geometry and topology (Vilenkin, 1986; Coleman et al, 1991).

For a description of fluctuations, at least two factors, the number of samples taken and the time of observation, usually enter into the consideration: For $N$ samples of a system in equilibrium, the fluctuations of physical quantities associated with the system are of the magnitude $N^{-1/2}$ and can be made arbitrarily small by making $N$ large. Thus in taking the thermodynamic limit of the system, i.e., letting $N$ and $V$ large but keeping $N/V$ constant, or, by looking at the system at longer time spans, the occurance of large fluctuations is statistically suppressed. The former operation forfeits the Poincaré recurrence, while the latter operation (made equivalent to averaging over a large number of copies) assumes the validity of ergodicity. By contrast, for finite nonequilibrium systems, large fluctuations can arise more readily. Because non-equilibrium systems have intrinsic time-scales, one cannot hope to get an ensemble-averaged suppression by taking a long enough waiting time, as in the equilibrium cases. As for the issues of time-reversibility of events involving fluctuations, although the appearance of a fluctuation and its disappearance are time-symmetric, the set-up of problems involving fluctuations is often such that the chronicle of interesting events starts at the time when the fluctuation first comes into existence, or becomes eventful. This imparts the subsequent history an apparent arrow of time. Thus we talk about the 'beginning' of a new phase, or the 'genesis' of the Universe, as if time only exists after that particular moment.

Irreversibility and thermal fluctuations are studied in many textbooks of non-equilibrium statistical mechanics. Here I want to focus on the statistical properties of vacuum fluctuations, especially in cosmological processes involving vacuum fluctuations. Let us first analyze entropy-generation and irreversibility in a simple but basic process, particle creation from the vacuum.

Pair creation involves the spontaneous or stimulated release of energy in the amount of the threshold or above from the vacuum or from existing particles. Note that the mechanism according to the basic physical laws is time-symmetric. Thus, given equal initial and final conditions, pair annihilation should be equally probable. However, the initial condition is usually arranged differently from the final conditions, and this is where the problem arises. It is easier for a pair to be created than for them to annihilate, because only particles-antiparticle pairs with $\pm k$ can do it and the two have to be brought together at the same point in spacetime for this to happen. (This is what is usually refered to as the phase space factor difference).

One of the reasons for our interest in vacuum particle creation processes is to try to get a handle on the nature of the ubiquitous, omnipotent, but mysterious and often ambiguous entity called the vacuum. Note that by comparison with the particles it creates, which carry precise and reproducible information content, the vacuum understood in a naive way contains little information. However, the vacuum is far more complex than a simple 'nothing'. It is made to play many different roles and perform many difficult tasks: The vacuum is every rich man's garbage dump (witness all the divergences) and every poor man's Messiah ("The Universe is a free lunch", Guth, 1981) It is far from devoid of information, because everything can in principle be obtained from it, given some viable mechanism (e.g., pair production) and some luck (probability and stochasticity). Therefore the mechanisms which transform the vacuum into physical reality is of special interest. It is for this reason that some understanding of the statistical properties of the vacuum is essential to launching the adventurous but noble quest to 'get everything from nothing', otherwise known as Don Quixote's 'free lunch'.

Cosmological particle creation adds into consideration an additional factor of the influence of background spacetimes on the vacuum (Parker, 1969). We shall look at just the dynamical effects here but not those effects associated with the global structures of spacetime such as the event horizon (Hawking, 1975; Unruh, 1976; Sciama, 1979). We have in earlier work analyzed the problem of entropy generation from cosmological particle creation and interaction processes. Let us try to understand the different nature of irreversibility in these processes.

Assuming that at an initial time $t_0$ the system is in an eigenstate of the number operator, then the number of particles in mode $k$ in a unit volume at a later time $t$ is given by (Parker, 1969)

$$< N_k(t) > = |\beta_k(t)|^2 + a_k < N_k(t_0) > \qquad (34.5)$$

where $\beta_k$ is the Bogolubov coefficient measuring the mixing of the positive and negative frequency components, and $a_k = 1 + 2|\beta_k(t)|^2$ is the parametric amplification factor for mode $k$. The two parts in this expression can be understood as the parametric amplification of vacuum fluctuations and that of particles already present in mode $k$. The first part (spontaneous creation) always increases while the second part (induced creation) can increase or decrease depending on the correlation and phase relation of the initial state and on whether the particle is a boson or a fermion.

Are these processes time-asymmetric? Is there entropy generation in a vacuum particle creation processes? The search for an answer to these seemingly simple questions teaches us something interesting. Let us separate the time-asymmetry question into two parts: one referring to the time-reversed process of pair annihilation, the other referring to the probability of particle creation in the Universe's contraction phase.

Assume the Universe is in the expansion phase. Consider first the more complicated but conceptually easier case of particle creation with interaction. If we measure only the one-particle distribution, the entropy function constructed from the reduced density matrix will under general conditions (assuming bosons with initial state an eigenstate of the number operator) increase. (For details see Hu and Kandrup, 1987). The primary reason is that one has ignored the information in the higher-order correlation functions. The presence of interaction is such that even if one starts with an initial state with no correlation between the relevant and irrelevant variables, interaction can change the correlations and bring about entropy generation. This case is similar in nature to our example above of dissipations in an interacting gas. These dissipative processes are irreversible, and their outcomes usually do not depend or depend only weakly on the initial conditions.

The other case of particle creation from the vacuum with no interaction is more subtle (Hu and Pavon, 1986; Kandrup, 1988). On the one hand we know that both the initial vacuum and the final particle pair are in a pure state, so there cannot be any entropy generation. On the other hand we clearly see an increase of particles in time, and one is tempted to use the particle number as a measure of entropy and conclude that entropy is generated in the process of particle creation. (Indeed, in the thermodynamic approximation, $S \sim N^3$, but this relation is only valid for collision-dominated gas, which assumes interaction, from which entropy generation is expected). The resolution of this paradox lies in the fact that usually in calculating particle creation one works in a Fock space representation where the initial state (e.g., the vacuum or the thermal state) is assumed to be an eigenstate of the number operator ($N$-representation). However, an uncertainty relation exists between the number and the phase information. It is at the sacrifice of the phase information that one sees an increase of the number in time. Had one chosen the initial state to be of definite phase ($P$-representation), particle number will not be monotonically increasing. Therefore it is only for the special choice of an eigenstate of the number operator as the initial state that the non-dissipative process of particle creation with no interaction appears to be irreversible. As in the case of phase mixing in Example D above, this apparent 'irreversibity' is also highly sensitive to the choice of the initial state.

Now consider the situation where these processes take place at the contraction stage of the Universe and ask the question whether they will take place with the same probability. Let us take the simplest case of cosmological particle creation, assuming that the *in*-vacuum and the *out*-vacuum are well defined (e.g., statically-bounded dynamics, or work with some conformal-vacuum) and symmetric. Since the Bogolubov transformations which relate a set of creation and annihilation operators at one time to another are time-reversal invariant, the process should be time-symmetric. That is, one should expect to see particle creation just as likely to happen in the contraction phase. However, except for steady state models, cosmological conditions are not symmetric between the *in* and the *out* states in the expanding and the contracting phases. In the expanding phase, the *in*-state for particle creation processes of any cosmological significance is usually taken to be at the singularity ('big bang') or at least around the Planck time, while the *out*-state is defined at late times before recontraction when curvature and field effects are weak. There is asymmetry in the *in* and *out* states between the expanding and the contracting phases which affects the production rates. Despite these differences, there is entropy generation associated with particle creation and interaction in both the expanding and the contracting phases. Thus the thermodynamic arrow of time defined by the direction of entropy increase will see no change at the turnaround point. To the extent that the thermodynamic arrow of time can be traced to be the root of many other arrows of time (including the psychological), entropy generation in particle creation can play a fundamental role in the problem of time-asymmetry.

We see in the above cosmological examples the workings of the differences between irreversible and dissipative processes as manifested in vacuum fluctuations and particle creation. We shall see next how these processes can affect the dynamics of the early Universe, and manifest as a relation between fluctuation in the quantum fields and dissipation in the dynamics of spacetime.

### 34.4 Fluctuation and Dissipation: Example from Cosmological Backreaction Processes

Cosmological particle creation comes from the amplification of vacuum fluctuations by the dynamics of the background spacetime. It is the transformation of a microscopic random process into macroscopic proportions. At late times like today's Universe this process is rather insignificant (Parker, 1969). However, near the Planck time $t_{pl} \sim (10^{-43}$ sec from the Big Bang), for non-conformal fields, or for non-conformally flat universes, production of particles might have been so copious that they could have exerted a strong influence on the dynamics of the early Universe (Zel'dovich 1970). In particular, anisotropies in the early Universe can be dissipated away in fractions of $t_{Pl}$ (Zel'dovich and Starobinsky, 1971; Hu and Parker, 1978; Hartle and Hu, 1980). Backreaction processes like these have been studied extensively for cosmological (origin of isotropy in the Universe), philosophical (chaotic cosmology program), and theoretical (quantum to classical transition) inquiries. Here we'd like to view it as an example of the fluctuation-dissipation relation relating the fluctuations of the vacuum to the dissipative dynamics of the Universe (Hu, 1989). Take for example a massless conformal scalar field in an anisotropic, homogeneous Bianchi Type-1 Universe with line element

$$ds^2 = a^2(\eta)[d\eta^2 - \sum_{i,j=1}^{3} e^{2\beta_{ij}(\eta)}dx_i dx_j].\tag{34.6}$$

The equation of motion for the anisotropic expansion rates $q_{ij} \equiv \beta'_{ij} \equiv d\beta_{ij}/d\eta$ calculated in the Schwinger (1961) - Keldysh (1964) (or closed time-path, or *in-in*) formalism is given by (Calzetta and Hu, 1987)

$$\frac{d}{d\eta}(Mq'_{ij}) + 3(2880\pi^2)^{-1}Kq'_{ij} + kq_{ij} = c_{ij},\tag{34.7}$$

where $c_{ij}$ is a constant measuring the initial anisotropy, $M$ and $k$ are funtions of $a, a'$ and $a''$. The nonlocal kernel $K(\eta - \eta')$

$$K q'_{ij} = \int_{-\infty}^{\eta} d\eta' (\frac{d^3}{d\eta^3} q'_{ij}) ln(\eta - \eta'). \tag{34.8}$$

linking the 'velocities' $q'_{ij}$ at different times gives a non-local viscosity function $\gamma$ (in Fourier space)

$$\gamma(\omega) = \frac{\pi}{60(4\pi)^2} \mid \omega \mid^3 . \tag{34.9}$$

which is responsible for the dissipation of anisotropy in the background dynamics. The energy density dissipated in the background dynamics is shown to be exacly equal to the energy density of the particles created:

$$\rho(\text{particle creation }) = \rho(\text{anisotropy dissipation}) \tag{34.10}$$

This relation, as we pointed out earlier (Hu, 1989), embodies the fluctuation-dissipation relation in the cosmological context, but does not yet have the correct form (F-D relation for black holes and de Sitter spacetimes have been proposed by Candelas and Sciama, 1977; Sorkin, 1986; and by Mottola, 1986 respectively).

Notice that velocity $\beta'$ enters in the equation of motion (34.7) instead of displacements $\beta$. This is because the coupling between the field and the background dynamics via the Laplace-Beltrami operator is of a derivative kind. This equation is in the form of a Langevin equation, except for the absence of explicit random forces. The search for this missing noise and a deeper statistical-mechanical meaning of dissipation in quantum fields led me to adopting the more general open-system viewpoint to treat these quantum field processes. This search resulted in the discovery of a generalised framework of semiclassical gravity based on an Einstein–Langevin equation (Calzetta and Hu, 1993b; Hu and Matacz, 1993b). Technically the ordinary effective action which takes into account the averaged effect of quantum fluctuations can be generalized to a coarsed-grained one, where the spacetime and the scalar field play the roles of the system and the environment respectively. The coarse-grained closed time-path effective action (Hu and Zhang, 1990; Hu, 1991b; Sinha and Hu, 1991) is intimately related to the influence action (Feynman and Vernon, 1963) which is needed for a full display of both dissipation and noise effects in quantum statistical systems (Hu, 1991a, 1993). The influence action has a real part which is responsible for particle production, and an imaginary part responsible for noise. The equation of motion derived from the influence action is the master equation for the reduced density matrix of the system after details of the environment are traced out. In the semiclassical limit the Wigner function associated with the reduced density matrix obeys the Fokker-Planck equation, while, equivalently, the system variable obeys a Langevin equation with an explicit noise term whose distribution function depends on the nature of the environment and its coupling with the system. It can be shown that the noise is related to the fluctuation in the number of particles created (Calzetta and Hu, 1993b).

Many physical processes in the macroscopic world manifest dissipative behavior, which is time-asymmetric. This is at variance with the basic laws governing the microscopic world, which are time-symmetric. To resolve this difference is one of the central tasks of statistical mechanics. One way is to conceive of a natural transformation (or spontaneous evolution) of a closed system to an open system involving the procedures outlined in Example C, i.e., separation of the system (the relevant variables) from the environment (the irrelevant variables), choice of boundary conditions, and averaging (coarse-graining) of the irrelevant variables. Backreaction of the averaged effect of the irrelevant variables modifies the dynamics of the relevant variables with a dissipative contribution. It is through this means that random microscopic reversible processes can bring forth irreversible behavior in the systematic macroscopic dynamics. The connection between these two aspects is best captured in the fluctuation-dissipation (FD) relation. In a concrete form, it provides a microscopic derivation of the kinetic coefficients (e.g. viscosity function). It is also one of the means that the quantum world described by wave functions and interference effects can be related to the classical world described by the classical equations of motion. We will discuss only the meaning of the fluctuation-dissipation relation but leave the decoherence effect in quantum to classical transition (Zurek, 1981, 1982, 1993; Joos and Zeh, 1985; Zeh, 1986; Goldeira and Leggett, 1985; Unruh and Zurek, 1989; Hu, Paz and Zhang, 1992, 1993; Paz, Habib and Zurek, 1993; Paz and Zurek, 1993; Zurek, Habib and Paz, 1993; Griffiths 1984; Omnès, 1992; Gell-Mann and Hartle, 1990: Dowker and Halliwell, 1992; Calzetta and Hu, 1993a), and the relation of noise and classical structure elsewhere (Hu, 1991a, 1993; Gell-Mann and Hartle, 1993). They are interrelated.

The FD relation is often written for equilibrium (finite temperature $T$) conditions and derived via linear-response theories (Callen and Welton, 1951; Kubo, 1959). We believe that, owing to its general nature, a relation should exist for non-equilibrium, and for quantum ($T = 0$) processes. In a recent work (Hu, Paz and Zhang, 1992, 1993a; see also Sinha and Sorkin, 1992) we have proven at least the latter case in quantum Brownian motion models. This provides the theoretical basis for a statistical interpretation of quantum backreaction processes, which include the well-known radiation-reaction problem in electrodynamics, as well as the backreaction problems in semiclassical cosmology. For the purpose of extending the fluctuation-dissipation relation to quantum fields, we used path-integral methods. Our results are summarized as follows.

Consider a Brownian particle with mass $M$ interacting with a thermal bath at temperature $T = (k_B \beta)^{-1}$. The classical action of the Brownian particle is

$$S_S[x] = \int_0^t ds \left\{ \frac{1}{2} m\dot{x}^2 - V(x) \right\} \tag{34.11}$$

The bath consists of a set of harmonic oscillators with mass $m_n$ whose motion is described by the classical action

$$S_E[\{q_n\}] = \int_0^t ds \sum_n \left\{ \frac{1}{2} m_n \dot{q}_n^2 - \frac{1}{2} m_n \omega_n^2 q_n^2 \right\} \qquad (34.12)$$

Assume as an example that the system and environment interact via a biquadratic coupling with action

$$S_{int}[x, \{q_n\}] = \int_0^t ds \sum_n \left\{ -\lambda C_n x^2 q_n^2 \right\} \qquad (34.13)$$

Here $\lambda$ is a coupling constant multiplied to each $C_n$ which is assumed to be small for perturbation calculations. The case of linear coupling has been derived by many authors (e.g., Feynman and Vernon, 1963; Caldeira and Leggett, 1983; Unruh and Zurek, 1989; Hu, Paz and Zhang, 1992). For biquadratic coupling, the fluctuation-dissipation relation between the second noise kernel $\tilde{\nu}(s_1 - s_2)$ and the dissipation kernel $\gamma(s_1 - s_2)$ can be written down explicitly as (Hu, Paz and Zhang, 1993a)

$$\hbar\tilde{\nu}(s) = \int_0^{+\infty} ds' K(s - s')\gamma(s') \qquad (34.14)$$

where the time convolution kernel $K(s)$ is given by

$$K(s) = \hbar \int_0^{+\infty} \frac{d\omega}{\pi} \left\{ \frac{1 + \coth^2 \frac{1}{4}\beta\hbar\omega}{2\coth \frac{1}{4}\beta\hbar\omega} \right\} \omega \cos \omega s. \qquad (34.15)$$

Except for the temperature dependent factor (the term within the curly brackets), this has the same form as the linear coupling case (which is given by $\coth(\beta\hbar\omega/2)$). For higher order couplings with an action $\lambda C_n f(x) q_n^r$ the FD relation again has the same form as (34.14) and (34.15), only that the temperature-dependent factor is different. In the high temperature limit $k_B T \gg \hbar\Gamma$, where $\Gamma$ is the cutoff frequency of the bath oscillators, $K(s) = 2k_B T \delta(s)$ and the fluctuation-dissipation relation reduces to

$$\hbar\tilde{\nu}(s_1 - s_2) = 2k_B T \gamma(s_1 - s_2) \qquad (34.16)$$

which is the famous Einstein formula (Einstein, 1905).

In the zero temperature limit

$$K(s) = \hbar \int_0^{+\infty} \frac{d\omega}{\pi} \omega \cos \omega s \qquad (34.17)$$

It is interesting to note that the fluctuation-dissipation relations for the linear and the nonlinear coupling models we have studied are identical both in the high temperature and the zero temperature limits. This insensitiveness to the different system-bath couplings reflects that it is a categorical relation (backreaction) between the stochastic stimuli (fluctuation-noise) of the environment and the averaged response of a system (dissipation-relaxation) which has a much deeper and broader meaning than that associated with the special cases studied in the literature. We have also derived the influence action for field theory models with non-linear coupling and colored noise environments (Zhang, 1990; Hu, Paz and Zhang, 1993b) and found that a set of FD relations exist which are identical in form to the quantum mechanical results given above. This seems to confirm our earlier suggestion about the universality of such relations (Hu, 1989). The FD relation suggests how macroscopic irreversibility can arise from microscopic reversible processes. It is in this capacity that it is relevant to the time-asymmetry problem.

The extension of the quantum Brownian motion results to quantum cosmology is under investigation. This requires first an upgrading in the treatment of the cosmological backreaction problem described above from the semiclassical to the full quantum level (describing wave functions of the Universe). One also needs to generalize this problem to statistical ensembles (of quantum states of the Universe) and study the evolution of the reduced density matrix of the Universe obtained by tracing out, say, the scalar fields viewed as the environment variables (see, e.g., Paz and Sinha, 1991, 1992). Consideration of this cosmological backreaction problem in the statistical context pushes the domain of validity of the fluctuation-dissipation relation to a new level, that which involves fluctuations of quantum fields and dissipative spacetime dynamics (Hu and Sinha, 1993a). This relation viewed in the cosmological context has direct implications on the notion of gravitational entropy and the time-asymmetry issue, as we now show.

## 34.5 Coarse-Graining and Dissipation in Spacetime: Example in Minisuperspace Cosmology

In a statistical-mechanical interpretation of the problem of backreaction due to particle creation, the background spacetime plays the role of the system while the matter field that of the environment. The backreaction can be calculated by the effective action method in loop expansions. In the ordinary approach, a background-fluctuation field decomposition is assumed, and the backreaction is due to the radiative correction effects $O(\hbar)$ of the matter field like vacuum fluctuation and particle creation. One can generalize this method to treat quantum statistical processes involving coarse-graining. Suppose one separates the field of the combined system $\phi$ into two parts: the system field $\bar{\phi}$ and the environment field $\tilde{\phi}$, i.e., $\phi = \bar{\phi} + \tilde{\phi}$, and assumes that they are coupled weakly with a small parameter $\lambda$. One can then construct a coarse-grained effective action $\Gamma[\bar{\phi}]$ by integrating away the environment variables. This procedure has been used in a renormalization group theory treatment of critical phenomena in the inflationary Universe (Hu and Zhang,

1990, Hu, 1991b). For quantum cosmology, one can use this method to study the effect of truncation in the gravitational degrees of freedom, and discuss the validity of the minisuperspace approximation. (A more comprehensive discussion of viewing minisuperspace as a quantum open system in quantum cosmology is given in Hu, Paz and Sinha, 1993)

### 34.5.1 Minisuperspace Approximation

Those cosmological models most often studied, like the Robertson-Walker, de Sitter, and the Bianchi universes, which possess high symmetries are but a small class of a large set of possible cosmological solutions of the Einstein equations. In terms of superspace, the space of all three-geometries, (Wheeler, 1968; DeWitt, 1967) these are the lower-dimensional minisuperspaces (Misner, 1972) (e.g., the mixmaster Universe with parameters $\alpha, \beta_+, \beta_-$ is a three-dimensional minisuperspace). In quantizing just the few lowest modes, as is often done in quantum cosmology studies, one ignores by fiat all these other modes. Is the minisuperspace quantization justified? (Kuchar and Ryan, 1986, 1989) Under what conditions is it justified? What is the backreaction effect of the inhomogeneous modes on the homogeneous mode? One can view the homogeneous geometry as the system and the matter fields (or the inhomogeneous perturbations of spacetime, the gravitons) as the environment, and use the coarse-grained effective action to calculate the averaged effect of the environment on the system. Notice the similarity with the statistical mechanical problems we have treated above. In one illustrative calculation (Sinha and Hu, 1991) we used a model of self-interacting quantum fields to mimic the nonlinear coupling of the gravitational waves modes (WKB time is used as it provides correct semiclassical results) and obtained an effective Wheeler-DeWitt equation for the minisuperspace sector with a new term containing a nonlocal kernel. Similar in form to Eq. (34.7) in the particle creation backreaction problem, it signifies the appearance of dissipative effects in the dynamics of the minisuperspace variables due to their interaction with the inhomogeneous modes. Thus one can conclude that the minisuperspace approximation is valid only if this dissipation is small. In the same sense as the other statistical processes we have considered above, the appearance of dissipation creates an arrow of time in the minisuperspace sector. This also provides one way to define gravitational entropy. Notice that in this view, as long as one limits one's observation to a subset of all possible geometrodynamics, and allows for some special initial conditions, dissipative behavior and the emergent arrow of time are unavoidable consequences.

### 34.5.2 Gravitational Entropy

The entropy of gravitational fields has been studied in connection with self-gravitating matter (Lynden-Bell and Wood, 1967; Lynden-Bell and Lynden-Bell,

1977; Sorkin, Wald and Zhang, 1984), with black holes (Hawking, 1975; Sorkin, 1986), with cosmology (Penrose, 1979) and with gravitons (Smolin 1985). We shall consider it for cosmological spacetimes without event horizons (See Davies, Ford and Page, 1989 for the case of de Sitter universe, which has an event horizon). Gravitational entropy of the Universe has also been discussed before in conjunction with quantum dissipative processes in the early Universe (Hu, 1983, 1984). Here I want to discuss it in the context of quantum cosmology (see also Kandrup, 1989) and the theme of the present conference, time-asymmetry.

Following the idea of minisuperspace approximation in quantum cosmology discussed above as a backreaction problem and generalizing the wave functions of the Universe to density matrices of the Universe, we can work with the reduced density matrix of the Universe constructed by tracing out the matter fields or the higher gravitational modes and define a gravitational entropy of the homogeneous Universe as

$$S = -Tr\rho_{red}ln\rho_{red} \qquad (34.18)$$

From the theory of subdynamics, we know that $S$ increases with time. (Note again that some notion of time has to be introduced beforehand, e.g., the WKB time, or the 4-volume time of Sorkin, 1993). The arrow of time arises as the direction of information flow from the relevant (spacetime, or the homogeneous gravitational modes) to the irrelevant (the matter fields, or the inhomogeneous modes) degrees of freedom. (See Hu, 1993; Hu and Sinha, 1993b for details.)

In this and earlier sections I have only sketched the statistical nature of certain quantum processes in semiclassical gravity and quantum cosmology, but I hope this array of examples and questions – from billiard balls to ink drops to plasma waves to particle creation to anisotropy damping to density matrix of the Universe – has demonstrated to you, despite the great disparity of their context, the universality of the issues involved and the conceptual unity in our understanding.

Up to now I have only discussed HOW one can see dissipation and the arrow of time arising in the system from coarse-graining the environment. I have not mentioned anything about the more fundamental and difficult questions in the system-environment approach to these issues in statistical mechanics, i.e., WHY? Why should the system be regarded as such? Why should the separation be made as such? Why should a sector be viewed as the system and get preferential treatment over the others? The answer to these questions when raised in the cosmological context can be more meaningfully sought with the open-system perspective if the spacetime has some distinguished global or physical structures like event horizon, particle horizon, non-trivial topology, etc. One can then define an objectively meaningful domain for the system and study its effective dynamics. The outcome also depends on how the coarse-graining (measurement, observation, participation) is taken, and how effective it is in producing persistent robust structures (Woo, 1989). Consistency in the behavior of the system after these procedures are taken (such as how stable any level of structure is with respect to iteration of the same coarse-graining routines, and how sensitive the open system is with respect to variations of coarse-graining) is certainly an important criterion in any consideration. I will now say a few things on these issues to conclude my talk.

### 34.6  Coarse-Graining and Persistent Structure in the Physical World

Let me summarize the main points of this talk and suggest a few questions to explore on the issue of irreversibility in cosmology.

On the whole, there are two different causes for the appearance of irreversibility: one due to special initial conditions, the other due to dissipation. † The first class is *a priori* determined by the initial conditions, the other is *a posteriori* rather insensitive to the initial conditions. Of the examples we have given, the first class includes chaotic dynamics, Landau damping, vacuum particle creation, the second class includes molecular dynamics, diffusion, particle creation with interaction, anisotropy dissipation, decoherence. Appearance of dissipation is accompanied by a degradation of information via coarse graining (such as the molecular chaos assumption in kinetic theory, restriction to one-particle distribution in particle creation with interaction, 'integrating out' some class of histories in decoherence). An arrow of time appears either because of some special prearranged conditions or as a consequence of coarse-graining introduced to the system. The issues we have touched on involve the transformation of a closed to an open system, the relation between the microscopic and the macroscopic world, and the transition from quantum appearances to classical realities. Many perceived phenomena in the observable physical world, including the phenomenon of time-asymmetry, can indeed be understood in the open-system viewpoint via the approximations introduced to the objective microscopic world by a macroscopic observer. We have discussed the procedures which can bring about these results. However, what to me seems more important and challenging is to explore under what conditions the outcomes become less subjective and less sensitive to these procedures, such as the system-environment split and the coarse-graining of the environment. These procedures provide one with a viable prescription to get certain general qualitative results, but are still not specific and robust enough to explain how and why the variety of observed phenomena in the physical world arise and stay in their particular ways. To address these issues one should ask a different set of questions:

*1) By what criteria are the system variables chosen? – collectivity and hierarachy of structure and interactions*

In a model problem, one picks out the system variables – be it the Brownian particle or the minisuperspace variables – by fiat. One defines one's system in

---

† As discussed earlier, dissipation also requires the stipulation of a somewhat special initial condition, i.e., that the system is not in an equilibrium state; but 'not more special than it needs to be' –in the words of R. Sorkin.

a particular way because one wants to calculate the properties of that particular system. But in the real world, certain variables distinguish themselves from others because they possess a relatively well-defined, stable, and meaningful set of properties for which the observer can carry out measurements and derive meaningful results. Its meaningfulness is defined by the range of validity or degree of precision or the level of relevance to what the observer chooses to extract information from. In this sense, it clearly carries a certain degree of subjectivity– not in the sense of arbitrariness in the will of the observer, but in the specification of the parameters of observation and measurement. For example, the thermodynamic and hydrodynamic variables are only good for systems close to equilibrium; in other regimes one needs to describe the system in terms of kinetic-theoretical or statistical-mechanical variables.

The soundness in the choice of a system in this example thus depends on the time scale of measurement compared to the relaxation time. As another example, contrast the variables used in nuclear collective model and the independent nucleon models. One can use the rotational-vibrational degrees of freedom to depict some macroscopic properties of the motion of the nucleus, and one can carry out meaningful calculations of the dissipation of the collective trajectories (in the phase space of the nucleons) due to stochastic forces. In such cases, the non-collective degrees of freedom can be taken as the noise source. However, if one is interested in how the independent nucleons contribute to the properties of the nucleus, such as the shell structure, one's system variable should, barring some simple cases, not be the elements of the $SO(3)$ group, or the $SU(6)$ group. At a higher still energy where the attributes of the quarks and the gluons become apparent, the system variables for the calculation of, say, the stability of the quark-gluon plasma should change accordingly. The level of relevance which defines one's system changes with the level of structure of matter and the relative importance of the forces at work at that level. The improvement of the Weinberg-Salam model with $W, Z$ intermediate bosons over the Fermi model of four-point interactions is what is needed in probing a deeper level of interaction and structure which puts the electromagnetic and weak forces on the same footing. Therefore, one needs to explore the rules for the formation of such relatively distinct and stable levels, before one can sensibly define one's system (and the environment) to carry out meaningful inquiries of a statistical nature.

What is interesting here is that these levels of structures and interactions come in approximate hierarchical order (so one doesn't need QCD to calculate the rotational spectrum of a nucleus, and the Einstein spacetime manifold picture will hopefully provide most of what we need in the post-Planckian era). One needs both some knowledge of the hierarchy of interactions (e.g., Weinberg 1980) and the way effective theories emerge– from 'integrating out' variables at very different energy scales in the hierarchical structure (e.g., ordinary gravity plus grand-unified theory regarded as a low energy effective Kaluza-Klein theory) The first part involves fundamental constituents and interactions and the second part the application of statistical methods. One should also keep in mind that what is viewed as funda-

mental at one level can be a composite or statistical mixture at a finer level. There
are system-environment separation schemes which are designed to accomodate or
reflect these more intricate structures, such as the mean field-fluctuation field split,
the dynamics of correlations (Balescu, 1975; Calzetta and Hu, 1988) and the multiple
source formalism (Cornwall, Jackiw and Tomboulis 1974; Calzetta and Hu, 1993a).
The validity of these approximations depends on where exactly one wants to probe
in between any two levels of structure. Statistical properties of the system such as
the appearance of dissipative effects and the associated irreversibility character of
the dynamics in an open system certainly depend on this separation.

*2) How does the behavior of the subsystem depend on coarse-graining?– sensitivity
and variability of coarse-graining, stability and robustness of structure*

Does there exist a common asymptotic regime as the result of including successively higher order iterations in the same coarse-graining routine? This measures the
sensitivity of the end result to a particular kind of coarse-graining. How well can
different kinds of coarse-graining measure produce and preserve the same result?
This is measured by its variability. Based on these properties of coarse-graining,
one can discuss the relative stability of the behavior of the resultant open system
after a sequence of coarse-grainings within the same routine, and its robustness with
respect to changes to slightly different coarse-graining routines.

Let me use some simple examples to illustrate what this problem is about. When
we present a microscopic derivation of the transport coefficients (viscosity, heat
conductivity, etc) in kinetic theory via the system-environment scheme, we usually
get the same correct answer independent of the way the environment is chosen or
coarse-grained. Have we ever wondered why? It turns out that this is the case only
if we operate in the linear-response regime (Feynman and Vernon 1963). The linear
coupling between the system and the environment makes this dependence simple.
This is something we usually take for granted, but has some deeper meaning. For
nonlinear coupling, the above problem becomes nontrivial. Another aspect of this
problem can be brought out in the following consideration (Balian and Veneroni,
1987). Compare these two levels of structure and interaction: hydrodynamic regime
and kinetic regime. Construct the relevant entropy (in the information theory sense)
from the one-particle distribution $\rho$ under the constraint that the average of any
physical variable $O$ is given by $< O >= Tr\rho O$. $\rho$ changes with different levels of
coarse-graining. In terms of the one-particle classical distribution function $f_1$ the
entropy function $S$ is given by

$$S_B = \int d\vec{r}d\vec{p}f_1(\vec{r},\vec{p})[1 - \ln h^3 f_1(\vec{r},\vec{p})] \tag{34.19}$$

in Boltzmann's kinetic theories, and

$$S_H \sim N^3, \quad N = \int d\vec{r} d\vec{p} f_1(\vec{r}, \vec{p}) \tag{34.20}$$

in hydrodynamics. Notice that $S_H > S_B$ is a maximum in the sequence of different coarse-graining procedures. In the terminology we introduced above, by comparison with the other regimes, the hydrodynamic regime is more robust in its structure and interactions with respect to varying levels of coarse-graining. The reason for this is, as we know, because the hydrodynamic variables describe systems in equilibrium. Further coarse-graining on these systems is expected to produce the same result. Therefore, a kind of 'maximal entropy principle' with respect to variability of coarse-graining is one way where thermodynamically robust systems can be located.

While including successively higher orders of the same coarse-graining measure usually gives rise to quantitative differences (if there is a convergent result, that is, but this condition is not gauranteed, especially if a phase transition intervenes), coarse-graining of a different nature will in general result in very different behavior in the dynamics of the open system. Let us look further at the relation of variability of coarse-graining and robustness of structure.

Sometimes the stability of a system with respect to coarse-graining is an implicit criterion behind the proper choice of a system. For example, Boltzmann's equation governing the one-particle distribution function which gives a very adequate depiction of the physical world is, as we have seen, only the lowest order equation in an infinite (BBGKY) hierarchy. If coarse-graining is by the order of the hierarchy – e.g., if the second and higher order correlations are ignored, then one can calculate without ambiguity the error introduced by such a truncation. The dynamics of the open system which includes dissipation effects and irreversible behavior will change little as one coarse-grains further to higher and higher order (if the series convergences, see, e.g., Dorfman, 1981). In another approximation, for a binary gas of large mass discrepancy, if one considers the system as the heavy mass particles, ignore their mutual interactions and coarse-grain the effect of the light molecules on the heavy ones, one can turn the Boltzmann equation into a Fokker-Planck equation for Brownian motion, and get qualitatively very different results in the behavior of the system.

In general the variability of different coarse-grainings in producing a qualitatively similar result is higher (more variations allowed) when the system one works with is closer to a stable level in the interaction range or in the hierarchical order of structure of matter. The result is more sensitive to different coarse-graining measures if it is far away from a stable structure, usually falling in between two stable levels.

One tentative analogy may help to fix these concepts: robust systems are like the stable fixed points in a parameter space in the renormalization group theory description of critical phenomena: the points in a trajectory are the results of performing successive orders of the same coarse-graining routine on the system (e.g., the Kadanoff-Migdal scaling), a trajectory will form if the coarse graining routine is stable. An unstable routine will produce in the most radical situations a random set of points. Different trajectories arise from different coarse-graining routines. Neighboring trajectories will converge if the system is robust, and diverge if not. Therefore the existence of a stable fixed point where trajectories converge to is an indication that the system is robust. Only robust systems survive in nature and carry definite meaning in terms of their persistent structure and systematic evolutions. This is where the relation of coarse-graining and persistent structures enters.

So far we have only discussed the activity around one level of robust structure. To investigate the domain lying in-between two levels of structures (e.g., between nucleons and quark-gluons) one needs to first know the basic constituents and interactions of the two levels. This brings back our consideration of levels of structures above. Studies in the properties of coarse-graining can provide a useful guide to venture into the often nebulous and evasive area between the two levels and extract meaningful results pertaining to the collective behavior of the underlying structure. But one probably cannot gain new information about the fine structure and the new interactions from the old just by these statistical measures. (cf. the old bootstrapping idea in particle physics versus the quark model). In this sense, one should not expect to gain new fundamental information about quantum gravity just by extrapolating what we know about the semiclasical theory, although studying the way how the semiclassical theory takes shape (viewed as an effective theory ) from a more basic quantum theory is useful. It may also be sufficient for what we can understand or care about in this later stage of the Universe we now live in.

There are immediate consequences from these theoretical discussions for cosmology. Questions like, why the Universe should in its later stage settle into the highly symmetric state of isotropy and homogeneity? Is this a particular choice of the 'system' from the beginning, or is it a consequence of coarse-graining an initial larger set of possibilities both in the spacetime and the matter degrees of freedom? What are the stable coarse-graining routines? How different can the coarse-graining routines be to still produce robust results? I have just begun to explore these questions in a number of ways. They are,

(i) Viewing the homogeneus cosmology as the infrared sector of spacetime excitations, and using the rules of dimensional reduction as possible explanation for its prevalance. (Hu, 1990)

(ii) Gravity as an effective theory and geometric structure as collective degrees of freedom (Sahkarov, 1968; Adler and Zee, 1984).

(iii) Einstein's gravity as the hydrodynamic limit of a nonlinear and nonlocal theory, drawing on the insight from the behavior of the Boltzmann equation, the BBGKY hiearachy and the long-wavelength hydrodynamic approximations.

I hope the discussions on the properties and origins of irreversible processes in cosmology, sketchy as they may appear, can help us gain a better perspective of the universality of these issues in physics and provide some theoretical basis for further discussions of their meanings.

## Acknowledgements

I thank Raphael Sorkin for a careful and thoughtful reading of this manuscript and for evoking many interesting discussions on various issues raised in this paper. This research is supported in part by the National Science Foundation under grant PHY91-19726.

## Discussion

**Cover**    In his question, P. Davies has suggested that the entropy of a gravitational field might be replaced by Kolmogorov (or algorithmic) complexity. It should be noted that entropy, as well as algorithmic complexity, are descriptive complexities. Moreover, they usually agree. And in the special case of equipartition of energy (or probability), entropy and Kolmogorov complexity equal the logarithm of the number of microstates of the given macrostate.

**Hartle**    If we are going to consider complexity then we are going to have to ask "whose complexity is it?" that is, what coarse-graining is going to be used to compute it?

**Hu**    Although coarse-graining has a strong element of subjectivity, those classes which lead to physical reality (including complexity) which is agreed upon by a large class of observers (including us) merit special attention. It is important to study the *criteria* and *conditions* for these coarse-grainings to be favorably selected in the evolutionary process which give rise to persistent structures (persistent at least to the degree we can perceive them).

## References

Adler, S. L. (1982) *Rev. Mod. Phys.* **54**, 719.
Albrecht, A. and Steinhardt, P. J. (1982) *Phys. Rev. Lett.* **48**, 1220.
Anderson, A. and Halliwell, J. J. (1993) *Phys. Rev.* **D48**, 2753.
Ashtekar, A. and Stachel, J. eds (1991) *Conceptual Problems in Quantum Gravity* (Birkhauser, Boston)
Balescu, R. (1975) *Equilibrium and Nonequilibrium Statistical Mechanics,* (Wiley, New York).
Balian, R. and Veneroni, M. (1987) *Ann. Phys. (N.Y.)* **174**, 229-244.
Bardeen, J. (1992) Talk at the ITP Conference, Beijing.
Barrow, J. D. (1982) *Phys. Rep.* **85**, 1.
Bekenstein, J. D. (1973) *Phys. Rev.* **D7**, 2333.
Bekenstein, J. D. (1974) *Phys. Rev.* **D9**, 3292.
Berger, B. K. (1992) in *Proc. GR13*, Cordoba, Argentina
Bogoiavlenskii, O. I. (1985) *Methods in the Qualitative Theory of Dynamical Systems in Astrophysics and Gas Dynamics* (Springer-Verlag, Berlin).
Caldeira, A. O. and Leggett, A. J. (1983) *Physica* **A121**, 587.
Caldeira, A. O. and Leggett, A. J. (1983) *Phys. Rev.* **A31**, 1059.
Callen, H. B. and Welton, T. A. (1951) *Phys. Rev.* **83**, 34.
Calzetta, E. (1989) *Class. Quantum Grav.* **6**, L227.
Calzetta, E. (1991) *Phys. Rev.* **D43**, 2498.

Calzetta, E. and Hu, B. L. (1987) *Phys. Rev.* **D35**, 495.

Calzetta, E. and Hu, B. L. (1988) *Phys. Rev.* **D37**, 2878.

Calzetta, E. and Hu, B. L. (1989) *Phys. Rev.* **D40**, 656.

Calzetta, E. and Hu, B. L. (1993a) "Decoherence of Correlation Histories" in *Directions in General Relativity* Vol 2 (Brill Festschrift) eds. B. L. Hu and T. A. Jacobson (Cambridge Univ., Cambridge)

Calzetta, E. and Hu, B. L. (1993b) "Noise and fluctuations in Semiclassical Gravity" Univ. Maryland preprint 93-216.

Calzetta, E. and Hu, B. L. (1993b) "From Kinetic Theory to Brownian Motion" unpublished.

Calzetta, E. and Mazzitelli, F. (1991) *Phys. Rev.* **D42**, 4066.

Candelas, P. and Sciama, D. W. (1977) *Phys. Rev. Lett.* **38**, 1372.

Coleman, S., Hartle, J., Piran, T., and Weinberg, S. eds (1990) *Quantum Cosmology and Baby Universes* (World Scientific, Singapore).

Cornwall, J. M., Jackiw, R., and Tomboulis, E. (1974) *Phys. Rev.* **D10**, 2428.

Davies, P. C. W. (1975) *J. Phys.* **A8**, 609.

Davies, P. C. W., Ford L. and Page, D. (1987) *Phys. Rev.* **D34**, 1700.

De Groot, S. R., van Leeuwen, W. A. and van Weert, Ch. G. (1980) *Relativistic Kinetic Theory* (North-Holland, Amsterdam)

DeWitt, B. S. (1967) *Phys. Rev.* **160**, 1113.

Dorfman, R. (1981) in *Perspectives in Statistical Physics* Vol. IX, Eds. H.J. Raveche (North-Holland, Amsterdam).

Dowker, H. F. and Halliwell, J. J. (1982) *Phys. Rev.* **D46**, 1580.

Einstein, A. (1905) *Ann. Phys. (Leipzig)* **17**, 549.

Feynman, R. P. and Vernon, F. L. (1963) *Ann. Phys. (N.Y.)*, **24**, 118.

Ford, G. W., Kac, M. and Mazur, P. (1963) *J. Math. Phys.* **6**, 504

Gell-Mann, M. and Hartle, J. B. (1990) in *Complexity, Entropy and the Physics of Information* ed. W. H. Zurek (Addison-Wesley, N.Y.).

Gell-Mann, M. and Hartle, J. B. (1993) *Phys. Rev.* **D47**, 3345

Goldwirth, D. S. (1991) *Phys. Lett.* **B256**, 354.

Grabert, H. (1982) *Projection Operator Techniques in Nonequilibrium Statistical Mechanics* (Springer Verlag, Berlin).

Grabert H., Schramm, P. and Ingold, G. (1988) *Phys. Rep.* **168**, 115.

Griffiths, R. B. (1984) *J. Stat. Phys.* **36**, 219.

Guth, A., (1981) *Phys. Rev.* **D23**, 347.

Halliwell, J. J. (1993) *Quantum Cosmology* (Cambridge Univ. Press, Cambridge, in preparation).

Hartle, J. B. (1983) in *The Very Early Universe,* eds. G. Gibbons, S. W. Hawking and S. Siklos (Cambridge Univ. Press, Cambridge)

Hartle, J. B. and Hawking, S. W. (1983) *Phys. Rev.* **D28**, 2960.

Hartle, J. B. and Hu, B. L. (1980) *Phys. Rev.* **D21**, 2756.

Hawking, S. W. (1975) *Commun. Math. Phys.* **87**, 395.

Heller, J. P. (1960) *Am. J. Phys.* **28**, 348-353.

Hu, B. L. (1982) in *Proc. Second Marcel Grossmann Meeting 1979,* Ed. R. Ruffini, (North-Holland, Amsterdam).

Hu, B. L. (1983) *Phys. Lett.* **97A**, 368.

Hu, B. L. (1984) in *Cosmology of the Early Universe,* Ed. L. Z. Fang and R. Ruffini (World Scientific, Singapore).

Hu, B. L. (1987) "Recent Development in Cosmological Theories", IAS Preprint, Princeton, IASSNS-HEP87/15.

Hu, B. L. (1989) *Physica,* **A158**, 399.

Hu, B. L. (1990) "Quantum and Statistical Effects in Superspace Cosmology" in *Quantum Mechanics in Curved Spacetime*, ed. J. Audretsch and V. de Sabbata (Plenum, London, 1990)

Hu, B. L. (1991a) "Statistical Mechanics and Quantum Cosmology", in *Proc. Second International Workshop on Thermal Fields and Their Applications*, eds. H. Ezawa et al (North-Holland, Amsterdam, 1991)

Hu, B. L. (1991b) "Coarse-Graining and Backreaction in Inflationary and Minisuperspace Cosmology" in *Relativity and Gravitation: Classical and Quantum*, Proc. SILARG VII, Cocoyoc, Mexico 1990, eds. J. C. D' Olivo et al (World Scientific, Singapore, 1991)

Hu, B. L. (1993) "Quantum Statistical Processes in the Early Universe" in *Quantum Physics and the Universe*, Proc. Waseda Conference, Aug. 1992 ed. M. Namiki, K. Maeda, et al (Pergamon Press, Tokyo, 1993)

Hu, B. L. and Kandrup, H. E. (1987) *Phys. Rev.* **D35**, 1776.

Hu, B. L. and Matacz, A. (1993a) "Quantum Brownian Motion in a Bath of Parametric Oscillators" Univ. Maryland preprint 93-210.

Hu, B. L. and Matacz, A. (1993b) "Einstein–Langevin Equation for Backreactions in Semiclassical Cosmology" Univ. Maryland preprint 94-31.

Hu, B. L. and Parker, L. (1978) *Phys. Rev.* **D17**, 933.

Hu, B. L. and Pavon, D. (1986) *Phys. Lett.* **180B**, 329.

Hu, B. L., Paz, J. P., and Sinha, S. (1993) "Minisuperspace as a Quantum Open System" in *Directions in General Relativity* Vol. 1, (Misner Festschrift) eds B. L. Hu, M. P. Ryan and C. V. Vishveswara (Cambridge Univ., Cambridge)

Hu, B. L., Paz, J. P. and Zhang, Y. (1992) *Phys. Rev.* **D45**, 2843.

Hu, B. L., Paz, J. P. and Zhang, Y. (1993a) "Quantum Brownian Motion in a General Environment II. Nonlinear coupling and perturbative approach" *Phys. Rev.* **D47** (1993)

Hu, B. L., Paz, J. P. and Zhang, Y. (1993b) "Stochastic Dynamics of Interacting Quantum Fields" *Phys. Rev.* D (1993)

Hu, B. L., Paz, J. P. and Zhang, Y. (1993c) "Quantum Origin of Noise and Fluctuation in Cosmology" in *Proc. Conference on the Origin of Structure in the Universe* Chateau du Pont d'Oye, Belgium, April, 1992, ed. E. Gunzig and P. Nardone (NATO ASI Series) (Kluwer, Dordrecht, 1993)

Hu, B. L. and Sinha, Sukanya (1993a) "Fluctuation-Dissipation Relation in Cosmology" Univ. Maryland preprint 93-164.

Hu, B. L. and Sinha, Sukanya (1993b) "Spacetime Coarse-Graining and Gravitaional Entropy" unpublished.

Hu, B. L. and Zhang, Y. (1990) "Coarse-Graining, Scaling, and Inflation" Univ. Maryland Preprint 90-186 (1990)

Hu, B. L. and Zhang, Y. (1992) "Uncertainty Principle at Finite Temperature" Univ. Maryland preprint 93-161.

Hu, B. L. and Zhang, Y. (1993) "Quantum and Thermal Fluctuations, Uncertainty Principle, Decoherence and Classicality" in *Quantum Dynamics of Chaotic Systems*: Proc. Third International Workshop on Quantum Nonintegrability, Drexel University, Philadelphia, May 1992, ed. J. M. Yuan, D. H. Feng, and G. M. Zaslavsky (Gordon and Breach, Langhorne, 1993)

Isham, C. J. (1991) "Conceptual and Geometrical Problems in Quantum Gravity" Lectures at the Schladming Winter School, Imperial College preprint TP/90-91/14

Joos, E. and Zeh, H. D. (1985) *Z. Phys.* **B59**, 223.

Kandrup, H. E. (1988) *Phys. Rev.* **D37**, 3505.

Kandrup, H. E. (1988) *Class. Quanatum Grav.* **5**, 903

Keldysh, L. V. (1964) *Zh. Eksp. Teor. Fiz.* **47**, 1515 [*Sov. Phys. JEPT* **20**, 1018 (1965)]

Kreuzer, H. J. (1981) *Nonequilibrium Thermodynamics and Its Statistical Foundations* (Oxford Univ., Oxford).

Kubo, R. (1959) *Lectures in Theoretical Physics*, Vol 1, pp 120-203 (Interscience, N. Y. 1959)

Kuchar, K. (1992) "Time and Interpretaions in Quantum Gravity" in *Proceedings of the 4th Canadian Conference on General Relativity and Relativistic Astrophysics*, eds. G. Kunstatter, D. Vincent and J. Williams (World Scientific, Singapore).

Kuchar, K. and Ryan, M. P., Jr., (1986) In *Gravitational Collapse and Relativity* Ed. H. Sato and T. Nakamura (World Scientific, Singapore).

Kuchar, K. and Ryan, M. P. Jr., (1989) *Phys. Rev.* **D40**, 3982.

Linde, A. (1982) *Phys. Lett.* **108B**, 389.

Lynden-Bell D. and Wood, R. (1967) *MNRAS* **136**, 101.

Lynden-Bell D. and Lynden-Bell, R. M. (1977) *MNRAS* **181**, 405.

Ma, S. K. (1985) *Statistical Mechanics* (World Scientific, Singapore).

Misner, C. W. (1969) *Phys. Rev. Lett.* **22**, 1071.

Misner, C. W. (1972) in *Magic Without Magic,* Ed. J. Klauder (Freeman, San Francisco).

Morikawa, M. (1989) *Phys. Rev.* **D40**, 4023.

Mottola, E. (1986) *Phys. Rev.* **D33**, 2126.

Mukhanov, V., Feldman, H. and Brandenberger, R. (1992) *Phys. Rep.* **215**, 203.

Omnès, R. (1992) *Rev. Mod. Phys.* **64**, 339.

Page, D. M. (1984) private communication

Parker, L. (1969) *Phys. Rev.* **183**, 1057.

Parker, L. (1986) in: *The Quantum Theory of Gravity,* S. Christensen, Ed. (Adam Hilger, S. Bristol, 1986).

Paz, J.P. (1990) *Phys. Rev.* **D40**, 1054.

Paz, J. P., Habib, S. and Zurek, W. H. (1993) *Phys. Rev.* **D47**, 488.

Paz, J. P. and Zurek, W. H. (1993) *Phys. Rev.* **D48**, 728.

Paz, J. P. and Sinha, Sukanya (1991) *Phys. Rev.* **D44**, 1038.

Paz, J. P. and Sinha, Sukanya (1992) *Phys. Rev.* **D45**, 2823.

Peebles, P. J. E. (1971) *Physical Cosmology* (Princeton Univ. Press, Princeton).

Penrose, R. (1979) "Singularities and Time-Asymmetry" in *General Relativity: an Einstein Centenary Survey*, eds. S.W. Hawking and W. Israel (Cambridge University Press, Cambridge, 1979)

Prigogine, I. (1962) *Introduction to Thermodynamics of Irreversible Processes* 2nd ed. (Wiley, New York).

Rubin, R. (1960) *J. Math. Phys.* **1**, 309.

Ryan, M. P., Jr., and Shepley, L.C. (1975) *Homogeneous Relativistic Cosmologies* (Princeton Univ. Press, Princeton).

Sakharov, A. D. (1967) *Dok. Akad. Nauk. SSR,* **177**, 70 [*Sov. Phys. Dokl.* **12** (1968) 1040].

Sato, K. (1981) *Phys. Lett.* **99B**, 66.

Schwinger, J. S. (1961) *J. Math. Phys.* **2**, 407.

Sciama, D. W. (1979) in *Centenario di Einstein* Editrice Giunti Barbara-Universitaria.

Sexl, R. U. and Urbantke, H. K. (1969) *Phys. Rev.,* **179**, 1247

Sinha, Sukanya (1991) Ph. D. Thesis, University of Maryland.

Sinha, Sukanya and Hu, B. L. (1991) *Phys. Rev.* **D44**, 1028-1037.

Sinha, Supurna and Sorkin, R. D. (1992) *Phys. Rev.* **B45**, 8123-8126.

Smolin, L. (1985) *Gen. Rel. Grav.* **7**, 417-437.

Sorkin, R. D. (1986) *Phys. Rev. Lett.* **56**, 1885-1888.

Sorkin, R. D. (1993) *Int. J. Theor. Phys.*

Sorkin, R. D., Wald, R. M. and Zhang, Z. J. (1981) *Gen. Rel. Grav.* **12**, 1127.

Unruh, W. G. (1976) *Phys. Rev.* **D14**, 870.

Unruh, W. and Zurek, W. H. (1989) *Phys. Rev.* **D40**, 1071.

Vilenkin, A. (1986) *Phys. Rev.* **D33**, 3560.

Weinberg, S. (1972) *Gravitation and Cosmology* (John Wiley, N.Y.).

Weinberg, S. (1980) *Phys. Lett.* **B91**, 51.

Wheeler, J. A. (1968) in *Battelle Recontres*, eds. C. DeWitt and J. A. Wheeler (Benjamin, New York).

Woo, C. H. (1989) *Phys. Rev.* **D39**, 3174.

Zee, A. (1979) Phys. *Rev. Lett.* **42** 417.

Zeh, H. D. (1986) *Phys. Lett.* **A116**, 9.

Zel'dovich, Ya. B. and Starobinsky, A. A. (1971) *Zh. Eksp. Teor. Fiz* **61**, 2161 [*Sov. Phys. JETP*, **34**, 1159 (1972)].

Zel'dovich, Ya. B. (1970) *Pis'ma Zh Eksp. Teor. Fiz.* **12**, 443 [*JETP Lett.* **12**, (1970) 307].

Zhang, Yuhong (1990) Ph. D. Thesis, University of Maryland.

Zurek, W. H. (1981) *Phys. Rev.* **D24**, 1516.

Zurek, W. H. (1982) *Phys. Rev.* **D26**, 1861.

Zurek, W. H. (1993 *Prog. Theor. Phys.* **89**, 281.

Zurek, W. H., Habib, S. and Paz, J. P. (1993) *Phys. Rev. Lett.* **47**, 1187.

Zwanzig, R. (1961) in *Lectures in Theoretical Physics, Vol. III,* Eds. W.E. Britten, B.W. Downes and J. Downes (Interscience, New York).

# 35

# Fluctuation-Dissipation Theorem in General Relativity and the Cosmological Constant

Emil Mottola

*T-8 Los Alamos National Laboratory M. S. B285*
*Los Alamos, New Mexico 87545, USA*

## Abstract

Vacuum fluctuations are an essential feature of quantum field theory. Yet, the smallness of the scalar curvature of our universe suggests that the zero-point energy associated with these fluctuations does not curve spacetime. A possible way out of this paradox is suggested by the fact that microscopic fluctuations are generally accompanied by dissipative behavior in macroscopic systems. The intimate relation between the two is expressed by a fluctuation-dissipation theorem which extends to general relativity. The connection between quantum fluctuations and dissipation suggests a mechanism for the conversion of coherent stresses in the curvature of space into ordinary matter or radiation, thereby relaxing the effective cosmological "constant" to zero over time. The expansion of the universe may be the effect of this time-asymmetric relaxation process.

In classical general relativity, the requirement that the field equations involve no more than two derivatives of the metric tensor allows for the possible addition of a constant term, the cosmological term, to the Einstein equations:

$$R^a_{\ b} - \frac{1}{2}R\delta^a_{\ b} + \Lambda\delta^a_{\ b} = \kappa T^a_{\ b}. \qquad (35.1)$$

If transposed to the right side of the equation this term corresponds to a constant energy density $\frac{\Lambda}{\kappa}$ and isotropic pressure $-\frac{\Lambda}{\kappa}$ permeating all of space uniformly, and independently of any localized matter sources. Hence, even if $T^a_{\ b} = 0$, such a (positive) cosmological term causes spacetime to become curved with a radius of curvature of order $|\Lambda|^{-\frac{1}{2}}$. From the mere fact that the universe is at least $10^{27}$ cm. large (and perhaps a great deal larger), we immediately infer an extremely stringent bound on $\Lambda$ from observations, *viz.*

$$|\Lambda_{obs}| < 10^{-54} \text{ cm}^{-2} \approx 2 \times 10^{-118}(\hbar\kappa)^{-1}. \qquad (35.2)$$

† Electronic Mail: EMIL@PION.LANL.GOV

The fact that this limit is some 118 orders of magnitude smaller than unity in the Planck units, $(\hbar\kappa)^{-1}$ is often called the "cosmological constant problem" (although as I shall argue this name is somewhat misleading).

In purely classical physics there is no natural scale for $\Lambda$, and it is perfectly consistent to set it equal to zero. In quantum theory this is no longer the case, because each mode of a field theory contributes $\frac{1}{2}\hbar\omega$ to the vacuum energy, and the sum over modes is quartically divergent. Renormalization is required, which means that the value of $\Lambda$ is a free parameter of the quantum theory, which cannot be set equal to zero without fine tuning. Fine tuning is a technically precise term, in the sense that there is no symmetry forbidding the generation of a $\Lambda$ term of order unity in Planck units. Supersymmetry (or superstrings) would forbid such a $\Lambda$ term only if the supersymmetry is unbroken. Since it must be broken to agree with particle physics as we know it, the problem of fine tuning remains. As Steven Weinberg has remarked: "Just because something is infinite, does not necessarily mean that it's zero." [1]

The point is brought home most forcefully if one considers a theory with symmetry breaking, such as the Weinberg-Salam theory of electroweak interactions. The energy density difference between the symmetric vacuum of the theory, in which it finds itself in the very early universe at temperatures above $1\text{TeV}$, and the symmetry broken vacuum of the present epoch is of order $1\text{TeV}^4 \approx 10\text{cm}^{-2}$, or some 55 orders of magnitude above the observational bound (35.2). In other words, if the standard model is correct, the universe passed through a phase transition in which a latent heat 55 orders of magnitude larger than the present vacuum energy density was dumped into the system, so efficiently that there is no presently observable effect on the curvature of spacetime! If only a few parts in $10^{55}$ of this latent heat did not convert to ordinary matter and radiation but remained as vacuum energy density today, then it would exceed the observational bound. If the universe passed through several such phase transitions at still higher temperatures, the fine tuning becomes even more incredible.

Aside from the difficulties that this fine tuning thrusts upon theoretical cosmology and in particular, the inflationary universe scenario, we have direct evidence in the laboratory of the existence of zero point fluctuations and the stresses they cause, in the Casimir effect between two conducting layers [2]. Of course, every success of quantum theory at the microscopic level is indirect evidence for the entire framework, which requires that the problem of vacuum energy be taken seriously.

Since the vacuum energy problem (as I would prefer to call it) does not arise in classical relativity, one might simply lay the problem at the door of quantum gravity, by which is usually meant the physics of the Planck scale. However, there is a problem with energy densities enormously larger than the observational bound (35.2) being generated at distance scales very much larger than the Planck scale, as the exercise with the Weinberg-Salam phase transition above demonstrates. In other words, there is a problem at distance scales where quantum gravitational effects *per se* are thought to be negligible, and where it should be perfectly permissable to treat spacetime as a classical continuous manifold. So it is far from clear that Planck scale physics is what is required to resolve the problem. On the contrary, it is the physics of the correct vacuum state of quantum gravity that is really the issue, and this is controlled by the long-distance or *infrared* properties of the theory, not those of the ultra-short Planck scale.

An analogy to the vacuum energy problem exists also in electromagnetism. I will call it the "cosmological electric field problem." It consists of the elementary observation that Maxwell's equations *in vacuo* admit a solution with constant, uniform electric field $\mathbf{E}_{cosm}$ of arbitrary magnitude and direction. Why then are all electric fields found in nature always associated with localized electrically charged sources? Why do we not observe some huge uniform electric field pointing in the direction of the magellanic clouds, for example? Classically, we just set $\mathbf{E}_{cosm} = 0$ by a boundary condition, but this explains nothing. Is there a deeper reason in the quantum theory?

The answer is almost certainly yes, because of vacuum polarization effects. Suppose that $\mathbf{E}_{cosm}$ were not zero. Then the quantum vacuum of any charged matter fields interacting with it is polarized. This is described by a polarization tensor,

$$\Pi^{ab}(x, x'; \mathbf{E}) = i\langle \mathscr{T} j^a(x) j^b(x') \rangle_{\mathbf{E}}, \tag{35.3}$$

where $j^a$ is the charge current operator and the expectation value is evaluated in a vacuum state with classical background field $\mathbf{E}$. The time ordering symbol $\mathscr{T}$ enforces the Feynman boundary conditions on the polarization operator, *i. e.* positive frequency modes propagate forward in time, while negative frequency modes propagate backward in time. Notice that these boundary conditions are *not* time symmetric. Time-ordering (with an $i\epsilon$ prescription in the propagator) defines a different polarization tensor or Feynman Green's function from anti-time-ordering (with a $-i\epsilon$ prescription). This time-asymmetry is built into quantum theory by the demands of causality and the existence of a lowest energy ground state, or vacuum.

The polarization operator $\Pi^{ab}$ contains two pieces, an even and an odd piece under time reversal. The even piece describes the polarizability of the vacuum, since the vacuum fluctuations of virtual charge pairs may be thought of as giving rise to an effective polarizability of the vacuum (dielectric constant). The time reversal odd piece describes the creation of *real* particle anti-particle pairs from the vacuum by the Schwinger mechanism [3]. The creation of real charged pairs means that a real current flows $\mathbf{j} \geq 0$ in the direction of the electric field at later times, even if none existed initially. This implies that the electric field does work at a rate, $\mathbf{j} \cdot \mathbf{E}$. To the extent that this power cannot be recovered because the created particles interact and lose the coherence they may have had in their initial state, this is the rate of energy *dissipation*. The vacuum behaves like a normal conductor with a finite conductivity and resistivity, due to the random, uncorrelated motions of its fluctuating charge carriers. At the same time, the electric field is diminished by the Maxwell equation,

$$\frac{\partial \mathbf{E}}{\partial t} = -\langle \mathbf{j} \rangle, \tag{35.4}$$

in the case of exact spatial homogeneity of the average current.

An important question is that of the time scale of the dissipation. Does the degradation of the coherent electric field take place rapidly enough to effectively explain why there is no observed $\mathbf{E}_{cosm}$ today? In this connection it must be remembered that the original Schwinger calculation of the decay rate of the vacuum into charged pairs involves a tunneling factor, $\exp(-\pi m^2/eE\hbar)$ for the creation of the first pair from the vacuum. If some charged matter is present initially, the charges are accelerated, radiate, and pair produce without any tunneling suppression factor. Hence one should expect that the time scale for an *induced* cascade of particle pairs to develop and degrade the electric field energy will be very much faster than the Schwinger spontaneous vacuum rate.

I have discussed this electromagnetic analogy in some detail because one can readily form some intuition of the physics involved. It is that of a giant capacitor discharging. Any cosmological electric field initially present in the universe eventually shorts itself out, and degrades to zero, when the vacuum polarization effects described by $\Pi^{ab}$ are taken into account. The real and imaginary parts of $\Pi^{ab}$ are related by a Kramers-Kronig dispersion relation which is one form of a fluctuation-dissipation theorem for the electrically polarized quantum vacuum. It is simply a consequence of causality, and the existence of a positive Hilbert space of quantum states that quantum vacuum fluctuations and quantum vacuum dissipation are inseparably related. One necessarily implies the other. This is the modern, relativistically covariant formulation for quantum fields of the relation between an equilibrium quantity (mean square displacement) and a time asymmetric dissipative quantity (diffusion coefficient or viscosity) first discussed by Einstein in his theory of Brownian movement [4]. That the "vacuum" of quantum field theory can be described in terms closely parallel to that of a fluid medium has been emphasized by B. Dewitt [5], D. Sciama et al. [6], and more recently by B. L. Hu and coworkers [7].

These observations are of a very general nature, and apply equally well to vacuum fluctuations in a gravitational background field. Since the geometry couples to the energy-momentum stress tensor, it is the fluctuations in this quantity which govern the dynamics of the gravitational field, and we are led to consider the corresponding polarization tensor,

$$\Pi^{abcd}(x, x'; \bar{g}) = i \langle \mathscr{T} T^{ab}(x) T^{cd}(x') \rangle_{\bar{g}}, \tag{35.5}$$

where $\bar{g}_{ab}(x)$ represents some classical background metric. This polarization tensor may be handled by exactly the same techniques as (35.3), and the analogous fluctuation-dissipation theorem relating its real and imaginary parts may be proven. If the background $\bar{g}_{ab}(x)$ possesses a timelike Killing field, and therefore a Euclidean continuation with periodicity $\beta$, it is natural to introduce the Fourier transform with respect to the corresponding ststic coordinate time difference $t - t'$. Defining the Fourier transforms of the symmetric and anti-symmetric parts of (35.5) by:

$$\int_{-\infty}^{\infty} dt \, \langle \{ T^{ab}(x), T^{cd}(x') \}_+ \rangle e^{i\omega(t-t')} = S^{abcd}(\mathbf{r}, \mathbf{r}'; \omega),$$

$$\int_{-\infty}^{\infty} dt \, \langle [ T^{ab}(x), T^{cd}(x') ]_- \rangle e^{i\omega(t-t')} = D^{abcd}(\mathbf{r}, \mathbf{r}'; \omega), \qquad (35.6)$$

we find that the two pieces are related via [8]:

$$D^{abcd}(\mathbf{r}, \mathbf{r}'; \omega) = \tanh\left(\frac{\beta\omega}{2}\right) S^{abcd}(\mathbf{r}, \mathbf{r}'; \omega). \qquad (35.7)$$

One such background spacetime is the Schwarzschild metric of a black hole with mass $M$. Classically, any positive value of $M$ is allowed, and all such spacetimes are stable. Quantum mechanically, the situation is quite different. If initial conditions corresponding to no flux of particles in the past is imposed, then an outward flux of Hawking radiation is observed from the hole at late times [9]. Moreover, this radiation has a thermal spectrum with temperature equal to the Hawking temperature,

$$T_H = \frac{\hbar}{k_B \kappa M}. \qquad (35.8)$$

This Hawking effect in a fixed Schwarzschild background precisely parallels the Schwinger mechanism of $e^+e^-$ pair creation in a fixed electric field background. In each case the temperature is equal to $\frac{\hbar a}{2\pi k_B}$, where $a$ is the proper acceleration of a particle of mass $m$ (and charge $e$) at the event horizon.

Although pair creation is described by a completely unitary evolution in the full quantum theory, coherence is maintained by precise correlation of all phase relations between the separated pairs, as they move out of causal contact with each other. If the pairs never again causally interact, so that these phase correlations can never again reconstruct and interfere, it is reasonable to average over the correlations. Then the process may be described in time asymmetric terms with entropy increasing as one half of a particle pair disappears down the hole, or is accelerated to infinity by the electric field, carrying information with it. There is then an *effective time asymmetry* in the decay of the black hole or of the electric field, at least over macroscopically long time scales, similar to the effective time asymmetry we observe every day in macroscopic systems, even though we believe the microphysical dynamics is fully time reversible. This time asymmetry is described naturally by the time reversal odd piece of the polarization tensor (35.5), namely $D^{abcd}(\mathbf{r}, \mathbf{r}'; \omega)$.

There is one very important difference between the vacuum dissipative mechanism of relativistic systems described here and that of ordinary (non-relativistic) media. In thermodynamics one usually considers systems which are stable to linear perturbations. Then the standard linear response theory [10] gives a response function analogous to the polarization tensors (35.3) and (35.5), the time reversal odd piece of which describes the relaxation of the system back to its stable thermodynamic state, after it has been perturbed by a small external force. The perturbation of such a stable system decays with time; it does not grow. Obviously, this cannot be the case for the electric field or the black hole, which is *unstable* to the inclusion of quantum fluctuation and pair creation effects. This is due to the rather different behavior of the polarization tensors (35.3) and (35.5) when event horizons are present. One can show that the condition for particle creation effects to occur in the background electric or gravitational field is just the condition that the time asymmetric piece of the polarization function *diverges* for large $t - t'$, in particular that:

$$D^{abcd}(\mathbf{r}, \mathbf{r}'; \omega) \propto \omega^{-1}, \qquad \omega \to 0. \qquad (35.9)$$

In fact it is precisely the residue of this $\omega^{-1}$ pole which determines the particle creation rate in the adiabatic limit of slowly varying backgrounds [8]. This *singular* behavior at low frequencies means that the background is unstable to small perturbations, and this is why the inclusion of the effects of quantum fluctuations changes the behavior of the system dramatically over long times. The electric field or the black hole, both completely stable classically are not stable quantum mechanically.

Hawking has given a simple thermodynamic argument for the quantum instability of a black hole due to fluctuations of matter or radiation in its vicinity [11]. The key observation is that the relation (35.8) implies that a black hole has *negative* specific heat. This means that if we place the black hole in a cavity with reflecting boundary condition in equilibrium with its own Hawking radiation, then this equilibrium is unstable. For if in some small time interval $\delta t$, the black hole should emit slightly more than its average emittance, due to a quantum/thermal fluctuation, its mass $M$ will suddenly be a tiny bit smaller than its equilibrium value, due to the net energy loss. But from (35.8) this implies that it will be slightly *hotter* than its equilibrium temperature, and will therefore radiate all the more rapidly, driving it further from equilibrium. The corresponding argument applies if the fluctuation is of the opposite sign.

This behavior is quite different from more familiar systems with positive specific heat, and reflects itself in the singular $\omega^{-1}$ behavior of (35.9), which describes the response of the system to perturbations (or fluctuations) on length scales of order of the horizon size or larger. Notice that this instability could in no way be observed unless we consider small disturbances or fluctuations *away* from the mean equilibrium state. The mathematical existence of the equilibrium configuration (by Euclidean time continuation for example) tells us nothing about the stability of that configuration. For that the detailed behavior of the fluctuation spectrum, as described by the Fourier transform of the polarization tensor (35.5) is absolutely essential.

With this discussion of the electric field and Schwarzschild backgrounds, we are finally prepared to consider vacuum fluctuations in a cosmological setting, and return to our starting point, the cosmological constant or vacuum energy problem. Consider the Einstein equation for a spatially homogeneous and isotropic spacetime with Robertson-Walker scale factor, $a(t)$:

$$H^2 \equiv \left(\frac{\dot{a}}{a}\right)^2 = \frac{\kappa}{3}\rho, \qquad (35.10)$$

where any cosmological term is included in the right hand side as a vacuum energy contribution with equation of state,

$$\rho_V = -p_V. \qquad (35.11)$$

The Einstein equation (35.10), together with the equation of covariant energy conservation,

$$\dot{\rho} + 3\,H\,(\rho + p) = 0, \qquad (35.12)$$

imply that:

$$\dot{H} = -\frac{\kappa}{2}\,(\rho + p). \qquad (35.13)$$

This equation is to be compared to the Maxwell Eq. (35.4), or the equation for the change of the Schwarzschild mass parameter in Eddington-Finkelstein-Vaidya coordinates [12]. In each case there is a classical static background that solves the Maxwell or Einstein equations, with zero source terms on the right hand side. In the case of (35.13) it is de Sitter spacetime with $\rho_V = -p_V = \frac{\Lambda}{\kappa}$, constant. In each case this background field is stable to classical charge or matter perturbations. The matter is simply accelerated, and swept out by the electric or gravitational field. In the de Sitter case classical matter (obeying $\rho + p > 0$) is redshifted away by (35.12). In the first two cases we have seen that there are good reasons to believe that these classically stable backgrounds with event horizons are unstable when quantum fluctuations are considered, and that the evolution of the system is inevitably driven away from the quasi-static initial state towards a final state in which the classical field energy has been dissipated into matter or radiation field modes. The signal of this instability is the singular behavior of the Fourier transform of the polarization tensor (35.9). In Ref. [8] I have argued that the polarization function corresponding to scalar (*i. e.* metric trace) perturbations of the de Sitter background has precisely the required singular $\omega^{-1}$ behavior. Particle creation with time-asymmetric initial conditions occurs [13], as in the Schwarzschild case. Thus, de Sitter spacetime satisfies the condition for dissipation of curvature stress-energy into matter and radiation modes. Notice also that whereas the Schwinger suppression factor for vacuum tunnelling occurs in electrodynamics because there are no massless charged particles, massless particles do couple gravitationally, and would be expected to dominate the dissipative process.

There is also an analog of the Hawking thermodynamic instability argument in de Sitter spacetime. Consider as the Euclidean defined "vacuum," the Bunch-Davies state which is invariant under the $SO(4,1)$ de Sitter symmetry group [14]. Because of that symmetry, the expectation value of the energy-momentum tensor of any matter fields in this state must itself be of the form $\rho = -p = $ constant. This is the state usually assumed, tacitly or explicitly by inflationary model builders. However, as should be clear from the previous examples, the mere existence of such a state tells us nothing about its stability against small fluctuations. In fact, both the energy within one horizon volume, and the entropy of the de Sitter horizon are *decreasing* functions of the Hawking-de Sitter temperature:

$$E_H = \rho V = \frac{4\pi}{\kappa H} = \frac{2}{\kappa T_H}$$

$$S_H = \frac{1}{4} A_H = \frac{\pi}{H^2} = \frac{1}{4\pi T_H^2} \tag{35.14}$$

Hence, by considering a small fluctuation in the Hawking temperature of the horizon which (like the black hole case) causes a small net heat exchange between the region interior to the horizon and its surroundings, we find that this interior region behaves like a system with negative specific heat [15]. Since this statement applies to a region centered anywhere in de Sitter space, the entire space is unstable against quantum/thermal fluctuations in its Hawking temperature. The same conclusion is reached by an analysis of the polarization tensor in this background, because of the singular $\frac{1}{\omega}$ dependence in its trace sector. Finally, as if this is not persuasive enough, the Feynman propagator for the linearized metric tensor field in a de Sitter background diverges for large spacelike separations [16].

All of these considerations imply that de Sitter spacetime *cannot* be the stable ground state spacetime of the consistent quantum theory, any more than Schwarzschild spacetime can describe the ground state of matter with finite total energy in an infinite box. In the latter case, the system is unstable to Hawking evaporation, converting the background curvature into matter and radiation which disperse to $\mathscr{I}^+$. If the corresponding evaporation proceeds to completion in the cosmological case, then the universe approaches flat Minkowski spacetime in the asymptotic future, regardless of the value of the cosmological term. One can see the analogy with the electric field discharging by comparing (35.13) with (35.4). The creation of real matter with $\rho + p > 0$ causes the scalar curvature parameter $H$ to decrease continuously. In fact, unlike the electric current which is a vector and may change sign, $\rho + p$ is always positive, so we should expect $H$ to decrease monotonically to zero without the plasma oscillations that can occur in the electrodynamic case [17].

It is a very attractive self-consistent picture that emerges from these observations. We began by posing a question, *viz.* why do the zero-point energies necessarily associated with quantum fluctuations not curve spacetime? By recognizing that fluctuations are closely connected with dissipation in ordinary (non-relativistic) statistical mechanics, and extending the fluctuation-dissipation analysis to general relativity, we come to the realization that dissipative processes are necessarily associated with the fluctuations of the quantum "vacuum," which behaves in many respects like a reactive medium. Then these dissipative particle creation processes provide a dynamical mechanism to relax the scalar curvature (*i.e.* the effective cosmological "constant," as measured by the time-dependent energy density of the universe) to zero. The expansion of the universe may be precisely this relaxation to asymptotically flat space taking place before our eyes, with perhaps some component of vacuum energy (with $p_V = -\rho_V$) of order of the upper bound (35.2) still present today. If this is the case, it would certainly have consequences for observational cosmology, and imply that searches for non-baryonic dark matter are probably futile. The dark "matter" may be all vacuum fluctuation energy, and so be completely undetectable by ordinary particle detectors.

To see if these ideas really work, I have been following two distinct but parallel lines of thought. The first is to develop the necessary field theory formalism to carry out numerical simulations of the effect (back reaction) of particle creation processes on classical background fields. All of the main technical difficulties are solved by a systematic $1/N$ approximation [18]. Here $N$ is an artificial, but convenient gauge-invariant parameter, equal to the number of identical copies of the quantum matter field interacting with the classical electric or gravitational field. The expansion of the Schwinger-Dyson equations of field theory in powers of $1/N$ bears a great deal of similarity to the moment expansion in non-equilibrium statistical mechanics. Leading order in $1/N$ exactly reproduces the semi-classical mean-field Maxwell or Einstein equations (after an appropriate rescaling of the coupling constants). However, not until we go beyond this leading mean-field approximation to the next order in $1/N$ have we included the dynamical back reaction and dissipative effects of the polarization tensor on the classical background. As one might expect, just as things become interesting, they also become difficult. Including the effects of the polarization tensor, which is a non-local function of two spacetime points in the correct, causal way requires the closed time path formalism of Schwinger and Keldysh [19]. It also requires numerically solving *integro*-differential equations for the photon or graviton propagator functions in the dynamically evolving classical background. We are just beginning to tackle the details of implementing this in the QED case, and will move on to the gravitational case only after gaining some experience with the QED problem (which incidentally is quite interesting in its own right). The outstanding question of the time scale for relaxation should be answerable by this direct approach to the problem.

The second line of thought begins with the observation that the most infrared divergent fluctuations in de Sitter spacetime are the *scalar* fluctuations, associated with the trace of the metric tensor or the polarization tensor. Also the effective cosmological constant is measured by the Ricci scalar in Einstein's equations (35.1). This suggests that one should try to treat the scalar spin-0 sector of the theory exactly, and include the transverse, traceless parts of the metric (which contain the usual spin-2 graviton degrees of freedom) later as small perturbations. In the last few years a great deal of progress has been made in understanding non-critical string theories, which involve the consistent covariant quantization of the conformal part $\sigma$ of the metric tensor in two dimensions:

$$g_{ab}(x) = e^{2\sigma(x)}\bar{g}_{ab}(x), \tag{35.15}$$

where $\bar{g}_{ab}(x)$ is a fixed fiducial metric. It turns out that a great deal of the technology developed to study these non-critical string theories (which one may understand as simply 2D quantum gravity coupled to matter) may be carried over directly into four dimensions [20]. The scalar $\sigma$ field in 4D gravity acquires dynamics through the quantum trace anomaly of matter fields, which is altogether different from the Einstein theory (where $\sigma$ is constrained). It is the effective action and dynamics of this field which is responsible for the large effects of quantum fluctuations in de Sitter space, and which we conjecture will control the infrared, or long time behavior of the metric in four dimensions. This line of research may afford us some analytic insight into the relaxation processes we only know how to study numerically in the first approach.

To conclude, it seems quite possible that the problem of vacuum zero point fluctuations and the curvature of spacetime has within it the seed for its own solution in the dissipative, time-asymmetric properties of the "vacuum," necessarily associated with those very fluctuations.

### Acknowlegement

I would like to thank the sponsor, Fundación Banco Bilbao Vizcaya, and the organizers of the Workshop on the Physical Origins of Time Asymmetry, J. Halliwell and J. Pérez-Mercader, for a very thought provoking, enjoyable, and *timely* meeting.

### Discussion

**Paz**  In electromagnetism, the back reaction is amplified by a cascade effect. Is the same effect present for gravity?

**Mottola**  On very general grounds, spontaneous emission processes are always accompanied by induced emission processes, or a cascade effect. The corresponding Bremsstrahlung Feynman diagrams are present in gravity as well.

**Hartle**  Can't your arrow of time associated with particle production be traced to the same source as the other arrows, namely, a special initial condition in this case one in which there were no particles?

**Mottola**  Yes. Indeed, I tried to emphasize the close connection between this relativistic particle creation process and more familiar non- relativistic statistical thermodynamics. The mechanism I have described is another form of the thermodynamic arrow of time, albeit in a cosmological context.

**Starobinsky**  It is rather dangerous to speak about infrared particles in de Sitter space because the notion of particles with wavelength exceeding the de Sitter horizon is not well defined. In this case it is better to speak about amplitudes of a quantum field itself, just as in the case of generation of adiabatic perturbations and gravitational waves during inflation.

**Mottola**  I completely agree. I have used the word *particle* synonymously with *field amplitude*, as a common but imprecise term. When one deals with the back reaction equations (35.4) or (35.13), only physical currents or energy-momenta need to be considered, and the particle concept is superfluous.

**Unruh**  The cosmological constant produces a curvature far less than the curvatures around the earth say. If this process shorts out the cosmological constant, why doesn't it also short out the curvature which cause our gravitational fields? How can this process produce at best a $\Lambda$ which is say $10^{50}$ times its present value?

**Mottola**  It is not just the strength of the field, but also the distance over which it acts that is relevant, as should be clear from the electric field example. In all three cases which I discussed, there was an event horizon and a singular behavior of the polarization tensor for low frequencies. Particles accelerated by the field must be able to achieve velocities of order of $c$. That is what the horizon means. The gravitational field of the earth does not have this behavior because it falls off rapidly with distance. In cosmology, the field grows weaker but the distance grows larger as the universe expands, so the characteristic scale is always the size of the universe, which can become arbitrarily large. Hence $\Lambda_{eff}$ may become arbitrarily small.

**Vilkovisky**  If there are massive particles in the vacuum, then, unless some fine tuning is done, the cosmological term is produced even when all effects of particle creation (including back reaction) are taken into account. Do you mean that the two-point function will grow enormously, showing that this is wrong?

**Mottola**  There is nothing wrong with the first part of your statement. Massive particles do generate a large cosmological term. However, this tells us *nothing* about the effects of this term on the curvature of space, much less about back reaction effects of particle creation in curved space. In fact, the calculation you have in mind may be performed in *flat* space. It has no dynamical content, until you couple it to Einstein's equations in some consistent way. Most of the time this is done in a semi-classical approximation, by including the expectation value (one-point function) of $T^{ab}$. To go further and include back reaction, you *must* consider the quantum fluctuations described by the two-point polarization function $\Pi^{abcd}$. Only this can tell you if the fluctuations are negligible, or if they grow enormously over time. If the latter is the case, then the dynamical *assumption* that a large cosmological term, however produced, leads to a highly curved, but stable spacetime is wrong.

## References

[1] Weinberg, S. (1989) *Rev. Mod. Phys.* **61**, 1.

[2] Casimir, H. B. G. (1948) *Proc. Kon. Ned. Acad. Wetenschap.* **51**, 793; Sparnaay, M. J. (1958) *Physica* **24**, 751; Boyer, T. M. (1970) *Ann. Phys.* **56**, 474.

[3] Schwinger, J. (1951) *Phys. Rev.* **82**, 664.

[4] Einstein, A. (1905) *Ann. Phys. (Leipzig)* **17**, 549; English translation reprinted in (1956) *Investigations on the Theory of Brownian Movement*, Ed. R. Fürth Dover, New York; Callen, H. B. and Welton, T. H. (1951) *Phys. Rev.* **83**, 34.

[5] DeWitt, B. (1979) Quantum gravity: the new synthesis. In *General Relativity: An Einstein Centenary Survey*, Eds. S. W. Hawking and W. Israel, Cambridge University Press, Cambridge.

[6] Sciama, D. W., Candelas, P. and Deutsch, D. (1981) *Adv. Phys.* **30**, 327.

[7] Hu, B. L. (1982) *Phys. Lett.* **A90** 375; Hu, B. L. (1983) *Phys. Lett.* **A97**, 368; Parker, L. (1983) *Phys. Rev. Lett.* **50** 1009. Calzetta, E. and Hu, B. L. (1988) *Phys. Rev.* **37** 2878;

[8] Mottola, E. (1986) *Phys. Rev* **D33**, 2136.

[9] Hawking, S. (1974) *Nature*, **248**, 30; Hawking, S. (1975) *Comm. Math. Phys.* **43**, 199.

[10] Martin, P. C. and Schwinger, J. (1959) *Phys. Rev.* **115**, 1342; Kadanoff, L. P. and Baym, G. (1962) *Quantum Statistical Mechanics*, W. A. Benjamin, Menlo Park, Calif.

[11] Hawking, S. W. (1976) *Phys. Rev.* **D13**, 191; Hartle, J. B. and Hawking, S. W. (1976) *Phys. Rev.* **D13**, 2188.

[12] Bardeen, J. M. (1981) *Phys. Rev. Lett.* **46**, 382.

[13] Mottola, E. (1985) *Phys. Rev.* **D31**, 754.

[14] Tagirov, E. A. (1973) *Ann. Phys. (N. Y.)* **76**, 561; Bunch, T. S. and Davies, P. C. W., (1978) *Proc. R. Soc. London* **A360**, 117.

[15] Mottola, E. (1986) *Phys. Rev.* **D33**, 1616.

[16] Antoniadis, I. and Mottola, E. (1991) *Jour. Math. Phys.* **32**, 1037.

[17] Cooper, F., Eisenberg, J. M., Kluger, Y., Mottola, E., and Svetitsky, B. (1991) *Phys. Rev. Lett.* **67**, 2427.

[18] Cooper, F. and Mottola, E. (1989) *Phys. Rev.* **D40**, 456.

[19] Schwinger, J. (1961) *Jour. Math. Phys.* **2**, 407; Keldysh, L. V. (1965) *Sov. Phys. JETP* **20**, 1018 [(1964) *J. Exptl. Theor. Phys. (U. S. S. R.)* **47**, 1515].

[20] Antoniadis, I. and Mottola, E. (1992) *Phys. Rev.* **D45**, 2013.